A Modern Approach to Physical Chemistry

Volume I

A Modern Approach to Physical Chemistry

Volume I

Edited by **Bernard Wilde**

NY RESEARCH PRESS

New York

Published by NY Research Press,
23 West, 55th Street, Suite 816,
New York, NY 10019, USA
www.nyresearchpress.com

A Modern Approach to Physical Chemistry: Volume I
Edited by Bernard Wilde

International Standard Book Number: 978-1-63238-001-2 (Hardback)

Printed in China.

Contents

Preface

Physical chemistry is a multi-disciplinary field of science that studies about macroscopic, atomic, subatomic, and particulate phenomenology of chemical systems with respect to laws of physics, its principles, practices, and concepts such as motion, force, quantum chemistry, equilibrium among numerous others. Essentially, Physical Chemistry deals with the manner in which fundamental physics finds its applications in chemical problems.

The term 'Physical Chemistry' was coined in the year 1752 by a scientist named, Mikhail Lomonosov, when he presented a lecture titled similarly. Most of the principles in Physical Chemistry are premised on bulk instead of atomic structures alone. Colloids and chemical equilibrium are a good example of this. The origins of modern physical chemistry can be traced to experiments related to electrolytes in solutions, chemical kinetics and chemical thermodynamics, which were conducted between the years 1860s to 1880s. It becomes significant to note that in the field of Physical Chemistry, theoretical developments have gone in tandem with experiments.

Some of the relationships that are resolved and studied by physical chemistry are; discussion of physical properties of different materials which are acted upon by intermolecular forces, concepts related to kinetics of reaction on reaction rate, the identification of ions including electrical conductivity, surface chemistry of membrane and electrochemistry of membrane etc.

I especially wish to acknowledge the contributing authors, without whom a work of this magnitude would clearly not have been realizable. I thank all our authors for allocating much of their scarce time to this project. Not only do I appreciate their participation, but also their adherence as a group to the time parameters set for this publication. Finally, I hope that this book will achieve success, serving as a support to the readers and prove to be very useful to the scientific community.

Editor

1

Microscopic Rate Constants of Crystal Growth from Molecular Dynamic Simulations Combined with Metadynamics

Dániel Kozma[1] and Gergely Tóth[2]

[1] *Institute of Enzymology, Research Centre for Natural Sciences, Hungarian Academy of Sciences, Karolina út 29, Budapest 1113, Hungary*
[2] *Institute of Chemistry, Eötvös Loránd University, P.O. Box 32, Budapest 1518, Hungary*

Correspondence should be addressed to Gergely Tóth, toth@chem.elte.hu

Academic Editor: Dennis Salahub

Atomistic simulation of crystal growth can be decomposed into two steps: the determination of the microscopic rate constants and a mesoscopic kinetic Monte Carlo simulation. We proposed a method to determine kinetic rate constants of crystal growth. We performed classical molecular dynamics on the equilibrium liquid/crystal interface of argon. Metadynamics was used to explore the free energy surface of crystal growth. A crystalline atom was selected at the interface, and it was displaced to the liquid phase by adding repulsive Gaussian potentials. The activation free energy of this process was calculated as the maximal potential energy density of the Gaussian potentials. We calculated the rate constants at different interfacial structures using the transition state theory. In order to mimic real crystallization, we applied a temperature difference in the calculations of the two opposite rate constants, and they were applied in kinetic Monte Carlo simulation. The novelty of our technique is that it can be used for slow crystallization processes, while the simple following of trajectories can be applied only for fast reactions. Our method is a possibility for determination of elementary rate constants of crystal growth that seems to be necessary for the long-time goal of computer-aided crystal design.

1. Introduction

The macroscopic shape of a crystal reflects the net effect of complex processes. The number of the important processes and the dependence of the results on many factors mean a large task for theoretical description and computational modelling. The most important factors are the properties of the solute and the solvent, the temperature, the pressure, the impurities, and the form of the possible seeds. Crystallization is usually divided into two steps, the formation of the seeds and the growth of the crystal [1]. There are numerous ways to model both steps. A possible grouping of the methods is according to the size scale of the methods terming as microscopic, mesoscopic, and macroscopic ones. In detailed modelling, we can use quantum mechanics where the electrons of the system are explicitly included. If the effect of the electrons is averaged in classical mechanical interaction potentials among the atoms or molecules, we can use atomic/molecular modelling with classical mechanics. We can use mesoscopic lattice models, where the atoms or molecules occupy lattice points defined in advance. We may even simplify the model up to limit, where the model includes only the information, that a lattice point is occupied or it is vacant. If the basics of the model are planes of the crystal, we approach to macroscopic description of the processes [2]. In these macroscopic cases, the increase or decrease of the interfacial planes is mostly described by differential equations or by stochastic methods, and thermodynamic and kinetic properties of the crystal planes are the parameters of the model. Recently, classical density functional method is used in the simulations as well [3, 4].

Despite the recent development of computational power, the microscopic modelling is still not feasible for complex systems in the case of crystal growth from solution. Here, three points have to be fulfilled simultaneously: accurate description of the system, reasonable simulation time, and large system size. Most crystallization processes need longer simulation time than the accessible 10^{-9}–10^{-6} s in atomistic approaches. Although there are few systems [5–7] that can be reasonablly studied within this time domain, the most practically important crystallizations take hours or days in the reality. Considering the size limitation, quantum mechanical calculations are restricted to systems, where only very simple interfaces can be found, and the solution phase can be highly biased by artificial periodicity.

These model systems are usually too small to be treated as an adequate representation of the interfacial problem [1]. In the case of classical mechanical models, it is seldom possible to include the whole crystal with the surrounding solution into the simulation. Usually, only the modelling of clearly defined interfaces are performed one at a time and some interfacial structures, like kinks, edges, or even screw dislocations are included. While the time domain of these simulations can be extended by crude simplification of the classical mechanical interaction potential, it leads to a wrong representation of reality. In the case of the classical density functional method, up to now, the system consists of only simple model particles, and the application on realistic molecules has not been elaborated yet [3, 4].

The time and size range restrictions can be eliminated, if we use mesoscopic methods. A feasible method is the kinetic Monte Carlo simulation [1, 8–11]. Here, elementary kinetic processes were performed at randomly selected crystal positions of a supercell. The supercell is constructed from the three-dimensional periodic images of the unit cell. The supercell contains hundred thousands to several millions of unit cells. Initially, a small portion of the lattice points in the middle of the supercell are occupied creating a seed (crystal phase), while the other lattice points are unoccupied (solution phase). We take into consideration two elementary reactions, a crystallization process where an unoccupied lattice point changes to be occupied, and a dissolution process, where an occupied lattice point becomes unoccupied. If these processes are performed at randomly selected positions of the supercell with a probability reflecting the rate constant of the process in the given lattice point environment, we can reasonablly simulate the growth of the crystal. The exact knowledge of the rate constants is crucial. Using the transition state theory, we can calculate these values from the free energy surface of the system. Free energy calculations are not straightforward, but they can be approximated by potential-energy-surface calculations extended with vibration data of the important phase-space points. In the case of quantum mechanical calculations, we are able to calculate it for a few phase space points, but we need a very good guess, which are the important ones. There are different methods to determine the transition pathways, but their adaptation to complex systems containing both crystal and solution is not trivial. The given kinetic parameter set is seldom complete, because the small systems are inadequate representations of the reality. In the case of classical mechanical models, where reasonable system size can be used, the vibration calculations are not trustworthy, because classical mechanical models are seldom suitable for accurate vibration calculations.

Piana et al. [5–7] applied another way to determine the rate constants. They performed classical mechanical molecular dynamics on different interfaces of urea crystal and urea solution in water and in methanol. They count the number of the elementary processes (crystallization and dissolution) at different interfacial positions. They differentiated the positions by the occupation of the neighbouring lattice points. If we take into account only the first 6 hydrogen-bonded neighbours in a urea crystal, and we use some symmetry considerations, there are 36 different positions. It means 72 rate constants for the two elementary processes. Piana et al. counted the visually detectable crystallization and dissolution processes for each type of positions. They got rate constants with acceptable statistics, because the crystal growth of urea is fast. Thereafter, they performed kinetic Monte Carlo simulations with the rate constant sets for both solvents (water, methanol). They obtained micrometer size crystals with macroscopic shapes corresponding well to experimental data. Their investigation clearly showed in a numerical experiment, that microscopic interactions determine the macroscopic shape of the crystals. The reverse direction was investigated in our previous paper: how can we estimate the microscopic rate constant on the basis of the macroscopic shape of the crystal [12]. Using a combined technique of kinetic Monte Carlo simulation and genetic algorithm for parameter estimation, we obtained kinetic rate constant sets for urea in water and in methanol. The sets were qualitatively similar to the ones of Piana et al. Our investigation showed that macroscopic shape contains important and relevant information to clarify the microscopic crystal growth processes. There is a lack of complete rate constant sets on crystal growth except the studies of Piana et al. and our reverse work in the literature. The absence of data does not mean disinterest, because these sets are maybe mandatory to develop computer aided crystal design on mesoscopic scale. The lack of the data is due to the difficulties of obtaining the rate constants. Up to now, these microscopic rate constants cannot be determined experimentally. The method of counting the events in a molecular dynamic simulation can be performed for very fast crystal growth processes, but the most of the scientifically and technically important processes are slow. The parameterization on the macroscopic shape supplies uncontrollable data, because the physical background is missing from our multiparameter fitting algorithm. It turned out to be feasible for urea, but it should be tested on other systems as well. The determination of the potential energy surface augmented with vibration data is usable for a few phase space points and for the corresponding few rate constants, but the determination of all rate constants exceeds the actual computation capacity. Furthermore, the systems are very small in the quantum mechanical calculations.

During the preparation of our paper, two interesting papers have been published on combined techniques to identify microscopic rate constants of crystallization processes [13, 14]. Liu et al. used classical and quantum mechanical simulations with metadynamics and transition path sampling to investigate the free energy surface for the initial steps of dissolution of NaCl in water [13]. They focused on the possible reaction pathways. Stack et al. [14] described combination of MD simulation and rare event methods to find the reaction pathway and rate constants of attachment and detachment of a barium ion onto a stepped barite surface. They determined the rate constants of elementary steps, like kink to bidentate, bidentate to inner shell, inner shell to outer shell, and outer shell to dissolution processes.

In our paper, we present a similar way to determine the rate constants of crystal growth in microscopic simulations, but here we perform a kMC simulation also using our microscopic rate constants. We combined classical mechanical molecular dynamics with metadynamics [15] to explore the free energy surface of crystal/liquid interfaces. Our method keeps the advantage of classical mechanical simulations, that the simulated system is large enough to be adequate representation of the interface. It explores directly the free energy surface: there is no need of classical mechanical vibration calculations. The main issue is that our method can be used for slow processes as well, because it uses the transition state theory as opposed to the counting method of Piana et al. In this feasibility study, we applied the method on the liquid/crystal interface of liquid argon modelled by Lennard-Jones interaction. The crystallization of Lennard-Jones systems was studied several times with computer simulations [16–18], but their aim was different than the determination of rate constants. This one-component study is only our first step. Here, we were able to mimic temperature-induced crystallization instead of the more important concentration-induced cases. The rate constants obtained in the combined molecular dynamics and metadynamics method were used as inputs of mesoscopic kinetic Monte Carlo simulations. We set up this simplified system only for studying the usability of our method, but unfortunately it turned to be more complicated than the original concentration-induced two-component situation. Despite the unexpected difficulties, we could show that rate constants can be determined this way.

2. Details of Calculation/Methods

2.1. Molecular Dynamics. The basic system of the simulations was a three-dimensional periodic Lennard-Jones model of argon containing two stable crystal/liquid interfaces (Figure 1). The crystal was hexagonal close-packed and the interfaces were the HCP(0001) ones. Each interfacial crystal atom had 6 neighbours in the plane of the interface and three neighbours in the lower layer of the crystal at equal distances. The space of the three missing crystal neighbours was filled by the liquid. The density of the phases, the temperature, the simulation box size, and the crystal parameters were set to obtain stable interfaces in equilibrium.

We performed a systematic search close to the triple-point of Ar [19, 20]. The parameters were determined and optimized in several simulations, the reliability of the parameters was checked by comparison to data (Figure 2) in the literature [21]. The stability of the system was checked visually on the intact structure of the solid phase and by

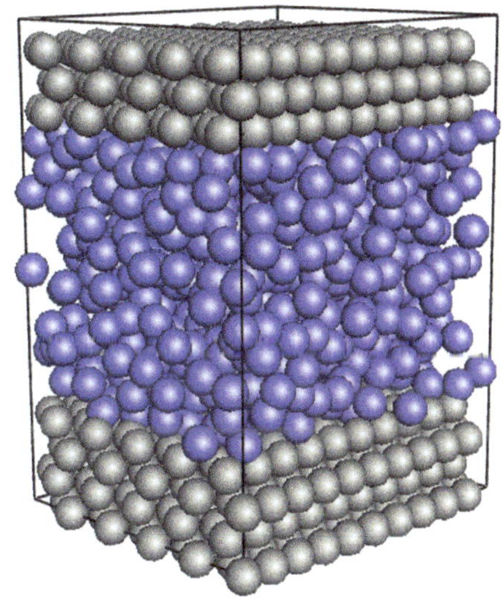

FIGURE 1: Simulation cell. Atoms: grey—crystal phase, blue—liquid phase.

calculating the square of the individual displacement of the atoms in the different phases [21], because this value was characteristically larger for the atoms in the liquid phase than for those ones in the solid phase. The isotropic feature of the pressure was checked also, because the cell contained a crystalline phase. It means the three axes of the simulation cell were set to obtain the same pressure in the three directions. The system consisted of 1400 atoms (700-700 atoms in the crystal and in the liquid phases). The final simulation parameters were: $T = 82$ K, $\rho_{\text{cryst}} = 0.021837\,\text{Å}^{-3}$, $\rho_{\text{liquid}} = 0.021519\,\text{Å}^{-3}$. There were 10-10 rows of atoms at the interfacial crystal in the x and y directions, and altogether 7 (4 + 3) layers of crystal in the z direction. The Lennard-Jones parameters were $\varepsilon = 956.11$ J/mol and $\sigma = 3.405$ Å [19]. The equilibrium distance in the crystal was 3.822 Å between two neighbours. It corresponded to the minimum of the Lennard-Jones potential. The equation of motions was integrated with the velocity Verlet method using a time step of 10^{-15} s. Nose-Hoover thermostat was used to maintain the temperature of the simulations in the canonical ensemble [19]. We used an *NVT* ensemble instead of *NPT*, that would be a more sophisticated solution, but the "phase transition" of a single atom does not change the density values significantly, so using *NVT* is acceptable.

The equilibrated basic system was used in the construction of different interfacial structures. Usually, atoms were taken out from the crystalline part at the interface to create vacancies, kinks, and steps. The density was maintained by addition of particles onto the other solid/liquid interface.

2.2. Metadynamics. Metadynamics is a numerical technique to determine free energy surfaces in computer simulations [15]. We define an s collective variable, which depends on the phase space variables of the system. The dynamics of the system is biased by auxiliary potentials placed on the trajectory of the collective variable. These cumulatively

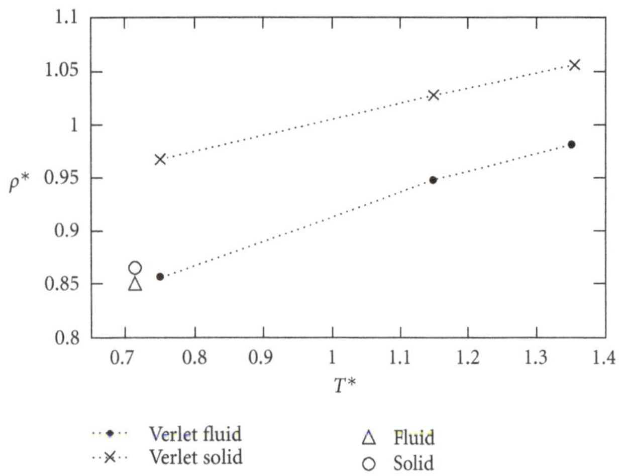

FIGURE 2: State variables of our system and the coexistence curve of Ar liquid/solid phase [19]. The values are in reduced units, where T * is the reduced temperature, ρ^* is the reduced density. The dotted line with × denotes the solid, the dotted line with black square denotes the fluid system. Δ and ○ represent the state variables of our fluid and solid phase.

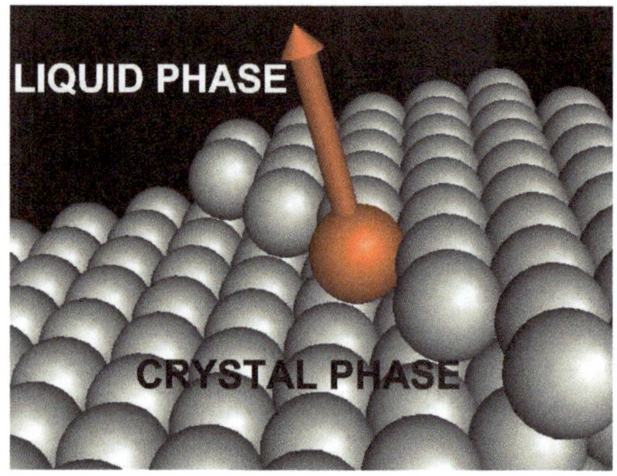

FIGURE 3: Stimulated displacement of an atom from the crystal to the liquid phase. Atoms: grey—crystal phase, red—atom to be displaced, black background—liquid phase.

introduced potentials repel the system of the previously occupied regions of the collective variable. The dynamics of the system are artificial, but the auxiliary potentials contain information on the free energy surface of the system. Usually, the auxiliary potentials are Gaussian functions centered on the average value of s in a previous time interval. The functions are cumulatively added to the system at each T period. It can be shown, that the s-dependent free energy of the system can be approximated for an infinitely long simulation as

$$F(s) = -w \sum_{t'=T,2T,3T,\ldots} \exp\left(-\frac{(s-s(t'))^2}{2\delta^2}\right). \quad (1)$$

w is the height and δ is the standard deviation parameter of the Gaussian. Metadynamics is a widely used technique for the determination of different free energy barriers and possible reaction pathways. There are many technical articles as well about the algorithm, the three fundamental articles of this method can be found in [15, 22, 23].

We used metadynamics to calculate the free energy barrier of displacing an atom from the crystal to the liquid phase (Figure 3). We defined the s variable as the actual Cartesian coordinates of the selected interfacial atom in the crystal. During the simulation, the repelling Gaussian potentials were added to the average position of this particle in every T period. A simulation was finished, if the chosen particle left the crystal phase and entered into the liquid phase, or it was finished after 70000 time steps that time was long enough to displace the investigated particle. We approximated the activation energy of the dissolution to the maxima of $F(s)$.

The activation energy of the crystallization was approximated by this one as well, because our two phases were in equilibrium. We note here that the obtained free energies were maybe lower limits of the activation energies, because the kinetic energy of the given particles contributed to pass the activation barrier. For the sake of simplicity, we did not apply any correction term. The parameters of the metadynamics were determined in test calculations. An important issue was that the total calculation had to be short enough to avoid spontaneous changes at the interface, like dissolution of other interfacial atoms or crystallization of atoms from the liquid phase. At first, we optimized the parameters of the auxiliary Gaussian functions. In our final choice, the height of the Gaussian (w) was $0.4\,k_BT$ and the standard deviation (δ) was 0.2 Å, where k_B denotes the Boltzmann constant. The relatively large height and standard deviation of the Gaussians were necessary, because there was a fast rearrangement of the interface. The height of the Gaussian was comparable to the average kinetic energy of one degree of freedom. We note here, that despite these large values, the auxiliary potential surface was relatively smooth, because the few hundred Gaussians were put along a three-dimensional s and the probability of putting two Gaussians very close to each other was small. Anyway, to reduce the uncertainty of the average free energy barrier, we performed at least 30 simulations for every case. The height of the Gaussians was kept fixed during the simulations. Perhaps, the well-tempered metadynamics method [24] would increase the accuracy of the free energy barriers, but the possible gain would be unimportant due to the large uncertainty caused by the interface reconstruction (detailed later). The time period (T) of adding the repelling potential and the time length of averaging were determined by checking the periodic motions of the phase at the interface. We found that the periodicity of an average in cage motion of the selected atom was about 600–800 time steps. We used different setups for the optimization of the number of time steps to wait after the addition of Gaussian and also for the position averaging of the selected particle to calculate $s(t')$. Finally, we concluded that $T = 300$ time steps was a feasible choice to add a new Gaussian. We did not apply waiting period, so all of the

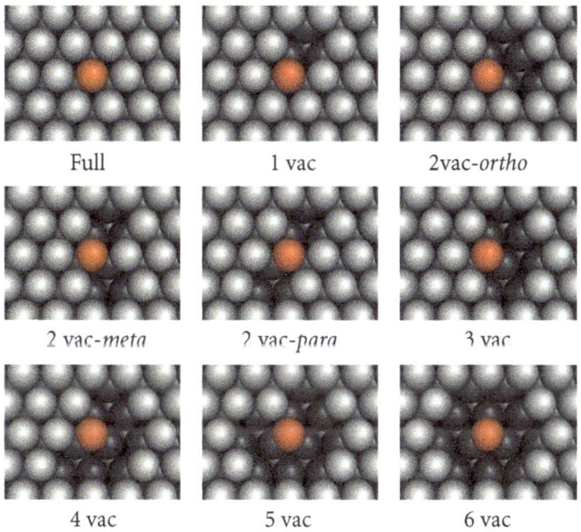

FIGURE 4: Interfacial positions with vacancies at the interface. Atoms: light grey: upper layer in the crystal.

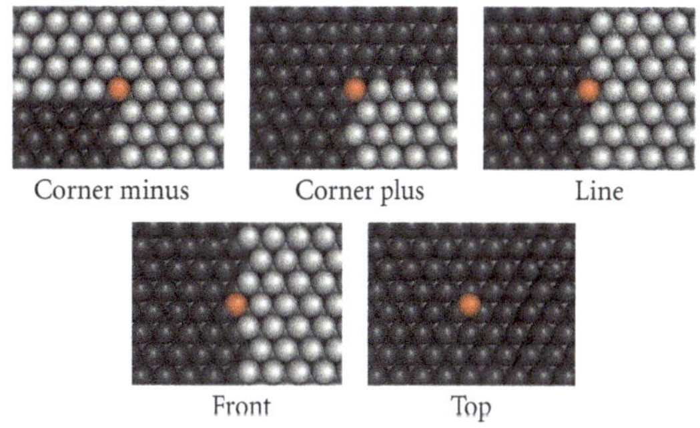

FIGURE 5: Interfacial structures. Atoms: light grey—upper layer in the crystal phase at the interface, dark grey—second layer in the crystal phase, red—atom be displaced.

positions of the selected particles were used during this T time steps to calculate the position of the next Gaussian.

The activation energy for a given simulation was defined as the maximal potential density of the auxiliary Gaussians after finishing the simulation. The points were identified with global extreme search adapted for the task. At first, we calculated at each $s(t')$ points, how many other Gaussian functions could be found within 3δ radius. Then we started a randomly driven simulated annealing [25] search from the $s(t')$ coordinate having the highest number of points. In this way, we usually found a global or local maxima around this $s(t')$. We repeated the annealing process several times, and we started simulating annealing also for all other Gaussian centers, where the number of the neighbouring Gaussians was more than 90% of the maximal one. The activation energy was calculated, where the highest Gaussian density was found.

2.3. The Simulated Lattice Environments. The lattice points were distinguished according the occupation of the neighbouring lattice points. In the case of anisotropy, taking into account the 12 first neighbours and demanding at least one occupied neighbour position, there were $2^{11} = 2048$ cases. If we supposed isotropic symmetry relations, this number reduced to 328. For example, the 12 cases with one missing neighbour were identical and the cases with only one existing neighbour were identical, as well. In the case of two missing neighbours (or 10 missing neighbours), there were 5 different subcases. We did not calculate all of the possible cases, but we simplly selected some examples.

One set of different lattice environments were created by deleting one or more neighbouring crystal atoms at the interface of the selected atom. A schematic view of this can be seen in Figure 4. The second set of lattice environments were created by forming different structures at the interface (Figure 5). Some details of the systems and the calculations can be found in Table 1. Few structures of the first and the second set had identical first-neighbour occupation, but their second-neighbour occupations were not the same. This redundancy was used to check the effect of the second neighbours.

2.4. A Simple Probabilistic Model for Evaluating the Calculated Free Energies. The probability distribution of the activation free energies at a type of interfacial structure can be supposed to be unimodal, as it can be seen in Figure 6 for some positions. If reconstruction of the interface occurred during the simulation, the histogram split to bi- or multimodal distribution. The different peaks might be identified as other structures, especially for the cases of the planar interfaces with different number of vacancies. The observed activation free energy histogram of an initial structure can be treated as a linear combination of the activation free energy histograms of the ideal cases of all of the structures developed during the interfacial reconstruction. If we assumed Gaussian distributions for the ideal cases of the planar interfacial structures with vacancies, the observed histograms of the full and of 1 to 6 vacancies could be described as linear combinations of the same 7 ideal Gaussian distributions. Numerically, it meant the decomposition of the curves to linear combinations of 7 Gaussian functions, where the fitted parameters were the means and standard deviations of the Gaussian functions and the linear coefficients. Our choice to use Gaussians as bases functions of the decompositions was according to its success in spectroscopic curve decompositions and that the surface rearrangement of free full and top positions looked like normally distributed.

There were too many independent linear coefficients; therefore, we applied a simple time-independent, probabilistic approach to reduce the number of the free parameters. The model consisted of two free parameters: probability of the occupation of a vacancy and probability of the dissolution of a neighbour. We used the formula of binomial distribution (2) to calculate the other linear coefficients using the fitted probability parameters. The shape of the equation is the same in the direction of crystallization and dissolution as well.

$$\zeta_{\text{process}}(n,k) = \binom{n}{k} p^k (1-p)^{(n-k)}. \quad (2)$$

Table 1: Details of the interfacial positions and activation energies. Columns are the following: interfacial structure, number of neighbours of the pushed particle, number of calculation performed on a given interfacial type, calculated average activation free energies with 95% confidence interval, median, minima, and maxima of the calculated activation free energy values, and the activation free energy values determined by the probabilistic method.

	No. of neighbours	No. of calculations	$\langle F^{\ddagger} \rangle$ (J/mol + 95% conf.)	$F^{\ddagger}_{\text{median}}$ (J/mol)	$F^{\ddagger}_{\text{min}}$ (J/mol)	$F^{\ddagger}_{\text{max}}$ (J/mol)	$F^{\ddagger}_{\text{Gaussian}}$ (J/mol)
Full	9	30	4653 ± 246	4749	2976	6109	**4757**
Interfacial positions with vacancies at the interface							
1 Vacancy	8	59	4517 ± 232	4542	2649	6620	**4600**
2 Vac—orto	7	30	4177 ± 357	4030	2501	5886	
2 Vac—meta	7	60	4405 ± 226	4389	2344	6713	**4259**
2 Vac—para	7	30	4409 ± 312	4324	2369	6412	
3 Vac—side by side	6	60	4412 ± 205	4484	2594	5853	**4090**
4 Vac—side by side	5	60	4203 ± 266	4162	1897	6756	**3957**
5 Vacancies	4	60	3758 ± 284	4006	1387	5603	**3652**
6 Vacancies	3	60	3770 ± 246	3726	1852	6193	**3295**
Interfacial structures							
Corner minus	8	90	**4127** ± 188	4235	2007	5950	
Line	6	90	**3488** ± 176	3446	1574	5483	
Front	6	90	**3293** ± 186	3114	1425	5339	
Corner plus	5	90	**2014** ± 96	1918	711	3423	
Top	3	90	**2021** ± 120	1964	847	3512	

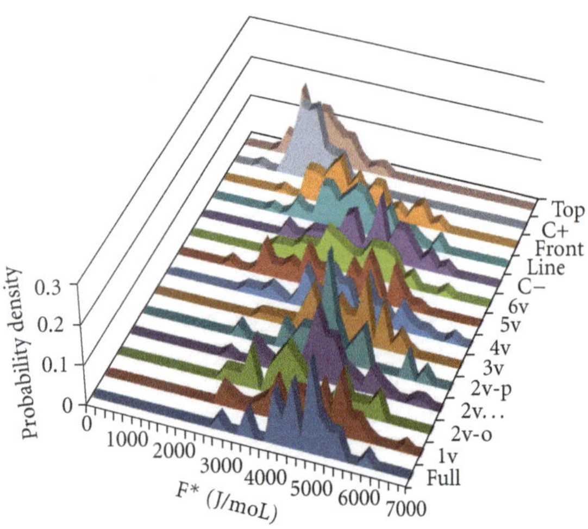

Figure 6: Free energy distributions for the different interfacial structures.

In the case of crystallization p denotes the probability of occupation of a vacancy and n denotes the number of vacancies, k is the number of crystallized particles. In the case of dissolution, these values are the probability of dissolution, number of occupied positions, number of dissolved neighbours. So, using this formula we can build a coefficient matrix, the upper and lower triangle contains the coefficients of the reconstructed interfaces with more or less neighbours, while the diagonal is the proportion of the original surface structure.

Using the free energy values determined in the numerous calculations of each surface type, we created a normalized energy histogram between 100 and 7100 J/mol with a 200 J/mol grid size. We fitted these histograms with the corresponding linear combination of Gaussian functions in the reduced parameter space. The parameters were the means and the widths of the 7 ideal Gaussian functions and the two p-s (for occupation and dissolution). The criteria of the fitting was to minimize the sum of the squared deviation of histograms and linear combination of the pure Gaussian functions. To ensure the convergence, we predefined the order of the activation energies and kept the probability values in the positive region. We got interpretable free energy values for the interfacial structures shown in Table 1.

3. Results and Discussion

A typical trajectory of s is shown for the full structure in Figure 7. More exactly, the coordinates of the Gaussian potentials are shown corresponding to the averaged coordinates of the selected particle in the subsequent T time periods. The particles left the initial position after the placement of 50–150 Gaussians. The particles entered to the liquid phase in many cases, but rather often they squeezed out a neighbouring particle to the liquid phase and annexed the neighbouring position. Sometimes the selected particle left to the direction of the bulk crystal and destroyed locally the crystal structure. In these cases, the interface decomposed fast. We express here that the trajectories of metadynamics are not related to real trajectories. The main goal of metadynamics is the determination of the $F(s)$ free energy surface.

The main aim of our investigation was to develop a method to determine kinetic constants for crystal growth

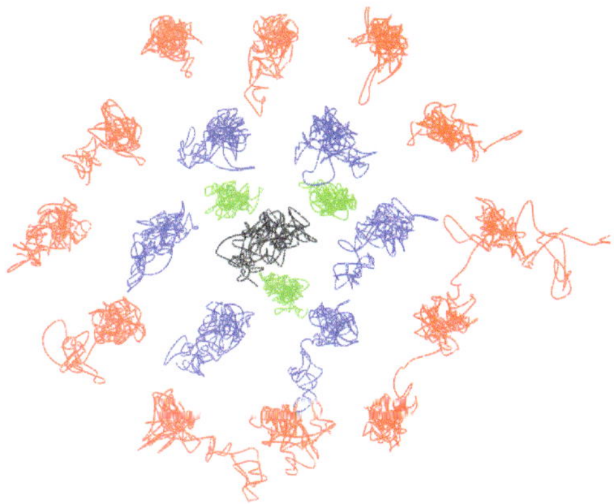

FIGURE 7: Typical "trajectory" of the Gaussians. Trajectories: black—particle to be displaced, blue—neighbouring atoms in the same crystal layer, green—neighbouring atoms in the next crystal layer, red—second neighbours in the same crystal layer.

processes for crystallization from solution. We chose a one-component system in our feasibility study. Here, instead of investigating concentration-induced crystallization, we studied temperature-induced crystallization. We considered it would be a simplification, but unfortunately it turned to be more complicated than the original concentration-induced two-component situation. In the case of two components, the empty lattice points of the crystal at the interface are mostly occupied by solvent molecules. In our one-component case, the unoccupied lattice points could be real voids or they were filled by the liquid. If they were voids, the atoms of the liquid tended to occupy the cavities. In the case of simple spherical model of Lennard-Jones argon, it was hard to distinguish an atom at a given position, that it was a part of the crystal or the liquid phase. The occupation of voids and rearrangement of the interface seemed to be a process comparable to the time scale of our simulation with metadynamics. We got difficulties in the interpretation of our data, and these difficulties could have been avoided, if we had chosen a two-component solvent + solute system.

In the case of the full interface, the activation energies were reliable. In the case of the simulations started with one vacancy at the surface, a large part of our simulations finished with different neighbourhood occupation than the initial one. We were able to remove the selected particle from the central position, but in the meanwhile the initially vacant neighbouring position was often occupied by a particle. Usually, an atom of the liquid phase filled up the hole. In the case of the planar interface with two neighbouring vacancies, the final neighbouring occupation was identical to the initial one only in 10 percent of the simulations. In the case of the other group of different interfacial structures (5 line, corner plus, etc), the main characteristics remained mostly intact during the simulations. Crystallization at the vacant positions was observed mostly only in the 5 corner minus situation. In the other cases, especially in the 5 corner plus case, the dissolution of the crystal was more probable. The structure of the interface far from the vacancies or interfacial structures remained usually unaltered. It means the differentiation of the initial liquid and crystalline phases remained still possible.

We tried different simulation constraints to avoid the reconstruction of the interface, but we were unable to find a method which did not influence the numerical results drastically. Therefore, we postponed the problem to the data evaluation process. At first, we selected visually the cases, where there was no surface reconstruction. We built a sphere with 3δ around the s_{max} value of the maximal free energy density in each simulation, and we searched the simulation time of the last added Gauss potential within this sphere. The snapshot of the particles at this time was checked for the initial condition. The simulations satisfying the condition were used in the calculation of average activation energies. The others with surface reconstruction were not used. It was a rather time-consuming method and did not provide clear cut results. The distinction of the altered/unaltered structures was fuzzy, and we got quantitatively different results depending on the subjective selecting. Therefore, we elaborated a method (see Details of Calculation Methods) on the histograms of the activation free energies. The results of the calculations are summarized in Table 1. The calculations provided data on interfacial structures, where the number of the occupied neighbour positions was between 3 and 9. The simple average of the activation free energies does not correspond well to a simplified chemical intuition based on a linear relation to the number of the neighbours. Let us discuss at first the planar interfacial structures with vacancies. Here the simple averages were not meaningfully caused by the enhanced reconstruction of the initial surface, mostly by the occupation of the vacancies. If we used medians for the estimation of the expected values, we obtained similarly biased activation free energies as for the averages. The distributions are rather broad (Figure 6) as numerically can be seen in the minimum and maximum values of the activation free energies (Table 1). If we used the Gaussian curve deconvolution described previously, we obtained meaningful activation energies for the ideal (pure) surface situations. On the other hand, also these activation energies could not be approximated by simple linear relationship to respect with the number of the neighbours. The activation process was not a clear cut energetic process based on the interaction among the neighbours. The entropic terms unpredictable differed for each structure. If we compared the case of the interfacial structures to the planar interfacial structures with vacancies, there was a large difference in the activation energies.

As it was mentioned, here the interfaces remained mostly unaltered causing more relevant activation free energies. More exactly, in some cases (e.g., 5 line, front, and corner plus) the opposite process, the melting of neighbours, took place. In the case of the planar structures with vacancies, the mean activation energies were larger than the ideal ones, but here the activation energies were maybe smaller than the ideal ones.

There was a significant difference among the data belonging to the same number of occupied neighbours. This difference could be easily explained by the difference in the second neighbour shells. In the case of the interfacial structures, there were missing second neighbours as well in the plane. There were lower activation energies (meaning faster crystal growth) for the crystallization or melting of partially existing planes in the parallel direction. These numerical data corresponded well to the general view of crystal growth processes. We did not perform the Gaussian curve decomposition for the interfacial structures. The necessary incorporation of the second neighbours would increase the number of the Gaussian functions up to a numerically underdetermined task.

The rate constants of crystal growth and melting were calculated using the transition state theory as follows:

$$k = \frac{RT}{h}\exp\left(-\frac{F^{\ddagger}}{RT}\right), \quad (3)$$

where h is the Planck's constant. We applied the simplest RT/h transmission coefficient, although it is not correct in the case of our reactions due to the liquid and solid phases and the non-elementary feature of the reactions. In the case of the kMC simulations, the difference of the opposite processes and their variability on the crystalline positions determine the shape of a crystal. If the same transmission coefficient is supposed in all rates, the magnitude of the transmission coefficient does not change the shape of the crystals, it scales only the timescale. We did not put effort in a better estimation of transition coefficient, because it would extend the study extremely. Therefore, our rates can be termed as relatively accurate rates only. If T was the temperature of the molecular dynamic simulation of the two-phase equilibrium system, we supposed that the two constants (k_{cryst} and k_{melt}) were equal. If we wanted to mimic temperature-induced crystallization, we defined different temperatures for the crystallization and the melting process (i.e., in the case of cooling the solid phase). In this approach, secondary nucleations can be neglected, because at this temperature, seeds cannot be formed in the liquid phase. In Table 2, we show rate constant of a system, where the temperature of the liquid was equal to the temperature of the equilibrium system, but the temperature of the crystal was 5 K less. This setup caused that the k_{cryst}-s were equal to the equilibrium ones, but the rate constant of melting decreased. The net effect can be interpreted as a temperature-induced crystal growth. In the calculation of the rate constants, we used the activation free energies printed in boldface in Table 1. It means for the interfacial positions with vacancies, we used the mean values of the Gaussian values of the curve resolution process. In the case of the interfacial structures, we used the average activation energies.

To present the usability of the rate constants, we performed kinetic Monte Carlo simulations. The detailed algorithm of the simulations can be found in [1, 12]. It was briefly described in the introduction as well. The supercell of the kMC simulation contained $1.5625 \cdot 10^7$ lattice points. The form of the supercell was close to a cube. The initial crystal occupied 0.5% of the supercell. The initially occupied lattice points were put in the middle of the supercell in a cubic form. The simulations were finished, if the crystal reached the border of the supercell or completely dissolved.

TABLE 2: Rate constants for temperature-induced crystallization of $\Delta T = 5\,\mathrm{K}$.

	k_{cryst} $T = 82\,\mathrm{K}\,(\mathrm{mol}^{-1}\,\mathrm{s}^{-1})$	k_{melt} $T = 77\,\mathrm{K}\,(\mathrm{mol}^{-1}\,\mathrm{s}^{-1})$
Full	$9.61 \cdot 10^{32}$	$5.73 \cdot 10^{32}$
1 Vacancy	$1.21 \cdot 10^{33}$	$7.33 \cdot 10^{32}$
2 Vac—orto	$1.99 \cdot 10^{33}$	$1.25 \cdot 10^{33}$
2 Vac—meta	$1.99 \cdot 10^{33}$	$1.25 \cdot 10^{33}$
2 Vac—para	$1.99 \cdot 10^{33}$	$1.25 \cdot 10^{33}$
3 Vac—side by side	$2.56 \cdot 10^{33}$	$1.63 \cdot 10^{33}$
4 Vac—side by side	$3.11 \cdot 10^{33}$	$2.00 \cdot 10^{33}$
5 Vacancies	$4.86 \cdot 10^{33}$	$3.22 \cdot 10^{33}$
6 Vacancies	$8.20 \cdot 10^{33}$	$5.63 \cdot 10^{33}$
Corner minus	$2.42 \cdot 10^{33}$	$1.53 \cdot 10^{33}$
Line	$6.18 \cdot 10^{33}$	$4.16 \cdot 10^{33}$
Front	$8.23 \cdot 10^{33}$	$5.65 \cdot 10^{33}$
Corner plus	$5.37 \cdot 10^{34}$	$4.17 \cdot 10^{34}$
Top	$5.32 \cdot 10^{34}$	$4.12 \cdot 10^{34}$

FIGURE 8: Argon crystal obtained in the kinetic Monte Carlo simulation.

In the case of argon crystal, the number of the lattice points with different first-neighbourhood occupancy is 2^{12}. If we use symmetry relations and isotropic approximation, we might reduce it to few hundreds of different environments. We did not calculate all of the different environments in this study. We calculated some typical ones and used them for the similar lattice positions. This similarity search was performed with comparing distance histograms calculated among the occupied neighbours for all type of lattice environments. These histograms were calculated for the lattice points for which the rate constants were determined in the simulations and for the other type of lattice points as well. The rate constants of a calculated case were assigned to that uncalculated environment which provided the highest overlap integral between the two histograms.

A typical kMC simulation is shown in Figure 8 with the set of Table 2. The temperature-induced simulation provided a featureless (close to spherical) crystal. It was not surprising because the simple spherical interactions of liquid argon and the HCP crystal structure do not cause necessarily characteristic crystals. Furthermore, it seems to be requisite to include second neighbourhood information in the simulation, for example, to mimic the preferred in plane growth of planes. Without the incorporation of the second neighbours in the classification of the positions, it was impossible to compare our data to other results on argon crystal growth [16–18].

4. Conclusion

The atomistic simulation of crystal growth can be decomposed into two steps: the determination of the microscopic rate constants and a mesoscopic kinetic Monte Carlo simulation. In our study, we proposed a method to determine kinetic rate constants of crystal growth. We performed classical mechanical molecular dynamic simulation on the liquid/crystal interface of argon at the equilibrium. The technique of metadynamics was used during the simulation to explore the free energy surface of crystal growth. A crystalline atom was selected at the interface, and it was displaced to the liquid phase by adding repulsive Gaussian potentials. The activation free energy of this process was calculated as the maximal potential-energy density of the Gaussian potentials in accordance with the basic idea of metadynamics. The calculations were performed for different interfacial structures and vacancies at the neighbouring positions of the selected particle.

The activation free energies were transformed to kinetic rate constants using the transition state theory. We supposed equal rate constants for the crystallization and the melting processes at a crystal position, since our system was in equilibrium. In order to mimic real crystallization, we applied a temperature difference in the transition state theory calculation of the two opposite reactions. The temperature perturbed set of kinetic rate constants was applied in mesoscopic kinetic Monte Carlo simulation of argon crystal growth.

The main advantage of this determination of rate constants is that it can be used for slow processes, while the simple following of trajectories in molecular dynamics and counting of elementary reactions taken place can be applied only for fast reactions. Our method provides a possibility for routine determination of elementary rate constants of crystal growth that seems to be necessary for the long-time goal of computer aided crystal design.

We should admit that our system choice was awkward for a feasibility study. We chose a one-component spherical system with high number of first neighbours both in the crystal and liquid phase. Here we were able to model crystal growth and crystal melting elementary processes of one selected particle instead of crystallization and dissolution processes. The determination of the activation free energy was not a simple task for some interfacial structures. The initial structure was often destroyed during the simulations with metadynamics. Therefore, the initial characterization of the lattice environment changed. It caused uncertainty in the interpretation of the obtained activation free energies. We think in the more important crystal growth from solutions, there is no need to take care with this bias, because in the two-component solute-solvent case, the solvent occupies the most of the liquid phase at the interface providing stable metadynamics [13]. The natural continuation of this feasibility study is the application of the method on scientifically important systems.

Acknowledgment

G. Tóth thanks the financial support of TÁMOP 4.2.1B-09/1/KMR-2010-0003.

References

[1] H. Jónsson, “Theoretical studies of atomic-scale processes relevant to crystal growth,” *Annual Review of Physical Chemistry*, vol. 51, pp. 623–653, 2000.

[2] S. X. M. Boerrigter, G. P. H. Josten, J. Van De Streek et al., “MONTY: Monte Carlo crystal growth on any crystal structure in any crystallographic orientation; Application to fats,” *Journal of Physical Chemistry A*, vol. 108, no. 27, pp. 5894–5902, 2004.

[3] K. R. Elder, K. Katakowski, M. Haataja, and M. Grant, “Modeling elasticity in crystal growth,” *Physical Review Letters*, vol. 88, no. 24, Article ID 245701, 4 pages, 2002.

[4] G. I. Tóth, G. Tegze, T. Pusztai, G. Tóth, and L. Gránásy, “Polymorphism, crystal nucleation and growth in the phase-field crystal model in 2D and 3D,” *Journal of Physics*, vol. 22, no. 36, Article ID 364101, 2010.

[5] S. Piana and J. D. Gale, “Interfaces between a saturated aqueous urea solution and crystalline urea: a molecular dynamics study,” *Journal of the American Chemical Society*, vol. 98, no. 6, pp. 1674–1681, 1994.

[6] S. Piana, M. Reyhani, and J. D. Gale, “Simulating micrometre-scale crystal growth from solution,” *Nature*, vol. 438, no. 7064, pp. 70–73, 2005.

[7] S. Piana and J. D. Gale, “Three-dimensional kinetic Monte Carlo simulation of crystal growth from solution,” *Journal of Crystal Growth*, vol. 294, no. 1, pp. 46–52, 2006.

[8] D. T. Gillespie, “A general method for numerically simulating the stochastic time evolution of coupled chemical reactions,” *Journal of Computational Physics*, vol. 22, no. 4, pp. 403–434, 1976.

[9] G. H. Gilmer, “Computer models of crystal growth,” *Science*, vol. 208, no. 4442, p. 355, 1980.

[10] A. F. Voter, “Classically exact overlayer dynamics: Diffusion of rhodium clusters on Rh(100),” *Physical Review B*, vol. 34, no. 10, pp. 6819–6829, 1986.

[11] K. A. Fichthorn and W. H. Weinberg, “Theoretical foundations of dynamical Monte Carlo simulations,” *The Journal of Chemical Physics*, vol. 95, no. 2, pp. 1090–1096, 1991.

[12] G. Tóth, “Microscopic kinetic data on crystal growth from macroscopic morphology,” *Crystal Growth & Design*, vol. 8, no. 11, pp. 3959–3964, 2008.

[13] L. M. Liu, A. Laio, and A. Michaelides, “Initial stages of salt crystal dissolution determined with ab initio molecular dynamics,” *Physical Chemistry Chemical Physics*, vol. 13, no. 29, pp. 13162–13166, 2011.

[14] A. G. Stack, P. Raiteri, and J. D. Gale, “Accurate rates of the complex mechanisms for growth and dissolution of minerals

using a combination of rare-event theories," *Journal of the American Chemical Society*, vol. 134, no. 1, pp. 11–14, 2012.
[15] A. Laio and M. Parrinello, "Escaping free-energy minima," *Proceedings of the National Academy of Sciences of the United States of America*, vol. 99, no. 20, pp. 12562–12566, 2002.
[16] W. C. Swope and H. C. Andersen, "106-particle molecular-dynamics study of homogeneous nucleation of crystals in a supercooled atomic liquid," *Physical Review B*, vol. 41, no. 10, pp. 7042–7054, 1990.
[17] P. R. Ten Wolde, M. J. Ruiz-Montero, and D. Frenkel, "Numerical evidence for bcc ordering at the surface of a critical fcc nucleus," *Physical Review Letters*, vol. 75, no. 14, pp. 2714–2717, 1995.
[18] P. R. ten Wolde, M. J. Ruiz-Montero, and D. Frenkel, "Numerical calculation of the rate of crystal nucleation in a Lennard-Jones system at moderate undercooling," *Journal of Chemical Physics*, vol. 104, no. 24, pp. 9932–9947, 1996.
[19] D. Frenkel and B. Smit, *Understanding Molecular Simulation*, Academic Press, London, UK, 1996.
[20] M. P. Allen and J. d. Tildesley, *Computer Simmulation of Liquids*, Oxford University Press, London, UK, 1991.
[21] J. P. Hansen and L. Verlet, "Phase transitions of the Lennard-Jones system," *Physical Review*, vol. 184, no. 1, pp. 151–161, 1969.
[22] A. Laio and F. L. Gervasio, "Metadynamics: a method to simulate rare events and reconstruct the free energy in biophysics, chemistry and material science," *Reports on Progress in Physics*, vol. 72, no. 12, Article ID 126601, 2008.
[23] G. Bussi, A. Laio, and M. Parrinello, "Equilibrium free energies from nonequilibrium metadynamics," *Physical Review Letters*, vol. 96, no. 9, Article ID 090601, 2006.
[24] A Barducci, G. Bussi, and M. Parrinello, "Well-tempered metadynamics: a smoothly converging and tunable free-energy method," *Physical Review Letters*, vol. 100, no. 2, Article ID 020603, 2008.
[25] W. H. Press, S. A. Teukolsky, W. T. Vetterling, and B. P. Flannery, *Numerical recipes in Fortran*, Cambridge University Press, Cambridge, UK, 1992.

2

Kinetics and Mechanism of Oxidation of L-Ascorbic Acid by Pt(IV)(aq) in Aqueous Hydrochloric Acid Medium

Sadhana Senapati,[1] S. P. Das,[2] and A. K. Patnaik[2]

[1] *PG Department of Chemistry, Utkal University, Odisha, Bhubaneswar 754004, India*
[2] *PG Department of Chemistry, Ravenshaw University, Odisha, Cuttack 753003, India*

Correspondence should be addressed to S. P. Das, dassmrutiprava@yahoo.in

Academic Editor: Marc Koper

Reduction of $[PtCl_6]^{2-}$ by L-ascorbic acid (H_2ASc) in 0.1 M aqueous acid medium has been investigated spectrophotometrically under pseudo-first order condition at $[PtCl_6]^{2-} = 0.005$–$0.007\,\text{mol dm}^{-3}$, $0.05 \le [H_2ASc]/\text{mol dm}^{-3} \le 0.3$, $298\,\text{K} \le T \le 308\,\text{K}$, $[H^+] = 0.14\,\text{mol dm}^{-3}$, $I = 0.5\,\text{mol dm}^{-3}$. The redox reaction follows the rate law: $d[\text{Pt(IV)}]/dt = k[H_2ASc][\text{Pt(IV)}]$, where k is the second-order rate constant and $[H_2ASc]$ is the total concentration of ascorbic acid. Electron transfer from $[H_2ASc]$ to Pt(IV) center leading to the release of two halide ions and formation of the reaction products, square planner Pt(II) halide complex, and dehydrated ascorbic acid is suggested. This redox reaction follows an outersphere mechanism as Pt(IV) complex is substituted inert. Activation parameters were calculated corresponding to rate of electron transfer reaction k. Activation parameters favor the electron transfer reaction.

1. Introduction

The redox reaction of L-ascorbic acid is of fundamental interest in chemistry, biochemistry, pharmacology, and several areas of medicine. It is used as a reducing agent with one or two electron reductants in chemical and biological systems [1–7]. Rate of reduction by ascorbic acid depends on nature of oxidant and pH of the medium. Different mechanisms have been proposed for the reaction between ascorbic acid and metal complexes [5–13].

In recent years there has been significant interest in the chemistry of square planar platinum (II) complex for remarkable anticancer properties [14–16]. Several thousand potentially biological active Pt(IV) complexes have been reported for anticancer properties [16].

We report here the interaction of ascorbic acid with Pt(IV)(aq). Ascorbic acid is a good biological reducing agent. Pt(IV)(aq) appeared attractive because these octahedral complexes are usually substitution inert and require reduction to Pt(II) species to act as potential anticancer drugs. The anticancer activity of Pt(IV) complexes is likely to be due to effective platinum (IV) transport into the cell followed by reduction to the more reactive platinum (II) compounds. The knowledge of the reactivity of Pt(IV) compounds towards the reduction by potential bioreductant like ascorbic acid may be important for understanding the mechanism of where antitumor activity as well as for designing new compounds with the least side effect. Chois et al. [17] reported that there is a correlation between the rate of the reduction and anticancer activity in a series of homologous Pt(IV) complex without going into the details of reaction mechanism. The present study was undertaken with the aim of extending the knowledge of the kinetics and thermodynamic parameters that characterized the reduction of Pt(IV) complex by ascorbic acid and clarifying the mechanism details using the model complex $PtCl_6{}^{2-}$.

2. Experimental

2.1. Materials. All chemicals used were analaR grade. L-ascorbic acid (Merck) and sodium hexachloro-platinate (Merck) were used as received. Sodium hexachloroplatinate stock solution was prepared with 0.1 M HCL. Solutions of different concentrations were prepared by proper dilution of

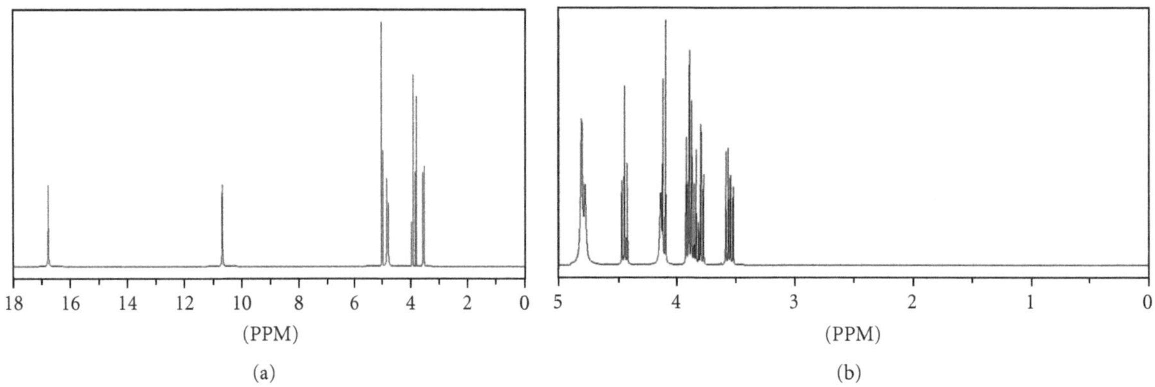

FIGURE 1: (a) ¹H NMR spectra of L-ascorbic acid. (b) ¹H NMR spectra of dehydroascorbic acid.

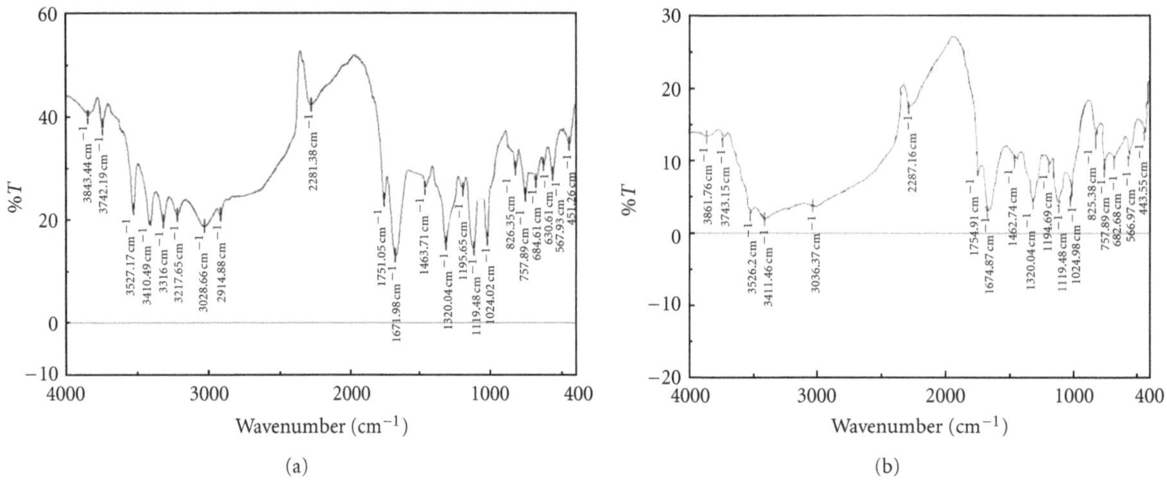

FIGURE 2: (a) IR spectra of ascorbic acid. (b) IR spectra of dehydroascorbic acid.

stock solution. Fresh solutions were used for kinetics measurements. Ionic strength was maintained with $NaClO_4$. All solutions were prepared in freshly prepared double distilled water using an all-glass distillation apparatus containing $KMnO_4$.

2.2. Kinetics. The kinetics of oxidation of L-ascorbic acid by sodium hexachloroplatinate in aqueous acid medium was studied spectrophotometrically under pseudo-first order conditions by keeping the concentration of ascorbic acid at least ten times in excess over Pt(IV) complex. The decrease of absorbance with time was monitored at 457 nm using conventional mixing technique in Systronic 2202 UV-VIS spectrophotometer equipped with a thermostatic bath for temperature control. The rate constant (k_{obs}) was calculated from the slope of $\ln(A_t - A_\infty)$ versus t(s) plot from the relationship

$$\ln(A_t - A_\infty) = \ln(A_0 - A_\infty) - k_{obs} \cdot t, \quad (1)$$

where A_0, A_t, A_∞ denote optical density of the reaction mixture at zero time, time t, and at infinity, respectively. A_∞ was measured after completion of the reaction mixture (after 24 hrs of mixing), were absorbance becomes almost constant. The correlation coefficients (R^2) of the plots used to determine k_{obs} were found to be 0.99 in most of the cases. All calculations were made on a PC using least square program.

2.3. Stoichiometry and Identification of Reaction Products. The stoichiometry for the reduction of $[PtCl_6]^{2-}$ by ascorbic acid was determined by iodine titration method. Excess of ascorbic acid is allowed to react with fixed amount of $[PtCl_6]^{2-}$. Titrating the ascorbic acid with iodine before and after the reaction indicates a 1 : 1 stoichiometry between Pt(IV) and L-ascorbic acid

$$H_2ASc + PtCl_6^{-2} \longrightarrow DHA + [PtCl_4]^{2-} + 2H^+ + 2Cl^- \quad (DHA = \text{dehydroascorbate}). \quad (2)$$

The product of redox reaction was identified as dehydroascorbate along with Pt(II). The product DHA was confirmed by ¹H NMR (see Figures 1(a) and 1(b)) and FTIR (see Figures 2(a) and 2(b)). Pt(II) was identified by UV-VIS spectra after 24 h (Figure 3(b)). Figure 1(a) shows peaks at δ4.78, 4.81, 10.68(enol), 16.77(enol) corresponding to four OH groups, δ5.0 for CH of furanone, δ3.93 corresponding to methine (CH) group in ascorbic acid, whereas

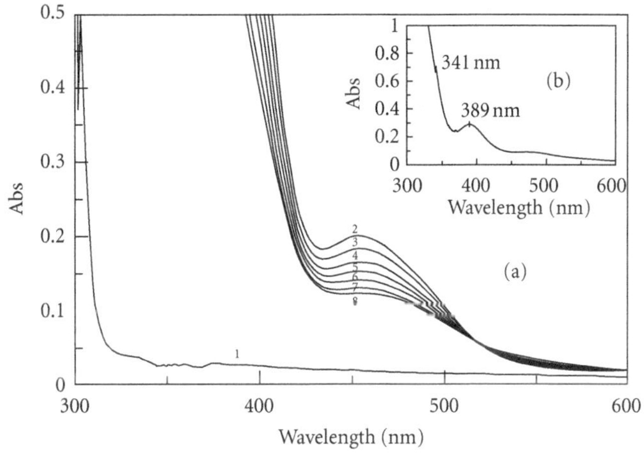

FIGURE 3: (a) UV-VIS spectral scan of reaction mixture of L-ascorbic acid and sodium hexachloroplatinate, $[H^+] = 0.14\,mol\,dm^{-3}$, $I = 0.5\,mol\,dm^{-3}$ at 303 K. (1) (L-ascorbic acid) = $1.0\times10^{-1}\,mol\,dm^{-3}$, (2) (sodium hexachloroplatinate) = $5.0\times10^{-3}\,mol\,dm^{-3}$, (3) immediately after mixing, (4) after 15 minutes, curve (5–8) $\Delta t = 5$ minutes. (b) Inset: spectra of the above reaction mixture after 24 hr.

TABLE 1: Pseudo-first order rate constant (k_{obs}) data at different temperatures.

$[H^+]$ ($mol\,dm^{-3}$)	$[H_2ASc]$ ($mol\,dm^{-3}$)	$10^3\,k_{obs}$ (pseudo-first order rate constant) (s^{-1}) 293 K	298 K	303 K	308 K
	0.05	1.215	2.000	4.000	5.700
	0.10	1.831	3.000	7.000	9.000
0.14	0.15	2.303	4.000	9.800	13.400
	0.20	2.817	5.000	12.500	17.500
	0.25	3.111	6.000	15.500	24.000
	0.30	3.560	7.000	18.000	29.000

Figure 1(b) shows peaks at $\delta 4.81$ and 4.78, peak at $\delta 4.45$, 4.11 corresponding to CH group in butyrolactone, and peak at $\delta 3.90$ corresponding to methine (CH). All peaks of ^1H NMR Spectra of Figure 1(b) corresponds to dehydroascorbic acid. IR spectra of Figure 2(b) show the presence of two OH groups due to the broad peaks at $3411.5\,cm^{-1}$ and $3526\,cm^{-1}$, whereas there are four OH groups in Figure 2(a) corresponding to peaks $3216\,cm^{-1}$, $3316\,cm^{-1}$, $3411.5\,cm^{-1}$, and $3527\,cm^{-1}$. In Figure 2(b), two peaks at $1675\,cm^{-1}$ corresponds to C–O, $1755\,cm^{-1}$ corresponds to C=O groups. All the IR spectra of Figure 2(b) corresponds to dehydroascorbic acid.

3. Results and Discussion

L-ascorbic acid (H_2ASc) has two acidic protons ($pK_1 = 4.04$ and $pK_2 = 11.34$) [18] and is a strong reducing agent ($E^0 = 0.390\,V$ NHE) in aqueous solutions. The electron transfer reaction between $PtCl_6^{2-}$ and L-ascorbic acid was investigated by changing (L-ascorbic acid) in the range 0.05 to $0.3\,mol\,dm^{-3}$ at $[H^+] = 0.14\,mol\,dm^{-3}$ and $I = 0.5\,mol\,dm^{-3}$. Temperature variation was carried out in the range 293 K to 308 K. The UV-VIS spectral scan (see Figure 3(a)) of the reaction mixture was over the wavelength range $350 \le \lambda$ (nm) ≤ 600. Gradual decrease of absorbance with time was observed without the formation of isosbestic point and without shifting of λ_{max} (maximum wavelength) indicating the fact that there is no formation of intermediate species. After a long interval of time (after 24 h), the peak at 457 nm disappeared completely and two new peaks (see Figure 3(b)) at 341 nm and 389 nm appeared corresponding to square plannar Pt(II) complex. In this redox process, octahedral Pt(IV) is reduced to square plannar Pt(II). Therefore this reaction is better classified as reductive elimination reaction [19]. IR and ^1H NMR spectra showed the formation of dehydroascorbate along with Pt(II). Similar product was reported by others [20–22] for the reaction between Pt(IV) and different reductants. Plots of pseudo-first order rate constant (k_{obs}) versus $[H_2ASc]_{total}$ were linear with zero intercept (see Figure 4). The fact indicates no reverse electron transfer reaction. The redox reaction follows the second order rate law given by (3)

$$\frac{-d[Pt(IV)]}{dt} = k_{obs}[Pt(IV)] = k[H_2ASc][Pt(IV)]_{total} \quad (3)$$

$$k_{obs} = k[H_2ASc]_{total}, \quad (4)$$

where k denotes second order rate constant. We arrive at the reaction shown in (4)

$$H_2ASc \rightleftharpoons HASc^- + H^+ \quad (pk_1 = 4.04)$$
$$Pt(IV) + HASc^- \xrightarrow{k} Pt(II) + DHA \quad (5)$$

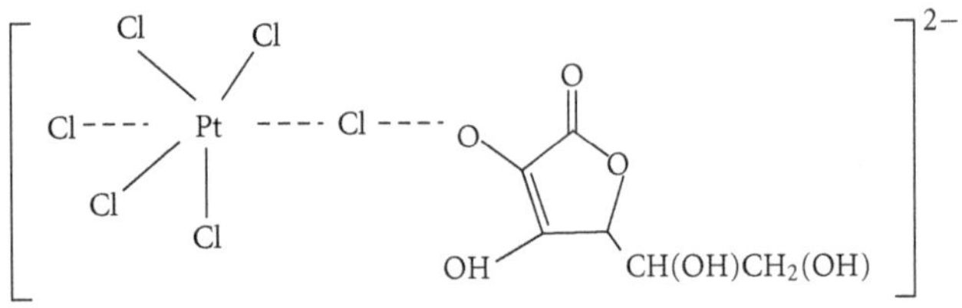

Scheme 1

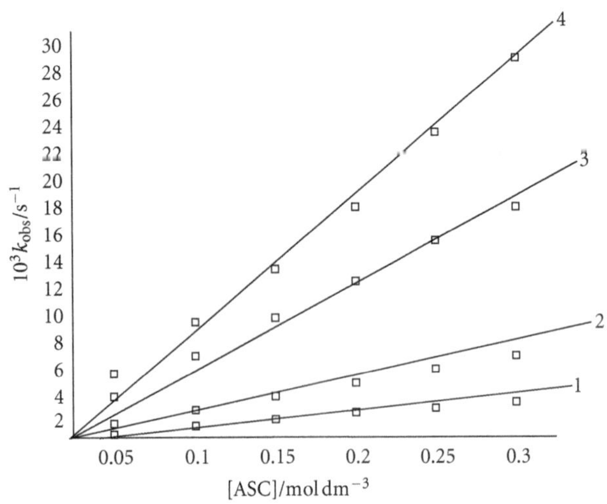

Figure 4: The plot of k_{obs} versus $[H_2Asc]$ at temperature (1) 293 K, (2) 298 K, (3) 303 K, (4) 308 K.

Table 2: Calculation of second order rate constant k and activation parameters.

Temperature (K)	$10^2\,k$ (rate of electron transfer reaction) ($mol^{-1}\,dm^3\,s^{-1}$)
293	0.9
298	2.0
303	5.6
308	9.4
$\Delta H^\ddagger$ (kilo joule per mole)	11.88 ± 1.31
$\Delta S^\ddagger$ (Joule per Kelvin per mole)	−51.66 ± 4.39

The second-order rate constant (k) for reduction of Pt(IV) was calculated using (4) by plotting k_{obs} versus $[H_2ASc]$ as collected in Table 1 at four different temperatures from 293 K to 308 K. k data were used to calculate activation parameters as $\Delta H^\ddagger = 11.88 \pm 1.31\,kJ\,mol^{-1}$ and $\Delta S^\ddagger = -51.66 \pm 4.39\,Jk\,mol^{-1}$ (see Table 2).

They are close to the reported value for similar systems [18, 19, 23]. The value of activation parameters is favorable for electron transfer reaction.

It is concluded that the redox reaction follows an outer sphere mechanism with the fact that Pt(IV) octahedral complex is substitution inert. The mechanism for reduction of Pt(IV) halide complex by biological reductant involve an attack by the reductant as a halide coordinated trans to good leaving groups forming a halide bridge activated in Pt(IV) complex which facilitates such rapid reductive elimination reaction [20, 23] (see Scheme 1).

A concerted two-electron transfer from ascorbate to Pt(IV) center leading to the release of two halide ions and formation of reaction products, Pt(II), dehydroascorbic acid are observed.

Acknowledgment

The authors are thankful to HOD Chemistry, Utkal University, for providing research facilities and also thankful to Prof. P. Mohanty (Rtd) for his help.

References

[1] E. Pelizzetti, E. Mentasti, E. Pramauro, and I. Ihaliz, *Chemistry*, vol. 15, p. 2829, 1976.

[2] E. Pelizzetti, E. Mentasti, and E. Pramauro, "Outer-sphere oxidation of ascorbic acid," *Inorganic Chemistry*, vol. 17, no. 5, pp. 1181–1186, 1978.

[3] D. H. Macartiney and A. Mc Auley, *Canadian Journal of Chemistry*, vol. 59, p. 132, 1981.

[4] Z. Amjad, J. C. Brodovitch, and A. McAuley, "Metal-ion oxidations in solution. Part XXI. Kinetics and mechanism of the reaction of ascorbic acid, hydroquinone, and catechol with 12-tungstocobaltoate (III)," *Canadian Journal of Chemistry*, vol. 55, no. 20, pp. 3581–3586, 1977.

[5] K. Tsuleahara and Y. Yamamota, *Bulletin of the Chemical Society of Japan*, vol. 54, p. 2642, 1981.

[6] N. H. Williams and J. K. Yamdell, "Outer-sphere electron-transfer reactions of ascorbate anions," *Australian Journal of Chemistry*, vol. 35, no. 6, pp. 1133–1144, 1982.

[7] M. Kimura and S. Yamab, *Journal of the Chemical Society, Dalton Transactions*, p. 423, 1982.

[8] P. Martinez, J. Zuluaga, J. Kralt, and R. Van Eldik, "Kinetics and mechanism of the oxidation of L-ascorbic acid by trisoxalatocobaltate(III) in basic aqueous solution," *Inorganica Chimica Acta*, vol. 146, no. 1, pp. 9–12, 1988.

[9] D. Pinnell and R. B. Jordan, "Kinetics of reduction of cobalt(III)-ammine complexes by dithionite," *Inorganic Chemistry*, vol. 18, no. 11, pp. 3191–3194, 1979.

[10] S. Abey Okada, H. Horic, and S. Taniguchr, *Journal of the Chemical Society, Perkin Transactions*, vol. 7, p. 5, 1987.

[11] D. H. Macartney and N. Sutin, *Inorganica Chimica Acta*, vol. 74, p. 221, 1983.

[12] E. Pelizzetti and E. Mentasti, *Journal of the Chemical Society, Dalton Transactions*, p. 61, 1978.

[13] M. M. Taqui khan and R. S. Sukla, *Inorganica Chimica Acta*, p. 149, 1989.

[14] N. Farrel, R. Ugo, and B. R. James, Eds., *Catalysis By Metal Complexes*, vol. 11, Kluwer, Dordrecht, The Netherlands, 1989.

[15] R. B. Weiss and M. C. Christian, "New cisplatin analogues in development: a review," *Drugs*, vol. 46, no. 3, pp. 360–377, 1993.

[16] M. C. Keage, M. J. Kelland, L. R. Neidles, and M. J. Warning, Eds., *Molecular Aspects of Anticover Drug DNA Interactions*, vol. 1, CRC Press, New York, NY, USA, 1993.

[17] Chois, C. Filotto, M. Bisanzo et al., "Reduction and anticancer activity of platinum(IV) complexes," *Inorganic Chemistry*, vol. 37, pp. 2500–2504, 1988.

[18] K. Moribe, W. Limwikrant, K. Higashi, and K. Yamamoto, "Drug Nanoparticle Formulation Using Ascorbic Acid Derivatives," *Journal of Drug Delivery*, vol. 2011, Article ID 138929, pp. 1–9, 2011.

[19] K. Lemma, A. Sargeson, and L. I. Elding, "Kinetics and mechanism for reduction of oral anticancer platinum(IV) dicarboxylate compounds by L-ascorbate ions," *Journal of the Chemical Society, Dalton Transactions*, no. 7, pp. 1167–1172, 2000.

[20] T. Shi, J. Berglund, and L. I. Elding, "Kinetics and mechanism for reduction of trans-dichlorotetracyanoplatinate(IV) by thioglycolic acid, l-cysteine, dl-penicillamine, and glutathione in aqueous solution," *Inorganic Chemistry*, vol. 35, no. 12, pp. 3498–3503, 1996.

[21] T. Shi, J. Berglund, and L. I. Elding, "Reduction of trans-dichloro- and trans-dibromo-tetracyano-platinate(IV) by L-methionine," *Journal of the Chemical Society, Dalton Transactions*, no. 12, pp. 2073–2077, 1997.

[22] K. Lemma, T. Shi, and L. I. Elding, "Kinetics and mechanism for reduction of the anticancer prodrug *trans,trans,trans*-$[PtCl_2(OH)_2(c\text{-}C_6H_{11}NH_2)(NH_3)]$ (JM335) by thiols," *Inorganic Chemistry*, vol. 39, no. 8, pp. 1728–1734, 2000.

[23] L. Kelemu, D. A. House, N. Retta, and L. I. Eldirs, "Kinetics and mechanism for reduction of halo- and haloam(m)ine platinum(IV) complexes by L-ascorbate," *Inorganica Chimica Acta*, vol. 331, pp. 98–108, 2002.

3

Polyethylene Glycols as Efficient Media for Decarboxylative Nitration of α,β-Unsaturated Aromatic Carboxylic Acids by Ceric Ammonium Nitrate in Acetonitrile Medium: A Kinetic and Mechanistic Study

K. Ramesh, S. Shylaja, K. C. Rajanna, P. Giridhar Reddy, and P. K. Saiprakash

Department of Chemistry, Osmania University, Hyderabad, Andhra Pradesh 500 007, India

Correspondence should be addressed to K. C. Rajanna; kcrajannaou@yahoo.com

Academic Editor: Leonardo Palmisano

Polyethylene glycols (PEGs) were found to be efficient media for decarboxylative nitration of α,β-unsaturated aromatic carboxylic acids by ceric ammonium nitrate (CAN) in acetonitrile to give β-nitrostyrene derivatives. Kinetics of the reaction exhibited second order kinetics with a first order dependence on [CAN] and [substrate]. Reactions were too sluggish to be studied in the absence of PEG; therefore detailed kinetics were not taken up. Reaction times were reduced from 24 hrs to few hours. The catalytic activity was found to be in the increasing order PEG-300 > PEG-400 > PEG-600 > PEG-200. Mechanism of PEG-mediated reactions was explained by Menger-Portnoy's scheme as applied in micellar kinetics.

1. Introduction

Cerium (IV) ammonium nitrate (CAN) is one of the most important reagents used for organic synthesis [1–4]. The formula of cerium (IV) ammonium nitrate, $[(NH_4)_2[Ce(NO_3)_6]$, reflects that the cerium (IV) ion is surrounded by six nitrate groups and the ammonium ions are counterions to compensate for the negative charge of the hexanitratocerate (IV) coordinating unit. CAN is a one-electron oxidation reagent. Cerium (IV) ammonium nitrate can also be used as a nitrating agent [5, 6], initiator for radical polymerization reactions [7], and also as a reagent to remove protecting groups [8]. Even though cerium (IV) reagents are milder oxidation reagents than the other metal-based oxidation reagents such as Mn(VII) and Cr(VI) salts, they are relatively much less toxic. Because of their high molecular mass, large quantities of cerium (IV) salts are required for stoichiometric reactions. Therefore, indirect and catalytic reactions using Ce(IV) have been developed [9, 10]. The main advantage of CAN over other cerium (IV) reagents is its higher solubility in organic solvents. The most popular solvents are (in decreasing order of importance) water, acetonitrile, dichloromethane, THF, and methanol [11]. Often mixtures of these solvents are also used as reaction medium. Other solvents have found only marginal use for this type of reactions.

Being good Michael acceptors, α,β-unsaturated nitroalkenes are widely applied in organic synthesis [12]. Among various methods reported for their preparation, a method involving the use of CAN provides a practical way to the synthesis of α,β-unsaturated "nitroalkenes" in good-to-excellent yields [13–23]. Recently polyethylene glycols (PEGs) have been used as catalysts and catalyst supports and also have been found to be an inexpensive, nontoxic, and environmentally friendly reaction medium, which avoid the use of acid or base catalysts. Moreover PEG can be recovered after completion of the reactions and recycled/reused [24–29] in another batch. Inspired by the striking features of PEG the author wants to use it as a catalyst by avoiding the use of acid in the present study; namely, ceric ammonium

TABLE 1: Binding constants of [CAN-PEG] at 303°K using Benesi-Hildebrand method.

S. N	PEG	Benesi-Hildebrand equation	K dm^3/mol	ε $dm^3/mol/cm$	$-\Delta G$ (kJ/mol)
1	PEG-200	$y = 7E-05x + 0.052$	743	19.23	16.7
2	PEG-300	$y = 9E-05x + 0.055$	611	18.18	16.2
3	PEG-400	$y = 5E-05x + 0.071$	1420	14.08	18.3

TABLE 2: Activation parameters of cinnamic acid in different PEG media. Units of $k'' = dm^3\ mol^{-1}\ min^{-1}$.

Type of PEG	PEG % (V/V)	k'' at 300 K	Equation	R^2	$\Delta H^{\#}$ kJ/mol	$\Delta G^{\#}$ kJ/mol	$\Delta S^{\#}$ J/K/mol
PEG-200	0.5	0.01	$y = -13.32x + 34.09$	0.999	110	84.2	85.9
	1.0	0.02	$y = -9.181x + 20.97$	0.999	76.3	83.3	−23.2
	2.0	0.03	$y = -8.399x + 18.75$	0.995	70.0	95.5	−41.7
	3.0	0.04	$y = -8.625x + 19.84$	0.998	91.8	81.8	−32.6
	4.0	0.08	$y = -7.047x + 15.06$	0.999	58.6	55.6	−72.3
	5.0	0.1	$y = -8.112x + 19.01$	0.995	67.5	39.7	39.5
PEG-300	0.5	0.01	$y = -22.79x + 64.43$	0.997	189	87.6	338
	1.0	0.03	$y = -4.655x + 6.308$	0.996	38.7	82.3	−145
	2.0	0.04	$y = -5.385x + 8.930$	0.999	44.8	81.8	−123
	3.0	0.05	$y = -2.684x + 0.252$	0.998	22.4	81.0	−195
	4.0	0.06	$y = -8.743x + 20.63$	0.999	72.7	80.5	−20.0
	5.0	0.08	$y = -7.750x + 17.60$	0.999	64.5	80.0	−51.3
PEG-400	0.5	0.02	$y = -6.979x + 13.65$	0.999	58.0	83.3	−84.0
	1.0	0.03	$y = -6.106x + 11.05$	0.997	50.8	82.5	−105
	2.0	0.04	$y = -9.224x + 21.84$	0.999	76.7	81.5	−16.0
	3.0	0.06	$y = -8.718x + 20.54$	0.999	72.5	38.0	−80.5
	4.0	0.07	$y = -7.054x + 15.38$	0.999	58.7	80.1	−69.7
	5.0	0.1	$y = -7.687x + 17.63$	0.999	64.0	79.3	−51.0
PEG-600	0.5	0.02	$y = -5.421x + 8.550$	0.998	45.0	83.0	−126
	1.0	0.03	$y = -4.644x + 6.274$	0.999	38.6	82.2	−145
	2.0	0.04	$y = -3.131x + 1.542$	0.999	26.0	81.4	−184
	3.0	0.05	$y = -5.555x + 9.813$	0.999	46.2	81.0	−115
	4.0	0.07	$y = -3.622x + 3.753$	0.998	30.1	80.0	−166
	5.0	0.1	$y = -4.247x + 5.956$	0.999	35.3	80.0	−148

nitrate (CAN) triggered decarboxylative nitration of α,β-unsaturated aromatic carboxylic acids. Acetonitrile is used as solvent in order to facilitate kinetic studies.

2. Experimental Details

2.1. Materials. Cinnamic acid, ceric ammonium nitrate (CAN), and polyethylene glycols were obtained from SD Fine Chemicals or Loba. Substituted cinnamic acids were prepared by Perkins reaction as cited in the literature [30, 31].

2.2. General Procedure for PEG-Mediated β-Nitrostyrenes Synthesis. In a typical solid state synthesis, cinnamic acid (0.01 mol), PEG (0.02 mmol), and CAN (0.012 mmol) are placed in a clean two-necked R. B. flask containing acetonitrile (MeCN) and stirred for certain time. Progress of the reaction is periodically monitored by thin layer chromatography (TLC). After completion, the reaction mixture is treated with 2% sodium carbonate solution, followed by the addition of "dichloromethane" (DCM) or "dichloroethane" (DCE). The organic layer was separated, dried over Na_2SO_4 and evaporated under vacuum, and purified with column chromatography using ethyl acetate : hexane (3 : 7) as eluent to get pure product. The products were characterized by IR, 1H-NMR, mass spectra, and physical data with authentic samples and found to agree well with earlier reports (Table 1). The yields of major products are compiled in Table 2.

2.3. Kinetic Method of Following the Reaction. Thermostat was adjusted to desired reaction temperature. Flask containing known amount of ceric ammonium nitrate (CAN) in acetonitrile solvent and another flask containing the substrate (α,β-unsaturated acid) and suitable amount of PEG solutions were clamped in a thermostatic bath. Reaction was initiated by mixing requisite amount of CAN with the other contents of the reaction vessel. The entire reaction mixture was

Table 3: Activation parameters of crotonic acid in different PEG media. Units of k'' = $dm^3\ mol^{-1}\ min^{-1}$.

Type of PEG	PEG % (V/V)	k'' at 300 K	Equation	R^2	$\Delta H^{\#}$ (kJ/mol)	$\Delta G^{\#}$ (kJ/mol)	$\Delta S^{\#}$ J/K/mol
PEG-200	0.5	0.01	$y = -9.907x + 22.7$	0.999	82.4	85.1	−9.0
	1.0	0.02	$y = -6.742x + 12.86$	0.999	56.0	83.3	−91.0
	2.0	0.03	$y = -5.069x + 7.688$	0.999	42.1	82.3	−134
	3.0	0.04	$y = -6.020x + 11.13$	0.999	50.0	81.5	105
	4.0	0.05	$y = -7.664x + 16.87$	0.999	64.0	81.2	−57.3
	5.0	0.06	$y = -7.062x + 14.92$	0.999	59.0	81.0	−73.5
PEG-300	0.5	0.01	$y = -9.797x + 22.35$	0.999	81.4	85.0	−12.0
	1.0	0.02	$y = -8.070x + 17.25$	0.999	67.0	83.2	−54.2
	2.0	0.04	$y = -4.233x + 5.180$	0.998	35.2	81.5	−154
	3.0	0.05	$y = -6.852x + 14.13$	0.999	54.6	79.0	−80.0
	4.0	0.06	$y = -6.564x + 13.36$	0.999	54.5	80.4	−86.5
	5.0	0.08	$y = -9.440x + 23.13$	0.999	78.5	80.0	−5.3
PEG-400	0.5	0.01	$y = -9.759x + 22.23$	0.999	81.2	85.0	−13.0
	1.0	0.02	$y = -7.864x + 16.59$	0.999	65.4	83.3	−57.0
	2.0	0.04	$y = -4.189x + 5.034$	0.999	35.0	82.0	−155
	3.0	0.05	$y = -8.018x + 17.93$	0.999	67.0	81.5	−48.5
	4.0	0.06	$y = -5.394x + 9.454$	0.999	45.0	80.7	−119
	5.0	0.08	$y = -8.250x + 19.25$	0.998	68.6	80.0	−37.5
PEG-600	0.5	0.01	$y = -9.630x + 21.81$	0.998	80.0	85.0	−16.3
	1.0	0.02	$y = -5.541x + 8.852$	0.998	46.0	83.2	−123
	2.0	0.03	$y = -5.256x + 8.404$	0.999	44.0	82.3	−128
	3.0	0.04	$y = -5.421x + 9.243$	0.998	45.0	81.2	−120
	4.0	0.06	$y = -4.492x + 6.466$	0.997	37.4	80.5	−144
	5.0	0.09	$y = -6.254x + 12.73$	0.999	52.0	80.0	−91.8

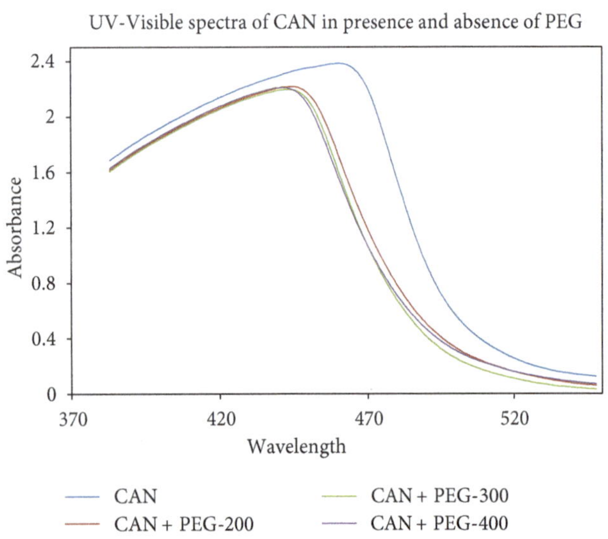

Figure 1: UV-Visible spectra of CAN in presence and absence of PEG in MeCN medium.

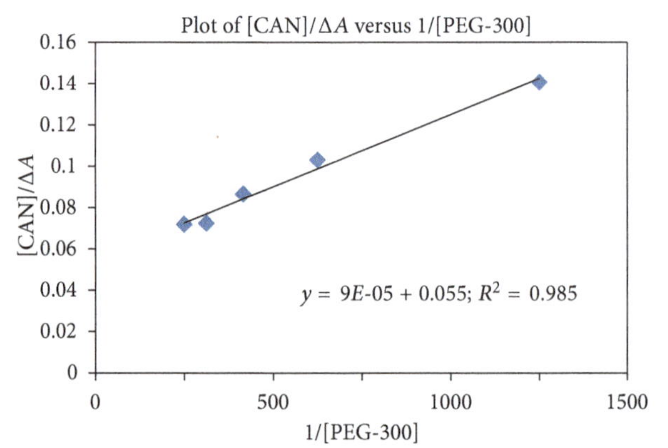

Figure 2: Benesi-Hildebrand plot of PEG-300 interaction with CAN.

mixed thoroughly. Flask was coated black from the outside to prevent photochemical effects. Aliquots of the reaction mixture were withdrawn into a cuvette and placed in the cell compartment of the laboratory visible spectrophotometer. Cell compartment was provided with an inlet and an outlet for circulation of thermostatic liquid at a desired temperature. The CAN content could be estimated from the previously constructed calibration curve showing absorbance (A) versus [CAN]. Absorbance values were in agreement with each other with an accuracy of ±3% error.

TABLE 4: Activation parameters of methoxy cinnamic acid in different PEG media. Units of $k'' = dm^3 mol^{-1} min^{-1}$.

Type of PEG	PEG % (V/V)	k'' at 300 K	Equation	R^2	$\Delta H^\#$ (kJ/mol)	$\Delta G^\#$ (kJ/mol)	$\Delta S^\#$ J/K/mol
PEG-200	0.5	0.01	$y = -12.85x + 32.55$	0.996	107	85.0	73.0
	1.0	0.02	$y = -13.90x + 36.01$	0.998	115	84.4	102
	2.0	0.04	$y = -6.038x + 11.21$	0.999	50.2	81.50	−104
	3.0	0.06	$y = -5.027x + 8.448$	0.999	42.0	80.2	−127
	4.0	0.08	$y = -3.844x + 4.748$	0.999	32.0	80.0	−158
	5.0	0.12	$y = -4.403x + 6.945$	0.999	37.0	79.0	−140
PEG-300	0.5	0.01	$y = -10.36x + 14.11$	0.997	86.0	35.3	110
	1.0	0.03	$y = -3.532x + 2.653$	0.999	46.0	98.5	−175
	2.0	0.04	$y = -5.385x + 8.930$	0.959	45.0	82.0	−123
	3.0	0.05	$y = -2.423x - 0.582$	0.999	20.2	78.0	−192
	4.0	0.06	$y = -8.743x + 20.63$	0.999	73.0	81.0	−26.0
	5.0	0.08	$y = -7.965x + 18.28$	0.999	66.2	80.0	−46.0
PEG-400	0.5	0.02	$y = -7.638x + 15.83$	0.999	63.5	83.3	−66.0
	1.0	0.03	$y = -10.01x + 24.07$	0.999	83.2	77.0	21.3
	2.0	0.04	$y = -9.756x + 23.58$	0.999	81.0	81.4	−2.0
	3.0	0.06	$y = -8.022x + 18.16$	0.999	67.0	80.0	−41.0
	4.0	0.08	$y = -6.000x + 11.86$	0.999	50.0	80.0	−99.0
	5.0	0.10	$y = -9.302x + 22.89$	0.999	77.3	79.5	−7.3
PEG-600	0.5	0.02	$y = -7.936x + 16.83$	0.999	66.0	83.3	−58.0
	1.0	0.03	$y = -5.955x + 10.61$	0.998	49.5	82.3	−110
	2.0	0.05	$y = -4.894x + 7.616$	0.999	41.0	81.3	−134
	3.0	0.06	$y = -6.259x + 12.33$	0.999	52.0	80.5	−95.0
	4.0	0.08	$y = -5.659x + 10.63$	0.999	47.0	38.72	−110
	5.0	0.1	$y = -6.719x + 14.05$	0.998	56.0	80.3	−81.0

2.4. [CAN-PEG] Binding Study. Spectrophotometric studies were performed in order to throw light on CAN binding with PEG (polyethylene glycol). UV-Visible spectrum of CAN in MeCN indicated a band at 459 nm. It underwent a hypsochromic shift (blue shift) of about 17 to 18 nm (band-shifted from 441 to 442 nm) in presence of PEG (Figure 1), followed by hypochromic shift. This observation indicates the formation of [CAN-PEG] complex. The CAN-PEG binding constants (K) were evaluated by the method of Benesi-Hildebrand [32]. When the equilibrium for molecular complexation is taken into consideration,

$$\text{PEG} + \text{CAN} \overset{K}{\rightleftharpoons} \text{C} \quad (1)$$

The equilibrium constant $K = [C]/[CAN][PEG]$, where [CAN], [PEG], and [C] are equilibrium concentrations of acceptor (CAN), donor (PEG), and complex, respectively. If $[CAN]_0$ and $[PEG]_0$ represent initial concentrations of CAN and PEG, respectively, then

$$K = \frac{[C]}{([CAN]_0 - [C])([PEG]_0 - [C])}. \quad (2)$$

Under condition that $[PEG]_0 \gg [CAN]_0 \Rightarrow [PEG]_0 \gg [C]$ and $([PEG]_0 - [C]) \sim [PEG]_0$

$$K = \frac{[C]}{([CAN]_0 - [C])[PEG]_0}$$

$$\Longrightarrow K = \frac{[C]}{([CAN]_0 [PEG]_0 - [C][PEG]_0)},$$

$$[C](1 + K[PEG]_0) = K[CAN]_0[PEG]_0$$

$$\Longrightarrow [C] = \frac{K[CAN]_0[PEG]_0}{(1 + K[PEG]_0)}. \quad (3)$$

But according to Lambert-Beer's law absorbance $(A) = \epsilon cl$.

In the previous equations, l is path length, A is absorbance, ϵ is the molar extinction coefficient, and K is formation constant of the complex, respectively,

$$[C] = \frac{A}{\epsilon l} = \frac{K[CAN]_0[PEG]_0}{(1 + K[PEG]_0)}. \quad (4)$$

According to this plot of absorbance (A) as a function of donor concentration ($[PEG]_0$) the plot indicated a linear increase in the value of (A) with an increase in ($[PEG]_0$) with

TABLE 5: Activation parameters of nitrocinnamic acid in different PEG media. Units of $k'' = dm^3\ mol^{-1}\ min^{-1}$.

Type of PEG	PEG % (V/V)	k'' at 300 K	Equation	R^2	$\Delta H^{\#}$ kJ/mol	$\Delta G^{\#}$ kJ/mol	$\Delta S^{\#}$ J/K/mol
PEG-200	0.5	0.01	$y = -12.00x + 29.73$	0.999	100	85.0	50.0
	1.0	0.02	$y = -10.87x + 26.63$	0.999	90.3	83.1	24.0
	2.0	0.04	$y = -4.711x + 6.807$	0.999	39.2	37.05	−141
	3.0	0.06	$y = -5.506x + 9.842$	0.999	46.0	38.04	−116
	4.0	0.07	$y = -5.012x + 8.347$	0.999	42.0	38.40	−128
	5.0	0.09	$y = -3.923x + 4.963$	0.998	33.0	38.99	−156
PEG-300	0.5	0.01	$y = -9.764x + 22.25$	0.999	81.2	85.0	−13.0
	1.0	0.02	$y = -7.861x + 16.58$	0.999	65.4	83.3	−60.0
	2.0	0.04	$y = -4.270x + 5.297$	0.999	35.5	37.0	−153
	3.0	0.05	$y = -6.852x + 14.13$	0.999	57.0	81.0	−80.0
	4.0	0.06	$y = -6.545x + 13.3$	0.998	54.4	81.2	−89.5
	5.0	0.08	$y = -9.042x + 21.84$	0.999	75.1	80.0	−16.0
PEG-400	0.5	0.02	$y = -7.717x + 16.10$	0.977	64.1	83.2	−64.0
	1.0	0.03	$y = -5.321x + 8.612$	0.999	44.2	82.0	−126
	2.0	0.04	$y = -8.683x + 20.01$	0.999	72.1	81.4	−31.2
	3.0	0.06	$y = -8.718x + 20.54$	0.999	72.4	80.4	−27.0
	4.0	0.07	$y = -7.947x + 18.23$	0.998	66.0	80.0	−46.0
	5.0	0.10	$y = -7.687x + 17.63$	0.999	64.0	79.3	−51.0
PEG-600	0.5	0.01	$y = -9.635x + 21.82$	0.998	80.1	85.0	−16.2
	1.0	0.02	$y = -8.534x + 18.80$	0.997	71.0	83.4	−41.3
	2.0	0.04	$y = -4.071x + 4.648$	0.999	34.0	82.0	−159
	3.0	0.06	$y = -4.703x + 7.167$	0.999	39.1	80.5	−138
	4.0	0.08	$y = -3.635x + 3.891$	0.999	30.2	80.0	−165
	5.0	0.10	$y = -5.600x + 10.65$	0.999	46.5	79.2	−109

Time (sec)	$\ln(A_o/A_t)$
0	0
60	0.1344
120	0.2636
180	0.4150
240	0.5422
300	0.6805
360	0.8176
420	0.9852
450	1.0497

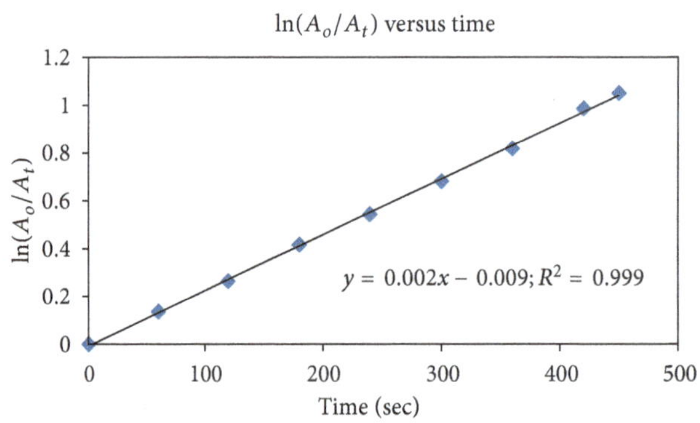

FIGURE 3: Pseudo first order plot of cinnamic acid ($\ln(A_o/A_t)$ versus time) at 323 K. $[CA] = 1.00 \times 10^{-2}\ mol\ dm^{-3}$; $[CAN] = 4.00 \times 10^{-3}\ mol\ dm^{-3}$; $[PEG\text{-}300] = 9.99 \times 10^{-1}\ mol\ dm^{-3}$.

a limiting value at its higher concentrations. Further, taking the reciprocals to the previous equation, it rearranges to

$$\left(\frac{[CAN]_0}{Al}\right) = \left(\frac{1}{K[PEG]_0\epsilon}\right) + \left(\frac{1}{\epsilon}\right). \quad (5)$$

For one cm path length, the aforementioned equation can be written as

$$\left(\frac{[CAN]_0}{A}\right) = \left(\frac{1}{K[PEG]_0\epsilon}\right) + \left(\frac{1}{\epsilon}\right). \quad (6)$$

But the absorbance of CAN and [CAN-PEG] absorbs in the same region significantly; therefore the observed absorbance (A) could be written as

$$A = A_{(CAN)} + A_{(Complex)}. \quad (7)$$

Now the true absorbance of [CAN-PEG] adduct is the difference between observed absorbance and absorbance of CAN:

$$\Longrightarrow A_{(Complex)} = \Delta A = A \sim A_{(CAN)}. \quad (8)$$

TABLE 6: Activation parameters of acrylic acid in different PEG media. Units of $k'' = dm^3\ mol^{-1}\ min^{-1}$.

Type of PEG	PEG % (V/V)	k'' at 300 K	Equation	R^2	$\Delta H^\#$ (kJ/mol)	$\Delta G^\#$ (kJ/mol)	$\Delta S^\#$ J/K/mol
PEG-200	0.5	0.01	$y = -11.85x + 29.21$	0.999	98.5	85.0	45.3
	1.0	0.02	$y = -8.384x + 18.31$	0.998	70.0	84.0	−45.3
	2.0	0.03	$y = -7.860x + 16.98$	0.998	65.3	82.2	−57.0
	3.0	0.04	$y = -8.164x + 18.27$	0.997	68.0	82.0	−47.0
	4.0	0.06	$y = -6.581x + 13.42$	0.999	55.0	81.0	−86.0
	5.0	0.08	$y = -6.061x + 11.96$	0.998	51.0	80.5	−98.2
PEG-300	0.5	0.01	$y = -11.85x + 29.21$	0.999	98.5	85.0	46.0
	1.0	0.02	$y = -11.43x + 28.49$	0.998	95.0	83.2	39.3
	2.0	0.03	$y = -10.67x + 26.25$	0.997	89.0	83.0	21.0
	3.0	0.04	$y = -9.163x + 21.61$	0.999	76.2	82.0	−18.0
	4.0	0.05	$y = -6.125x + 11.72$	0.999	51.0	81.0	−100
	5.0	0.07	$y = -5.443x + 9.792$	0.998	45.3	80.2	−116
PEG-400	0.5	0.02	$y = -8.314x + 18.07$	0.997	69.2	83.4	−47.4
	1.0	0.03	$y = -9.364x + 21.98$	0.998	78.0	82.5	−15.0
	2.0	0.04	$y = -9.756x + 23.58$	0.999	81.2	82.0	−1.6
	3.0	0.06	$y = -7.889x + 17.76$	0.997	66.0	81.0	−50.0
	4.0	0.08	$y = -6.573x + 13.69$	0.998	55.0	80.2	−84.0
	5.0	0.10	$y = -8.507x + 20.35$	0.999	71.0	80.0	−28.4
PEG-600	0.5	0.01	$y = -9.947x + 22.82$	0.998	83.0	85.4	−8.0
	1.0	0.02	$y = -6.653x + 12.57$	0.997	55.4	83.3	−93.0
	2.0	0.03	$y = -6.334x + 11.88$	0.997	53.0	83.0	−99.0
	3.0	0.05	$y = -2.362x - 0.780$	0.998	20.0	77.4	−191
	4.0	0.07	$y = -3.078x + 1.893$	0.998	26.0	80.5	−182
	5.0	0.09	$y = -4.666x + 7.455$	0.998	39.0	80.0	−135

Time (sec)	$\ln(A_o/A_t)$
0	0
5	0.068
10	0.1431
15	0.2025
25	0.34015
30	0.4054
35	0.4834
40	0.5679
45	0.6286
55	0.7621
65	0.9162

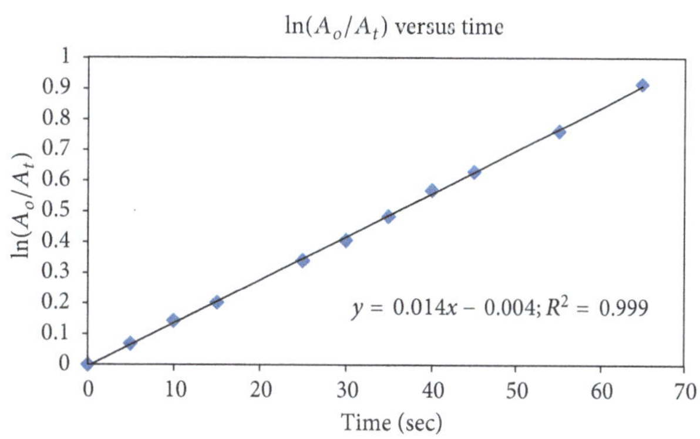

FIGURE 4: Pseudo first order plot of acrylic acid ($\ln(A_o/A_t)$ versus time) at 315 K. [AA] $= 1.00 \times 10^{-2}$ mol dm^{-3}; [CAN] $= 4.00 \times 10^{-3}$ mol dm^{-3}; [PEG-300] $= 13.32 \times 10^{-2}$ mol dm^{-3}.

Therefore, a plot of $([CAN]_0/\Delta A)$ versus $1/[PEG]_0$ should give a straight line according to the previous equation. These plots (Figure 2) have been realized in the present study. Formation constant (K) has been calculated from the ratio of intercept to slope, while inverse of the intercept gave molar extinction coefficient (ϵ) and is represented in Table 1.

3. Results and Discussion

3.1. Determination of the Order of Reaction. General form of the rate law for a nitrodecarboxylation (nitro-Hunsdiecker reaction) reaction could be represented by considering the following general scheme:

$$CAN + S \longrightarrow Products \tag{9}$$

where CAN = ceric ammonium nitrate and S = α,β-unsaturated carboxylic acid:

$$V = k_1[CAN]^x[CAN]^y. \tag{10}$$

Reactions were conducted under two different conditions. Under pseudo first order conditions $[CAN] \gg [CAN]$,

Time (sec)	$\ln(A_o/A_t)$
0	0
10	0.047
20	0.1140
30	0.1490
40	0.1854
50	0.2425
60	0.2985
70	0.3457
80	0.3902
90	0.4467
100	0.5108
110	0.5634
120	0.6008

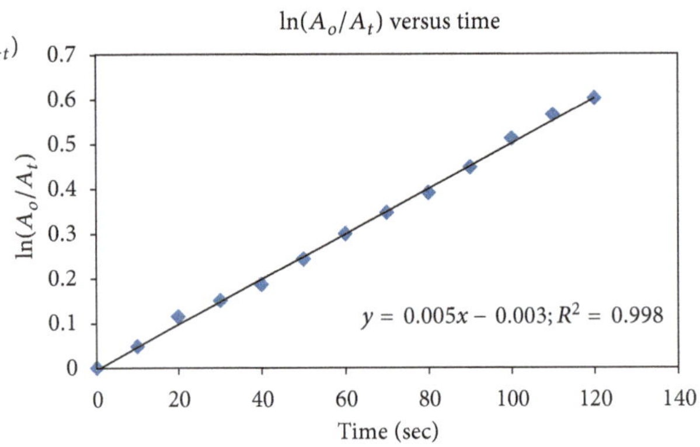

FIGURE 5: Pseudo first order plot of nitrocinnamic acid ($\ln(A_o/A_t)$ versus time) at 315 K. [NCA] = 1.00×10^{-2} mol dm^{-3}; [CAN] = 4.00×10^{-3} mol dm^{-3}; [PEG-400]= 1.25×10^{-1} mol dm^{-3}.

Time (min)	$1/A_t$
0	1.1543
10	1.8408
20	2.3255
30	2.9727
40	3.6433
50	4.2761
70	4.8236
80	5.4823
90	6.1428

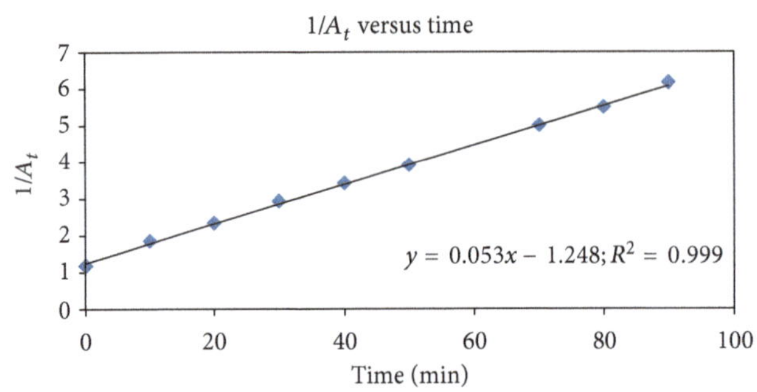

FIGURE 6: Second order kinetic plot of cinnamic acid ($1/A_t$ versus Time) at 310 K. [CA] = 4.00×10^{-2} mol dm^{-3}; [CAN] = 4.00×10^{-2} mol dm^{-3}; [PEG-200] = 5.00×10^{-1} mol dm^{-3}.

plots of $\ln(A_o/A_t)$ versus time were straight lines with a negative slope indicating order (x) with respect to [CAN] to be unity. As typical examples pseudo first order plots are shown in Figures 3, 4, and 5. First order rate constant (k') was obtained from the slopes of line plots cited (k').

The reaction is also conducted under second order conditions with equal concentrations of $[\text{CAN}]_0$ = $[\text{CAN}]_0$. Kinetic plots of $[1/(a-x)]$ or $[1/(A_t)]$ versus time (Figures 6, 7, and 8) have been found to be linear with a positive gradient and definite intercept on ordinate (vertical axis), indicating overall second order kinetics.

In the present study, kinetic data have been collected at three to four different temperatures within the range of 300–315 K. Activation parameters such as $\Delta H^{\#}$ and $\Delta S^{\#}$ have been evaluated from Eyring's plots (Figures 9 and 10). Free energy of activation ($\Delta G^{\#}$) is obtained from Gibbs-Helmholtz equation:

$$\Delta G^{\#} = \Delta H^{\#} - T\Delta S^{\#}. \quad (11)$$

3.2. Mechanism of Nitrodecarboxylation of Unsaturated Acids by Ce(IV) Nitrate (CAN) in MeCN Medium. In order to gain an insight into the mechanistic aspects of CAN-CA reaction in MeCN medium, the knowledge of distribution of CAN species in HNO_3 medium could be useful. In HNO_3 Ce(IV) mainly exists as $Ce(NO_3)_6^{2-}$, $Ce(NO_3)_5^{-}$, $Ce(OH)(NO_3)_4^{-}$, $Ce(NO_3)_4$, which could arise from the following equilibria as cited in the literature reports [33–36]:

$$\begin{gathered} Ce(NO_3)_4 + NO_3^- \rightleftharpoons Ce(NO_3)_5^- \\ Ce(NO_3)_5^- + NO_3^- \rightleftharpoons Ce(NO_3)_6^{2-} \\ Ce(NO_3)_4 + H_2O \rightleftharpoons Ce(OH)(NO_3)_4^- + H^+ \end{gathered} \quad (12)$$

But the CAN species in the present study could be entirely different because the reactions are conducted in MeCN medium, which is large excess over [CAN]. MeCN may penetrate into the coordination spheres of Ce(IV) and form solvated CAN species according to the following equilibrium:

$$Ce(NO_3)_4 + 2ACN \rightleftharpoons [Ce(NO_3)_4(ACN)_2] \quad (13)$$

When unsaturated acid (USA) is added to the reaction mixture containing solvated CAN, it may form [Ce(IV) carboxylate] precursor which dissociates and releases HNO_3 *in situ*. The HNO_3 thus released in turn may combine with another solvated CAN to form nitronium ion (NO_2^+) which

$$\underset{\text{(PEG)}}{\text{H-(OCH}_2\text{-CH}_2)_n\text{-OH}} + \underset{\text{(CAN)}}{[\text{Ce(NO}_3)_4(\text{ACN})_2]} \underset{-\text{HNO}_3}{\overset{K}{\rightleftharpoons}} \underset{\text{[PEG-CAN]}}{[\text{H-(OCH}_2\text{-CH}_2)_n\text{-O-Ce(NO}_3)_3(\text{ACN})_2]}$$

Scheme 1

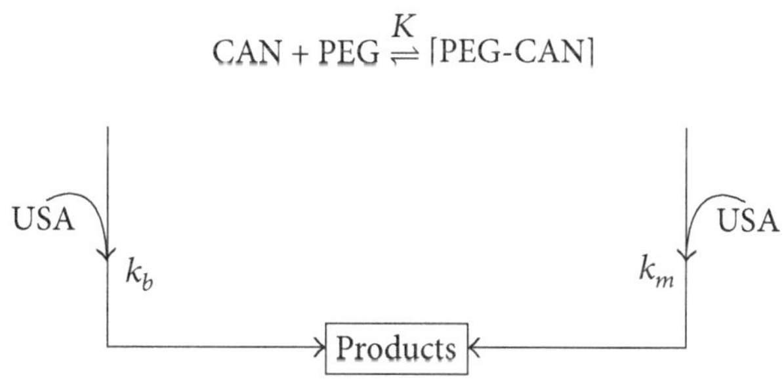

Scheme 2

finally reacts with coordinated USA to afford β-nitrostyrene as shown in the following sequence of steps:

$$\text{RCH=CH-COOH} + [\text{Ce(NO}_3)_4(\text{CH}_3\text{CHN})_2] \rightleftharpoons \underset{\text{(Precursor)}}{[\text{Ce(NO}_3)_4\,(\text{ACN})\,(\text{RCH=CH-COOH})]}$$
$$\text{Precursor} \overset{\text{(Slow)}}{\longrightarrow} \text{HNO}_3 + [\text{Ce(NO}_3)_3\,(\text{ACN})\,(\text{RCH=CH-COOH})]$$
$$\text{HNO}_3 + [\text{Ce(NO}_3)_4(\text{ACN})_2] \longrightarrow \text{NO}_2{}^+ + \text{NO}_3{}^- + [\text{Ce(NO}_3)_3\,(\text{OH})\,(\text{ACN})_2]$$
$$\text{NO}_2{}^+ + [\text{Ce(NO}_3)_3\,(\text{ACN})\,(\text{RCH=CH-COOH})] \longrightarrow \text{RCH=CHNO}_2 + \text{CO}_2 + [\text{Ce(NO}_3)_3\,(\text{ACN})]^+$$
$$[\text{Ce(NO}_3)_3\,(\text{ACN})]^+ + \text{NO}_3{}^- + \text{ACN} \longrightarrow [\text{Ce(NO}_3)_4(\text{ACN})_2] \qquad (14)$$

3.3. Mechanism of Nitrodecarboxylation in PEG Media. Progress of the reaction has been studied in the presence of a set of polyoxyethylene compounds (PEGs) with varied molecular weights ranging from 200 to 600 units, and it was found that the reaction is enhanced remarkably in all PEGs. Reaction times were reduced from 24 hrs to few hours. The catalytic activity was found to be in the increasing order PEG-300 > PEG-400 > PEG-600 > PEG-200. Further, it is also interesting to note that the absorbance of solvated Ce(IV) species is increased when PEG is added to [CAN] solution. This observation may indicate that solvated [CAN] species could bind with PEG to form *PEG supported CAN species* according to the equilibrium shown in Scheme 1.

The plots of k_m (rate constant of PEG reaction) versus C_{PEG} (concentration of PEG) indicated a rate maxima nearly in the vicinity of 1.50 mol dm^{-3} PEG-200, 0.99 mol dm^{-3} PEG-300, 0.99 mol dm^{-3} PEG-400, 0.500 mol dm^{-3} PEG-600. Efforts have been made to interpret PEG effect on the mechanism of CAN-USA reaction. Mechanism of PEG-mediated reactions was explained in the lines of micellar catalysis because PEG resembles the structure of Triton-X. The first and widely applied model is that of Menger and Portnoy [37] which closely resembles that of an enzymatic Catalysis. According to this model, formation of PEG bound reagent (PEG-CAN) could occur in the preequilibrium step due to the interaction of Ce(IV) with PEG. The complex thus formed may possess higher or lower reactivity to give products. A general mechanism is proposed by considering the bulk phase and micellar phase reactions as shown in Scheme 2, where k_m and k_0 or (k_b) represents rate constants for PEG and bulk phases, respectively, and K is the [PEG-CAN] binding constant. For the previous mechanism, rate law could be derived according to the following sequence of steps in the lines of micellar catalyzed reactions.

Considering the total concentration of (C_s) as the algebraic sum of free species and PEG bound substrate complex (CAN-PEG) species,

$$C_S = [\text{CAN}] + [\text{CAN-PEG}]. \qquad (15)$$

From micelle-substrate binding equilibrium,

$$K = \frac{[\text{CAN-PEG}]}{[\text{PEG}]\,[\text{CAN}]}, \quad \text{or}$$
$$[\text{CAN}] = \frac{[\text{CAN-PEG}]}{K\,[\text{PEG}]}. \qquad (16)$$

Time (min)	$1/A_t$
0	1.3457
10	1.5518
20	1.8230
30	2.0408
40	2.2222
50	2.5000
60	2.8027
70	3.0303
80	3.3333
90	3.5037
100	3.7454

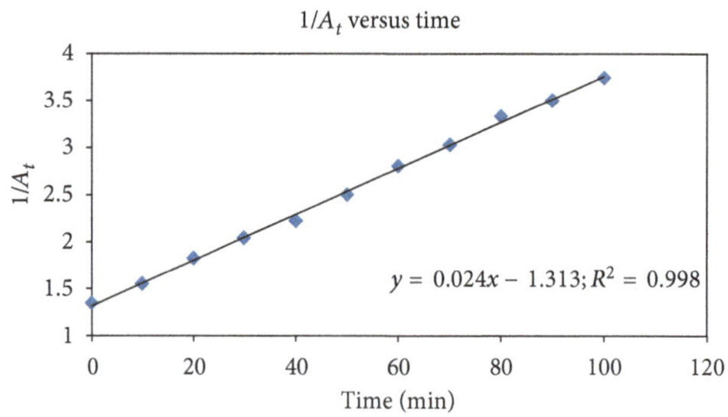

FIGURE 7: Second order kinetic plot of acrylic acid ($1/A_t$ versus Time) at 310 K. [AA] = 4.00×10^{-2} mol dm^{-3}; [CAN] = 4.00×10^{-2} mol dm^{-3}; [PEG-300] = 3.33×10^{-1} mol dm^{-3}.

Time (min)	$1/A_t$
0	1.1473
10	1.3666
20	1.5543
30	1.7867
40	1.9607
50	2.2408
60	2.4255
70	2.6641
80	2.8571
90	3.1250
100	3.2714
120	3.7562

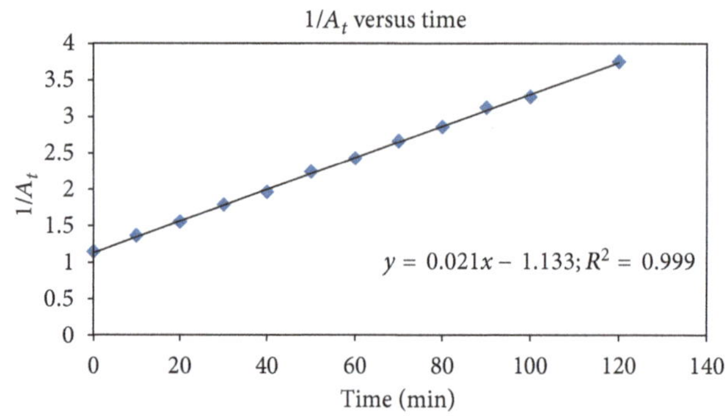

FIGURE 8: Second order kinetic plot of nitrocinnamic acid ($1/A_t$ versus time) at 310 K. [NCA] = 4.00×10^{-2} mol dm^{-3}; [CAN] = 4.00×10^{-2} mol dm^{-3}; [PEG-600] = 3.32×10^{-1} mol dm^{-3}.

Substitution of [CAN] in (15) gives

$$C_S = \frac{[\text{CAN-PEG}]}{K[\text{PEG}]} + [\text{CAN-PEG}] = \frac{[\text{CAN-PEG}] + K[\text{PEG}][\text{CAN-PEG}]}{K[\text{PEG}]} \quad \text{or} \qquad (17)$$

$$[\text{CAN-PEG}] = \frac{K[\text{PEG}]C_S}{1 + K[\text{PEG}]}.$$

Similarly free substrate [CAN] is written as [CAN] = C_S − [CAN-PEG]

$$[\text{CAN}] = C_S - \frac{K[\text{PEG}]C_S}{1 + K[\text{PEG}]}. \qquad (18)$$

After simplification, the previous equation reduces to

$$[\text{CAN}] = \frac{C_S}{1 + K[\text{PEG}]}. \qquad (19)$$

Substitution of [CAN-PEG] and [CAN] in (15) gives

$$k' = \frac{k_0 C_S}{1 + K[\text{PEG}]} + \frac{k_m K[\text{PEG}]C_S}{1 + K[\text{PEG}]} \quad \text{or}$$

$$k_\varphi = \frac{(k_0 + k_m K[\text{PEG}])}{1 + K[\text{PEG}]} \qquad (20)$$

$$k_\varphi = \frac{(k_0 + k_m K[\text{PEG}])}{1 + K[\text{PEG}]},$$

where $k_\varphi = (k'/[\text{CAN}])$, the second order rate constant in PEG media. Subtracting k_0 from both sides of equation and rearranging

$$k_\varphi - k_0 = \frac{(k_m - k_w)K[\text{PEG}]}{1 + K[\text{PEG}]}. \qquad (21)$$

But, since the reactions are too sluggish in the absence of [PEG], the rate constant (k_0) would be much smaller than ($k_m K$[PEG]); that is, ($k_0 \lll k_m K$[PEG]). Therefore the (k_0) term could be neglected in the previous equation. The rate law (21) could be then considered as

$$k_\varphi = \frac{k_m K[\text{PEG}]}{1 + K[\text{PEG}]}. \qquad (22)$$

$10^3/T$	$\ln(k''/T)$
3.3333	−8.9226
3.2786	−8.4334
3.225	−7.9491
3.1746	−7.562

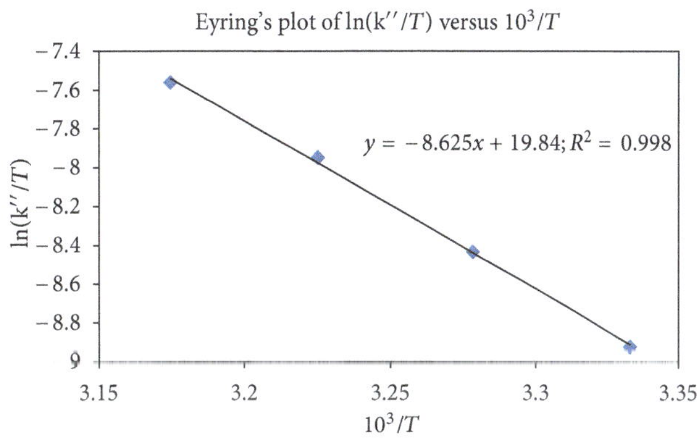

FIGURE 9: Eyring's plot of cinnamic acid ($\ln(k''/T)$ versus $10^3/T$). Kinetic study with [PEG-200] = 2.00 mol dm^{-3}.

$10^3/T$	$\ln(k''/T)$
3.3333	−10.3089
3.2786	−9.6323
3.225	−9.0431
3.1746	−8.4118

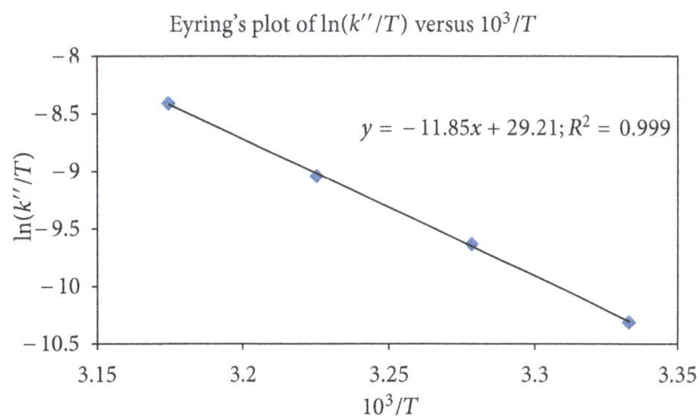

FIGURE 10: Eyring's plot of acrylic acid ($\ln(k''/T)$ versus $10^3/T$). Kinetic study [PEG-300] = 13.32×10^{-1} mol dm^{-3}.

This rate-law resembles Michaelis-Menten type rate law that is used for enzyme kinetics. Interestingly the plots of rate constant (k_φ), that is, second order rate constant of PEG-mediated reaction versus [PEG], indicated either Hill type curve (i.e., a gradual increase with an increase in [PEG] passing through a maximum point in the profile). This observation could be explained due to the CAN-PEG binding interactions to give [CAN-PEG] species which is responsible for rate accelerations in PEG-mediated reactions. Formation of [CAN-PEG] species has also been established by UV-Visible spectroscopic studies, and binding constants are compiled in Table 1. These observations strengthened us to take up detailed reaction kinetics at various PEG concentrations in order to have an insight into the variation in the enthalpies and entropies of activation with [PEG].

3.4. Effect of Structure on the Reactivity, Enthalpy, and Entropy Changes. The enthalpy ($\Delta H^{\#}$) and entropy of activation ($\Delta S^{\#}$) are the two parameters typically obtained from the temperature dependence of a reaction rate, when these data are analyzed using Eyring's equation. Both of these values are obtained from Eyring's plots according to standard procedures [37–39]. Typical plots are shown in Figures 9 and 10. The entropy of activation ($\Delta S^{\#}$) provides clues about the molecularity of the rate determining step in a reaction, that is, whether the reactants are bonded to each other or not. Even though structural change in the cinnamic acid exhibits slight variation on its reactivity, no quantitative conclusions could be obtained from rate constant data. However, the observed positive $\Delta S^{\#}$ values in lower PEG concentration (0.1 M PEG) decreased with an increase in [PEG] and are becoming more negative for all the unsaturated acids which probably indicates that PEG is tightly bound to CAN and lends support that PEG bound CAN is the active species in PEG-mediated reactions. The observed negative $\Delta S^{\#}$ values are in consonance with the concept that for associative mechanism, entropy decreases upon achieving the transition state, and values near zero are difficult to interpret [38–40]. By and large similar magnitude of observed free energy of activation ($\Delta G^{\#}$) in this reaction series (Tables 2, 3, 4, 5, 6) indicates a similar type of mechanism which is operative in the present study. The observed negative magnitude of ΔG (Table 1) indicates the spontaneity of complexation of [CAN-PEG] due to PEG binding with CAN.

4. Conclusions

Being a versatile chemical reagent CAN has been applied to organic reactions in catalytic or stoichiometric amounts. Our research group has succeeded in using CAN to perform nitrodecarboxylation of α,β-unsaturated aromatic carboxylic acids to give β-nitrostyrene derivatives. Furthermore, use of PEG supported CAN could effectively initiate the decarboxylation of carboxyl group followed by nitration. Because the CAN reagent is cost effective, we believe that the present protocol offers a rapid and clean alternative and reduces reaction times.

References

[1] T. L. Ho, "Ceric ion oxidation in organic chemistry," *Synthesis*, vol. 6, pp. 347–354, 1973.

[2] V. Nair, L. Balagopal, R. Rajan, and J. Mathew, "Recent advances in synthetic transformations mediated by cerium(IV) ammonium nitrate," *Accounts of Chemical Research*, vol. 37, no. 1, pp. 21–30, 2004.

[3] J. R. Hwu and K. Y. King, "Versatile reagent ceric ammonium nitrate in modern chemical synthesis," *Current Science*, vol. 81, no. 8, pp. 1043–1053, 2001.

[4] V. Nair and A. Deepthi, "Cerium(IV) ammonium nitrate—a versatile single-electron oxidant," *Chemical Reviews*, vol. 107, no. 5, pp. 1862–1891, 2007.

[5] V. Singh, V. Sapehiyia, and G. L. Kad, "Ultrasonically activated oxidation of hydroquinones to quinones catalysed by ceric ammonium nitrate doped on metal exchanged K-10 clay," *Synthesis*, no. 2, pp. 198–200, 2003.

[6] H. M. Chawla and R. S. Mittal, "Oxidative nitration by silica gel-supported cerium(IV) ammonium nitrate," *Synthesis*, vol. 1, pp. 70–72, 1985.

[7] S. Dincturk and J. H. J. Ridd, "Nitration of an electron-rich aromatic compound by CAN-SiO_2," *Journal of the Chemical Society, Perkin Transactions*, vol. 2, pp. 965–969, 1982.

[8] A. S. Sarac, "Redox polymerization," *Progress in Polymer Science*, vol. 24, no. 8, pp. 1149–1204, 1999.

[9] I. E. Marko, A. Ates, A. Gautier et al., "Cerium(IV)-catalyzed deprotection of acetals and ketals under mildly basic conditions," *Angewandte Chemie International Edition*, vol. 38, no. 21, pp. 3207–3209, 1999.

[10] N. Maulide, J. C. Vanherck, A. Gautier, and I. E. Markó, "Mild and neutral deprotections catalyzed by cerium(IV) ammonium nitrate," *Accounts of Chemical Research*, vol. 40, no. 6, pp. 381–392, 2007.

[11] J. Christoffers, T. Werner, and M. Rössle, "Cerium-catalyzed oxidative C–C bond forming reactions," *Catalysis Today*, vol. 121, no. 1-2, pp. 22–26, 2007.

[12] M. Rössle, T. Werner, W. Frey, and J. Christoffers, "Cerium-catalyzed, aerobic oxidative synthesis of 1,2-dioxane derivatives from styrene and their fragmentation into 1,4-dicarbonyl compounds," *European Journal of Organic Chemistry*, no. 23, pp. 5031–5038, 2005.

[13] E. Ganin and I. Amer, "Cerium-catalyzed selective oxidation of alkylbenzenes with bromate salts," *Synthetic Communications*, vol. 25, no. 20, pp. 3149–3154, 1995.

[14] K. Auty, B. C. Gilbert, C. B. Thomas, S. W. Brown, C. W. Jones, and W. R. Sanderson, "The selective oxidation of toluenes to benzaldehydes by cerium(III), hydrogen peroxide and bromide ion," *Journal of Molecular Catalysis A*, vol. 117, no. 1–3, pp. 279–287, 1997.

[15] R. P. Kreh, R. M. Spotnitz, and J. T. Lundquist, "Mediated electrochemical synthesis of aromatic aldehydes, ketones, and quinones using ceric methanesulfonate," *Journal of Organic Chemistry*, vol. 54, no. 7, pp. 1526–1531, 1989.

[16] H. Feuer and A. T. Nielsen, *Nitro Compounds Recent Advances in Synthesis and Chemistry*, vol. 9, VCH Publishers, New York, NY, USA, 1990.

[17] A. G. M. Barrett, "Heterosubstituted nitroalkenes in synthesis," *Chemical Society Reviews*, vol. 20, no. 1, pp. 95–127, 1991.

[18] R. S. Varma and G. W. Kabalka, "Nitroalkenes in the synthesis of heterocyclic compounds," *Heterocycles*, vol. 24, no. 9, pp. 2645–2677, 1986.

[19] S. E. Denmark and L. R. Marcin, "A general method for the preparation of 2,2-disubstituted 1-nitroalkenes," *Journal of Organic Chemistry*, vol. 58, no. 15, pp. 3850–3856, 1993.

[20] M. Node, A. Itoh, K. Nishide et al., "Preparation of nitroalkenes: Substitution reaction via addition elimination using β nitrovinyl sulfoxides," *Synthesis*, no. 11, pp. 1119–1124, 1992.

[21] R. Ballini, R. Castagnani, and M. Petrini, "Chemoselective synthesis of functionalized conjugated nitroalkenes," *Journal of Organic Chemistry*, vol. 57, no. 7, pp. 2160–2162, 1992.

[22] S. S. Jew, H. D. Kim, Y. S. Cho, and C. H. Cook, "A practical preparations of conjugated nitroalkenes," *Chemistry Letters*, vol. 1986, no. 10, pp. 1747–1748.

[23] W. W. Sy and A. W. By, "Nitration of substituted styrenes with nitryl iodide," *Tetrahedron Letters*, vol. 26, no. 9, pp. 1193–1196, 1985.

[24] V. V. Namboodiri and R. S. Varma, "Microwave-accelerated Suzuki cross-coupling reaction in polyethylene glycol (PEG)," *Green Chemistry*, vol. 3, no. 3, pp. 146–148, 2001.

[25] A. Haimov and R. Neumann, "Polyethylene glycol as a non-ionic liquid solvent for polyoxometalate catalyzed aerobic oxidation," *Chemical Communications*, no. 8, pp. 876–877, 2002.

[26] L. Heiss, "Polyethylene glycol monomethyl ether-modified pig liver esterase: Preparation, characterization and catalysis of enantioselective hydrolysis in water and acylation in organic solvents," *Tetrahedron Letters*, vol. 36, no. 22, pp. 3833–3836, 1995.

[27] S. Chandrasekhar, C. Narsihmulu, S. S. Sultana, and N. R. Reddy, "Osmium tetroxide in poly(ethylene glycol) (PEG): a recyclable reaction medium for rapid asymmetric dihydroxylation under sharpless conditions," *Chemical Communications*, vol. 9, no. 14, pp. 1716–1717, 2003.

[28] K. Tanemura, T. Suzuki, Y. Nishida, and T. Horaguchi, "Aldol condensation in water using polyethylene glycol 400," *Chemistry Letters*, vol. 34, no. 4, pp. 576–577, 2005.

[29] R. Kumar, P. Chaudhary, S. Nimesh, and R. Chandra, "Polyethylene glycol as a non-ionic liquid solvent for Michael addition reaction of amines to conjugated alkenes," *Green Chemistry*, vol. 8, no. 4, pp. 356–358, 2006.

[30] J. R. Johnson, "The perkin reaction and related reactions," *Organic Reactions*, vol. 1, p. 210, 1942.

[31] C. I. Chiriac, F. Tanasa, and M. Onciu, "A novel approach in cinnamic acid synthesis: direct synthesis of cinnamic acids from aromatic aldehydes and aliphatic carboxylic acids in the presence of boron tribromide," *Molecules*, vol. 10, no. 2, pp. 481–487, 2005.

[32] H. A. Benesi and J. H. Hildebrand, "Ultraviolet absorption bands of iodine and aromatic hydrocarbons," *Journal of the American Chemical Society*, vol. 71, pp. 1832–1833, 1949.

[33] M. Santappa and B. Sethuram, "Oxidation studies. IV. Kinetics of oxidation of HCHO and some alcohols by ceric salts in HNO_3 medium," *Proceedings of the Indian Academy of Sciences Section A*, vol. 67, no. 2, pp. 78–89, 1968.

[34] K. V. Rao and S. S. Muhammed, "Oxidation of alcohols by cerium(IV). I. Oxidation of methanol by ceric perchlorate," *Bulletin of the Chemical Society of Japan*, vol. 36, no. 8, pp. 943–949, 1963.

[35] S. S. Muhammed and B. Sethuram, "Oxidation of alcohols by cerium(IV). I. Oxidation of isopropanol and secondary butanol by ceric nitrate," *Acta Chimica Hungarica*, vol. 46, pp. 115–124, 1965.

[36] N. Dutt, R. R. Nagori, and R. N. Mehrotra, "Kinetics and mechanisms of oxidations by metal ions. Part VI. Oxidation of α-hydroxyl acids by cerium (IV) in aqueous nitric acid," *Canadian Journal of Chemistry*, vol. 64, no. 1, pp. 19–23, 1986.

[37] F. M. Menger and C. E. Portnoy, "Chemistry of reactions proceeding inside molecular aggregates," *Journal of the American Chemical Society*, vol. 89, no. 18, pp. 4698–4730, 1967.

[38] K. A. Connors, *Chemical Kinetics: The Study of Reaction Rates in Solution*, VCH, New York, NY, USA, 1990.

[39] J. H. Espenson, *Chemical Kinetics and Reaction Mechanism*, McGraw-Hill, New York, NY, USA, 1981.

[40] V. Jagannadham, "The change in entropy of activation due to solvation (hydration) of ions: proton versus carbocation," *Chemistry (the Bulgarian Journal of Chemical Education)*, vol. 18, p. 89, 2009.

4

Comparative Biosorption Studies of Hexavalent Chromium Ion onto Raw and Modified Palm Branches

Mona A. Shouman, Nady A. Fathy, Soheir A. Khedr, and Amina A. Attia

Surface Chemistry and Catalysis Laboratory, National Research Centre, El-Dokki, Cairo 12622, Egypt

Correspondence should be addressed to Amina A. Attia; mollyattia@hotmail.com

Academic Editor: Marc Koper

The waste of palm branches (PB) was tested for its ability to remove chromium (VI) from aqueous solution by batch and column experiments. Palm branches chemically modified with an oxidizing agent (sulphuric acid) then coated with chitosan and surfactant (hexadecyl trimethyl ammonium bromide surfactant, HDTMA), respectively, were carried out to improve the removal performance of PB. The results of their Cr (VI) removal performances are pH dependent. The adsorption data could be well interpreted by the Langmuir, Freundlich, and Flory-Huggins isotherm models. The maximum adsorption capacity obtained from the Langmuir model for the chitosan coated oxidized palm branches is 55 mg/mg. The adsorption process could be described by pseudo-second-order kinetic model. The intraparticle diffusion study revealed that film diffusion might be involved. The biosorbents were successfully regenerated using 1 M HCL solution.

1. Introduction

Increased industrialization and human activities have created impact on the environment through the disposal of waste containing heavy metals. Heavy metals are toxic even in extremely minute quantities on human health and to the fauna and flora of receiving water [1]. It is known that legal standards on environment control are becoming strict and, as a result, the discharge of heavy metals into aquatic bodies and sources of potable water is being rigorously controlled. Most heavy metals exist in a cationic form, but Cr (VI) combined with oxygen is anionic in natural aquatic system. The maximum permissible limit of chromium in natural water is only 0.05 mg/L by the US Environmental Protection Agency [2]. Acute exposure to Cr (VI) causes nausea, diarrhea, liver, and kidney damage, internal hemorrhage, and respiratory problems [3, 4]. Chromium compounds are widely used in electroplating, metal finishing, magnetic tapes, pigments, leather tanning, wood protection, chemical manufacturing, electrical and electronic equipment, and catalysis [5]. Several treatment technologies have been developed to remove chromium from aqueous phase. Common methods include biosorption [6], ion exchange [7], chemical reduction, ultrafiltration [8], and adsorption [9]. The adsorption process is an effective technique for the treatment of anionic metals. Activated carbon, a representative adsorbent, has been widely used to remove diverse pollutants, including organic pollutants, reactive dyes, cationic heavy metals, and anionic metals from the aqueous phase. Although it is a preferred adsorbent, and its widespread use, it is restricted due to its high cost production. In order to decrease the cost of treatment, attempts also have been made to find inexpensive alternative adsorbents [10]. Recently, there is a great interest to improve the adsorption capacity of low-cost adsorbents for remediation of wastewater [11–13]. Nowadays, surface modification technique of diverse biosorbents has been implied to increase the surface charge and to enhance their adsorption capacity for ionic pollutants [11, 13]. Surfactants are chemicals that have an amphipathic structure along with a hydrophobic tail and a hydrophilic head. The surfactants are adsorbed onto solid surface and change the surface properties of solids, especially at the interface. Up to date, the performance of surfactants in water and wastewater treatment has been regarded for the separation of metal

TABLE 1: Physicochemical properties of the biosorbents.

Properties	PB	COPB	SOPB
pH_{PZC}	6	7.3	7.5
Surface area, m^2/g	2.8	6.3	5.9
Total pore volume, cm^3/g	0.004	0.012	0.012
Average pore diameter, Å	9.9	12.5	19.5

ions and other toxic substance [14, 15]. Also, the use of chitosan over commercially available agricultural byproducts for the removal of heavy metals is becoming popular due to its low cost, abundant availability, non toxicity, potential for ion exchange, and it has been recently reported in some papers that chitosan adsorbent has high adsorption capacities towards many classes of heavy metals [16], dyes [17], and other ionic molecules [18]. Chitosan, a naturally occurring linear polymer of glucosamine, can be obtained from crustacean chitin on an industrial scale due to a large amount of crustacean's exoskeleton is available as a byproduct of food processing industries.

Palm branches (PB) are produced in great quantities in our country, especially in Siwa Oasis. Palm tree wastes include trunks, fronds, fruit waste, and empty fruit branches. The farmers trim about 12 branches yearly from each tree, so the authors have decided that a fraction of this wood is believed to be enough to supply the growing demand for preparing potential biosorbents. The palm tree branches have the characteristic of exchange sorption properties due to the presence of some functional groups such as carboxylic, hydroxyl, and lactone which have a high affinity for metal ions [19]. In the present investigation an attempt was made to explore the effect of both the chitosan and cationic surfactant modifying agents on the adsorption performance of palm branches for removing Cr (VI) ions by using batch and column systems. In order to better understand the adsorption characteristics, isotherms and kinetics were employed to evaluate the sorption process.

2. Material and Method

2.1. Adsorbate Solution. The stock solution of Cr (VI) was made by dissolving 0.2827 g $K_2Cr_2O_7$ (AR grade from Merck, Germany) in 1 L double distilled water. The solutions of different concentrations used in various experiments were obtained by dilution of the stock solutions.

2.2. Adsorbent Materials. The palm tree branches (PB) were provided from the farm of our institute. The precursor was first washed to remove dirt from its surface and then dried in an oven overnight at 100°C. The dried PB were crushed and sieved to desired mesh size of 2-3 mm. Then the dried PB were treated with 2% H_2SO_4 (v/v) in an incubator at 100°C for 24 h and soaked with deionized water until the solution pH was stable. Afterwards, the adsorbent was soaked in 3% $NaHCO_3$ (w/v) till any residual acid left was removed. Finally, the sample was dried overnight in an oven at 100°C, then cooled at room temperature, and stored in a desiccator [16]. Hereafter, it was called OPB (oxidized palm tree branches).

Surface modification of OPB by the chitosan and cationic surfactant was carried out as follows. About 20 g of chitosan flakes (98% acetylation degree, Figure 1(a)) were immersed in 10% of acetic acid (v/v) with constant stirring for 24 h to form a homogenized whitish viscous gel. Afterwards, OPB was dipped into the gel (dose ratio of chitosan to OPB is 1:5) and shaken overnight at 120 rpm. This sample was then washed with deionized water and dried at 100°C. It was called COPB. The process was repeated for four times to form a thick coating of chitosan onto oxidized palm branches. In case of modification with a cationic surfactant, a 10 g of OPB was added to a solution of hexadecyltrimethyl ammonium bromide (HDTMA) of 0.197 g/L (Figure 1(b)). This concentration corresponds to the 0.5 critical micelle concentration (CMC), which is a measure of surfactant efficiency. Above the CMC, a surfactant forms a micelle that is very stable and remains in the bulk zone. For that reason, 0.5 CMC was selected to maximize the amount of surfactant adsorbed onto the OPB. The mixture was agitated for 24 h and then filtered. The filtered OPB was dried at 100°C for 4 h. The surfactant modified OPB was called SOPB. The characteristics of the three raw PB, COPB, and SOPB biosorbents are summarized in Table 1.

2.3. Characterization of the Biosorbents. The determination of pH_{PZC} for the biosorbents was performed in 50 mL of 0.01 M NaCL solution and placed in a closed Erlenmeyer flask. The pH was adjusted to a value between 2 and 12 by adding 0.1 M NaOH or HCl solutions. Then, 0.1 g of biosorbents was added and the final pH was measured after 48 h under agitation at 30°C. The pH_{PZC} is the point where the curve pH_{final} versus $pH_{initial}$ crosses the line $pH_{initial} = pH_{final}$ [20].

The morphology of biosorbents was analyzed by scanning electron microscope using JEOL 560LV SEM at 20 KeV with background subtraction and a summation of 60 scans. The functional surface groups present on the prepared biosorbents were determined by FTIR analysis. The FTIR spectra of the biosorbents were recorded using JASCO-410 FTIR instrument in the region 4000–400 cm^{-1}. The specific surface area and total pore volume of the biosorbents were calculated from the adsorption data of nitrogen at 77 K by BET analyzer (Quantachrome AS1Win, Version 2.01, USA).

2.4. Batch Studies. All batch experiments were carried out with biosorbent samples in 250 mL conical flasks with 100 mL Cr (VI) aqueous solutions on a rotary shaker at 250 rpm/min. The concentration of chromium ion was determined spectrophotometrically at 540 nm using 1,5-diphenyl carbazide (Sigma-Aldrich, USA) as the a complexing agent [21]. The study of the pH (2–12) dependency of Cr (VI) adsorption onto PB, COPB, and SOPB was carried out in 100 mL of Cr (VI) solution with initial concentration of 100 mg/L at room temperature. The pH value was adjusted by 1 M NaOH or HCl solution.

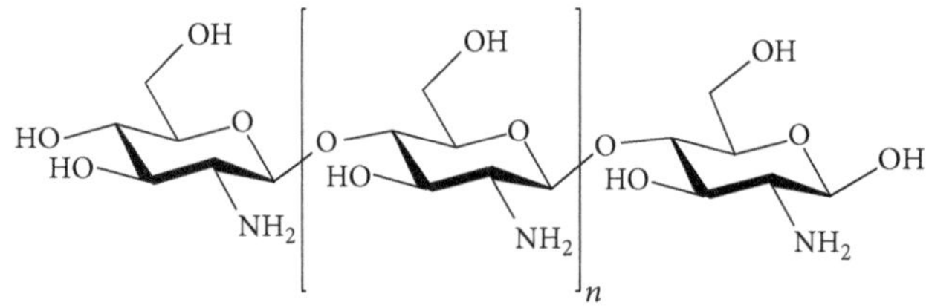

(a)

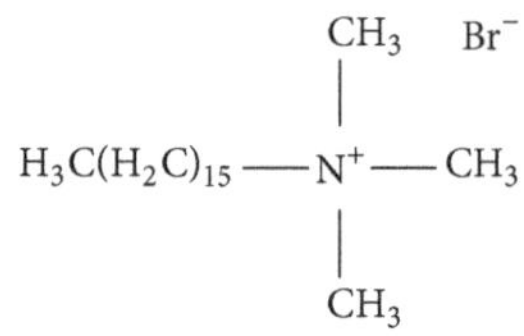

(b)

FIGURE 1: Chemical structures of (a) chitosan and (b) hexadecyltrimethyl ammonium bromide surfactant.

Isotherm studies were conducted by contacting 0.1 g of PB, COPB, and SOPB samples with 50 mL of Cr (VI) solution at different initial concentrations (20, 40, 60, 80, and 100 mg/L) shaking for 24 h at room temperature (30°C). The maximum removal of Cr (VI) solution was found at pH 2 for PB while for COPB and SOPB adsorbents at pH 6, respectively. The percentage dye adsorbed (%R) and the amount of residual metal ions adsorbed on biosorbents (q_e, mg/g) was calculated by the following equations:

$$\%R = \frac{(C_0 - C_e)\,100}{C_0}, \qquad q_e = \frac{(C_0 - C_e)\,V}{W}, \tag{1}$$

where q_e is the equilibrium adsorption capacity (mg/g) and C_0 and C_e are the initial and equilibrium liquid-phase concentrations of solute in aqueous solution (mg/L), respectively. V is the liquid phase volume (L) and W is the mass of biosorbent used (g).

To investigate the kinetic characteristics of the adsorption, 0.1 g of biosorbents was added to 100 mL of heavy metal solution with initial concentration of 100 mg/L and the samples were agitated for designated time periods (5, 10, 20, 30, 40, 50, and 60 min).

2.5. Column Studies. The fixed-bed column was made of Pyrex glass tube of 2 cm inner diameter and 30 cm height. The column performance of Cr (VI) adsorption onto biosorbents was studied at initial concentration of 100 mg/L, bed height 60 mm, and flow rate 2 mL//min. The Cr (VI) solution was pumped to the column in a down-flow direction by a peristaltic pump. Samples were collected at regular intervals at 30°C and the initial pH was 2.

2.6. Column Regeneration Studies. Regeneration studies help to elucidate the recycling nature of the spent biosorbent. The regeneration method was performed in the same column packed with 1 g of each of the three different biosorbents. Then the column was loaded with the 100 mg/L concentration of Cr (VI) solution at the flow rate of 2 mL/min. The column was operated until the effluent concentration matched the concentration of the loaded chromium (VI). After complete saturation of the adsorbent, distilled water was passed through the column to remove traces of undesorbed Cr (VI). Afterwards, the efficiency of 1 M HCl was tested as an eluent to desorb Cr (VI) from the three biosorbents.

3. Results and Discussion

3.1. Surface Features of the Prepared Biosorbents. Figures 2(a), 2(b), and 2(c) show the SEM micrographs for the raw palm branches (PB) and its modified biosorbents (COPB and SOPB). In Figure 2(a), the surface of PB was rough and dented, besides small pores are also evident. For COPB and SOPB biosorbents, Figures 2(b) and 2(c) show highly defined pores and cavities resulted in the modification process. This explains why the BET surface area value for PB was lower than that of COPB and SOPB (cf. Table 1).

To confirm the type of functional groups, FTIR spectra of the three biosorbents are shown in Figure 3. When comparing the FTIR spectra of palm tree branches precursor with the modified samples, significant shifts in absorption peaks were obtained which may be assigned to the effect of modification process on the surface of PB. As shown in Figure 3, characteristic peaks obtained for PB are 3452 cm^{-1} (stretching vibration of the –OH groups), 2926 and 2874 cm^{-1} (C–H sp^3 asymmetric and symmetric stretching), 1633 cm^{-1} (C=O stretching of acetyl groups), 1446 cm^{-1} (O–H bending), and 1174 and 1090 cm^{-1} (alcoholic C–O) [22]. After modification, the peaks obtained for COPB biosorbent are found to be 3423 cm^{-1} (wide peak of the –NH groups stretching), 2926 cm^{-1} (C–H stretching of polymer backbone), 1628 cm^{-1} (N–H bending and C=O stretching of acetyl groups), 1320 cm^{-1} (C–N stretching), and 617–560 cm^{-1} (nitro compounds) [23]. This confirms that the chitosan copolymer is successfully loaded onto palm tree branches. For SOPB (Figure 3(c)), their absorption peaks were observed at 3751 cm^{-1} (N–H groups stretching of secondary amides). Depending on the configuration of surfactant molecules, the vibration bands of HDTMA micelles possess vibration at 2927 cm^{-1} that have been assigned to the symmetric and asymmetric stretching vibration of methyl and methylene groups. Bands at about 1600–1200 cm^{-1} have been ascribed to vibration of trimethyl ammonium quaternary group CN $(CH)_3^+$ [24]. This result indicates that the HDTMA modified the surface properties of palm tree branches.

3.2. Effect of pH. The effect of initial solution pH on Cr (VI) removal by three biosorbents was performed over a range 2–12 and the studies were carried out for 24 h. Figure 4 indicates

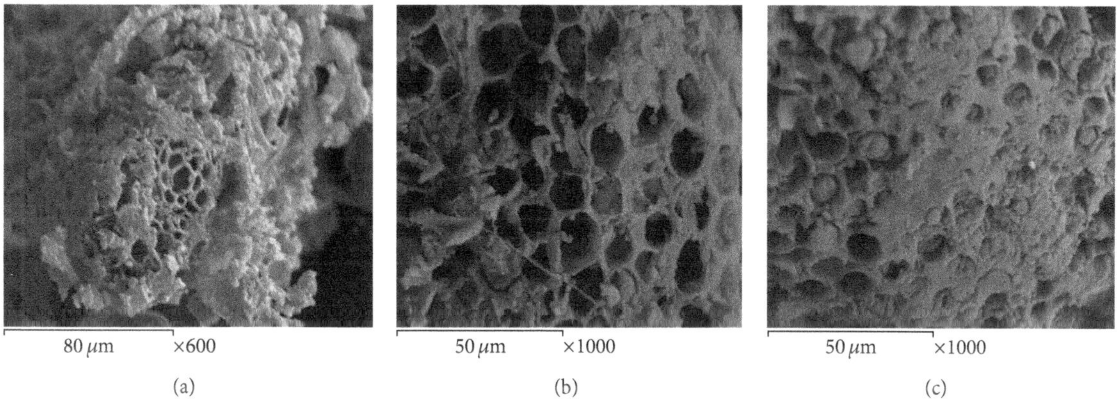

FIGURE 2: SEM images of (a) PB, (b) COPB, and (c) SOPB biosorbents.

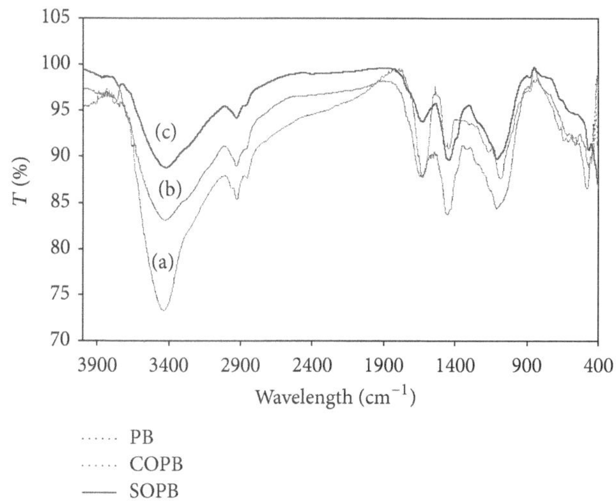

FIGURE 3: FTIR spectra of (a) PB, (b) COPB, and (c) SOPB biosorbents.

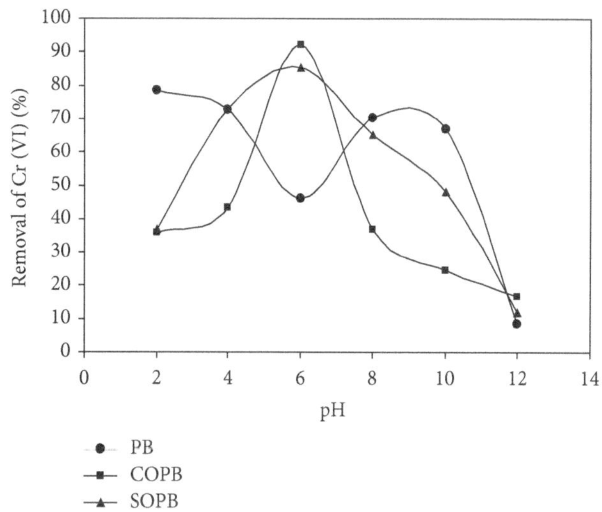

FIGURE 4: Effect of pH on the removal of Cr (VI) by the biosorbents.

that the Cr (VI) uptakes were much higher in acidic solution than those in alkaline conditions. Optimal metal removal efficiency occurs at pH 6 for COPB and SOPB while for PB the optimal is pH 2. It can be explained that the pH of the aqueous solution affects the speciation of chromium and the surface charge of the biosorbent [24, 25].

Cr (VI) exists in different forms in aqueous solution and the stability of these forms is dependent on the pH of the system. In the pH range of 1 to 6 different forms of chromium ions such as $Cr_2O_7^{2-}$, $HCrO_4^{2-}$, $Cr_3O_{10}^{2-}$, and $Cr_4O_{13}^{2-}$, while $HCrO_4^{2-}$ predominates in the solution [25]. Thus Cr (VI) ions exist predominantly as $HCrO_4^{2-}$ in acidic medium and the amino groups ($-NH_2$) of chitosan (COPB) would be in protonated cationic form ($-NH_3^+$) to a higher extent in acidic solution which cause an electrostatic interaction between the biosorbent and $HCrO_4^{2-}$ resulting in high chromium removal [26]. The same trend was also observed for SOPB as depending on the configuration of surfactant molecules, the surfactant head (trimethyl ammonium group) has a lot of positive charge in acidic medium [27]. With respect for PB, originally the palm branches have a lot of positively charged groups on the surface as depicted from FTIR, for this reason its adsorption capacity is relatively high in acidic medium.

At higher pH, the decrease in removal of chromium may be explained by the dual competition of both anions [$HCrO_4^{2-}$ and OH^-] to be adsorbed on the surface of the biosorbents of which OH^- predominates in the solution [27, 28]. The pH_{PZC} obtained for the three biosorbents were in the range 6–7.5. This is consistent with results reported in our study [27]. Accordingly, the biosorption of chromium ions performed at pH 2 for raw PB while for COPB and SOPB samples occurred at pH 6.

3.3. Adsorption Isotherms. Adsorption isotherm studies are important to determine the efficacy of adsorption. This provides a relationship between the concentration of Cr (VI) in the liquid phase and the amount of Cr (VI) adsorbed on the solid-phase when the two phases are at equilibrium. Langmuir, Freundlich, and Flory-Huggins adsorption isotherms were used in this study [29].

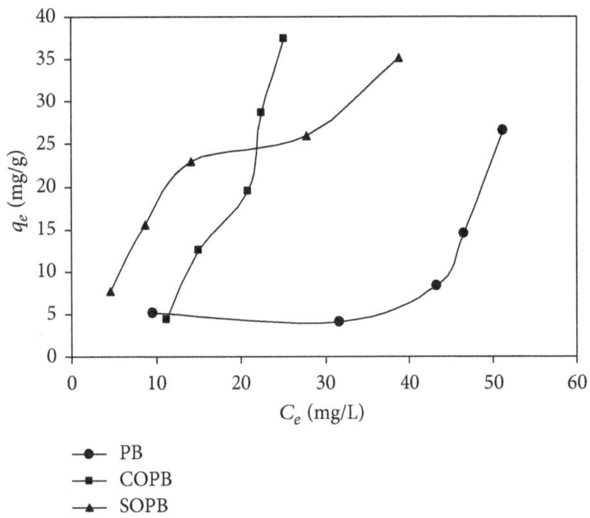

FIGURE 5: Adsorption isotherms of Cr (VI) on the biosorbents.

The Langmuir model assumes that a monomolecular layer is formed when adsorption takes part without any interaction between the adsorbed molecules [30]. The Langmuir model can be represented as linear form:

$$\frac{C_e}{q_e} = \frac{1}{K_L q_{\max}} + \frac{1}{q_{\max}} C_e, \quad (2)$$

where $q_{\max}$ is the maximum monolayer adsorption capacity (mg/g), and K_L is the Langmuir constant (g/L) related to the affinity of the binding sites. The experimental data were plotted as C_e/q_e versus C_e (figure not shown). The values of Langmuir constants $q_{\max}$ and K_L were obtained by linear regression method and are listed in Table 2.

The biosorption isotherm of Cr (VI) on the three biosorbents is presented in Figure 5. This figure shows that the biosorption of Cr (VI) within the three biosorbents corresponds to type-S adsorption isotherm in the Giles classification system [31]. In this type, the initial curvature shows that, as more sites in the adsorbent are filled, it becomes increasingly difficult for a bombarding solute molecule to find a vacant site available. This implies either that the adsorbed solute molecule is vertically oriented or that there is no strong competition from the solvent. The types of systems which give this curve do in fact fulfill these calculations. Thus they have one of the following characteristics: (a) are mono-functional, (b) have moderate intermolecular attraction, causing it to pack vertically in regular array in the adsorbed layer, and (c) meet strong competition for substrate sites from molecules of the solvent or of another species [31, 32]. The weakness of the adsorbate-adsorbent forces will cause the uptake at low concentrations to be small, but once a molecule has become adsorbed, the adsorbate-adsorbent forces will promote the adsorption of further molecules—a cooperative process. Therefore, the isotherm will become a convex to the concentration axis. A polar adsorbate of particular interest in this context is water because the dispersion contribution to its overall interaction energy is usually small compared with the polar contribution. Not surprisingly, water provided many examples of type-S isotherms. One of the palm branches is covered with a layer of adsorbed water; however, the adsorbate-adsorbent interaction would be virtually reduced to the weak dispersion energy of water with Cr (VI), so that a type-S isotherm is obtained. In a system that gives rise to a type-S isotherm, however, multiple layers are built upon some parts on the surface while a monomer is still incomplete on other parts.

The Freundlich isotherm is an empirical equation that assumes that the adsorption process takes place on heterogeneous surfaces and adsorption capacity is related to the concentration of Cr (VI) at equilibrium. This isotherm model is defined by the following linear form:

$$\log q_e = \log K_F + \frac{1}{n} \log C_e, \quad (3)$$

where K_F is the Freundlich constant $(\text{mg/g})/(\text{L/mg})^{1/n}$ and also referred to adsorption capacity, while n is the heterogeneity factor and related to adsorption intensity. The value of n can be used also to describe the adsorption whether is linear ($n = 1$) or a physical process ($n > 1$) is favorable, or a chemical process ($n < 1$). On the other hand, the value of $1/n < 1$ or $1/n > 1$ indicates a normal Langmuir isotherm and cooperative adsorption, respectively. A plot of $\ln q_e$ versus $\ln C_e$ gives a straight line, where the values of K_F and $1/n$ are determined from the intercept and the slope, respectively (figure not shown). As depicted from Table 2, the Cr (VI) adsorption characteristics of the three adsorbents are better explained by the Langmuir isotherm model. Also, the adsorption capacity of Cr (VI) onto COPB shows a higher amount, and it can be implied that the amine groups (NH_3^+) of chitosan are largely responsible for the adsorption of chromium ions from the solution, and the interaction is chiefly electrostatic attraction in nature [33].

The Florry-Huggins accounts for the degree of surface coverage characteristics of adsorbate on the biosorbent and the linear form of the equation is expressed as [34]

$$\log\left(\frac{\theta}{C_0}\right) = \log K_{\text{FH}} + n_{\text{FH}} \log(1-\theta), \quad (4)$$

where θ is the surface coverage of the adsorbent by adsorbate and K_{FH} and n_{FH} are the Florry-Huggins constants. These constants can be obtained from the plot of log (θ/C_0) versus $\log(1-\theta)$ (figure not shown). Table 2 showed that the apparent number n_{FH} of metal ionoccupying sorption sites is greater for COPB than that of SOPB and PB biosorbents. The values of correlation coefficient (R^2) were found to be ≤ 0.98. The value of K_{FH} was used for the calculation of the spontaneity Gibbs free energy ($\Delta G^0 = -RT \ln(K_{\text{FH}})$, J/mol). It was found that the ΔG^0 value was negative that pointed out the feasibility and spontaneity of the process.

3.4. Adsorption Kinetics. Figure 6 shows the effect of contact time on the adsorbed amount of Cr (VI) by the raw PB, COPB, and SOPB samples, respectively. It can be seen that the kinetics of Cr (VI) removal by modified palm branches

TABLE 2: Langmuir, Freundlich, and Flory-Huggins constants for the uptake of Cr (VI) on the biosorbents.

Adsorbents	Langmuir			Freundlich			Florry-Huggins			
	$q_{\max}$ (mg/g)	K_L (L/g)	R^2	K_F	n	R^2	K_{FH}	n_{FH}	ΔG^0 (kJ/mol)	R^2
PB	25	0.023	0.982	0.230	1.0	0.883	0.018	3.4	−32.2	0.969
COPB	55	0.047	0.988	10	1.3	0.912	0.139	4.0	−31.8	0.978
SOPB	41.7	0.660	0.990	6.3	3.5	0.887	0.103	2.4	−31.7	0.982

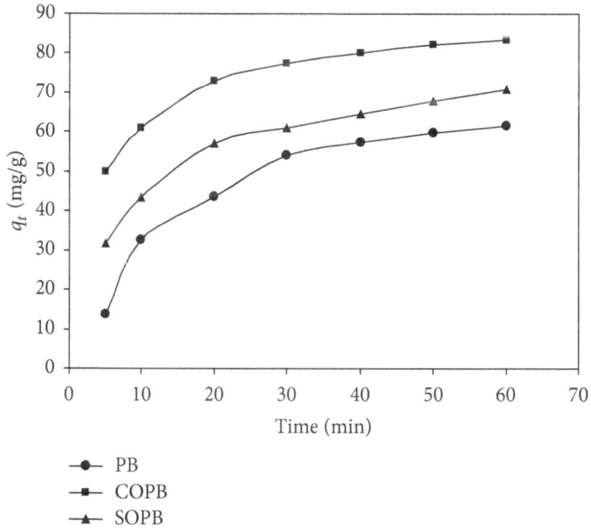

FIGURE 6: Effect of contact time on the adsorbed amount of Cr (VI) by the biosorbents.

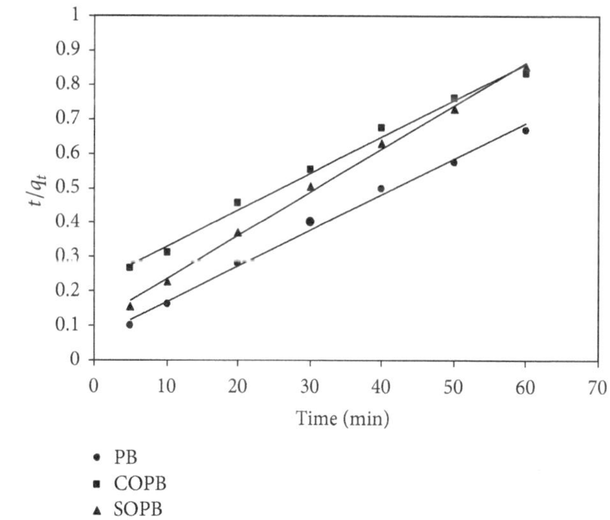

FIGURE 7: Pseudo-second-order plots for the biosorbents.

presents a shape characterized by a strong increase of the capacity of chromium removal during the first few minutes of contact between the solution and biosorbents, and followup of a slow increase until a state of equilibrium is reached.

In order to examine the mechanism of adsorption process, pseudo-first-order (Lagergren) and pseudo-second-order kinetic models were used, respectively, as follows:

$$\log(q_e - q_t) = \log q_e - \frac{k_1 t}{2.303},$$
$$\frac{t}{q_t} = \frac{1}{k_2 q_e^2} + \frac{t}{q_e}, \quad (5)$$

where q_e and q_t are the amounts of Cr (VI) adsorbed (mg/g) onto the three biosorbents at equilibrium and at time t (min), respectively. k_1 (min^{-1}) and k_2 (g/mg·min) are the adsorption rate constants of both pseudo-first and pseudo-second order models. A plot of (5) gave straight lines with very promising correlation. However, data was more linked when examined according to pseudo-second-order model as depicted by high value correlation coefficient (R^2) as shown in Table 3 and also Figure 7. In addition the calculated values from the pseudo-second order model ($q_{e,\mathrm{cal}}$) were also found to be very closed to experimental values ($q_{e,\mathrm{exp}}$). Thus experimental results support the assumption behind the model that the rate limiting step in adsorption of heavy metals is chemisorption involving valences forces through the sharing or exchange of electrons between biosorbent and metal ions. Some studies on the kinetic of Cr (VI) adsorption onto various adsorbents have also reported higher correlations for pseudo-second-order model [35].

Further, the intraparticle diffusion model can be added to determine the rate-determining step in the adsorption process of Cr (VI) by the prepared adsorbents. The equation is given as follows:

$$q_t = k_p t^{1/2} + C, \quad (6)$$

where k_p, $t^{1/2}$, and C are the intraparticle diffusion rate constant (mg/g·min$^{1/2}$), square root of the time (min)$^{1/2}$, and the intercept from the plot of q_t versus $t^{1/2}$ (not shown here), respectively. Intraparticle diffusion plots show multilinear, implying that more than one process affected the adsorption of chromium ions. The initial portion of the plot indicated an external mass transfer, whereas the second linear portion is due to the intraparticle or pore diffusion [36].

3.5. Column Adsorption Studies. The time for breakthrough appearance and the shape of the breakthrough curve very important characteristics for determining the operation and the dynamic response of an adsorption column [37–39]. The breakthrough curves show the loading behavior of metal to be removed from solution in a fixed bed and is usually expressed in terms of adsorbed metal concentration [C_{ad} = inlet metal concentration (C_0) − outlet metal concentration

Table 3: Kinetic parameters for Cr (VI) adsorption on the biosorbents.

Adsorbents	$q_{e,\exp}$ (mg/g)	Pseudo-first order			Pseudo-second order			Intraparticle diffusion	
		k_1	$q_{e,\text{cal}}$ (mg/g)	R^2	k_2	$q_{e,\text{cal}}$ (mg/g)	R^2	K_p	C
PB	80.5	0.013	38	0.9251	6.4×10^{-3}	82	0.997	7.2	19
COPB	96.2	0.017	79	0.7294	1.7×10^{-4}	98	0.999	6.0	20
SOPB	89.5	0.014	76	0.9225	7.1×10^{-4}	90	0.998	7.6	24

Table 4: Parameters of breakthrough curve for Cr (VI) adsorption on the biosorbents.

Adsorbents	Breakthrough parameters					
	C_o (mg/g)	t_{tot} (min)	m_{tot}	q_{tot} (mg/g)	Total % removal	q_e (mg/g)
PB	100	355	50	24	62.4	12
COPB	100	288	58	31	55.1	19.5
SOPB	100	250	71	39	41	15.6

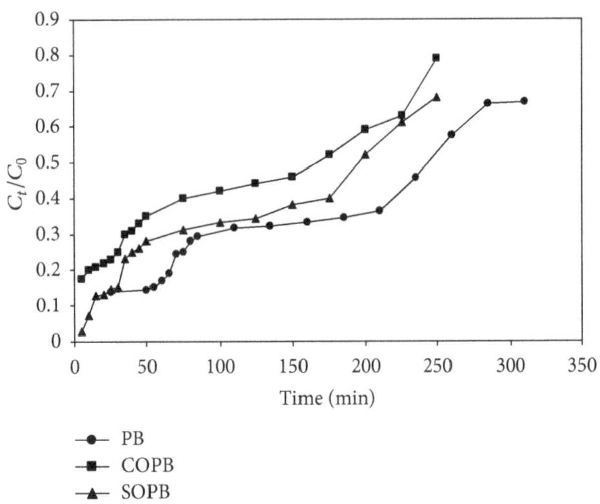

Figure 8: Breakthrough curves for Cr (VI) adsorption onto the biosorbents (flow rate = 2 mL/min, initial concentration = 100 mg/L, and pH = 2).

(C_t)] or normalized concentration defined as the ratio of effluent metal concentration to inlet metal concentration (C_t/C_0) as a function of time or volume of effluent for a given bed height [37]. Effluent volume (V_{eff}) can be calculated from the following equation:

$$V_{\text{eff}} = Q_t, \tag{7}$$

where t and Q_t are the total flow time (min) and volumetric flow rate (mL/min), respectively. The area under the breakthrough curve (A) is obtained by integrating the adsorbed concentration (C_{ad}) versus time (t), and plot can be used to find the total adsorbed metal quantity (maximum column capacity). Total adsorbed metal quantity (q_{tot}, mg/g) in the column for a given feed concentration and flow rate is calculated from the following equation:

$$q_{\text{tot}} = \frac{Q_t A}{1000} = \frac{Q_t}{1000} \int_{t=0}^{t=\text{total}} C_{\text{ad}} dt. \tag{8}$$

The total amount of adsorbed metal ions sent to column (m_{tot}) is calculated from the following equation [38]:

$$m_{\text{tot}} = \frac{C_0 Q_t}{1000}. \tag{9}$$

Total removal is calculated from the following equation [36]:

$$\text{Total removal\%} = \frac{q_{\text{tot}}}{m_{\text{tot}}} \times 100. \tag{10}$$

Equilibrium metal uptake (q_{eq}) in the column is defined by (11) as the total amount of metal adsorbed per gram of adsorbent (X) at the end of the total flow time:

$$q_{\text{eq}} = \frac{q_{\text{tot}}}{X}. \tag{11}$$

The breakthrough curves of Cr (VI) adsorption by PB, COPB, and SOPB at flow rate (2 mL/min), fixed bed height of 6 cm, and initial Cr (VI) concentration of 100 mg/L are shown in Figure 8. At the interval time of 80 min, the value of C_t/C_0 reached 0.29, 0.39, and 0.42 for PB, SOPB and COPB, respectively. The breakthrough curve of COPB becomes steeper than the other sorbents. The probable reason behind this is the presence of deriving force which represents the amine groups, these groups are responsible for interaction between COPB and Cr (VI) ions, which leads to an increase in the total adsorbed quantities and maximum Cr (VI) uptakes and removal percent [39] (cf. Table 4).

3.6. Desorption Column Studies. The possibility of adsorbent regeneration (desorption) and metal recovery was primarily studied based on the general assumption that the regeneration of adsorbent promotes economic adsorption treatment. The results showed 50% desorption for raw palm branches (PB), while 95% desorption for both COPB and SOPB with 1 M HCl solution. From this data it may be implied that the adsorbed chromium forms a film of $Cr(OH)_3$ on the external surface of modified palm branches and this film can be easily dissolved in 1 M HCl [1].

4. Conclusions

This study investigated the biosorption characteristics of Cr (VI) on raw and modified palm branches (PB). Adsorption of hexavalent chromium ion is found to be effective in the lower pH range. Equilibrium isotherm data were fitted using Langmuir and Flory-Huggins. Kinetic study showed that the pseudo-second-order kinetic model is appropriate and film diffusion might be involved in the sorption process. The chitosan coated palm branches (COPB) appear to be technically feasible and eco-friendly and with high efficacy compared to the other adsorbents. The adsorbents could be regenerated by using hydrochloric acid and therefore can be reused. These adsorbents could be good candidate for adsorption of Cr (VI) ions.

References

[1] R. Dobrowolski and M. Otto, "Study of chromium(VI) adsorption onto modified activated carbons with respect to analytical application," *Adsorption*, vol. 16, no. 4-5, pp. 279–286, 2010.

[2] U.S. Environmental Protection Agency, *Toxicological Review of Hexavalent Chromium*, National Center for Environmental Assessment, Office of Research and Development, Washington, DC, USA, 1998.

[3] R. J. Irwin, M. V. N. Mouwerik, L. Stevens, M. D. Seese, and W. Basham, *Environmental Contaminants Encyclopedia Chromium (VI) (Hexavalent Chromium) Entry*, National Park Service, Water Resources Division, Fort Collins, Colo, USA, 1971.

[4] D. Mohan, K. P. Singh, and V. K. Singh, "Trivalent chromium removal from wastewater using low cost activated carbon derived from agricultural waste material and activated carbon fabric cloth," *Journal of Hazardous Materials*, vol. 135, no. 1–3, pp. 280–295, 2006.

[5] M.-N. Zhang, X.-P. Liao, and B. Shi, "Adsorption of surfactants on chromium leather waste," *Journal of the Society of Leather Technologies and Chemists*, vol. 90, no. 1, pp. 1–6, 2006.

[6] L. Ramrakhiani, R. Majumder, and S. Khowala, "Removal of hexavalent chromium by heat inactivated fungal biomass of termitomyces clypeatus: surface characterization and mechanism of biosorption," *Chemical Engineering Journal*, vol. 171, no. 3, pp. 1060–1068, 2011.

[7] S. Rengaraj, C. K. Joo, Y. Kim, and J. Yi, "Kinetics of removal of chromium from water and electronic process wastewater by ion exchange resins: 1200H, 1500H and IRN97H," *Journal of Hazardous Materials*, vol. 102, no. 2-3, pp. 257–275, 2003.

[8] G. Ghosh and P. K. Bhattacharya, "Hexavalent chromium ion removal through micellar enhanced ultrafiltration," *Chemical Engineering Journal*, vol. 119, no. 1, pp. 45–53, 2006.

[9] A. A. Attia, S. A. Khedr, and S. A. Elkholy, "Adsorption of chromium ion (VI) by acid activated carbon," *Brazilian Journal of Chemical Engineering*, vol. 27, no. 1, pp. 183–193, 2010.

[10] J. Cui, N. Zhao, N. Wei, J. Li, Z. Qiao, and F. He, "Surface properties of chemically modified activated carbons for adsorption rate of Cr (VI)," *Chemical Engineering Journal*, vol. 115, no. 1-2, pp. 133–138, 2005.

[11] W. E. Marshall, L. H. Wartelle, D. E. Boler, M. M. Johns, and C. A. Toles, "Enhanced metal adsorption by soybean hulls modified with citric acid," *Bioresource Technology*, vol. 69, no. 3, pp. 263–268, 1999.

[12] H.-D. Choi, J.-M. Cho, K. Baek, J.-S. Yang, and J.-Y. Lee, "Influence of cationic surfactant on adsorption of Cr(VI) onto activated carbon," *Journal of Hazardous Materials*, vol. 161, no. 2-3, pp. 1565–1568, 2009.

[13] L. H. Wartelle and W. E. Marshall, "Citric acid modified agricultural by-products as copper ion adsorbents," *Advances in Environmental Research*, vol. 4, no. 1, pp. 1–7, 2000.

[14] C. A. Basar, C. Aydiner, S. Kara, and B. Keskinler, "Removal of CrO_4 anions from waters using surfactant enhanced hybrid PAC/MF process," *Separation and Purification Technology*, vol. 48, no. 3, pp. 270–280, 2006.

[15] K. Baek and T. W. Yang, "Competitive bind of anionic metals with cetylpyridinium chloride micelle in micellar-enhanced ultrafiltration," *Desalination*, vol. 167, no. 1–3, pp. 101–110, 2004.

[16] S. Babel and T. A. Kurniawan, "Cr(VI) removal from synthetic wastewater using coconut shell charcoal and commercial activated carbon modified with oxidizing agents and/or chitosan," *Chemosphere*, vol. 54, no. 7, pp. 951–967, 2004.

[17] Y. C. Wong, Y. S. Szeto, W. H. Cheung, and G. McKay, "Adsorption of acid dyes on chitosan-equilibrium isotherm analyses," *Process Biochemistry*, vol. 39, no. 6, pp. 693–702, 2004.

[18] S. Chatterjee and S. H. Woo, "The removal of nitrate from aqueous solutions by chitosan hydrogel beads," *Journal of Hazardous Materials*, vol. 164, no. 2-3, pp. 1012–1018, 2009.

[19] H.-D. Choi, M.-C. Shin, D.-H. Kim, C.-S. Jeon, and K. Baek, "Removal characteristics of reactive black 5 using surfactant-modified activated carbon," *Desalination*, vol. 223, no. 1–3, pp. 290–298, 2008.

[20] D. Park, Y.-S. Yun, D. S. Lee, S.-R. Lim, and J. M. Park, "Column study on Cr(VI)-reduction using the brown seaweed Ecklonia biomass," *Journal of Hazardous Materials*, vol. 137, no. 3, pp. 1377–1384, 2006.

[21] S. M. de Oliveira Brito, H. M. C. Andrade, L. F. Soares, and R. P. de Azevedo, "Brazil nut shells as a new biosorbent to remove methylene blue and indigo carmine from aqueous solutions," *Journal of Hazardous Materials*, vol. 174, no. 1–3, pp. 84–92, 2010.

[22] S. Panumati, K. Chudecha, P. Vankhaew et al., "Adsorption of phenol from diluted aqueous solutions by activated carbons obtained from bagasse, oil palm shell and pericarp of rubber fruit," *Songklanakarin Journal of Science and Technology*, vol. 30, no. 2, pp. 185–189, 2008.

[23] S. Chatterjee, M. W. Lee, and S. H. Woo, "Influence of impregnation of chitosan beads with cetyl trimethyl ammonium bromide on their structure and adsorption of congo red from aqueous solutions," *Chemical Engineering Journal*, vol. 155, no. 1-2, pp. 254–259, 2009.

[24] V. Swarnakar, N. Agrawal, and R. Tamar, "Sorption of Cr (VI) and As (V) on HDTMA-modified zeolites," *International Journal of Computational Science and Engineering* , vol. 2, no. 5, pp. 1–9, 2011.

[25] E. Malkoc and Y. Nuhoglu, "Fixed bed studies for the sorption of chromium(VI) onto tea factory waste," *Chemical Engineering Science*, vol. 61, no. 13, pp. 4363–4372, 2006.

[26] L. Dambies, C. Guimon, S. Yiacoumi, and E. Guibal, "Characterization of metal ion interactions with chitosan by X-ray photoelectron spectroscopy," *Colloids and Interface*, vol. 177, no. 2-3, pp. 203–214, 2001.

[27] H.-D. Choi, W.-S. Jung, J.-M. Cho, B.-G. Ryu, J.-S. Yang, and K. Baek, "Adsorption of Cr(VI) onto cationic surfactant-modified activated carbon," *Journal of Hazardous Materials*, vol. 166, no. 2-3, pp. 642–646, 2009.

[28] A. Sharma and K. G. Bhattacharyya, "Adsorption of chromium (VI) on azadirachta indica (Neem) leaf powder," *Adsorption*, vol. 10, no. 4, pp. 327–338, 2005.

[29] J. Febrianto, A. N. Kosasih, J. Sunarso, Y. H. Ju, N. Indraswati, and S. Ismadji, "Equilibrium and kinetic studies in adsorption of heavy metals using biosorbent: a summary of recent studies," *Journal of Hazardous Materials*, vol. 162, no. 2-3, pp. 616–645, 2009.

[30] Z. Aksu, "Determination of the equilibrium, kinetic and thermodynamic parameters of the batch biosorption of nickel(II) ions onto Chlorella vulgaris," *Process Biochemistry*, vol. 38, no. 1, pp. 89–99, 2002.

[31] E. Karadag, D. Saraydin, and O. Guven, "Cationic dye adsorption by acrylamide/itaconic acid hydrogels in aqueous solutions," *Polymers for Advanced Technologies*, vol. 8, no. 9, pp. 574–578, 1997.

[32] C. H. Giles, T. H. MacEwan, S. N. Nakhwa, and D. Smith, "Studies in adsorption. part XI. a system of classification of solution adsorption isotherms, and its use in diagnosis of adsorption mechanisms and in measurement of specific surface areas of solids," *Journal of the Chemical Society*, vol. 786, pp. 3973–3993, 1960.

[33] X.-J. Hu, J.-S. Wang, Y.-G. Liu et al., "Adsorption of chromium (VI) by ethylenediamine-modified cross-linked magnetic chitosan resin: isotherms, kinetics and thermodynamics," *Journal of Hazardous Materials*, vol. 185, no. 1, pp. 306–314, 2011.

[34] U. Shafique, A. Ijaz, M. Salman et al., "Removal of arsenic from water using pine leaves," *Journal of the Taiwan Institute of Chemical Engineers*, vol. 43, no. 2, pp. 256–263, 2012.

[35] Y. A. Aydin and N. D. Aksoy, "Adsorption of chromium on chitosan: optimization, kinetics and thermodynamics," *Chemical Engineering Journal*, vol. 151, no. 1–3, pp. 188–194, 2009.

[36] W. H. Cheung, Y. S. Szeto, and G. McKay, "Intraparticle diffusion processes during acid dye adsorption onto chitosan," *Bioresource Technology*, vol. 98, no. 15, pp. 2897–2904, 2007.

[37] Z. Aksu and F. Gonen, "Biosorption of phenol by immobilized activated sludge in a continuous packed bed: prediction of breakthrough curves," *Process Biochemistry*, vol. 39, no. 5, pp. 599–613, 2004.

[38] T. V. N. Padmesh, K. Vijayaraghavan, G. Sekaran, and M. Velan, "Batch and column studies on biosorption of acid dyes on fresh water macro alga *Azolla filiculoides*," *Journal of Hazardous Materials*, vol. 125, no. 1–3, pp. 121–129, 2005.

[39] E. Malkoc, Y. Nuhoglu, and M. Dundar, "Adsorption of chromium(VI) on pomace-an olive oil industry waste: batch and column studies," *Journal of Hazardous Materials*, vol. 138, no. 1, pp. 142–151, 2006.

5

Constructing Potential Energy Surfaces for Polyatomic Systems: Recent Progress and New Problems

J. Espinosa-Garcia, M. Monge-Palacios, and J. C. Corchado

Departamento de Química Física, Universidad de Extremadura, 06071 Badajoz, Spain

Correspondence should be addressed to J. Espinosa Garcia, joaquin@unex.es

Academic Editor: Laimutis Bytautas

Different methods of constructing potential energy surfaces in polyatomic systems are reviewed, with the emphasis put on fitting, interpolation, and analytical (defined by functional forms) approaches, based on quantum chemistry electronic structure calculations. The different approaches are reviewed first, followed by a comparison using the benchmark H + CH_4 and the H + NH_3 gas-phase hydrogen abstraction reactions. Different kinetics and dynamics properties are analyzed for these reactions and compared with the available experimental data, which permits one to estimate the advantages and disadvantages of each method. Finally, we analyze different problems with increasing difficulty in the potential energy construction: spin-orbit coupling, molecular size, and more complicated reactions with several maxima and minima, which test the soundness and general applicability of each method. We conclude that, although the field of small systems, typically atom-diatom, is mature, there still remains much work to be done in the field of polyatomic systems.

1. Introduction

At the heart of chemistry lies the knowledge of the reaction mechanism at atomic or molecular levels, that is, the motion of the nuclei in the potential field due to the electrons and nuclei. When the separation of the nuclear and electronic motions is possible, that is, within the Born-Oppenheimer (BO) approximation [1], the potential energy surface (PES) is the electronic energy (which includes electronic kinetic and Coulomb interaction energy) plus nuclear repulsion of a given adiabatic electronic state as a function of geometry. The PES is then the effective potential energy for the nuclear motion when the system is in that electronic state.

The actual knowledge of the chemical reactivity is based on quantum mechanical first principles, and the complete construction of the PES represents a very important challenge in theoretical chemistry. For small reactive systems (three or four atoms), the PES construction is a relatively mature field, although even today new surfaces are still being developed for atom-diatom systems [2–6] using high-level *ab initio* calculations. For instance, one of the latest atom-diatom surfaces has been developed in the present year by Jiang et al. [6] for the H + HBr hydrogen abstraction reaction. *Ab initio* calculations were performed based on the multireference configuration interaction (MRCI) method including Davidson's correction using augmented correlation-consistent-polarized valence X-tuple zeta basis sets, with X up to 5, extrapolating the energies to the complete basis set (CBS) limit, and taking into account the spin-orbit correction. In order to define the complete configuration space for the BrH_2 system, *ab initio* calculations were performed at more than 12000 geometries. Then, a three-dimensional cubic spline interpolation was employed to yield the ground-state PES for this system. As can be seen, the computational effort (correlation energy + basis set + number of points) is enormous, and doing this for larger systems is still clearly prohibitive.

The extension to larger reactive systems (more than 4 atoms) is an open and promising field, although computationally very costly. Thus, for polyatomic systems of n atoms, with $3n - 6$ internal degrees of freedom, if one needs m configurations for each degree of freedom, one would need a total of $m^{(3n-6)}$ evaluations of the energy. For instance, taking a typical value of m of 10, for the H + CH_4 hydrogen abstraction reaction 10^{12} evaluations would be needed. This implies too much computer time to perform *ab initio* calculations with chemical accuracy (1 kcal mol^{-1}) and becomes totally unaffordable if one aims to achieve spectroscopic accuracy (1 cm^{-1}).

The accuracy of the kinetics and dynamics description of a chemical reaction depends, aside from the quality of the PES, on the dynamics method used. If the motion of the nuclei on the PES is determined by the Hamilton equations in the phase-space configuration, the dynamics methods are classical or quasiclassical trajectory (QCT) approaches [7], and if the system is described by the Schrödinger equation (time dependent or time independent), the dynamics methods are quantum mechanical (QM) [8]. Finally, variational transition-state theory with multidimensional tunneling contribution (VTST/MT) can be derived from a dynamical approach by statistical mechanics and provides an excellent and affordable method to calculate thermal rate and state-selected constants [9].

In sum, electronic structure theory, dynamics methods, and constructed potential energy surfaces are the keystones of the theoretical study of chemical reactivity. Recent years have seen a spectacular development in these strongly related disciplines, paving the way towards chemical accuracy in polyatomic systems and toward spectroscopic accuracy in smaller problems. The scope of this paper is the construction of potential energy surfaces for bimolecular gas-phase polyatomic reactive systems in their electronic ground-state. Section 2 describes different methods of constructing potential energy surfaces, with especial focus on the methodological approach developed by our research group. Section 3 presents a library of PESs available for the scientific community, and Section 4 presents some results for the H + CH_4 benchmark polyatomic system and the H + NH_3 reaction, which permits us to compare results from different methods. Section 5 is devoted to analyzing the treatment of the peculiarities found in some reactions that entail additional challenges to the construction of potential energy surfaces, such as spin-orbit coupling, the molecular size, and the presence of various maxima and minima. Finally, Section 6 summarizes the main conclusions of the paper.

2. Constructing Potential Energy Surfaces

The construction of potential energy surfaces in reactive systems began with the dawn of Quantum Chemistry when different quantum approaches were used, basically empirical or semiempirical. In present days, the data needed for this construction are obtained from high-level *ab initio* calculations, although density functional theory (DFT) methods are also widely used, especially in the case of larger molecular systems.

As is well known, and this special issue is a clear indication of it, the construction of PESs has a long tradition in theoretical and computational chemistry, and an exhaustive review of the literature is beyond the scope of the present paper (see, for instance, a recent review in [10]). Here we only highlight some important contributions with the aim that the reader will get an overall perspective on this wide field of research.

The most straightforward procedure to describe a reactive system is from electronic structure calculations carried out "on the fly," sometimes also called "direct dynamics" [11], where every time the dynamics algorithm requires an energy, gradient, or Hessian, it is computed from electronic structure calculations or molecular mechanics methods. The advantage of this approach is the direct application of electronic structure calculations to dynamics problems without intermediaries, but when chemical accuracy is needed, very high-level *ab initio* calculations are required, and hence the time and computational cost are very high. For polyatomic systems, such an approach is still prohibitive. A recent example is the direct dynamics trajectory study of the formaldehyde cation with molecular hydrogen gas-phase reaction [12]. Even using a modest MP2/6-311G(d,p) *ab initio* level, which yields a barrier only 0.5 kcal mol^{-1} below the benchmark value, the ≈9600 trajectories required about 2.5 CPU years on a small Linux cluster. In this regard, one has to keep in mind that QCT calculations require a large number of trajectories to adequately sample the events and obtain results that from a purely statistical point of view are tolerable, that is, with very small errors. Note that the time devoted to the QCT calculation is negligible in this time scale.

The alternative to direct dynamics calculations is to construct a mathematical PES, that is, developing an algorithm that can provide the potential energy for any given geometry of the system without depending on "on-the-fly" electronic structure computations. In this sense, three basic approaches have been considered: fitting, interpolation, or analytical (defined by functional forms). In fact, the fitting and analytical procedures share the idea of functional form, and this analytical function needs to be fitted to theoretical information. This classification is therefore arbitrary, and other possibilities can be found in the literature [10]. We think, however, that it can help the reader get a better general view of this field. The interpolation methods have been widely applied in the construction of potential energy surfaces, and different strategies have been developed [9, 10, 13–15]. Thus, Collins et al. [14, 16, 17] developed and applied a method for generating potential energy surfaces using a modified Shepard interpolation. The PES at any configuration (Z) is represented by a weighted average of Taylor series $T_i(R)$:

$$V = \sum_{i=1}^{N\text{data}} w_i(Z)T_i(R), \tag{1}$$

where N_{data} is the number of molecular configurations whose energy and its first (gradient) and second (Hessian) derivatives have been evaluated, $w_i(Z)$ is the normalized weighting factor, R is a set of internal coordinates, and the Taylor series are expanded around the point i:

$$T_i(R) = V(R) + \sum_{k=1}^{3n-6} \frac{\partial V}{\partial R_k}[R_k - R_k(i)] + \frac{1}{2!}\sum_{k=1}^{3n-6}\sum_{j=1}^{3n-6}[R_k - R_k(i)]\frac{\partial^2 V}{\partial R_k \partial R_j}\left[R_j - R_j(i)\right]. \tag{2}$$

The use of the energy and its first and second derivatives makes the method very robust, but computationally very expensive, especially if very high-level *ab initio* calculations are needed for a correct description of the reactive system. For instance, very recently a new PES for the H + CH_4 gas-phase reaction has been developed by Collins by interpolation of 30000 data points obtained with high-level *ab initio* calculations [18]. The *ab initio* calculations took about 600 days on a workstation with 8 CPU cores in total, while the evaluations of the potential values on all the grids for this reaction took 400 days. Three advantages of this method are the following: first, that the interpolated function agrees precisely with the values of the data points; second, that new electronic structure points may be incorporated in the process to improve the PES; third, that the PES is invariant to the permutation or exchange of indistinguishable nuclei. However, Shepard interpolation often yields slight oscillations in equipotential contours and vibrational frequencies (second derivatives of the energy); that is, the potential energy surface is not smooth enough to provide smooth changes in the vibrational frequencies. This is a limitation when one wants to compute tunneling effects at low energies, since the barrier to tunnel through is unrealistic.

Fitting methods, such as the interpolated ones considered above, have been widely used in the construction of potential energy surfaces [19–21], and they have been recently reviewed [10]. Thus, for instance, a fitting procedure widely used is based on the many-body expansion (MBE) method [22], amplified by Varandas [23] in the so-called double many-body expansion (DMBE) method. The potential is given by a sum of terms corresponding to atoms, diatoms, triatoms, and tetra-atoms and a series of adjustable coefficients. Another fitting procedure has been developed and applied by Bowman et al. [24–27], in which the *ab initio* data are globally fitted to a permutational symmetry invariant polynomial. The function describing the PES has the form

$$V = p(x) + \sum_{i<j} q_{i,j}(x) \cdot y_{i,j}, \tag{3}$$

where x is an n-dimensional vector, depending on n internuclear distances, with components $x_{i,j} = \exp(-r_{ij})$, r_{ij} being the distance between nuclei i and j, and $y_{i,j}$ given by

$$y_{i,j} = \frac{e^{-r_{i,j}}}{r_{i,j}}, \tag{4}$$

and $p(x)$ and $q(x)$ are polynomials constructed to satisfy the permutation symmetry with respect to the indistinguishable nuclei. For instance, for the CH_5^+ polyatomic system [28], the PES was least-squares fitted to 20 639 *ab initio* energies, obtained at the MP2/cc-pVTZ level. This fit contains 2303 coefficients and an rms fitting error of 51 cm^{-1}. Bowman and coworkers have developed 16 of such PESs for polyatomic systems, and they have been recently revised [27]. In general, the fitting approach is linear least squares, and therefore these surfaces do not exactly reproduce the *ab initio* data.

The third alternative consists of the analytical surfaces defined by functional forms. In this method the *ab initio* data are fitted to a valence bond (VB) functional form, augmented with molecular mechanics (MM) terms which give great flexibility to the potential energy surface. These VB/MM functional forms have a long history in the development of potential energy surfaces [29–33], although at first the surfaces were semiempirical in the sense that theoretical and experimental data were used in the fitting procedure. This has been the methodological approach developed by our group [10, 34, 35].

The first surfaces were developed for the H + CH_4 hydrogen abstraction reaction, as a paradigm of polyatomic systems [29–33]. The potential energy for a given geometry, V, is given by the sum of three terms: stretching potential, V_{stretch}, harmonic bending term, V_{harm}, and anharmonic out-of-plane potential, V_{op},

$$V = V_{\text{stretch}} + V_{\text{harm}} + V_{\text{op}}. \tag{5}$$

The stretching potential is the sum of four London-Eyring-Polanyi (LEP) terms, each one corresponding to a permutation of the four methane hydrogens:

$$V_{\text{stretch}} = \sum_{i=1}^{4} V_3(R_{\text{CH}_i}, R_{\text{CH}_\text{B}}, R_{\text{H}_i\text{H}_\text{B}}), \tag{6}$$

where R is the distance between the two subscript atoms, H_i stands for one of the four methane hydrogens, and H_B is the attacking H atom. Although the functional form of the V_3 LEP potential is well known, we will recall it for the sake of completeness:

$$V_3(R_{\text{CH}_i}, R_{\text{CH}_\text{B}}, R_{\text{H}_i\text{H}_\text{B}}) = Q(R_{\text{CH}_i}) + Q(R_{\text{CH}_\text{B}}) + Q(R_{\text{H}_i\text{H}_\text{B}}) - \sqrt{\frac{1}{2}[J(R_{\text{CH}_i}) - J(R_{\text{H}_i\text{H}_\text{B}})]^2 + \frac{1}{2}[J(R_{\text{H}_i\text{H}_\text{B}}) - J(R_{\text{CH}_\text{B}})]^2 + \frac{1}{2}[J(R_{\text{CH}_\text{B}}) - J(R_{\text{CH}_i})]^2},$$

$$Q(R_{XY}) = \frac{E^1(R_{XY}) + E^3(R_{XY})}{2},$$

$$J(R_{XY}) = \frac{E^1(R_{XY}) - E^3(R_{XY})}{2},$$

$$E^1(R_{XY}) = D^1_{XY}\{\exp(-2\alpha_{XY}[R_{XY} - R^e_{XY}]) \\ -2\exp(-\alpha_{XY}[R_{XY} - R^e_{XY}])\},$$
$$E^3(R_{XY}) = D^3_{XY}\{\exp(-2\alpha_{XY}[R_{XY} - R^e_{XY}]) \\ +2\exp(-\alpha_{XY}[R_{XY} - R^e_{XY}])\},\tag{7}$$

where there are 12 fitting parameters, four for each of the three kinds of bond, R_{CH_i}, R_{CH_B}, and $R_{H_iH_B}$. In particular, these are the singlet and triplet dissociation energies, D^1_{XY} and D^3_{XY}, the equilibrium bond distance, R^e_{XY}, and the Morse parameter, α_{XY}. The Morse parameter for the CH_i bonds, α_{CH}, however, is not taken as a constant but rather as a function of the CH distances,

$$\alpha_{CH} = a_{CH} + b_{CH}\left(\frac{\tanh\left[c_{CH}\left(\overline{R} - R^e_{CH}\right)\right] + 1}{2}\right),\tag{8}$$

with $\overline{R}$ being the average R_{CH_i} distance,

$$\overline{R} = \frac{1}{4}\sum_{i=1}^{4} R_{CH_i}.\tag{9}$$

In this way, α_{CH} changes smoothly from its value at methane, $a_{CH} + (b_{CH}/2)$, to its value at the methyl radical, $a_{CH} + b_{CH}$, as the reaction evolves. Therefore, 14 parameters are required to describe the stretching potential.

One of the problems with this functional was that the equilibrium C–H distances for the reactants, saddle point, and products are the same, leading to a very rigid surface. Chakraborty et al. [36], for the H + C_2H_6 reaction, included a modification to endow the surface with greater flexibility. The reference C–H bond distance is transformed smoothly from reactant to product using the following equation:

$$R^o_{CH} = P_1 R^o_{CH,R} + (1 - P_1) R^o_{CH,P},\tag{10}$$

where P_1 is

$$P_1 = \prod_{i=1}^{4} T_1(R_{CH_i}),\tag{11}$$

which is symmetric with respect to all the four hydrogen atoms and goes to zero as one of the hydrogen atoms is abstracted, and T_1 is a geometry-dependent switching function, given by

$$T_1(R_{CH_i}) = 1 - \tanh[w_1(R_{CH_i} - w_2)],\tag{12}$$

where w_1 and w_2 are adjustable parameters. Therefore, this adds 2 new parameters (total 16 parameters) to describe the stretching potential.

The V_{harm} term is the sum of six harmonic terms, one for each bond angle in methane:

$$V_{harm} = \frac{1}{2}\sum_{i=1}^{3}\sum_{j=i+1}^{4} k^0_{ij} k_i k_j \left(\theta_{ij} - \theta^0_{ij}\right)^2,\tag{13}$$

where k^0_{ij} and k_i are force constants and θ^0_{ij} are the reference angles. The k^0_{ij} force constants are allowed to evolve from their value in methane, k^{CH_4}, to their value in methyl, k^{CH_3}, which are two parameters of the fit, by means of switching functions:

$$k^0_{ij} = k^{CH_4} + k^{CH_4}\left[S_1(R_{CH_i})S_1\left(R_{CH_j}\right) - 1\right] \\ + \left(k^{CH_4} - k^{CH_3}\right)\left[S_2(R_{CH_i})S_2\left(R_{CH_j}\right) - 1\right],\tag{14}$$

while k_i is a function of both the R_{CH_i} and $R_{H_iH_B}$ distances:

$$k_i = A_1 \exp\left[-A_2(R_{CH_i} - R^e_{CH})^2\right],$$
$$A_1 = 1 - \exp\left[-aa_1(R_{H_iH_B})^2\right],\tag{15}$$
$$A_2 = aa_2 + aa_3 \exp\left[-aa_4\left(R_{H_iH_B} - R^e_{H_iH_B}\right)^2\right].$$

Thus, four adjustable parameters are involved in the definition of k_i.

The reference angles are also allowed to change from their value at methane, τ = 109.47°, or arccosine (−1/3), to methyl, 120° or $2\pi/3$ radians, by means of switching functions [37]:

$$\theta^0_{ij} = \text{acos}\left(-\frac{1}{3}\right) + \left[\text{acos}\left(-\frac{1}{3}\right) - \frac{\pi}{2}\right]\left[S_\varphi(R_{CH_i})S_\varphi\left(R_{CH_j}\right) - 1\right] \\ + \left[\text{acos}\left(-\frac{1}{3}\right) - \frac{2\pi}{3}\right]\left[S_\theta(R_{CH_k})S_\theta(R_{CH_l}) - 1\right].\tag{16}$$

Finally, the switching functions are given by

$$S_1(R_{CH_i}) = 1 - \tanh\left[\alpha^s_1(R_{CH_i} - R^e_{CH})(R_{CH_i} - \beta^s_1)^8\right],$$
$$S_2(R_{CH_i}) = 1 - \tanh\left[\alpha^s_2(R_{CH_i} - R^e_{CH})(R_{CH_i} - \beta^s_2)^6\right],$$
$$S_\varphi(R_{CH_i}) = 1 - \tanh\left\{A_\varphi(R_{CH_i} - R^e_{CH})\exp\left[B_\varphi\left(R_{CH_i} - C_\varphi\right)^3\right]\right\},$$
$$S_\theta(R_{CH_i}) = 1 - \tanh\left\{A_\theta(R_{CH_i} - R^e_{CH})\exp\left[B_\theta(R_{CH_i} - C_\theta)^3\right]\right\},\tag{17}$$

involving 10 more adjustable parameters. In total, 16 terms need to be fitted for the calibration of the V_{harm} potential.

The V_{op} potential is a quadratic-quartic term whose aim is to correctly describe the out-of-plane motion of methyl:

$$V_{op} = \sum_{i=1}^{4} f_{\Delta_i}\sum_{\substack{j=1\\ j\neq i}}^{4}\left(\Delta_{ij}\right)^2 + \sum_{i=1}^{4} h_{\Delta_i}\sum_{\substack{j=1\\ j\neq i}}^{4}\left(\Delta_{ij}\right)^4.\tag{18}$$

The force constants, f_{Δ_i} and h_{Δ_i}, have been incorporated into a new switching function which is such that V_{op} vanishes at the methane limit and which directs the change of the methyl fragment from pyramidal to planar as the reaction evolves:

$$f_{\Delta_i} = [1 - S_3(R_{CH_i})] f_\Delta,$$
$$h_{\Delta_i} = [1 - S_3(R_{CH_i})] h_\Delta, \quad (19)$$
$$S_3(R_{CH_i}) = 1 - \tanh\left[\alpha_3^s (R_{CH_i} - R_{CH}^e)(R_{CH_i} - \beta_3^s)^2\right],$$

with f_Δ, h_Δ, α_3^s, and β_3^s being the only parameters of V_{op} that enter the fitting process. Δ_{ij} is the angle that measures the deviation from the reference angle:

$$\Delta_{ij} = \operatorname{acos}\left(\frac{(\vec{r_k} - \vec{r_j}) \times (\vec{r_l} - \vec{r_j})}{\left\|(\vec{r_k} - \vec{r_j}) \times (\vec{r_l} - \vec{r_j})\right\|} \frac{\vec{r_i}}{\|\vec{r_i}\|}\right) - \theta_{ij}^0, \quad (20)$$

where $\vec{r_i}$, $\vec{r_j}$, $\vec{r_k}$, and $\vec{r_l}$ are vectors going from the carbon atom to the i, j, k, and l hydrogen atoms, respectively, and θ_{ij}^0 are the reference angles defined in (16). The first term to the right of (20) is therefore the angle between the CH_i bond and a vector perpendicular to the plane described by the j, k, and l hydrogen atoms and centred at the j atom. To correctly calculate Δ_{ij}, the motion from k to l has to be clockwise from the point of view of the i atom.

In the case of the H + CH_4 reaction the CH_3 radical product is planar. What happens if the product presents a nonplanar geometry? In these cases of CX_3 products, we have modified the reference angle θ_{ij}^o in (13) and (20). Thus, the original expression,

$$\theta_{ij}^o = \tau + \left(\tau - \frac{\pi}{2}\right)\left[S_\varphi(R_{CX_i}) \cdot S_\varphi(R_{CX_j}) - 1\right] + \left(\tau - \frac{2\pi}{3}\right)\left[S_\vartheta(R_{CX_k}) \cdot S_\vartheta(R_{CX_l}) - 1\right], \quad (21)$$

where $\tau = 109.47°$, is replaced by

$$\theta_{ij}^o = \tau + (\tau - \tau_1)\left[S_\varphi(R_{CX_i}) \cdot S_\varphi(R_{CX_j}) - 1\right] + (\tau - \tau_2)[[S_\vartheta(R_{CX_k}) \cdot S_\vartheta(R_{CX_l}) - 1], \quad (22)$$

where τ_2 is the bending angle in the non-planar product, and τ_1 is related to τ_2 by the expression,

$$\tau_1 = \pi - \arcsin\left[\frac{\sin(\tau_2/2)}{\sin(\pi/3)}\right]. \quad (23)$$

In these cases of non-planar products, CX_3, this correction would add new parameters in the fitting procedure.

The PES, therefore, depends on at least 36 parameters, 16 for the stretching, 16 for the harmonic term, and 4 for the out-of-plane potential. These 36 parameters give great flexibility to the PES, while keeping the VB/MM functional form physically intuitive.

In the case of the H + CH_4 reaction the CH_4 reactant presents T_d symmetry. What happens if the reactant presents a different symmetry? For instance, ammonia presents symmetry C_{3v}, characterized by an inversion mechanism through a planar structure with symmetry D_{3h}. The functional form (5) used for the H + CH_4 reaction cannot be applied without modification to any kind of system . Thus, when this potential is applied to the study of the H + NH_3 → H_2 + NH_2 reaction, a major drawback was observed [38], namely that it wrongly describes the NH_3 inversion reaction, predicting that the planar ammonia (D_{3h} symmetry), which is a saddle point to the ammonia inversion, is about 9 kcal mol^{-1} more stable than the pyramidal structure (C_{3v} symmetry).

In the original expression for the H + CH_4 reaction (5), the V_{op} term was added to obtain a correct description of the out-of-plane bending in the methyl radical, 580 cm^{-1}. However, Yang and Corchado [38] noted that this term leads to unphysical behaviour along the ammonia inversion path. To avoid this drawback, our laboratory recently developed [39] a new PES where the V_{op} term is removed:

$$V = V_{stretch} + V_{harm}. \quad (24)$$

Consequently, each of the 16 parameters is required for the stretching terms and the harmonic bending terms of the PES which were fitted to high-level *ab initio* CCSD(T)/cc-pVTZ calculations.

In sum, starting from a basic functional form, this form must be adapted to each particular case, looking for the greatest flexibility and suitability for the problem under study. This has been the main aim of the modifications on the original functional form (5), described by (10)–(12), (22), (23), and (24). While this could represent a disadvantage with respect to the fitting or interpolation methods, because new functional forms must be developed in each case, it also represents an advantage, because simple stepwise modifications can give great flexibility to the surface.

Once the functional form is available, the fitting procedure is started. A very popular approach for fitting a function is the least-squares method, which, using some local optimization algorithm, gives values of the parameters that minimize (locally) the function

$$R = \sum_x |E(x) - F(x, p)|^2, \quad (25)$$

where $E(x)$ is the *ab initio* energy associated with a particular molecular configuration specified by "x" and $F(x, p)$ is the energy predicted by the analytical function at the same molecular configuration, which depends on a set of m parameters denoted as p.

One must note, however, that any fitting procedure has certain limitations. First, due to the large number of parameters, it is very hard to find a global minimum for the fit, which accurately describes the entire surface. Second, due to the nature of the linear least-square method, the result is dependent on the initial parameters; third, since we use a mathematical approach without physical intuition, one usually obtains a number of distinct sets of mathematical parameters, all equally probable and good at reproducing the complete system. To make matters worse, as noted by Banks and Clary [40], while the differences between these parameter sets may be small, the dynamics information obtained from them can vary notably. This is especially true because small changes in the energy derivatives can cause large changes in dynamical properties of the PES. One therefore requires an extremely accurate fit to the topology of the given points to obtain an acceptable set of parameters.

To solve at least partially some of the above problems, we adopt a different approach [41]. Firstly, we will try to obtain the values of the parameters that minimize the function

$$R = \sum_x w_x^e \left| E(x) - F(x,p) \right|^2 + \sum_x w_x^g \left| g(x) - \frac{\partial F(x,p)}{\partial x} \right|^2 + \sum_x w_x^H \left| H(x) - \frac{\partial^2 F(x,p)}{\partial x^2} \right|^2, \quad (26)$$

where $g(x)$ denotes the gradients (first derivative of the energy), $H(x)$ the Hessian elements (second derivatives of the energy), and w_x^e, w_x^g, and w_x^H are weights. In practice, the gradients enter the fitting process only at the stationary points, and the Hessian elements by means of the harmonic vibrational frequencies at selected points. Secondly, the choice of the number of points to be fitted is critical, and this strategy allows this number to be reduced drastically (see more details in the original paper [41]). We begin by choosing the stationary points (reactants, products, saddle point, and intermediate complexes) as data points. The geometry, energy, and frequencies are fitted and, thus, indirectly, the first and second derivatives of the energy. It is necessary to remember that if the information included was only the energy, a mesh of m^{3n-6} points would be required. However, when the information included is not only the energy at a stationary point but also the first and second derivatives, only one point is needed to reproduce each stationary point configuration. Moreover, additional points on the minimum energy path are included. Thus, with the inclusion of the stationary points and representative points on the reaction path, we search for a good reproduction of the topology of the path connecting reactants to products. Finally, we add the energy of a point not on the reaction path to describe zones of the reaction valley relevant to tunneling dynamics and rotational excitation of the products. The fitting procedure is currently automated on a computer with no user intervention except to analyze the results and, when called for, to change the weights.

This strategy presents certain advantages over other methods. The first is transferability of the functional form and the fitted parameters. Since one can regard VB/MM as some kind of highly specific MM force field, it can be expected that some of the fitting parameters are transferable to a similar system, although obviously the parameter values are system specific. In our group we have used this feature to obtain, for example, the PES for the F + CH_4 reaction [42] using as the starting point the analytical PES for H + CH_4 [43]. Secondly, the PESs are guaranteed to some extent to have the capability of reproducing the energetic interactions of the chemical system in region not included in the fit. For example, high-energy collisions might require knowledge of the PES at very high energies that may not have been sampled in the fitting process. An interpolation method cannot ensure the absence of spurious wells or unphysical repulsive regions when nearby points are absent, while MM force fields can. Thirdly, VB/MM PES parameters can be refitted so as to fine-tune the PES using additional data (e.g., higher *ab initio* calculations at selected points) at a low computational cost. Furthermore, VB/MM surfaces and their energy derivatives (when they can be analytically calculated) are usually smoother than interpolated surfaces. As was mentioned above, Shepard interpolation sometimes gives discontinuities in the derivatives, which is a highly undesirable feature for kinetics and dynamics studies.

Thus, our research group has developed economical alternatives for constructing analytical PESs of polyatomic systems, which basically are VB/MM-type surfaces. This strategy of seeking an optimal tradeoff of time and computational cost has been successfully used in several gas-phase hydrogen abstraction reactions of five [44, 45], six [42, 43, 46–52], and seven [53] atoms. In general, a good correspondence between kinetics and dynamics theoretical results and experimental measurements is found. In the first phase of our research, theoretical and experimental information was used in the fitting procedure, making the analytical surface semiempirical in nature, which was a serious problem and a limitation in kinetics and dynamics studies. However, in the last few years, our surfaces have been fitted exclusively to very high-level *ab initio* calculations, avoiding the aforementioned limitations.

3. Library of Potential Energy Surfaces

The FORTRAN codes and the fitted parameters of the polyatomic potential energy surfaces developed by our group are available, free of charge, for the scientific community, and can be downloaded from the POTLIB library [54]: http://comp.chem.umn.edu/potlib/ or http://users.ipfw.edu/DUCHOVIC/POTLIB2001, or obtained from the authors upon request.

4. Applications

(a) The H + CH_4 *Reaction.* The gas-phase H + CH_4 → H_2 + CH_3 hydrogen abstraction reaction, as well as its deuterated isotopomers, is the prototype polyatomic reactive system and has been widely studied both theoretically and experimentally [10]. The construction of its PES has a long history, and it is one of the few reactive systems for which different approaches to the construction have been developed, which permits a direct comparison.

We will focus attention, except otherwise stated, on the three most recent and accurate surfaces for this reactive system, which were constructed with different strategies. Chronologically, in 2006 Zhang et al. [55, 56] developed the family of ZBBi surfaces, using the invariant polynomial method, based on the fitting to more than 20000 *ab initio* energies at the RCCSD(T)/aug-cc-pVTZ level. In 2009 we developed [41] an analytical PES, CBE surface, which is a VB/MM functional form, and the 36 parameters are fitted using exclusively high-level electronic structure calculations at the CCSD(T)/cc-pVTZ level. Very recently, in 2011,

Table 1: CH_5 saddle point properties.[a]

	Fitting (ZBB3)[b]	Analytical (CBE)[c]	Interpolated (ZFWCZ)[d]
Barrier height	14.78	15.01	15.03
Geometry			
R(C–H′)	1.409	1.390	1.399
R(H′–H$_B$)	0.901	0.973	0.895
<H-C-H′	102.8	106.8	103.1
<C H′ H$_B$	180.0	180.0	180.0
Vibrational frequencies			
	3289	3173	3223
	3289	3173	3223
	3104	3036	3073
	1726	1833	1784
	1434	1443	1442
	1434	1443	1442
	1078	1173	1115
	1078	1173	1115
	1062	1085	1062
	440	542	522
	440	542	522
	1320i	1488i	1467i

[a]Energy in kcal mol^{-1}, geometry in Å and degrees, and vibrational frequency in cm^{-1}; [b][55]; [c][41]; [d][18].

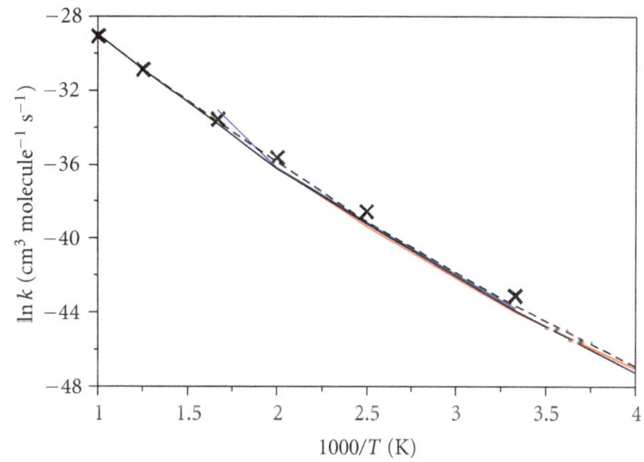

Figure 1: Arrhenius plots of ln k (cm^3 molecule^{-1} s^{-1}) for the forward thermal rate coefficients of the H + CH_4 reaction against the reciprocal of temperature (K), in the range 250–1000 K. Black line: MCDTH quantum calculations on the CBE surface; red line: MCDTH quantum calculations on the WWM surface; blue line: quantum calculations on the ZBB2 surface; black dashed line: VTST/MT calculations on the CBE surface; crosses: experimental values [60].

Zhou et al. [18] constructed a full-dimensional surface using the modified Shepard interpolation scheme based on 30000 data points at the CCSD(T)/aug-cc-pVTZ level—the ZFWCZ surface. The energy, geometry and vibrational frequencies of the saddle point are summarized in Table 1 for the three surfaces. Comparing the wide range of properties, we conclude that the three surfaces present excellent agreement, with small differences in the CBE with respect to the other two surfaces in the <H–C–H′ bending angle, ≈3°, and in the ZBB3 compared to the other two surfaces in the imaginary frequency, ≈100 cm^{-1}. The three surfaces present a colinear saddle point, 180.0°, with a similar barrier height, with small differences, ±0.2 kcal mol^{-1}, that is, within the chemical accuracy, and very close to the best estimation predicted at the CCSD(T)/aug-cc-pVQZ level, 14.87 kcal mol^{-1} [57].

The best macroscopic measure of the accuracy of a potential energy surface is probably the rate constant, at least in the thermal bottleneck region. When we compare different dynamics methods using the same surface, the dynamics approach is tested, and when we compare theoretical and experimental results, both the dynamics method and the surface are tested. Figure 1 plots the thermal rate coefficients computed with accurate fulldimensional quantum dynamics approaches, on the analytical CBE surface; a very accurate interpolation surface developed by Wu et al., WWM surface [57–59], which used the modified Shepard interpolated method developed by Collins et al., based on CCSD(T)/cc-pVQZ or CCSD(T)/aug-cc-pVQZ *ab initio* calculations; and a fitted surface, named ZBB2, an earlier version of the ZBB3 surface developed by Bowman et al. In the same figure, there also appear the results obtained with the variational transition-state theory with multidimensional tunneling effect [41] and the experimental data [60] for comparison. First, with quantum dynamics approaches, the rate coefficients obtained with the three surfaces agree almost perfectly in the common temperature range. Second, when the rate coefficients are obtained using different dynamics approaches, VTST/MT and MCDTH (multiconfigurational time-dependent Hartree approach [58, 59]) on the same CBE surface, excellent agreement is found, and both reproduce the experimental information in the common temperature range. These results indicate, first, that VTST/MT is a powerful and computationally economic tool for the kinetics study of polyatomic systems, with results comparable to those obtained with computationally more expensive quantum dynamical methods. Second, the CBE surface is accurate at least in the region of low energies, which is the most relevant region for thermal rate coefficient calculations and, therefore, provides a satisfactory description of the reaction path and the transition-state region.

Next, we analyze some dynamics properties where different surfaces have been used and compared with the sparse experimental data. We focus on the H + CD_4 → HD + CD_3 reaction because there is more experimental information available for comparison. The product energy partitioning has been experimentally obtained by Valentini's group. [61] only for the vibration and rotation of the HD product, and this same group found that more than 95% of the HD product is formed in the $v' = 0$ and $v' = 1$ vibrational states. These results appear in Table 2, together with the QCT results on the analytical CBE and the fitted ZBB1 [55] surfaces. The QCT calculations on these two PESs show excellent

Table 2: HD product energy partitioning and vibrational distribution (percentages) for the H + CD_4 reaction at 1.52 eV.

Surface	f'_{vib} (HD)	f'_{rot} (HD)	HD ($v' = 0$)	HD ($v' = 1$)
Analytical[a]	18	18	84	15
Fitted[b]	22	18	78	22
Exp.[c]	7	9	≥95	

[a]CBE surface, [41] [b]ZBB1 surface, [55] [c]Valentini et al. [61].

agreement for the HD product energy partitioning as well as for the HD vibrational distribution. In this latter case, the QCT results reproduce the experimental evidence, but, in the first case, they strongly contrast with the experimental measurements, which measure an internal excitation of the HD product of 7% and 9% for vibration and rotation, respectively. The agreement between the results from the two surfaces leads us to think that the discrepancies with experiment are mainly due to the dynamical method, that is, to limitations of the QCT approach. Experimental problems, however, cannot be totally ruled out, as was recently observed by Hu et al. [62], who suggested that the conclusions from Valentini et al.'s CARS experimental study might need to be reinterpreted.

The product angular distribution is, doubtless, one of the most sensitive dynamics features with which to test the quality of the potential energy surface, but experimentally it is very difficult to measure in some cases. When the Photoloc technique is used, the laboratory speed depends on both the scattering angle and the speed of the CD_3 product, which is influenced by the HD coproduct internal energy distribution. Uncertainties in this quantity could produce errors in the scattering angle.

Camden et al. [63] reported the first study of the state-to-state dynamics differential cross-section at high energies (1.95 eV) for the H + CD_4 gas-phase reaction using the Photoloc technique. They found that the CD_3 products are sideways/forward scattered with respect to the incident CD_4, suggesting a stripping mechanism (note that in the original papers [63–65] the CD_3 product is measured with respect to the incident H). Later, this same laboratory [64, 65] reported new experimental studies, also at high energy (1.2 eV), finding the same experimental behaviour. Experimentally, state-to-state dynamics studies are difficult to perform at low energies for the title reaction, because the H atoms, which are produced in a photolysis process, are hot. Only very recently have Zhang et al. [66] reported crossed molecular beam experiments for the H + CD_4 reaction at lower collision energies, ranging from 0.72 to 1.99 eV. Note that the lower value, 0.72 eV, is close to the barrier height, and consequently its dynamics will be influenced mainly by the transition-state region. Figure 2(a) plots these experimental results. A backward angle is clearly observed at low energies (rebound mechanism), changing towards sideways when the energy increases (stripping mechanism).

This is an excellent opportunity to test the quality of the PES and the dynamics methods (Figures 2(b)–2(d)). Only two surfaces have been used to study this problem theoretically: an analytical surface from our group (versions 2002 [43, 67] and 2008, labeled here as CBE [41, 68]), using both QCT and QM calculations, and an "on-the-fly" B3LYP/631G(d,p) density functional theory surface [65], using QCT calculations. At low energies, ≈0.7 eV, our analytical surfaces, PES-2002 and CBE, using QCT and QM methods (Figures 2(b) and 2(c)), show backward scattering, associated with a rebound mechanism, reproducing the recent experimental data [66]. The B3LYP "on-the-fly" surface using QCT calculations (Figure 2(d)) yields a more sideways scattering, with large uncertainties, in contrast with experiment. These differences could be due to the poor statistics on the B3LYP surface [65] and to the severe underestimation of the barrier heights, about 5 kcal mol^{-1} lower than the best *ab initio* calculations. This low barrier artificially permits reactive trajectories with larger impact parameters, favouring the sideways scattering region. When this error in the barrier height is corrected, our PES-2002 and, more noticeably in the better CBE surface, the low impact parameters are favoured, and this explains the rebound mechanism. Note that, interestingly, this experimental behaviour was already predicted by our group in 2006 [67] before the experimental data were available.

At higher collision energies, as they increase from 1.06 to 1.99 eV, Zhang et al. [66] found a shift of the product angular distribution from backwards to sideways. This behaviour agrees qualitatively with the previous observations of Camden et al. [63–65], although these latter workers found a clear sideways distribution at 1.21 and 1.95 eV, with practical extinction of the backward signal. This may have been due to the use of the Photoloc technique, which neglects the internal energy contribution of the HD co-product, and, as the authors themselves recognized in 2006, "clearly a more detailed picture of the differential cross section is desirable but it will have to await more experimental work." The QCT results using the analytical CBE surface (Figure 2(b)) reproduce the new experimental evidence. QM calculations on the same CBE surface [68] (Figure 2(c)) give more sideways scattering than the QCT calculations and experiment, although it is not clear whether this is due to significant quantum effects or simply are an artifact of the reduced dimensionality approach in the QM calculations.

Note that QCT calculations on the old PES-2002 surface also predicted the subsequently experimentally observed behaviour [66]—backwards-sideways scattering. However, they contradicted the experimental information available at the time the study was carried out—in 2006. In addition, since direct-dynamics QCT calculations at the B3LYP/631G(d,p) level showed sideways scattering, reproducing the experimental measurements [63–65], the validity of the analytical PES-2002 was questioned. The recent experiments of Zhang et al. [66], however, changed the whole picture and now the direct-dynamics QCT calculations at low collision energies are questioned because of, as noted above, the presence of reactive trajectories with erroneously large impact parameters due to the underestimation of the barrier height. In the case of higher collision energies, however, this error in the barrier should be of less concern than at lower energies.

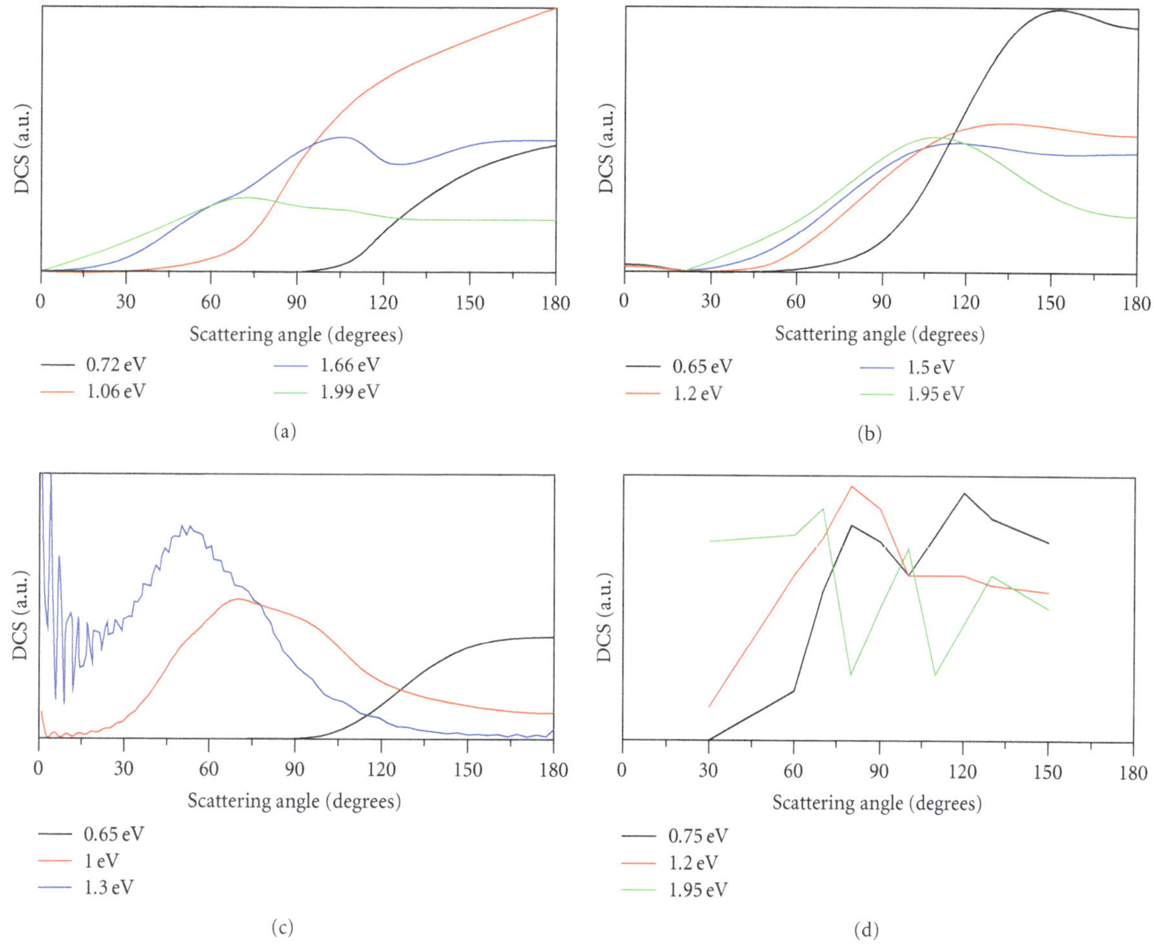

FIGURE 2: CD_3 product angular distribution (with respect to the incident CD_4) for the $H + CD_4 \rightarrow HD + CD_3$ reaction at different collision energies. (a) Experimental results from [64]; (b) QCT angular distribution on the CBE analytical surface [66]; (c) QM angular distribution on the CBE analytical surface [66]; (d) QCT angular distribution on the B3LYP "on-the-fly" surface [65].

Very recently, Zhou et al. [18] have performed an exhaustive analysis of the total reaction probabilities and integral cross-section using quantum dynamics calculations on the three most recent surfaces: fitted ZBB3, analytical CBE, and interpolated ZFWCZ surfaces, with collision energies up to 1.7 eV. Figure 3 plots the integral cross-section results. At collision energies up to 1.0 eV, the three surfaces show satisfactory agreement, while at higher energies of collision, the CBE surface overestimates this dynamics property, while the remaining two surfaces show good agreement. This could be attributed to deficiencies of the CBE surface at high energies. This is not surprising since the calibration of the CBE surface was done thinking of thermal behaviour. The information used during the fit focused on the reaction path and reaction valley, and higher energy areas were neither sampled nor weighted sufficiently. Therefore, as the collision energy increases, the accuracy of the CBE surface diminishes.

Finally, another severe test of the quality of the PES is the study of the effect of the vibrational excitation on the dynamics. In fact, the dynamics of a vibrationally excited polyatomic reaction presents a challenge both theoretically and experimentally. Camden et al. [69] carried out the

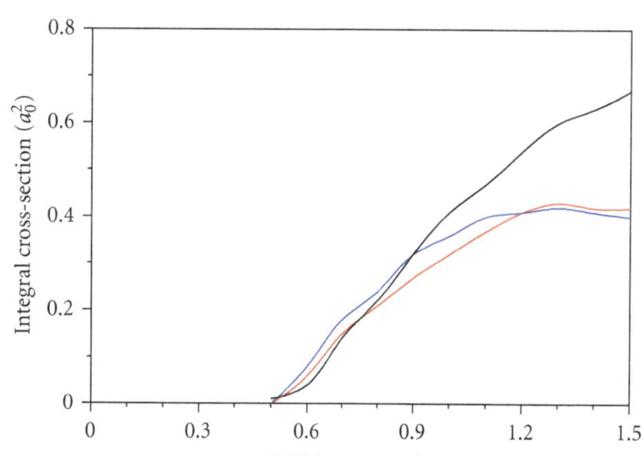

FIGURE 3: Quantum mechanical integral cross-section (a_o^2) versus collision energy (eV) for the $H + CH_4 \rightarrow H_2 + CH_3$ reaction on the CBE (black line), ZBB3 (blue line), and ZFWCZ (red line) surfaces.

first experimental study on the effect of the C–H stretch excitation on the gas-phase $H + CH_4$ hydrogen abstraction reaction. They found that the excitation of the asymmetric C–H stretch mode enhances the reaction cross-section by a factor of 3.0 ± 1.5 with respect to the ground-state methane, and this enhancement is practically independent of the collision energies for the three cases analyzed—1.52, 1.85, and 2.20 eV. In the following year, 2006, two theoretical papers on this issue were published: one by Xie et al. [56] using QCT calculations on the ZBB3 surface, and another from our laboratory [70] also using QCT calculations on the older analytical PES-2002 surface. At 1.52 eV, the theoretical results are close, with computed enhancement factors of 2.3 and 1.9, respectively, both of them within the experimental uncertainties. Note that our more recent CBE surface also predicts an enhancement factor of 1.9 (unpublished results).

(b) The $H + NH_3$ *Reaction.* The reaction of hydrogen atom with ammonia is a typical five-body reactive system, and presents a rare opportunity to study both intermolecular and intramolecular dynamics. For the intermolecular case, the gas-phase $H + NH_3 \rightarrow H_2 + NH_2$ hydrogen abstraction reaction is similar to the $H + CH_4$ reaction. It presents a barrier height of 14.5 kcal mol^{-1} and a reaction exoergicity of 5.0 kcal mol^{-1}, as compared to 14.87 and an endoergicity of 2.88 kcal mol^{-1}, respectively, for the $H + CH_4$ analogue. Also, the inversion of ammonia between two pyramidal structures (C_{3v} symmetry) passing through a planar structure (D_{3h} symmetry, which is a saddle point) is an example of intramolecular dynamics [71] with a barrier height in the range 5.20–5.94 kcal mol^{-1}, depending on the level of calculation.

Only three potential energy surfaces have been developed for the $H + NH_3$ system. In 2005, Moyano and Collins [72] developed an interpolation potential energy surface for the ammonia inversion, and the hydrogen abstraction and exchange reactions of $H + NH_3$ using a modified Shepard interpolated scheme based on 2000 data points calculated at the CCSD(T)/aug-cc-pVDZ level (PES1 version) or as single-point calculations at the CCSD(T)/aug-cc-pVTZ (PES2 version) level. The first and second derivatives of the energy were calculated by finite differences in the energy. Previously, in 1997, our group constructed the first surface for the hydrogen abstraction reaction exclusively, CE-1997 [44], which was fitted to a combination of experimental and theoretical information that is, it was semiempirical in nature, which represents a limitation for dynamics studies. Moreover, as was previously noted, recently Yang and Corchado [38] reported a major drawback of the CE-1997, namely that it describes incorrectly the NH_3 inversion motion, predicting incorrectly that the planar ammonia (D_{3h} symmetry) is about 9 kcal mol^{-1} more stable than the pyramidal structure (C_{3v} symmetry). To correct this behaviour of the NH_3 inversion, together with its semiempirical character, a new analytical potential energy surface, named EC-2009, was recently developed by our group [39], describing simultaneously the hydrogen abstraction and ammonia inversion reactions. This EC-2009 surface is basically a valence bond-molecular mechanics (VB/MM) surface, given by (24), and was fitted exclusively to very high level *ab initio* calculations

TABLE 3: NH_4 hydrogen abstraction saddle point properties.[a]

	Interpolated (MC)[b]	Analytical (EC-2009)[c]	*Ab initio*[d]
Barrier height	14.64	14.48	14.73
Geometry			
R(N–H′)	1.323	1.279	1.308
R(H′–H_B)	0.900	0.868	0.890
<C–H′–H_B	160.8	180.0	158.4
Vibrational frequencies			
	3450	3444	3478
	3350	3373	3384
	2057	1861	1888
	1527	1623	1566
	1174	1497	1280
	1047	1080	1063
	650	622	677
	629	581	506
	1371i	1602i	1662i

[a]Energy in kcal mol^{-1}, geometry in Å and degrees, and vibrational frequency in cm^{-1}; [b][72]; [c] [39]; [d] [39], at the CCSD(T)/cc-pVTZ level.

at the CCSD(T)/cc-pVTZ level. Note that the first derivatives of this surface are analytical, which implies a significant reduction in the computer time required for dynamical calculations as well as more accurate derivatives than is possible with numerical methods.

We begin by analyzing the hydrogen abstraction reaction. Table 3 lists the energy, geometry, and vibrational frequencies of the saddle point, with the interpolated MC (PES2 version) and the analytical EC-2009 surfaces. Using as target the CCSD(T)/cc-pVTZ *ab initio* level, the two surfaces present similarities, although important differences must be noted. First, while the interpolated MC surface reproduces the *ab initio* N–H′–H bend angle, the analytical surface yields a collinear approach. However, in previous papers [73, 74], we demonstrated that this was not a serious problem in the kinetics and dynamics description of the system. Second, the imaginary frequency obtained with the MC surface differs by about 300 cm^{-1} from the *ab initio* value.

For the two surfaces, Figure 4 shows the energy along the minimum energy path (V_{MEP}) and the ground-state vibrationally adiabatic potential curve (V_a^G), which is defined as the sum of the potential and vibrational zero-point energies along the reaction path. Note that s is the reaction coordinate, being zero at the saddle point, positive in the product channel, and negative in the reactant channel. Firstly, the V_{MEP} curve is smooth and shows no oscillations in either surface. However, V_a^G shows significant oscillations when computed for the interpolated surface. The reason is that the potential energy surface is not smooth enough to provide smooth changes in the frequencies (second derivatives of the energy). The situation becomes worse as one moves away from the saddle point and approaches areas where little *ab initio* information for interpolation is available. Thus, the bump in V_a^G at around $s = -1$ a.u. is due to wild oscillations

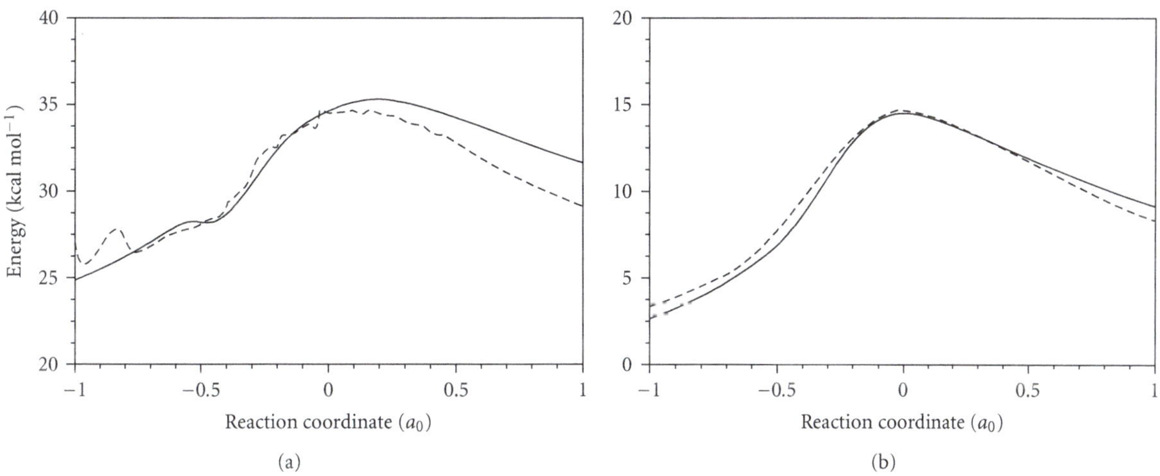

FIGURE 4: Potential energy (b) and vibrational ground-state energy (a) along the reaction path of the H + NH_3 → H_2 + NH_2 reaction computed using the EC-2009 (solid lines) and MC (dashed lines) surfaces. The zero of energy is set to the equilibrium potential energy of the reactants.

of the frequencies that lead to unphysical values of V_a^G for $|s| > 1$. This is a limitation when one wants to compute the tunneling effect at low energies (see below), since the barrier to tunnel through is totally unrealistic.

Figure 5 plots the thermal rate coefficients computed with the VTST/MT approach on both surfaces, where tunneling was estimated using the least-action tunneling (LAT) method [75, 76], together with experimental values [77] for comparison. In the common temperature range, 490–1780 K, both surfaces reproduce the experimental data, which is a test of both the surface and the dynamical method. The differences between the rates for the two surfaces are relatively small, 40% at 600 K, and diminishing as temperature increases. At low temperatures, the differences between the two surfaces increase, with the rates computed on the MC surface being 92% larger than the ones with the EC-2009 surface at 200 K. In this low temperature regime, where tunneling is important, the analytical EC-2009 surface is more accurate due to the unrealistic oscillations in the adiabatic reaction path on the MC surface, which negatively influence the tunneling.

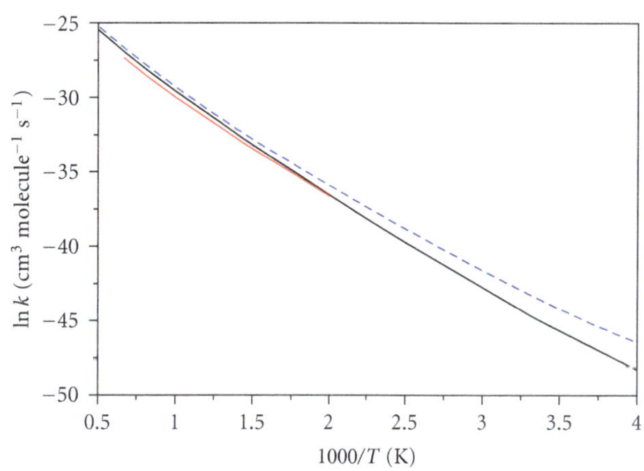

FIGURE 5: Arrhenius plots of ln k (cm^3 molecule^{-1} s^{-1}) for the forward thermal rate coefficients of the H + NH_3 → H_2 + NH_2 reaction against the reciprocal of temperature (K) in the range 200–2000 K. Solid black line: analytical EC-2009; dashed blue line: interpolated MC; dotted red line: experimental values from [77].

The integral cross-sections in the range 10–30 kcal mol^{-1} have been evaluated using quasi-classical trajectory (QCT) calculations on both surfaces [72, 78] and are plotted in Figure 6. As apparent from this figure, both surfaces agree reasonably, showing typical threshold behaviour, starting from 10 kcal mol^{-1} and increasing with the translational energy. Unfortunately, there is no experimental data for comparison, and we think that these theoretical results might stimulate experimental work on this little studied system.

Finally, the angular distributions of the H_2 product with respect to the incident H atom has only been determined using the EC-2009 surface [78]. Figure 7 plots this property for collision energies of 25 and 40 kcal mol^{-1}. At 25 kcal mol^{-1}, the scattering distribution is in the sideways-backward hemisphere, associated with a rebound mechanism and low impact parameters. When the collision energy increases, 40 kcal mol^{-1}, the scattering is shifted slightly towards the sideways hemisphere, due to larger impact parameters.

As was mentioned above, the EC-2009 surface describes, in addition to the aforementioned hydrogen abstraction reaction, the ammonia inversion, an example of interesting intramolecular dynamics. Figure 8 plots the equipotential contours in the two significant coordinates, r(N–H) bond length and <H–N–H angle, for the analytical EC-2009 PES and the CCSD(T)/cc-pVTZ *ab initio* level, for comparison. As seen, the contours are smooth and the analytical PES reproduces the fitted *ab initio* data points [39].

A very stringent test of the quality of this surface is the ammonia splitting, which demands spectroscopic accuracy. The inversion motion is represented by a symmetric double-well potential. As a consequence of the perturbation originating this double well, a splitting of each degenerate vibrational

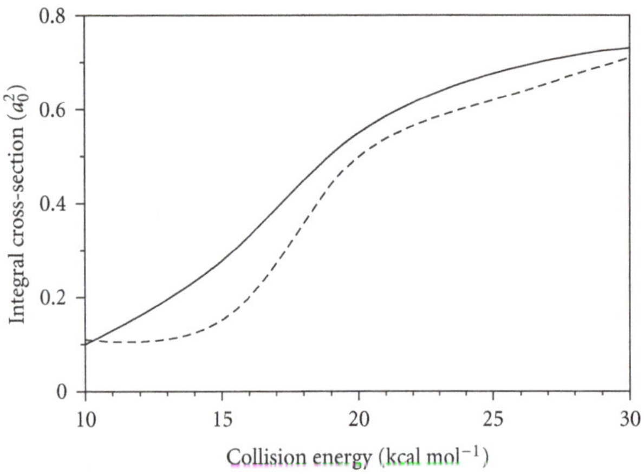

Figure 6: QCT reaction cross-section (a_0^2) versus the collision energy (kcal mol^{-1}) for the $H + NH_3 \rightarrow H_2 + NH_2$ reaction computed using the analytical EC-2009 (solid line) and interpolated MC (dashed line) surfaces.

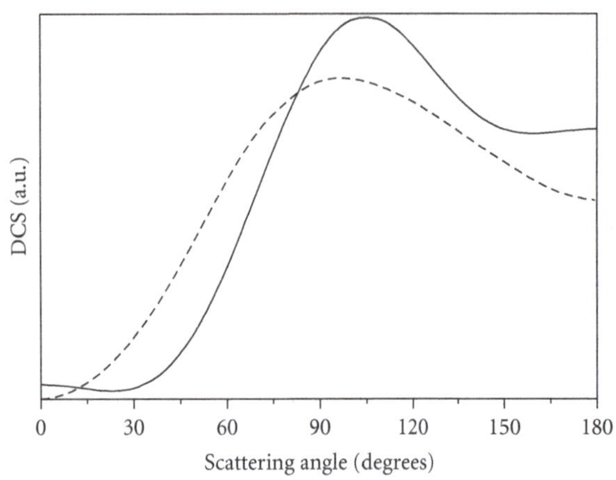

Figure 7: H_2 product angular distribution (with respect to the incident H) for the $H + NH_3 \rightarrow H_2 + NH_2$ reaction at 25 kcal·mol^{-1} (solid line) and 40 kcal mol^{-1} (dashed line), computed using QCT calculations on the EC-2009 surface.

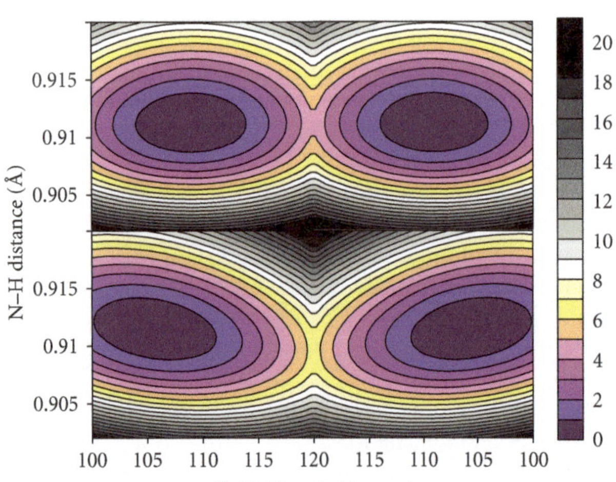

Figure 8: Ammonia inversion reaction. Contour plots of the analytical EC-2009 (upper panel) and CCSD(T)/ccpVTZ *ab initio* surface (lower panel).

level into two levels appears, ΔE, due to quantum mechanical tunneling [71], and the splitting increases rapidly with the vibrational number. The splitting ΔE can be computed from the tunneling rate of inversion, k_{tunn}, which is in turn obtained from the imaginary action integral, $\theta(E)$, using the WKB approximation [79]:

$$\Delta E_0 = \frac{k_{\text{tunn}}}{2c}, \qquad k_{\text{tunn}} = \frac{2c\nu_2}{\pi}\exp[-\theta(E)], \tag{27}$$

with c being the speed of light and ν_2 the eigenvalue associated with the inversion mode of ammonia, 1113 cm^{-1}. Figure 9 plots this path for ammonia where the first two pairs of split eigenvalues are superimposed. The computed splitting is listed in Table 4 together with experimental values [80] for comparison. The EC-2009 results overestimate the experimental values. The reason for the discrepancy mainly lies in the shape of the PES, although other factors such as the tunneling calculation cannot be discarded. Indeed, we found [39] that the overestimation we obtain in tunneling splitting is due to our barrier to inversion being slightly lower and thinner than those of other studies. Unfortunately, tunneling splitting is so sensitive to the shape of the PES that a few tenths of kcal·mol^{-1} give rise to a factor of four in the computed splitting. With respect to the effect of isotopic substitution, for the ND_3 case we obtain a value $\Delta E = 0.37$ cm^{-1}. Although this is greater than the experimentally reported value, 0.05 cm^{-1} [81], it correctly predicts about one order of magnitude reduction upon isotopic substitution. In sum, despite the enormous effort required for the construction of the potential energy surface, it does not suffice to obtain spectroscopic accuracy, and our surface can only give a qualitative description of the splitting in ammonia.

5. When the Problems Increase

The benchmark $H + CH_4$ hydrogen abstraction reaction, with five light atoms and a single heavy atom, which allows a large number of very high-level *ab initio* calculations to be performed, gives the impression of the process being a "piece of cake" regarding polyatomic bimolecular reactions. However, it still presents kinetics and dynamics differences depending on the potential energy used and discrepancies with the experimental measures. Then, what will the case be when more complicated systems are studied? In this section we will analyze some problems which can appear alone or in combination in the study of polyatomic systems and strongly complicate the construction of potential energy surfaces.

5.1. The Spin-Orbit Problem and Multisurface Dynamics. This is a typical problem, for instance, in reactions involving halogen atoms, $X(^2P)$, with molecular systems, R–H:

$$X(^2P) + H\text{–}R \longrightarrow XH + R \quad (X = F, Cl, Br). \tag{28}$$

The halogen atom presents two spin-orbit electronic states, $^2P_{3/2}$ and $^2P_{1/2}$, with a splitting of 404 cm^{-1} (1.1 kcal mol^{-1}),

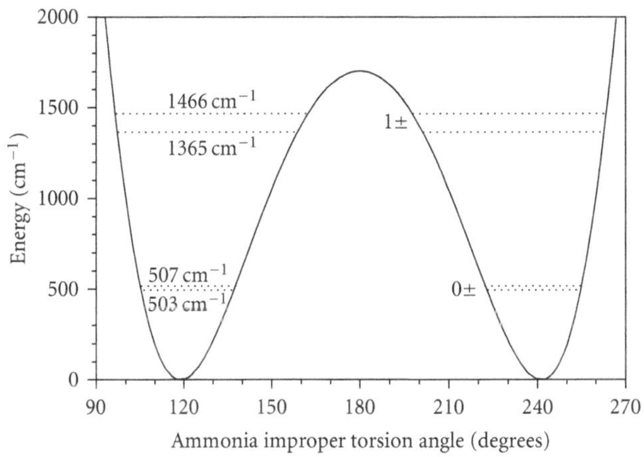

FIGURE 9: Classical potential for the ammonia inversion path obtained from the EC-2009 surface. The first two pairs of eigenvalues are shown.

TABLE 4: Eigenvalues of ammonia inversion (in cm^{-1}).

ν_2	EC-2009[a]	Exp.[b]
0^+	0.00	0.00
0^-	3.64	0.79
1^+	861.39	932.43
1^-	962.90	968.12
2^+	1533.36	1598.47
2^-	1942.29	1822.18

[a] [39]; [b] [80].

882 cm^{-1} (2.5 kcal mol^{-1}), and 3685 cm^{-1} (3.5 kcal mol^{-1}) for F, Cl, and Br, respectively. A priori, the smaller the separation, the greater the possibility of the reaction coming from the two states, which complicates the PES construction and the dynamics study.

This problem affects all the previously described theoretical methods to develop surfaces, because it is a problem intrinsic to the initial information required: the quantum mechanical calculations. For this problem, relativistic calculations would be needed, which would immensely increase the computational cost and would make these calculations impractical in polyatomic systems. In addition, new functional forms for the analytical functions need to be developed to include the coupling between states and its dependence on coordinates [82, 83], in order to make it possible to include nonadiabatic effects and hopping between surfaces in the dynamics study of these multistate systems.

In the case of atom-diatom reactions, some results have shed light on the spin-orbit problem, although some theory/experiment controversies still persist. For instance, for the well-studied $F(^2P_{3/2}, {}^2P_{1/2}) + H_2$ reaction, Alexander et al. [82, 84] found that the reactivity of the excited s-o state of F is small, 10–25% of the reactivity of the ground s-o state, and concluded that the overall dynamics of the F + H_2 reaction could be well described by calculations on a single, electronically adiabatic PES, although for a direct comparison with experiment the coupling between the ground and excited s-o surfaces must be considered.

For the analogue $Cl(^2P_{3/2}, {}^2P_{1/2}) + H_2$ reaction, because of the larger energy separation, one would expect that the reaction could evolve on the ground-state adiabatic surface, with the contribution of the chlorine excited state, $^2P_{1/2}$, being practically negligible according to the Born-Oppenheimer (BO) approach. However, recently a theory/experiment controversy has arisen on this issue. Thus, Lee and Liu [85–87] demonstrated experimentally the contrary for the Cl + H_2 reaction; that is, the excited chlorine atom (Cl*) is more reactive to H_2 than the ground-state chlorine (Cl) by a factor of at least ≈6. In any case, even taking into account experimental error bars, the authors were confident about the reactivity relationship Cl* > Cl, which could have a significant effect on the temperature dependence of the thermal rate coefficients. The authors interpreted this surprising result by postulating a nonadiabatic transition (breakdown of the BO approximation) from the excited Cl* to the ground-state Cl surface by either electrostatic or spin-orbit coupling in the entrance channel. This experimental study initiated a major theoretical and experimental debate on the influence of the excited Cl* in the reactivity of the Cl + H_2 reaction. Different laboratories [88–93] performed theoretical/experimental studies of the validity of the BO approximation in this reaction, concluding that the adiabatically allowed reaction ($Cl(^2P_{3/2}) + H_2$) will dominate the adiabatically forbidden reaction ($Cl(^2P_{1/2}) + H_2$). Hence, these results are in direct contrast with the experiment of Liu et al. [85–87], suggesting that this experiment claiming high reactivity of Cl* needs to be reexamined.

In the case of polyatomic systems, this level of sophistication has not been achieved and would still be computationally prohibitive. Thus, some approaches have been considered to take into account, indirectly, the s-o effect in the construction of the PES and the dynamics study. First, for thermochemical or rate coefficient calculations, the s-o effect on the multiple electronic states is taken into account by the electronic partition function of the reactants in the usual expression

$$Qe = 4 + 2\exp\left(-\frac{\varepsilon}{k_B T}\right), \tag{29}$$

where ε is the s-o splitting of the halogen atom. Second, there is an additional effect on the barrier height (Figure 10). In fact, if we assume that along the entire reaction path the states are fully quenched, considering the s-o effect would lower the energy of the s-o ground state of the halogen atom by $1/3\,\varepsilon$ below its nonrelativistic energy, increasing the barrier height by this amount. This represents 0.38, 0.83, and 1.17 kcal mol^{-1}, respectively, for F, Cl, and Br. Our group considered these approaches in the construction of the surface for the $F(^2P) + CH_4$ [42], $Cl(^2P) + CH_4$ [49], and $Cl(^2P) + NH_3$ [94, 95] reactions. Note, however, that this approach is to some extent incoherent in the sense that as we move away from the saddle point towards reactants the energy asymptotically tends to zero, the energy of the lower state of the halogen atom. However, as we move from the

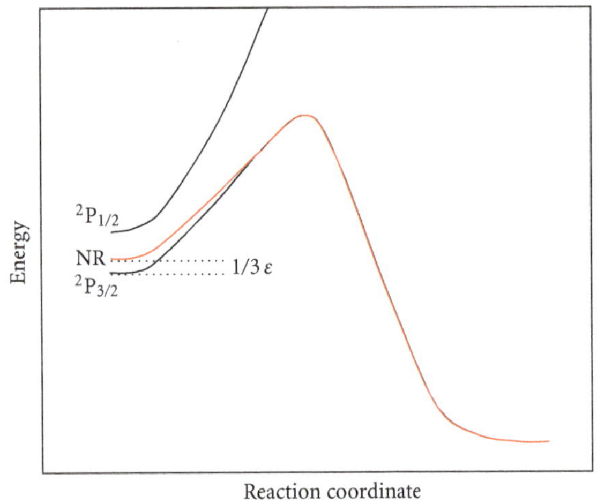

FIGURE 10: Schematic representation of the potential energy along the reaction path for a reaction with spin/orbit effects on the reactants. Red line: nonrelativistic calculations.

saddle point towards reactants, it should tend asymptotically to 1/3 ε until it reaches the point where the two states interact. From that point towards the reactants, the surface has to tend to zero. There is therefore a gap of up to 1/3 ε between the exact result and our approximation results. However, we can assume that this gap is sufficiently small and is located so far from the dynamically important regions of the surface that we can safely neglect it. Moreover, in our analytical surfaces, the changes in the energy can be fitted in order to change the slope of the reaction path so that this asymptotic behaviour can be corrected. In addition, in the latter case of the $Cl(^2P) + NH_3$ reaction, because of the presence of wells in the reactant channel (see below), this gap occurs before the system reaches the well in the region connecting the well with reactants, and its influence on the kinetics and dynamics is entirely negligible.

5.2. The Molecular "Size" Problem. Obviously, the cost of calculating the quantum chemical information needed to build the PES increases exponentially with the number of electrons involved, and it is still a prohibitive task for large molecules and heavy atoms. Thus, for instance, while the $H + CH_4$ benchmark reaction involves five light hydrogen atoms with only eleven electrons, when third-row atoms are considered, for instance, $H + SiH_4$, nineteen electrons are involved, or when larger systems are considered, for instance, $H + CCl_4$, four heavy chlorine atoms and 75 electrons must be included in the calculations. This represents an enormous computational effort, and that is prohibitive if high-level *ab initio* calculations are used to obtain chemical accuracy. Obviously, fitting or interpolation approaches, based on grids of 20000–30000 data points or direct dynamics calculations are still unaffordable, although more economical alternatives could be used to calculate the quantum mechanical data, such as the "dual level" technique, where the geometries and vibrational frequencies are calculated at a lower *ab initio* or DFT level and the energies are calculated as single points on these geometries at a higher quantum mechanical level. Even so, the computational effort would be enormous.

In these complicated cases with a large number of electrons, our strategy to build the PES, based on a smaller number of *ab initio* calculations, represents an interesting and practical alternative. Thus, our laboratory has constructed surfaces for several five-body systems, $H + NH_3$ [39, 44], 11 electrons, $F + NH_3$ [45], 19 electrons, and $Cl(^2P) + NH_3$ [94, 95], 27 electrons; six-body systems, $H + SiH_4$ [46], 19 electrons, $H + GeH_4$ [96], 37 electrons, $Cl(^2P) + CH_4$ [49], 27 electrons, $Br(^2P) + CH_4$ [51], 45 electrons, $H + CCl_4$ [52, 97], 75 electrons; one of seven bodies, $OH + CH_4$ [53], 19 electrons. In addition, we have also studied reactions with asymmetrically substituted methane, $H + CH_3Cl$ and $Cl + CHClF_2$ [98, 99], which represent another challenge in the PES construction for polyatomic systems, because in addition to the aforementioned problems, the possibility of several reaction channels needs to be taken into account. Although the older surfaces were semiempirical, that is, they combined theoretical and experimental information in the fitting procedure due to computational limitations at that time, the newer surfaces, those developed from 2007 onward, are based exclusively on quantum mechanical information: $H + CH_4$ [41], $H + CCl_4$ [97], $H + NH_3$ [39], and $Cl(^2P) + NH_3$ [95].

It is noteworthy that the functional form has remained almost unchanged, being that of the $H + CH_4$ reaction, adding different modifications following the requirements of the systems under study. Thus, for the five-body systems we removed the dependency on one of the hydrogen atoms of CH_4, and for the seven-body system, $OH + CH_4$, we added additional Morse and harmonic terms to describe the OH bond and <HOH angle, while for the asymmetric reactions we allowed the four atoms bonded to the carbon atom to vary independently. In addition, we improved the analytical form when building the $Cl + NH_3$ surface by allowing the equilibrium N–H distance to vary along the reaction path [95].

5.3. Reactions with More Complicated Topology. The last problem analyzed in this section is that associated with the presence of several maxima and minima in the polyatomic reaction. As was recently noted by Clary [100], "reactions with several maxima and minima in the potential energy surface present the severest challenge to calculating and fitting potential energy surfaces and carrying out quantum dynamics calculations."

Basically, taking into account the topology of the potential energy surface and independently of whether the reaction is exothermic, endothermic, or thermoneutral, the bimolecular reactions can be classified into three broad categories (Figure 11): reactions with a single barrier (Figure 11(a)); barrierless reactions (Figure 11(b)), and reactions with more complicated potentials (e.g., Figures 11(c) and 11(d)). We focus on this last case with polyatomic systems ($n > 4$). Obviously, the existence of wells in the entry and exit channels is associated with the presence of heavy atoms,

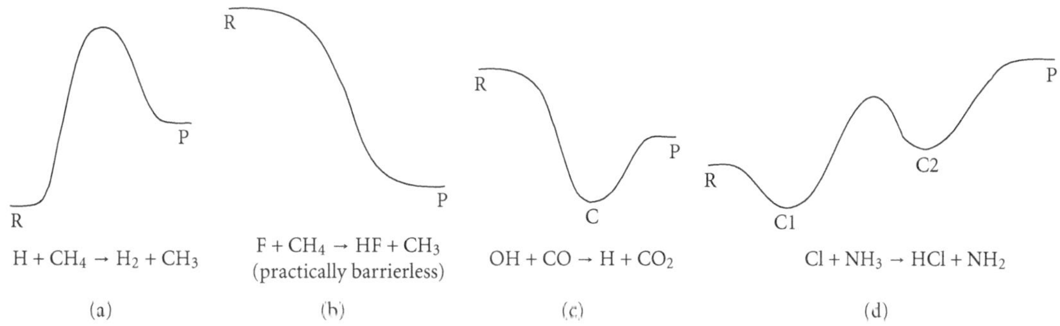

FIGURE 11: Schematic representation of the potential energy along the reaction path for different types of bimolecular reactions. R, P and C denote, respectively, reactants, products, and intermediate complexes. In all the cases, the vertical axis represents the energy and the horizontal axis the reaction coordinate.

especially halogen atoms, which favour the formation of intermediate complexes, either van der Waals or hydrogen-bonded complexes. Thus, bimolecular reactions with a complicated topology, as shown in Figures 11(c) or 11(d), also tend to pose several of the problems considered above, that is, the increase of the number of electrons involved and the spin-orbit problem. In addition, the presence of several maxima and minima requires a fine evaluation of the energies, gradients, and Hessians needed for a correct description of the PES. This, obviously, represents a complication in the construction of the PES, and only very few global surfaces have as yet been developed [26, 95].

We finish this section by describing the most complicated surface analyzed by our group [95]: the analytical potential energy surface for the $Cl(^2P) + NH_3$ polyatomic reaction, which presents several wells in the entry and exit channels, with a topology showed in Figure 11(d). Different intermediate complexes were found in the entry and exit channels at the CCSD(T)/cc-pVTZ level, and the intrinsic reaction path was calculated (energies, gradients, and Hessians) starting from the saddle point. With respect to the barrier height, one of the most difficult energy properties of a PES to estimate, different laboratories report electronic structure calculations using different levels (correlation energy and basis sets) [94, 95, 101–103]. Gao et al. [102] constructed the reaction path using the MPWB1K density functional (DFT) method [104], finding a barrier of 5.2 kcal mol^{-1}, and, using different correlation energy levels and basis sets, despite, they reported values in the range 4.8–6.2 kcal mol^{-1}. Xu and Lin [103] performed a computational study of the mechanisms and kinetics of the reaction. The geometries of the stationary points were optimized using the B3LYP DFT method, and their energies were refined with the modified Gaussian-2 (G2M) theory. They obtained a value of 7.2 kcal mol^{-1}, in contrast with the preceding values. Finally, we [94] obtained a barrier height of 7.6 kcal mol^{-1} at the CCSD(T)/cc-pVTZ level and of 5.8 kcal mol^{-1} [95] at a higher level, CCSD(T) = FULL/aug-cc-pVTZ, close to Gao et al.'s result. These results illustrate the dramatic influence of the electronic correlation and basis set on the correct description of the barrier, and it must be borne in mind that small differences in the saddle point produce large deviations in the kinetics and dynamics analysis. To further complicate the study of this reaction, the spin-orbit coupling must be considered, since the chlorine atom has two low-lying fine structure electronic states, $^2P_{1/2}$ and $^2P_{3/2}$, with a separation of $\varepsilon = 882\,cm^{-1} \approx 2.5\,kcal\,mol^{-1}$. As discussed above, the spin-orbit coupling was taken into account in our nonrelativistic calculations in two ways: first, in the electronic partition function of the reactant (29), and second, by adding one-third of the split between the two states, that is, 0.8 kcal mol^{-1}, to the barrier height, where we have assumed that the s-o coupling is essentially fully quenched at the saddle point. With this correction, our best estimate of the barrier height was 6.6 kcal mol^{-1}.

With all this information, an analytical PES was constructed and kinetics information was obtained using VTST/MT methods [95]. The results agree well with experimental values of the rate coefficients and equilibrium constants, showing that the wells have little influence on the kinetics of the reaction. This is mainly due to the fact that VTST/MT for this reaction, whose reaction path is lower in energy than the products, assumes that tunneling has no effect. Therefore, kinetics is controlled by the properties of a transition state, which is located near the saddle point. In addition, calculations of the recrossing of the transition state that the presence of wells could cause showed that it is very small. Therefore, only the saddle point region determines the reaction probabilities.

An exhaustive dynamics study using QCT and QM methods is currently in progress in our group and should be published soon. From the results it seems that the reaction cross-sections show significant values at very low collision energies using both methodologies. In order to analyze to what extent the presence of wells (especially the well on the reactant side) is responsible for this behaviour, further studies will be carried out with a model surface similar to the $Cl + NH_3$ from which the wells have been removed. Note that the latter study is possible because of the availability of an analytical surface that can be refitted to remove the wells without significantly modifying other regions of the PES, which is an added value to this kind of analytical PES. These kinds of study can help us to understand the dynamics of such a complicated system and whether transition-state theory, which (as noted above) assumes that only the saddle point region is significant, needs to be challenged.

6. Final Remarks

The construction of potential energy surfaces in polyatomic reactive systems represents a major theoretical challenge, with a very high computational and human time cost. Based on high-level electronic structure calculations, fitting, interpolation, and analytical (defined by functional forms) approaches have been developed and applied in the kinetics and dynamics (classical, quasi-classical, and quantum mechanical) study of these reactions. However, in spite of the enormous progress in the last 20–30 years in theoretical algorithms and computational power, the construction of potential energy surfaces in polyatomic systems has still not reached the level of accuracy achieved for the triatomic systems.

The quality of these surfaces is still an open and debatable question, and even for the benchmark H + CH_4 polyatomic reaction, which involves only five light hydrogen atoms and eleven electrons, small differences are found depending on the PES construction and the dynamics method. Unfortunately, the problems will increase for other important chemical systems, where some effects are present such as spin-orbit coupling, increase in the number of electrons and molecular size, or potentials with more complicated topology. In these cases, the kinetics and dynamics results will be even more strongly dependent on the quality of the PES. In the last few decades, there has been much theoretical effort on the part of various laboratories, but there is still much to do in this research field.

Throughout this paper, the emphasis has been put on the strategy developed by our group which has constructed about 15 surfaces for polyatomic systems of five, six and seven bodies. These are freely available for download from the POTLIB websites, http://comp.chem.umn.edu/POTLIB/ or http://users.ipfw.edu/DUCHOVIC/POTLIB2001/, or can be requested from the authors.

The analytical surfaces developed in our group are based on a basic functional form with slight modifications to suit it to each of the systems studied. In this sense, they are very specialized force fields, with mathematical functions that allow further improvements to be introduced. For example, adding terms to describe anharmonicity, mode-mode coupling, or an improved dependence of the constants of the fit on the reaction coordinate are possibilities, which can improve the functional form.

The inclusion of additional reaction channels, such as the exchange reactions in methane or ammonia, is another pending matter in work on our surfaces. This would open up the possibility of analyzing competitive channels and could lead to a better understanding of the behaviour of complex reactions.

An additional advantage of this approach is the negligible computational cost of the evaluation of the potential energy surface and its derivatives, which is a very desirable feature when one wants to apply expensive QM methods for the study of these reactions. Undoubtedly, with the evolution of computer resources, eventually a point will be reached when direct dynamics calculations using high *ab initio* levels will be feasible. However, until then, the possibility of having an analytical function that can describe the potential energy with reasonable accuracy is the fastest option. In recent years ad hoc surfaces have been constructed which only reproduce a single kinetics or dynamics property but do not provide a complete description of the reaction system. However, the aim of our work is to build a PES that reproduces at least qualitatively all the kinetics and dynamics features of the reactive system. Obviously a quantitative description of all kinetics and dynamics properties will be the final goal, and with this target in mind in our group we have adopted a flexible functional form that can be easily improved and adapted to other similar systems.

Acknowledgments

This work has been partially supported in recent years by the Junta de Extremadura, Spain, and Fondo Social Europeo (Projects nos. 2PR04A001, PRI07A009, and IB10001). One of the authors (M.Monge-Palacios) thanks Junta de Extremadura (Spain) for a scholarship.

References

[1] M. Born and R. Oppenheimer, "Zur quantentheorie der molekeln," *Annals of Physics*, vol. 84, pp. 457–484, 1927.

[2] C. N. Ramachandran, D. De Fazio, S. Cavalli, F. Tarantelli, and V. Aquilanti, "Revisiting the potential energy surface for the He + H_2^+ → HeH^+ + H reaction at the full configuration interaction level," *Chemical Physics Letters*, vol. 469, no. 1–3, pp. 26–30, 2009.

[3] F. George and S. Kumar, "Ab initio ground and the first excited adiabatic and quasidiabatic potential energy surfaces of H^+ + CO system," *Chemical Physics*, vol. 373, no. 3, pp. 211–218, 2010.

[4] T. Tanaka and T. Takayanagi, "Quantum reactive scattering calculations of H + F_2 and Mu + F_2 reactions on a new *ab initio* potential energy surface," *Chemical Physics Letters*, vol. 496, no. 4–6, pp. 248–253, 2010.

[5] W. Skomorowski, F. Pawlowski, T. Korona, R. Moszynski, P. S. Zuchowski, and J. M. Hutson, "Interaction between LiH molecule and Li atom from the state-of-the-art electronic structure calculations," *Journal of Chemical Physics*, vol. 134, p. 114109, 2011.

[6] B. Jiang, Ch. Xie, and D. Xie, "New ab iniito potential energy surface for BrH_2 and rate constants for the H + HBr → H_2 + Br abstraction reaction," *Journal of Chemical Physics*, vol. 134, p. 114301, 2011.

[7] M. Karplus, R. N. Porter, and R. D. Sharma, "Exchange reactions with activation energy. I. Simple barrier potential for (H, H_2)," *Journal of Chemical Physics*, vol. 43, no. 9, pp. 3259–3287, 1965.

[8] D.J. Kouri, Y. Sun et al., "New time-dependent and time-independent computational methods for molecular collisions," in *Mathematical Frontiers in Computational Chemical Physics*, D. G. Truhlar, Ed., pp. 207–244, Springer, New York, NY, USA, 1988.

[9] A. Fernandez-Ramos, J. A. Miller, S. J. Klippenstein, and D. G. Truhlar, "Modeling the kinetics of bimolecular reactions," *Chemical Reviews*, vol. 106, pp. 4518–4584, 2006.

[10] T. V. Albu, J. Espinosa-García, and D. G. Truhlar, "Computational chemistry of polyatomic reaction kinetics and

dynamics: the quest for an accurate CH_5 potential energy surface," *Chemical Reviews*, vol. 107, no. 11, pp. 5101–5132, 2007.

[11] W. L. Hase, K. Song, and M. S. Gordon, "Direct dynamics simulations," *Computing in Science and Engineering*, vol. 5, no. 4, pp. 36–44, 2003.

[12] J. Liu, K. Song, W. L. Hase, and S. L. Anderson, "Direct dynamics trajectory study of the reaction of formaldehyde cation with D_2: vibrational and zero-point energy effects on quasiclassical trajectories," *Journal of Physical Chemistry A*, vol. 109, no. 50, pp. 11376–11384, 2005.

[13] T.-S. Ho and H. Rabitz, "A general method for constructing multidimensional molecular potential energy surfaces from ab initio calculations," *Journal of Chemical Physics*, vol. 104, no. 7, pp. 2584–2597, 1996.

[14] M. A. Collins, "Molecular potential-energy surfaces for chemical reaction dynamics," *Theoretical Chemistry Accounts*, vol. 108, no. 6, pp. 313–324, 2002.

[15] T.-S. Ho and H. Rabitz, "Reproducing kernel Hilbert space interpolation methods as a paradigm of high dimensional model representations: application to multidimensional potential energy surface construction," *Journal of Chemical Physics*, vol. 119, no. 13, pp. 6433–6442, 2003.

[16] J. Ischtwan and M. A. Collins, "Molecular potential energy surfaces by interpolation," *Journal of Chemical Physics*, vol. 100, no. 11, pp. 8080–8088, 1994.

[17] M. A. Addicoat and M. A. Collins, "Potential energy surfaces: the forces of chemistry," in *Tutorials in Molecular Reaction Dynamics*, M. Brouard and C. Vallance, Eds., p. 28, RSC Publishing, Cambridge, UK, 2010.

[18] Y. Zhou, B. Fu, C. Wang, M. A. Collins, and D. H. Zhang, "Ab initio potential energy surface and quantum dynamics for the $H + CH_4 \rightarrow H_2 + CH_3$," *Journal of Chemical Physics*, vol. 134, pp. 64323–64330, 2011.

[19] J. N. Murrell, S. Carter, S. C. Farantos, P. Huxley, and A. J. C. Varandas, *Molecular Potential Energy Functions*, Wiley, Chistester, UK, 1984.

[20] G. C. Schatz, "The analytical representation of electronic potential-energy surfaces," *Reviews of Modern Physics*, vol. 61, no. 3, pp. 669–688, 1989.

[21] J. M. Bowman and G. C. Schatz, "Theoretical studies of polyatomic bimolecular reaction dynamics," *Annual Review of Physical Chemistry*, vol. 46, no. 1, pp. 169–195, 1995.

[22] K. Sorbie and J. N. Murrell, "Theoretical study of the $O(^1D)+H_2(^1S_g)$ reactive quenching process," *Molecular Physics*, vol. 31, pp. 905–920, 1976.

[23] A. J. C. Varandas, "Four-atom bimolecular reactions with relevance in environmental chemistry," *International Reviews in Physical Chemistry*, vol. 19, no. 2, pp. 199–245, 2000.

[24] A. Brown, B. J. Braams, K. Christoffel, Z. Jin, and J. M. Bowman, "Classical and quasiclassical spectral analysis of CH_5^+ using an ab initio potential energy surface," *Journal of Chemical Physics*, vol. 119, no. 17, pp. 8790–8793, 2003.

[25] Z. Jin, B. J. Braams, and J. M. Bowman, "An ab initio based global potential energy surface describing $CH_5^+ \rightarrow CH_3^+ + H_5$," *Journal of Physical Chemistry A*, vol. 110, no. 4, pp. 1569–1574, 2006.

[26] B. J. Braams and J. M. Bowman, "Permutationally invariant potential energy surfaces in high dimensionality," *International Reviews in Physical Chemistry*, vol. 28, no. 4, pp. 577–606, 2009.

[27] J. M. Bowman, G. Czako, and B. Fu, "High-dimensional ab initio potential energy surfaces for reaction dynamics calculations," *Physical Chemistry Chemical Physics*, vol. 13, no. 18, pp. 8094–8111, 2011.

[28] A. B. McCoy, B. J. Braams, A. Brown, X. Huang, Z. Jin, and J. M. Bowman, "Ab initio diffusion monte carlo calculations of the quantum behavior of CH_5^+ in full dimensionality," *Journal of Physical Chemistry A*, vol. 108, no. 23, pp. 4991–4994, 2004.

[29] T. Valencich and D. L. Bunker, "Energy-dependent cross sections for the tritium-methane hot atom reactions," *Chemical Physics Letters*, vol. 20, no. 1, pp. 50–52, 1973.

[30] L. M. Raff, "Theoretical investigations of the reaction dynamics of polyatomic systems: chemistry of the hot atom ($T^* + CH_4$) and ($T^* + CD_4$) systems," *Journal of Chemical Physics*, pp. 2220–2244, 1970.

[31] R. Steckler, K. J. Dykema, F. B. Brown, G. C. Hancock, D. G. Truhlar, and T. Valencich, "A comparative study of potential energy surfaces for $CH_3 + H_2 \leftrightarrow CH_4 + H$," *Journal of Chemical Physics*, vol. 87, no. 12, pp. 7024–7035, 1987.

[32] T. Joseph, R. Steckler, and D. G. Truhlar, "A new potential energy surface for the $CH_3 + H_2 \leftrightarrow CH_4 + H$ reaction: calibration and calculations of rate constants and kinetic isotope effects by variational transition state theory and semiclassical tunneling calculations," *Journal of Chemical Physics*, vol. 87, no. 12, pp. 7036–7049, 1987.

[33] M. J. T. Jordan and R. G. Gilbert, "Classical trajectory studies of the reaction $CH_4 + H \rightarrow CH_3 + H_2$," *Journal of Chemical Physics*, vol. 102, no. 14, pp. 5669–5682, 1995.

[34] J. Espinosa-García and J. C. Corchado, "Global surfaces and reaction-path potentials. Applications of variational transition-state theory," *Recent Research Development in Physical Chemistry*, vol. 1, pp. 165–192, 1997.

[35] J. Espinosa-Garcia, "Analysis of economical strategies for the construction of potential energy surfaces," *Trends in Physical Chemistry*, vol. 8, pp. 49–73, 2000.

[36] A. Chakraborty, Y. Zhao, H. Lin, and D. G. Truhlar, "Combined valence bond-molecular mechanics potential-energy surface and direct dynamics study of rate constants and kinetic isotope effects for the $H + C_2H_6$ reaction," *Journal of Chemical Physics*, vol. 124, no. 4, pp. 44315–44328, 2006.

[37] R. J. Duchovic, W. L. Hase, and H. B. Schlegel, "Analytic function for the $H + CH_3 \rightleftarrows CH_4$ potential energy surface," *Journal of Physical Chemistry*, vol. 88, no. 7, pp. 1339–1347, 1984.

[38] M. Yang and J. C. Corchado, "Seven-dimensional quantum dynamics study of the $H + NH_3 \rightarrow H_2 + NH_2$ reaction," *Journal of Chemical Physics*, vol. 126, no. 21, pp. 214312–214321, 2007.

[39] J. Espinosa-Garcia and J. C. Corchado, "Analytical potential energy surface and kinetics of the $NH_3 + H \rightarrow NH2 + H_2$ hydrogen abstraction and the ammonia inversion reactions," *Journal of Physical Chemistry A*, vol. 114, no. 12, pp. 4455–4463, 2010.

[40] S. T. Banks and D. C. Clary, "Reduced dimensionality quantum dynamics of $Cl + CH_4 \rightarrow HCl + CH_3$ on an ab initio potential," *Physical Chemistry Chemical Physics*, vol. 9, no. 8, pp. 933–943, 2007.

[41] J. C. Corchado, J. L. Bravo, and J. Espinosa-Garcia, "The hydrogen abstraction reaction $H + CH_4$. I. New analytical potential energy surface based on fitting to ab initio calculations," *Journal of Chemical Physics*, vol. 130, no. 18, pp. 184314–184323, 2009.

[42] C. Rángel, M. Navarrete, and J. Espinosa-García, "Potential energy surface for the $F(^2P_{3/2}, ^2P_{1/2}) + CH_4$ hydrogen abstraction reaction. Kinetics and dynamics study," *Journal of Physical Chemistry A*, vol. 109, no. 7, pp. 1441–1448, 2005.

[43] J. Espinosa-García, "New analytical potential energy surface for the CH_4 + H hydrogen abstraction reaction: thermal rate constants and kinetic isotope effects," *Journal of Chemical Physics*, vol. 116, no. 24, pp. 10664–10673, 2002.

[44] J. C. Corchado and J. Espinosa-García, "Analytical potential energy surface for the NH_3 + H ↔ NH_2 + H_2 reaction: application of variational transition-state theory and analysis of the equilibrium constants and kinetic isotope effects using curvilinear and rectilinear coordinates," *Journal of Chemical Physics*, vol. 106, no. 10, pp. 4013–4021, 1997.

[45] J. Espinosa-García and J. C. Corchado, "Analytical surface for the reaction with no saddle-point NH_3 + F → NH_2 + FH. Application of variational transition state theory," *Journal of Physical Chemistry A*, vol. 101, no. 40, pp. 7336–7344, 1997.

[46] J. Espinosa-García, J. Sansón, and J. C. Corchado, "The SiH_4 + H → SiH_3 + H_2 reaction: potential energy surface, rate constants, and kinetic isotope effects," *Journal of Chemical Physics*, vol. 109, no. 2, pp. 466–473, 1998.

[47] J. Espinosa-García and J. C. García-Bernáldez, "Analytical potential energy surface for the CH_4 + O(^{3}P) → CH_3 + OH reaction. Thermal rate constants and kinetic isotope effects," *Physical Chemistry Chemical Physics*, vol. 2, no. 10, pp. 2345–2351, 2000.

[48] J. Espinosa-García, J. L. Bravo, and C. Rangel, "New analytical potential energy surface for the F(^{2}P) + CH_4 hydrogen abstraction reaction: kinetics and dynamics," *Journal of Physical Chemistry A*, vol. 111, no. 14, pp. 2761–2771, 2007.

[49] J. C. Corchado, D. G. Truhlar, and J. Espinosa-García, "Potential energy surface, thermal, and state-selected rate coefficients, and kinetic isotope effects for Cl + CH_4 → HCl + CH_3," *Journal of Chemical Physics*, vol. 112, no. 21, pp. 9375–9389, 2000.

[50] C. Rangel, M. Navarrete, J. C. Corchado, and J. Espinosa-García, "Potential energy surface, kinetics, and dynamics study of the Cl + CH_4 → HCl + CH_3 reaction," *Journal of Chemical Physics*, vol. 124, no. 12, pp. 124306–124324, 2006.

[51] J. Espinosa-García, "Potential energy surface for the CH_3 + HBr → CH_4 + Br hydrogen abstraction reaction: thermal and state-selected rate constants, and kinetic isotope effects," *Journal of Chemical Physics*, vol. 117, no. 5, pp. 2076–2086, 2002.

[52] C. Rangel and J. Espinosa-García, "Potential energy surface for the CCl_4 + H → CCl_3 + ClH reaction: kinetics and dynamics study," *Journal of Chemical Physics*, vol. 122, no. 13, pp. 134315–134319, 2005.

[53] J. Espinosa-García and J. C. Corchado, "Potential energy surface for a seven-atom reaction. Thermal rate constants and kinetic isotope effects for CH_4 + OH," *Journal of Chemical Physics*, vol. 112, no. 13, pp. 5731–5739, 2000.

[54] R.J. Duchovic, Y.L. Volobuev, G.C. Lynch et al., "POLTLIB," 2001.

[55] X. Zhang, B. J. Braams, and J. M. Bowman, "An ab initio potential surface describing abstraction and exchange for H + CH_4," *Journal of Chemical Physics*, vol. 124, no. 2, pp. 021104–021108, 2006.

[56] Z. Xie, J. M. Bowman, and X. Zhang, "Quasiclassical trajectory study of the reaction H + CH_4 (v_3 =0,1) → CH_3 + H_2 using a new ab initio potential energy surface," *Journal of Chemical Physics*, vol. 125, no. 13, pp. 133120–133127, 2006.

[57] T. Wu, H. J. Werner, and U. Manthe, "Accurate potential energy surface and quantum reaction rate calculations for the H + CH_4 → H_2 + CH_3 reaction," *Journal of Chemical Physics*, vol. 124, no. 16, pp. 164307–164318, 2006.

[58] F. Huarte-Larrañaga and U. Manthe, "Quantum dynamics of the CH_4 + H → CH_3 + H_2 reaction: full-dimensional and reduced dimensionality rate constant calculations," *Journal of Physical Chemistry A*, vol. 105, no. 12, pp. 2522–2529, 2001.

[59] G. Schiffel, U. Manthe, and G. Nyman, "Full-dimensional quantum reaction rate calculations for H + CH_4 → H_2 + CH_3 on a recent potential energy surface," *Journal of Physical Chemistry A*, vol. 114, no. 36, pp. 9617–9622, 2010.

[60] J. W. Sutherland, M. C. Su, and J. V. Michael, "Rate constants for H + CH_4, CH_3 + H_2, and CH_4 dissociation at high temperature," *International Journal of Chemical Kinetics*, vol. 33, no. 11, pp. 669–684, 2001.

[61] G. J. Germann, Y. D. Huh, and J. J. Valentini, "State-to-state dynamics of atom + polyatom abstraction reactions. II. The H + CD_4 → HD(v',J') + CD_3 reaction," *Journal of Chemical Physics*, vol. 96, no. 8, pp. 1957–1966, 1992.

[62] W. Hu, G. Lendvay, D. Troya et al., "H + CD4 abstraction reaction dynamics: product energy partitioning," *Journal of Physical Chemistry A*, vol. 110, no. 9, pp. 3017–3027, 2006.

[63] J. P. Camden, H. A. Bechtel, and R. N. Zare, "Dynamics of the simplest reaction of a carbon atom in a tetrahedral environment," *Angewandte Chemie*, vol. 42, no. 42, pp. 5227–5230, 2003.

[64] J. P. Camden, H. A. Bechtel, D. J. A. Brown et al., "A reinterpretation of the mechanism of the simplest reaction at an sp 3-hybridized carbon atom: H + CD_4 → CD_3 + HD," *Journal of the American Chemical Society*, vol. 127, no. 34, pp. 11898–11899, 2005.

[65] J. P. Camden, W. Hu, H. A. Bechtel et al., "H + CD_4 abstraction reaction dynamics: excitation function and angular distributions," *Journal of Physical Chemistry A*, vol. 110, no. 2, pp. 677–686, 2006.

[66] W. Zhang, Y. Zhou, G. Wu et al., "Depression of reactivity by the collision energy in the single barrier H + CD_4 → HD + CD_3 reaction," *Proceedings of the National Academy of Sciences of the United States of America*, vol. 107, no. 29, pp. 12782–12785, 2010.

[67] C. Rangel, J. Sansón, J. C. Corchado, J. Espinosa-Garcia, and G. Nyman, "Product angular distribution for the H + CD_4 → HD + CD_3 reaction," *Journal of Physical Chemistry A*, vol. 110, no. 37, pp. 10715–10719, 2006.

[68] J. Espinosa-García, G. Nyman, and J. C. Corchado, "The hydrogen abstraction reaction H+ CH_4. II. Theoretical investigation of the kinetics and dynamics," *Journal of Chemical Physics*, vol. 130, no. 18, pp. 184315–184323, 2009.

[69] J. P. Camden, H. A. Bechtel, D. J. A. Brown, and R. N. Zare, "Effects of C-H stretch excitation on the H + CH_4 reaction," *Journal of Chemical Physics*, vol. 123, no. 13, pp. 134301–134309, 2005.

[70] C. Rangel, J. C. Corchado, and J. Espinosa-García, "Quasiclassical trajectory calculations analyzing the reactivity and dynamics of asymmetric stretch mode excitations of methane in the H + CH_4 reaction," *Journal of Physical Chemistry A*, vol. 110, no. 35, pp. 10375–10383, 2006.

[71] C. H. Townes and A. L. Schawlow, *Microwave Spectroscopy*, McGraw-Hill, New York, NY, USA, 1955.

[72] G. E. Moyano and M. A. Collins, "Interpolated potential energy surface for abstraction and exchange reactions of NH_3 + H and deuterated analogues," *Theoretical Chemistry Accounts*, vol. 113, no. 4, pp. 225–232, 2005.

[73] J. Espinosa-García, "Capability of LEPS surfaces to describe the kinetics and dynamics of non-collinear reactions," *Journal of Physical Chemistry A*, vol. 105, no. 1, pp. 134–139, 2001.

[74] J. Espinosa-García, "Capability of LEP-type surfaces to describe noncollinear reactions. 2. Polyatomic systems," *Journal of Physical Chemistry A*, vol. 105, no. 38, pp. 8748–8755, 2001.
[75] B. C. Garrett and D. G. Truhlar, "A least-action variational method for calculating multidimensional tunneling probabilities for chemical reactions," *Journal of Chemical Physics*, vol. 79, no. 10, pp. 4931–4938, 1983.
[76] R. Meana-Pañeda, D. G. Truhlar, and A. Fernández-Ramos, "Least-action tunnelling transmission coefficient for polyatomic reactions," *Journal of Chemical Theory and Computation*, vol. 6, no. 1, pp. 6–17, 2010.
[77] T. Ko, P. Marshall, and A. Fontijn, "Rate coefficients for the $H + NH_3$ reaction over a wide temperature range," *Journal of Physical Chemistry*, vol. 94, no. 4, pp. 1401–1404, 1990.
[78] J. Espinosa-García and J. C. Corchado, "Quasi-classical trajectory calculations of the hydrogen abstraction reaction $H + NH_3$," *Journal of Physical Chemistry A*, vol. 114, no. 21, pp. 6194–6200, 2010.
[79] A. M. Hapern, B. R. Ramachandran, and E. D. Glendening, "The inversion potential of ammonia: an intrinsic reaction coordinate calculation for student investigation," *Journal of Chemical Education*, vol. 84, no. 6, pp. 1067–1072, 2007.
[80] V. Špirko, "Vibrational anharmonicity and the inversion potential function of NH_3," *Journal of Molecular Spectroscopy*, vol. 101, no. 1, pp. 30–47, 1983.
[81] V. M. Devi, P. P. Das, K. N. Rao, S. Urban, D. Papouseck, and V. Spirko, "The diode laser spectrum of the V_2 band of $^{14}ND_3$ and $^{15}ND_3$," *Journal of Molecular Spectroscopy*, vol. 88, pp. 293–299, 1981.
[82] M. H. Alexander, D. E. Manolopoulos, and H. J. Werner, "An investigation of the $F + H_2$ reaction based on a full ab initio description of the open-shell character of the $F(^2P)$ atom," *Journal of Chemical Physics*, vol. 113, no. 24, pp. 11084–11100, 2000.
[83] G. Capecchi and H. J. Werner, "Ab initio calculations of coupled potential energy surfaces for the $Cl(^2P_{3/2},^2P_{1/2}) + H_2$ reaction," *Physical Chemistry Chemical Physics*, vol. 6, no. 21, pp. 4975–4983, 2004.
[84] M. H. Alexander, H. J. Werner, and D. E. Manolopoulos, "Spin-orbit effects in the reaction of $F(^2P)$ with H_2," *Journal of Chemical Physics*, vol. 109, no. 14, pp. 5710–5713, 1998.
[85] S. H. Lee, "State-specific excitation function for $Cl(^2P) + H_2$ (v = 0,j): effects of spin-orbit and rotational states," *Journal of Chemical Physics*, vol. 110, no. 17, pp. 8229–8232, 1999.
[86] S.-H. Lee and K. Liu, "Exploring the spin-orbit reactivity in the simplest chlorine atom reaction," *Journal of Chemical Physics*, vol. 111, no. 14, pp. 6253–6259, 1999.
[87] F. Dong, S. H. Lee, and K. Liu, "Direct determination of the spin-orbit reactivity in $CI(^2P_{3/2}, ^2P_{1/2}) + H_2/D_2/HD$ reactions," *Journal of Chemical Physics*, vol. 115, no. 3, pp. 1197–1204, 2001.
[88] M. H. Alexander, G. Capecchi, and H. J. Werner, "Theoretical study of the validity of the Born-Oppenheimer approximation in the $CL + H_2 \rightarrow HCL + H$ reaction," *Science*, vol. 296, no. 5568, pp. 715–718, 2002.
[89] N. Balucani, D. Skouteris, L. Cartechini et al., "Differential cross sections from quantum calculations on coupled ab initio potential energy surfaces and scattering experiments for $Cl(^2P) + H_2$ reactions," *Physical Review Letters*, vol. 91, no. 1, pp. 013201–013204, 2003.
[90] M. H. Alexander, G. Capecchi, and H.-J. Werner, "Details and consequences of the nonadiabatic coupling in the $Cl(^2P) + H^2$ reaction," *Faraday Discussions*, vol. 127, pp. 59–72, 2004.
[91] E. Garand, J. Zhou, D. E. Manolopoulos, M. H. Alexander, and D. M. Neumark, "Nonadiabatic interactions in the $Cl + H_2$ reaction probed by CIH_2^- and CID_2^- photoelectron imaging," *Science*, vol. 319, no. 5859, pp. 72–75, 2008.
[92] D. M. Neumark, "Slow electron velocity-map imaging of negative ions: applications to spectroscopy and dynamics," *Journal of Physical Chemistry A*, vol. 112, no. 51, pp. 13287–13301, 2008.
[93] X. Wang, W. Dong, C. Xiao et al., "The extent of non-born-oppenheimer coupling in the reaction of $Cl(^2P)$ with *para*-H_2," *Science*, vol. 322, no. 5901, pp. 573–576, 2008.
[94] M. Monge-Palacios and J. Espinosa-Garcia, "Reaction-path dynamics calculations of the $Cl + NH_3$ hydrogen abstraction reaction: the role of the intermediate complexes," *Journal of Physical Chemistry A*, vol. 114, no. 12, pp. 4418–4426, 2010.
[95] M. Monge-Palacios, C. Rangel, J. C. Corchado, and J. Espinosa-Garcia, "Analytical potential energy surface for the reaction with intermediate complexes $NH_3 + Cl \rightarrow NH_2 + HCl$: application to the kinetics study," *International Journal of Quantum Chemistry*. In press.
[96] J. Espinosa-García, "Analytical potential energy surface for the $GeH_4 + H \rightarrow GeH_3 + H_2$ reaction: thermal and vibrational-state selected rate constants and kinetic isotope effects," *Journal of Chemical Physics*, vol. 111, no. 20, pp. 9330–9336, 1999.
[97] J. Espinosa-García, C. Rangel, M. Monge-Palacios, and J. C. Corchado, "Kinetics and dynamics study of the $H + CCl_4 \rightarrow HCl(v', j') + CCl_3$ reaction," *Theoretical Chemistry Accounts*, vol. 128, pp. 743–755, 2010.
[98] C. Rangel and J. Espinosa-García, "Analytical potential energy surface describing abstraction reactions in asymmetrically substituted polyatomic systems of type $CX_3Y + A \rightarrow$ products," *Journal of Physical Chemistry A*, vol. 110, no. 2, pp. 537–547, 2006.
[99] C. Rangel and J. Espinosa-García, "Potential energy surface for asymmetrically substituted reactions of type CWXYZ + A. kinetics study," *Journal of Physical Chemistry A*, vol. 111, no. 23, pp. 5057–5062, 2007.
[100] D. C. Clary, "Theoretical studies on bimolecular reaction dynamics," *Proceedings of the National Academy of Sciences of the United States of America*, vol. 105, no. 35, pp. 12649–12653, 2008.
[101] S. Kondo, K. Tokuhashi, A. Takahashi, and M. Kaise, "Ab initio study of reactions between halogen atoms and various fuel molecules by Gaussian-2 theory," *Journal of Hazardous Materials*, vol. 79, no. 1-2, pp. 77–86, 2000.
[102] Y. Gao, I. M. Alecu, P.-C. Hsieh, B. P. Morgan, P. Marshall, and L. N. Krasnoperov, "Thermochemistry is not a lower bound to the activation energy of endothermic reactions: a kinetic study of the gas-phase reaction of atomi-chlorine with ammonia," *Journal of Physical Chemistry A*, vol. 110, no. 21, pp. 6844–6850, 2006.
[103] Z. F. Xu and M. C. Lin, "Computational studies on the kinetics and mechanisms for NH_3 reactions with ClOx (x = 0-4) radicals," *Journal of Physical Chemistry A*, vol. 111, no. 4, pp. 584–590, 2007.
[104] Y. Zhao and D. G. Truhlar, "Hybrid meta density functional theory methods for thermochemistry, thermochemical kinetics, and noncovalent interactions: the MPW1B95 and MPWB1K models and comparative assessments for hydrogen bonding and van der Waals interactions," *Journal of Physical Chemistry A*, vol. 108, no. 33, pp. 6908–6918, 2004.

A Theoretical Investigation of the Ring Strain Energy, Destabilization Energy, and Heat of Formation of CL-20

John A. Bumpus

Department of Chemistry and Biochemistry, University of Northern Iowa, Cedar Falls, IA 50614, USA

Correspondence should be addressed to John A. Bumpus, john.bumpus@uni.edu

Academic Editor: Dennis Salahub

The cage compound CL-20 (a.k.a., 2,4,6,8,10,12-hexanitro-2,4,6,8,10,12-hexaazaisowurtzitane, HNIW, or 2,4,6,8,10,12-hexanitro-2,4,6,8,10,12-hexaazatetracyclo[5.5.0.$0^{3,11}$.$0^{5,9}$]dodecane) is a well-studied high-energy-density material (HEDM). The high positive gas- ($\Delta_f H_g^\circ$) and solid- ($\Delta_f H_s^\circ$) phase heat of formation values for CL-20 conformers have often been attributed to the strain energy of this cage compound and, by implication, to the conventional ring strain energy (CRSE) inherent in isowurtzitane which may be viewed as a "parent compound" (although not the synthetic precursor) of CL-20. $\Delta_f H_g^\circ$ values and destabilization energies (DSEs), which include the contribution from CRSE, were determined by computation using a relatively new multilevel *ab intio* model chemistry. Compared to cubane, isowurtzitane does not have an exceptionally high CRSE. It is about the same as that of cyclopropane and cyclobutane. These investigations demonstrate that instead of the CRSE inherent in the isowurtzitane parent compound, the relatively high $\Delta_f H_g^\circ$ and DSE values of CL-20 conformers must be due, primarily, to torsional strain (Pitzer strain), transannular strain (Prelog strain), and van der Waals interactions that occur due to the presence of the six >N–NO_2 substituents that replace the six methylene (–CH_2–) groups in the isowurtzitane parent compound. These conclusions are even more pronounced when 2,4,6,8,10,12-hexaazaisowurtzitane is viewed as the "parent compound."

1. Introduction

The cage compound CL-20 (a.k.a., 2,4,6,8,10,12-hexanitro-2,4,6,8,10,12-hexaazaisowurtzitane, HNIW, or 2,4,6,8,10,12-hexanitro-2,4,6,8,10,12-hexaazatetracyclo [5.5.0.$0^{3,11}$.$0^{5,9}$]dodecane) is a high-energy-density material (HEDM) that has been developed and studied during the past several years (see Scheme 1).

This compound was first synthesized in 1987 by scientists at the China Lake Naval Weapons Center [1, 2]. There is substantial interest in the use of CL-20 as a high explosive and many of its properties (density = ~2 g cm^{-3}, detonation velocity = 9.4 mm ms^{-1}, oxygen balance = −11, detonation pressure = 420 kbar, etc.) suggest that it is superior to RDX (1,3,5-trinitroperhydro-1,3,5-triazine) and HMX (octahydro-1,3,5,7-tetranitro-1,3,5,7-tetrazocine), two high explosives that are widely used at the present time [3–6] (see Scheme 2).

In addition to its use as a high explosive compound, [3] CL-20 has potential use as a propellant [6].

CL-20 is often compared to octanitrocubane (ONC) which, like CL-20, is a polynitro cage compound. Although both have many properties that recommend them for use as high explosives [7–12], economic considerations appear to limit the large scale production and usefulness of ONC [11, 12]. On the other hand such problems for large-scale production of CL-20 appear to have been successfully addressed, although cost reduction is still an issue. CL-20 is produced commercially by ATK-Thiokol Propulsion (Brigham City, UT, USA) [13].

The relatively high heat of combustion and heat of detonation values of many HEDMs are due, in part, to their high positive heat of formation values. Octanitrocubane (ONC) is a case in point (see Scheme 3).

Estimates of gas-phase ($\Delta_f H_g^\circ$) and solid-phase ($\Delta_f H_s^\circ$) heat of formation values of this compound have been

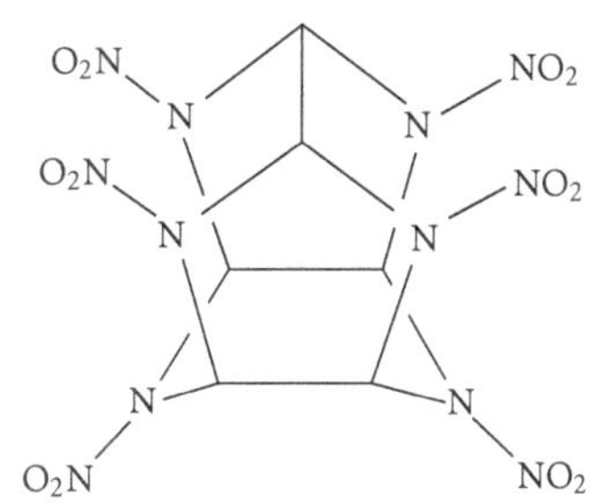

CL-20

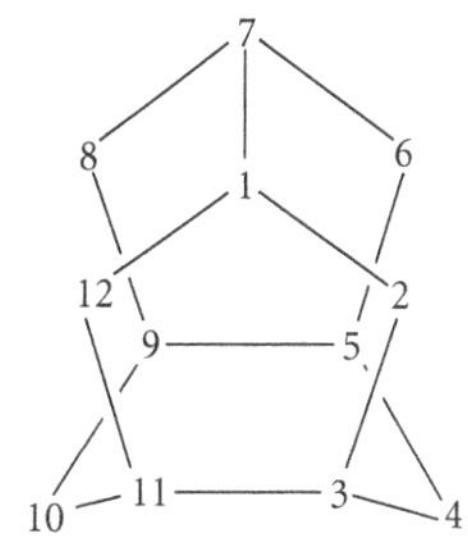

Numbering system for CL-20 and other substituted isowurtzitanes

SCHEME 1

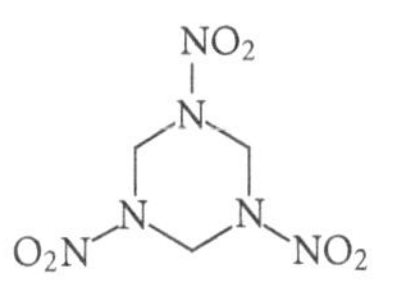

RDX

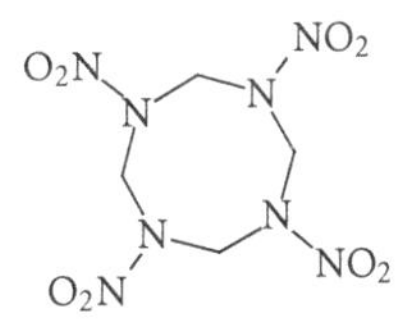

HMX

SCHEME 2: Structures of RDX and HMX.

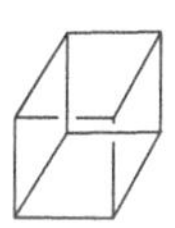

Cubane

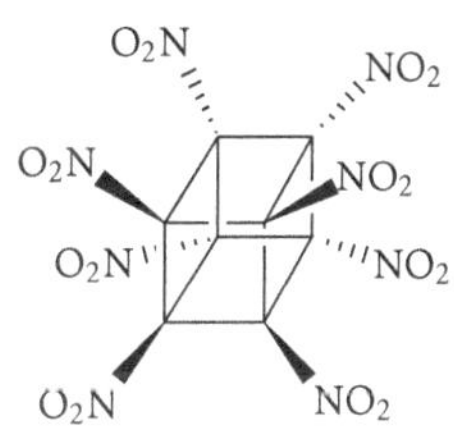

Octanitrocubane (ONC)

SCHEME 3: Structures of cubane and octanitrocubane.

calculated to be ~726 and ~594 kJ/mol, respectively [8, 9], and it is reasonable to conclude that such high heat of formation values are due, in large measure, to the considerable conventional ring strain energy (CRSE) inherent in the parent compound, cubane.

At first glance, it may appear that, like cubane, isowurtzitane and/or 2,4,6,8,10,12-hexaazaisowurtzitane may also have very high CRSEs that contribute to the high heat of formation of CL-20 (see Scheme 4).

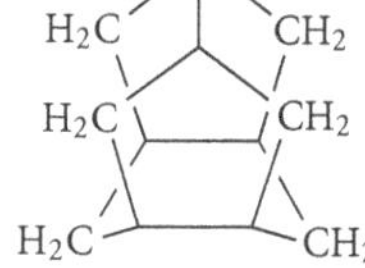

Isowurtzitane

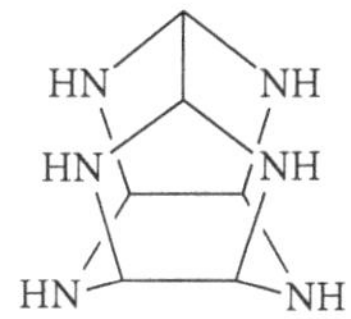

2,4,6,8,10,12-Hexaazaisowurtzitane

SCHEME 4: Structures of isowurtzitane and 2,4,6,8,10,12-hexaazaisowurtzitane.

In this study it is shown that this is not the case. Using gas-phase calculations, it was demonstrated that the CRSE inherent in isowurtzitane is relatively minor (compared to cubane, e.g., which is clearly highly strained) and the CRSEs of 2,4,6,8,10,12-hexaazaisowurtzitane conformers are exceptionally small. Thus the modestly high (compared to ONC) destabilization energy (DSE) and heat of formation values of CL-20 must be due to factors other than the CRSE inherent in the parent compound(s).

2. Computational Approach

2.1. Heat of Formation. Gas-phase heat of formation values were calculated using a recently described [15] multilevel *ab initio* model chemistry (known as T1) as implemented in the Spartan 10 (Wavefunction, Inc., Irvine, CA, USA) suite of programs. The T1 program has been described in substantial detail by Ohlinger et al. [15]. Briefly this program is based on the G3MP2 model chemistry. The major differences between T1 and G3MP2 model chemistries are those in the T1 program; the MP2/6-31G(d) equilibrium geometry is substituted with an HF/6-31G(d) equilibrium geometry. The HF/6-31G(d) frequency and QCISD(T)/6-31G(d) energy steps are removed and the MP2/G3MP2 large energy step is approximated using a dual basis set RI-MP2 technique [15].

Comparing computed results with experimental values from 1805 structurally diverse compounds from the NIST database, Ohlinger et al. [15] reported this procedure to have a mean absolute error of 8.5 kJ/mol and a root mean square error of 11.5 kJ/mol. Thus it is reasonable to suggest that the values computed in this investigation have similar accuracy.

2.2. Conventional Ring Strain Energy (CRSE) and Destabilization Energy (DSE). Conventional ring strain energies (CRSE) as described by Benson and associates [16, 17] were calculated by difference using gas-phase heat of formation ($\Delta_f H^\circ_{(\text{grp. add.})g}$) values determined by group additivity theory (without consideration of ring strain energy, etc.) and the gas-phase heat of formation ($\Delta_f H^\circ_g$) of the molecule determined by experiment or by computation, using theory (such as T1 or G3MP2) that is sufficiently robust to account for and include ring strain and other contributing factors. Thus, for gas-phase reactions, CRSE is calculated as follows:

$$\text{CRSE} = \Delta_f H_g^\circ - \Delta_f H^\circ{}_{(\text{grp. add.})g}. \quad (1)$$

Leroy [18] makes the point that for unsubstituted cyclic compounds, destabilization energy (Leroy uses the term negative stabilization energy) is equivalent to CRSE. *Thus in this study the term CRSE is used to describe strain inherent in the unsubstituted cyclic and cage compounds under investigation.* In contrast to CRSE, the term destabilization energy (DSE) is used here as a more inclusive term and is reserved for characterization of *substituted cyclic and cage compounds* although it is calculated in precisely the same way as CRSE:

$$\text{DSE} = \Delta_f H_g^\circ - \Delta_f H^\circ{}_{(\text{grp. add.})g}. \quad (2)$$

Although the distinction between CRSE and DSE is somewhat arbitrary, it is warranted as it should be appreciated that for substituted cyclic and cage compounds DSE includes not only the energy contribution from ring strain but also the effects that substituents impose on the structure. For example, substituents, such as nitro groups, may increase or decrease bond angles and bond lengths (depending on placement). Furthermore van der Waals interactions between adjacent or nearby substituents will affect DSE. For nitro groups, close interaction leads to an increase in DSE. Other factors such as torsional strain (Pitzer strain) and transannular strain (Prelog strain) also contribute to DSE.

It should also be appreciated that DSE may be calculated for acyclic compounds in which CRSE, by definition, is not a factor.

The ASTM CHETAH 8.0 software program (CHETAH (The Computer Program for Chemical Thermodynamics and Energy Release Evaluation) University of South Alabama, Mobile Alabama 36688-0002 (http://www.southalabama.edu/engineering/chemical/chetah/index.html) and ASTM International (http://www.astm.org/)) was used to calculate strainless gas-phase heat of formation values ($\Delta_f H^\circ{}_{(\text{grp. add.})g}$) that, in turn, were used to calculate CRSE and DSE, as described above. This software uses the Benson group additivity approach. Thus the Benson group values as well as other group values provided by the software manufacturer were used in these calculations [16, 17, 19].

2.3. Contribution of Ring Nitrogen and Nitro Group Substituents to the DSE, ($\Delta_f H^\circ{}_{(\text{grp. add.})g}$), and $\Delta_f H_g^\circ$ Values of CL-20. The contributions of ring nitrogen and nitro group substituents to the DSE and $\Delta_f H_g^\circ$ values of CL-20 were determined using the T1 multilevel *ab initio* model chemistry [15]. To do this, the $\Delta_f H_g^\circ$ value for CL-20 conformer HNIW-III (one of four known conformers of CL-20 See [14]) was calculated first. The $\Delta_f H_g^\circ$ value for 2,4,6,8,10-pentanitro-2,4,6,8,10,12-hexaazaisowurtzitane was then calculated following computational removal of one of the nitro groups from the equilibrium conformation of HNIW-III. $\Delta_f H_g^\circ$ values for the remaining nitrohexaazaisowurtzitanes were calculated in a similar manner. The same approach was used to assess the effect of ring nitrogen substitution on $\Delta_f H_g^\circ$. Nitro groups or ring nitrogens were computationally removed in the following order N-12, N-10, N-8, N-6, N-4, and N-2. $\Delta_f H^\circ{}_{(\text{grp. add.})g}$ values were also calculated using the group additivity approach described above and DSE values were calculated using (2). All $\Delta_f H_g^\circ$, $\Delta_f H^\circ{}_{(\text{grp. add.})g}$, and DSE values refer to such values at 298.15 K.

For comparison, the effect of nitro group substituents on $\Delta_f H_g^\circ$ and DSE values for nitrocubane and polynitrocubanes was determined in a similar manner. In these calculations, the $\Delta_f H_g^\circ$ and DSE values for octanitrocubane were calculated first. The $\Delta_f H_g^\circ$ and DSE values for heptanitrocubane were then calculated following computational removal of one of the nitro groups from the equilibrium conformation of octanitrocubane. $\Delta_f H_g^\circ$ and DSE values for the remaining nitrocubanes were calculated in a similar manner. In addition to octanitrocubane and heptanitrocubane, $\Delta_f H_g^\circ$ and DSE values were calculated for 1,2,3,4,5,6-hexanitrocubane, 1,2,3,4,5-pentanitrocubane, 1,2,3,8-tetranitrocubane, 1,2,3-trinitrocubane, 1,2-dinitrocubane, and nitrocubane.

3. Results and Discussion

3.1. Conventional Ring Strain Energy (CRSE) and Heat of Formation. Computed values for gas-phase heat of formation ($\Delta_f H_g^\circ$), strainless heat of formation ($\Delta_f H^\circ{}_{(\text{grp. add.})g}$), and destabilization energy (DSE) for four CL-20 conformations as well as several other compounds directly relevant to this investigation are presented in Table 1.

There are several polymorphs (conformers) possible for CL-20 [11, 14, 20–26]. To date five polymorphs have been isolated experimentally and four have been studied computationally. Zhou et al. [14] used density functional theory at the B3LYP/6-31G(d,p) level to characterize four CL-20 conformations. In Zhou et al.'s [14] terminology, conformations HNIW-I, HNIW-II, HNIW-III, and HNIW-IV represent local minima on the potential energy surface. Zhou et al.'s [14] terminology is used here (Figure 1).

The computed gas-phase heat of formation values of the four CL-20 conformations (HNIW-I, HNIW-II, HNIW-III, and HNIW-IV) range between 548.0 and 571.6 kJ/mol. Subtracting the heat of sublimation value (168.7 kJ/mol) for CL-20 calculated by Zeman and Krupka [27] yields solid phase heat of formation values for these conformations ranging from 379.3 to 402.9 kJ/mol. These values are clearly consistent with the experimentally determined solid phase heat of formation values (377.4 ± 13 and 431.0 ± 13 kJ/mol) reported for two of the crystalline polymorphs [5].

Isowurtzitane and 2,4,6,8,10,12-hexaazaisowurtzitane may be thought of as the parent compounds (although not the synthetic precursors) of CL-20. Similarly cubane may be thought of as the parent compound for ONC. Thus it was informative to compute the $\Delta_f H_g^\circ$ values, $\Delta_f H^\circ{}_{(\text{grp. add.})g}$ values, and CRSE values for these compounds to determine and compare how the cage structures of the parent

TABLE 1: Comparison of the strainless gas-phase heat of formation, destabilization energy, and gas-phase heat of formation of CL-20, ONC, and other relevant compounds of interest.

Compound	Strainless heat of formation $\Delta_f H^\circ_{(\text{grp. add.})g}$ (kJ/mol)	Heat of formation $\Delta_f H^\circ_g$ (kJ/mol)	Destabilization energy or CRSE (kJ/mol)	Destabilization energy per C + N present in the ring or the cage (kJ/mol)
		55.8	117.7	39.2
Cyclopropane	−61.9	(39.3)	(101.2)	(33.7)
		(53.3)	(115.2)	(38.4)
Cyclobutane	−82.5	29.7	112.2	28.1
		(27.7)	(110.2)	(27.6)
HAIW I	325.3	306.6	−18.7	−1.6
HAIW II	325.3	301.7	−23.6	−2.0
HAIW III	325.3	323.1	−2.2	−0.2
HAIW IV	325.3	290.5	−34.8	−2.9
Isowurtzitane	−171.5	−63.5	108.0	9.0
CL-20 (HNIW I)	240.0	558.8	318.8	26.6
CL-20 (HNIW II)	240.0	554.4	314.4	26.2
CL-20 (HNIW III)	240.0	571.6	331.6	27.6
CL-20 (HNIW IV)	240.0	548.0	308.0	25.7
Cubane	−63.6	622.2	685.8	85.7
		(622.2)	685.8	85.7
Octanitrocubane	−405.0	687.5	1,092.5	136.6
		(726)	(1,131)	(141)

The values in parentheses for octanitrocubane are the heat of formation value calculated by Zhang and Xiao [8] and values calculated using this heat of formation.
Other values in parentheses are those reported in, or calculated using, those referenced in the NIST Chemistry WebBook (http://webbook.nist.gov/chemistry/) or in the CRC's Standard Thermodynamic Properties of Chemical Substances.
HAIW refers to 2,4,6,8,10,12-hexaazaisowurtzitane.

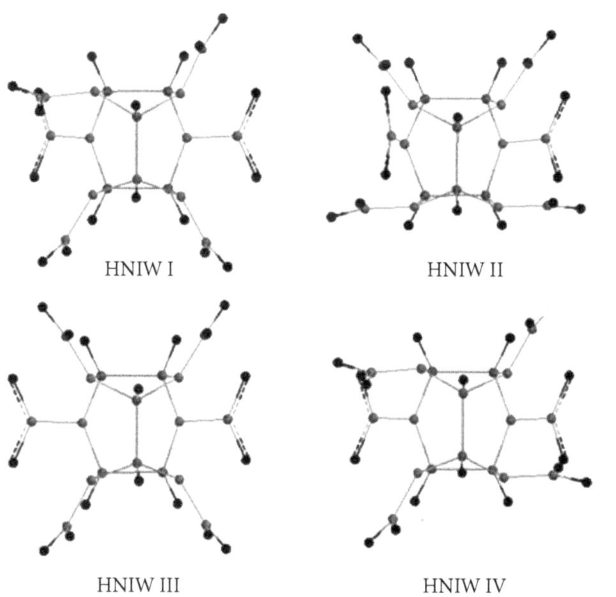

FIGURE 1: Conformations of CL-20 described by Zhou et al. [14].

compounds may contribute to the $\Delta_f H^\circ_g$ values of CL-20 and ONC. These values are also presented in Table 1.

One characteristic of ONC that makes it a high-energy compound is the fact that the cubane cage imposes tremendous ring strain on the compound and it is the ring strain that is inherent in the cubane cage that imparts a very substantial contribution to the $\Delta_f H^\circ_g$ and DSE of ONC. Several experimental and computed values for the CRSE (destabilization energy) of cubane have been reported (see [28] and references therein). These values range between 656.9 and 707.6 kJ/mol and are consistent with the value (685.8 kJ/mol) calculated in this investigation (Table 1). It should also be appreciated that substitution of the eight hydrogens in cubane by eight nitro groups to form ONC leads to a more negative $\Delta_f H^\circ{}_{(\text{grp. add.})g}$, and this contributes substantially to the exceptionally high DSE of ONC.

Because ONC and CL-20 are both polynitro cage compounds, it is tempting to assume that the high heat of formation values characteristic of both compounds are due, in large measure, to ring strain inherent in the parent cage compounds (i.e., their CRSE). This is clearly true for ONC. However, this is not the case for CL-20. Data presented in Table 1 shows that the calculated CRSE (destabilization energy) of isowurtzitane is only 108.0 kJ/mol. Although this value is not inconsequential (it is about the same as that observed for cyclopropane and cyclobutane), it is much less than the CRSE of cubane (CRSE = 685.8 kJ/mol), a compound that clearly deserves to be called "highly strained."

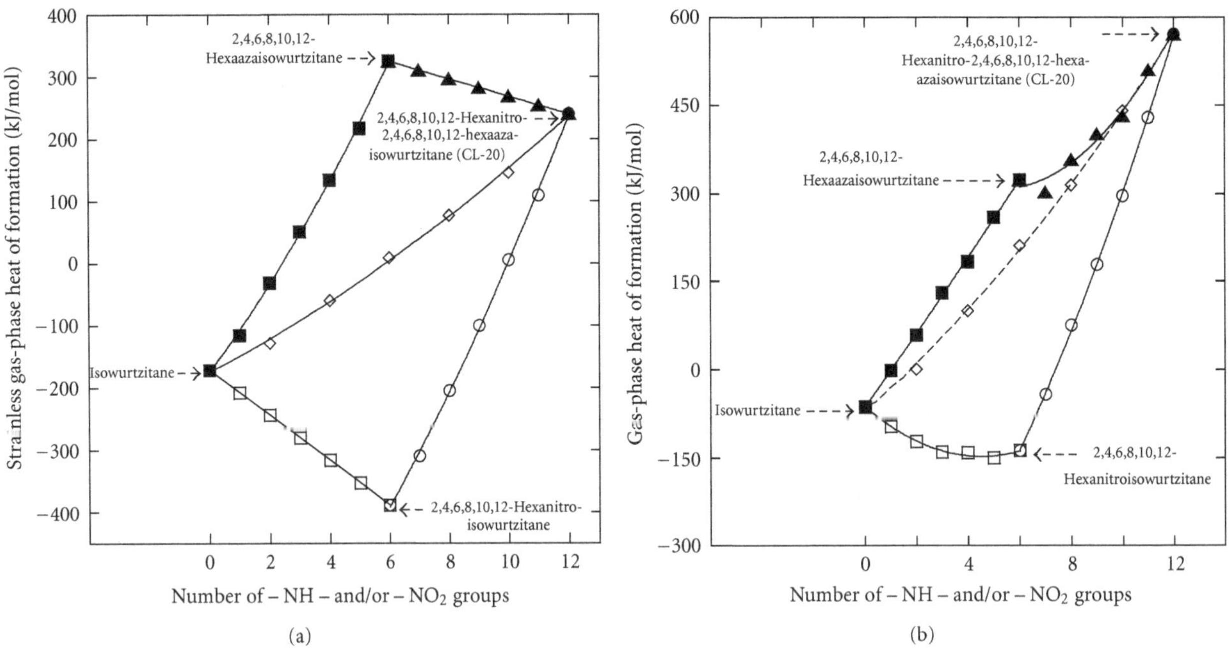

Figure 2: (a) Effect of –NH– and $-NO_2$ substitution on the strainless gas-phase heat of formation of isowurtzitane. Closed squares: effect on the strainless heat of formation of adding –NH– groups to isowurtzitane in the 2, 4, 6, 8, 10 and 12 positions, open squares: effect on the strainless heat of formation of adding $-NO_2$ groups to isowurtzitane in the 2, 4, 6, 8, 10 and 12 positions, open circles: effect on the strainless heat of formation of adding –NH– groups to 2,4,6,8,10,12-hexanitroisowurtzitane in the 2, 4, 6, 8, 10 and 12 positions, closed triangles: Effect on the strainless heat of formation of adding $-NO_2$ groups to 2,4,6,8,10,12-hexaazaisowurtzitane in the 2, 4, 6, 8, 10 and 12 positions, tilted open squares: effect on the strainless heat of formation of adding $>N-NO_2$ groups to isowurtzitane in the 2, 4, 6, 8, 10 and 12 positions. (b) Effect of –NH– and $-NO_2$ substitution on the gas phase heat of formation of isowurtzitane calculated using the T1 multilevel *ab initio* model chemistry approach described by Ohlinger et al. [15]. Closed squares: effect on the heat of formation of adding –NH– groups to isowurtzitane in the 2, 4, 6, 8, 10 and 12 positions, Open squares: Effect on the heat of formation of adding $-NO_2$ groups to isowurtzitane in the 2, 4, 6, 8, 10 and 12 positions, Open circles: Effect on the heat of formation of adding –NH– groups to 2,4,6,8,10,12-hexanitroisowurtzitane in the 2, 4, 6, 8, 10 and 12 positions, Closed triangles: effect on the heat of formation of adding $-NO_2$ groups to 2,4,6,8,10,12-hexaazaisowurtzitane in the 2, 4, 6, 8, 10 and 12 positions, tilted open squares: effect on the heat of formation of adding $>N-NO_2$ groups to isowurtzitane in the 2, 4, 6, 8, 10 and 12 positions.

This comparison is even more striking when calculated CRSEs are expressed on a per carbon basis. When analyzed in this manner, the CRSE for isowurtzitane per carbon is even less than that observed for cyclopropane and cyclobutane and almost ten times less than that observed for cubane (Table 1).

An even more appropriate comparison involves 2,4,6,8,10,12-hexaazaisowurtzitane. When CRSEs (destabilization energies) for four conformers of 2,4,6,8,10,12-hexaazaisowurtzitane (derived from HNIW-I, HNIW-II, HNIW-III, and HNIW-IV) were calculated, values were found to range from −34.8 to −2.2 kJ/mol. Negative CRSE or DSE values are more properly known as stabilization energies (SEs) and are clearly not responsible for the high $\Delta_f H_g^\circ$ and DSE of the CL-20 conformers.

3.2. Effect of –NH– and $-NO_2$ Substitutents on CRSE, DSE and Heat of Formation. Figures 2(a) and 2(b) present the relative effects of successive –NH–, $-NO_2$ and $>N-NO_2$ group substitutions on the heat of formation of isowurtzitane.

In Figure 2(a), only the strainless group contribution to gas phase heat of formation ($\Delta_f H^\circ{}_{(\text{grp. add.})g}$) values is considered. Substitution of hydrogens in isowurtzitane by $-NO_2$ groups (open squares) actually contributes to a decrease in $\Delta_f H^\circ{}_{(\text{grp. add.})g}$ values. Conversely, substitution of nitrogen for carbon atoms in isowurtzitane (closed squares) results in an increase of $\Delta_f H^\circ{}_{(\text{grp. add.})g}$ values. A similar analysis can also be accomplished by substituting methylene ($-CH_2-$) groups with $>N-NO_2$ groups. This, of course, is the equivalent of simultaneously substituting a carbon and two hydrogens with a nitrogen and an $-NO_2$ group, respectively. Figure 2(a) demonstrates that substitution of methylene ($-CH_2-$) groups with $>N-NO_2$ groups results in an increase in strainless $\Delta_f H^\circ{}_{(\text{grp. add.})g}$ values (tilted open square).

In Figure 2(b) a similar analysis is presented using $\Delta_f H_g^\circ$ values computed using the T1 *ab initio* procedure.

Results (Figure 2(b)) demonstrate that substitution of hydrogens in isowurtzitane by $-NO_2$ decreases $\Delta_f H_g^\circ$ values (open squares) while nitrogen substitution increases $\Delta_f H_g^\circ$ values (closed squares). When the hydrogens bonded to nitrogens in 2,4,6,8,10,12-hexaazaisowurtzitane are substituted by $-NO_2$ groups, the $\Delta_f H_g^\circ$ value of the first compound in this series is decreased. However, subsequent substitutions of hydrogen by $-NO_2$ result in increased $\Delta_f H_g^\circ$ values

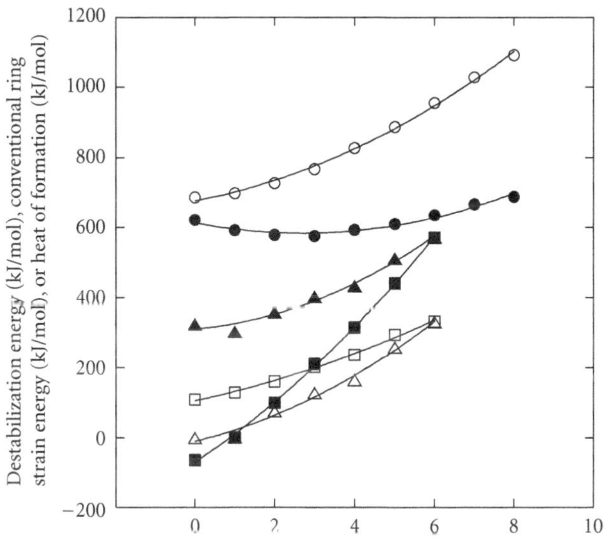

FIGURE 3: The effect on DSE and $\Delta_f H_g^\circ$ of substituting $-NO_2$ groups for hydrogens in >C–H groups in cubane compared with (1) the effect of substituting $-NO_2$ groups for hydrogens in >N–H groups in 2,4,6,8,10,12-hexaazaisowurtizane and (2) the effect of substituting $-CH_2$-groups in isowurtzitane with >N–NO_2 groups. Note: for compounds having no $-NO_2$ substituents, CRSE = DSE. Open circles: CRSE and DSE values for cubane and nitrocubanes, open triangles: CRSE and DSE values for 2,4,6,8,10,12-hexaazaisowurtzitane and $-NO_2$ substituted 2,4,6,8,10,12-hexaazaisowurtzitanes, open squares: CRSE and DSE values for isowurtzitane and >N–NO_2 substituted isowurtzitane (i.e., nitroazaisowurtzitanes). Closed circles: $\Delta_f H_g^\circ$ of cubane and nitrocubanes, closed triangles: $\Delta_f H_g^\circ$ of 2,4,6,8,10,12-hexaazaisowurtzitane and $-NO_2$-substituted 2,4,6,8,10,12-hexaazaisowurtzitanes, closed squares: $\Delta_f H_g^\circ$ of isowurtzitane and >N–NO_2-substituted isowurtzitanes (i.e., nitroaza-isowurtzitanes).

(closed triangles). When substitution of methylene groups by >N–NO_2 groups is considered, results demonstrate that $\Delta_f H_g^\circ$ values increase as a result of each substitution (tilted open squares).

The major point of comparison between Figures 2(a) and 2(b) involves the effect of substituting $-NO_2$ groups for hydrogens on nitrogens in 2,4,6,8,10,12-hexaazaisowurtzitane. In Figure 2(a), such substitutions result in a decrease in $\Delta_f H^\circ{}_{(\text{grp. add.})g}$ (closed triangles), whereas in Figure 2(b) this substitution results, initially, in a decrease of $\Delta_f H_g^\circ$ (closed triangles), but as more $-NO_2$ groups are added, a substantial increase is observed due to a combination of torsional strain (Pitzer strain), transannular strain (Prelog strain), and van der Waals interactions between the six >N–NO_2 substituents.

It should be pointed out that such strain and interactions result because, in contrast to noncage or acyclic compound, the parent cage compounds presents a rigid scaffold that substantially limits bond rotation to relieve or minimize strain, and van der Waals interactions.

Finally, the effect on DSE and $\Delta_f H_g^\circ$ of substituting $-NO_2$ groups for hydrogens in >C–H groups in cubane is compared with the effect of substituting $-NO_2$ groups for hydrogens in >N–H groups in 2,4,6,8,10,12-hexaazaisowurtizane (Figure 3).

Also presented and compared (Figure 3) is the effect on DSE and $\Delta_f H_g^\circ$ caused by substituting $-CH_2$-groups in isowurtzitane with >N–NO_2 groups. In all cases, these substitutions increase the DSE relative to the parent compounds and the preceding compound in these substitution series. Figure 3 also nicely demonstrates that, *relative to cubane*, the CRSE inherent in isowurtzitane and 2,4,6,8,10,12-hexaazaisowurtzitane have a comparatively minor effect on the DSE (and therefore on the $\Delta_f H_g^\circ$) of CL-20.

4. Conclusions

It is known that the high $\Delta_f H_g^\circ$ and DSE values characteristic of octanitrocubane are due, in large measure, to the high $\Delta_f H_g^\circ$ and CRSE values inherent in and characteristic of cubane, its highly strained parent compound.

The situation for CL-20, however, is different. Isowurtzitane and 2,4,6,8,10,12-hexaazaisowurtzitane can be viewed as parent compounds (but not synthetic precursors) of CL-20, but neither of these compounds have exceptionally high CRSE values. This is due to the fact that the bond angles in these compounds are not as highly strained as the 90° CCC bond angles in cubane and ONC.

Thus the high $\Delta_f H_g^\circ$ and modestly high DSE values characteristic of CL-20 conformers are not due to high CRSE values of the parent compound(s). Instead, the high $\Delta_f H_g^\circ$ and modestly high DSE values of CL-20 conformers must be due, in large measure, to torsional strain (Pitzer strain), transannular strain (Prelog strain), and van der Waals interactions that occur due to the presence of the six >N–NO_2 substituents that are present in this HEDM. Such strain and interactions result or are enhanced because the parent cage compounds presents a rigid scaffold that substantially limits bond rotation that would otherwise relieve or minimize strain and van der Waals interactions. It should be appreciated that the group contribution energies characteristic of the >N–NO_2 groups provide the greatest contribution to the $\Delta_f H_g^\circ$ and DSE of CL-20.

Acknowledgment

The author wishes to thank the Environmental Sciences Program at the University of Northern Iowa for its continuing support.

References

[1] A. T. Nielsen, "Caged polynitramine compound," *Chemical Abstracts*, vol. 128, article 36971t, 1998, U. S. Department of Navy, U. S. Patent Office Application Case No. 70631, June 1987, U. S. Patent Application No. 253, 106, September 1988, U. S. Patent No. 5, 693, 794 (CI 540-554, C07D259/00, December 1997.

[2] A. T. Nielsen, A. P. Chafin, S. L. Christian et al., "Synthesis of polyazapolycyclic caged polynitramines," *Tetrahedron*, vol. 54, no. 39, pp. 11793–11812, 1998.
[3] M. Geetha, U. R. Nair, D. B. Sarwade, G. M. Gore, S. N. Asthana, and H. Singh, "Studies on CL-20: the most powerful high energy material," *Journal of Thermal Analysis and Calorimetry*, vol. 73, no. 3, pp. 913–922, 2003.
[4] V. D. Ghule, P. M. Jadhav, R. S. Patil, S. Radhakrishnan, and T. Soman, "Quantum-chemical studies on hexaazaisowurtzitanes," *Journal of Physical Chemistry A*, vol. 114, no. 1, pp. 498–503, 2010.
[5] R. L. Simpson, P. A. Urtiew, D. L. Ornellas, G. L. Moody, K. J. Scribner, and D. M. Hoffman, "CL-20 performance exceeds that of HMX and its sensitivity is moderate," *Propellants, Explosives, Pyrotechnics*, vol. 22, no. 5, pp. 249–255, 1997.
[6] D. Mueller, "New gun propellant with CL-20," *Propellants, Explosives, Pyrotechnics*, vol. 24, no. 3, pp. 176–181, 1999.
[7] M.-X. Zhang, P. E. Eaton, and R. Gilardi, "Hepta- and octanitrocubanes," *Angewandte Chemie International Edition*, vol. 39, pp. 401–404, 2000.
[8] J. Zhang and H. Xiao, "Computational studies on the infrared vibrational spectra, thermodynamic properties, detonation properties, and pyrolysis mechanism of octanitrocubane," *Journal of Chemical Physics*, vol. 116, no. 24, pp. 10674–10683, 2002.
[9] A. M. Astakhov, R. S. Stepanov, and A. Y. Babushkin, "On the detonation parameters of octanitrocubane," *Combustion, Explosion and Shock Waves*, vol. 34, no. 1, pp. 85–87, 1998.
[10] P. E. Eaton, M.-X. Zhang, R. Gilardi, N. Gelber, S. Iyer, and R. Surapaneni, "Octanitrocubane: a new nitrocarbon," *Propellants, Explosives, Pyrotechnics*, vol. 27, pp. 1–6, 2000.
[11] U. R. Nair, R. Sivabalan, G. M. Gore, M. Geetha, S. N. Asthana, and H. Singh, "Hexanitrohexaazaisowurtzitane (CL-20) and CL-20-based formulations," *Combustion, Explosion and Shock Waves*, vol. 41, no. 2, pp. 121–132, 2005.
[12] P. E. Eaton, R. L. Gilardi, and M.-X. Zhang, "Polynitrocubanes: advanced high-density, high-energy materials," *Advanced Materials*, vol. 12, no. 15, pp. 1143–1148, 2000.
[13] National Research Council of the National Academies, *Advanced Energetic Materials*, Committee on Advanced Energetic Materials and Manufacturing Technologies, Board on Manufacturing and Engineering Design, Division on Engineering and Physical Sciences, The National Academies Press, Washington, DC, USA, 2004.
[14] G. Zhou, J. Wang, W. D. He, N. B. Wong, A. Tian, and W. K. Li, "Theoretical investigation of four conformations of HNIW by B3LYP method," *Journal of Molecular Structure*, vol. 589-590, pp. 273–280, 2002.
[15] W. S. Ohlinger, P. E. Klunzinger, B. J. Deppmeier, and W. J. Hehre, "Efficient calculation of heats of formation," *Journal of Physical Chemistry A*, vol. 113, no. 10, pp. 2165–2175, 2009.
[16] S. W. Benson, F. R. Cruickshank, D. M. Golden et al., "Additivity rules for the estimation of thermochemical properties," *Chemical Reviews*, vol. 69, no. 3, pp. 279–324, 1969.
[17] S. W. Benson, *Thermochemical Kinetics: Methods for the Estimation of Thermochemical Data and Rate Parameters*, John Wiley & Sons, New York, NY, USA, 2nd edition, 1976.
[18] G. Leroy, "The theoretical approach to some chemical problems," *Advances in Quantum Chemistry*, vol. 17, no. C, pp. 1–95, 1985.
[19] CHETAH- ASTM Computer Program for Chemical Thermodynamic and Energy Release Evaluation—CHETAH Version 8. 0, http://www.normas.com/ASTM/COMPs/CHETAH-9-0 .html.
[20] R. A. Pesce-Rodriguez, R. A. Fifer, K. L. McNesby et al., "Thermal decomposition of HNIW-based formulations," Technical Report BRL-TR-3402, AD-A255 613, 1991, http://oai.dtic.mil/oai/oai?verb=getRecord&metadataPrefix=html&identifier=ADA255613.
[21] M. F. Foltz, C. L. Coon, F. Garcia, and A. L. Nichols III, "Thermal stability of the polymorphs of hexanitrohexaazaisowurtzitane, Part I," *Propellants, Explosives, Pyrotechnics*, vol. 19, no. 1, pp. 19–25, 1994.
[22] M. F. Foltz, "Thermal stability of E-hexanitrohexaazaisowurtzitane in an Estane formulation," *Propellants, Explosives, Pyrotechnics*, vol. 19, no. 2, pp. 63–69, 1994.
[23] M. F. Foltz, C. L. Coon, F. Garcia, and A. L. Nichols III, "Thermal stability of the polymorphs of hexanitrohexaazaisowurtzitane, Part II," *Propellants, Explosives, Pyrotechnics*, vol. 19, no. 3, pp. 133–144, 1994.
[24] N. B. Bolotina, M. J. Hardie, R. L. Speer Jr., and A. A. Pinkerton, "Energetic materials: variable-temperature crystal structures of γ- and ε-HNIW polymorphs," *Journal of Applied Crystallography*, vol. 37, no. 5, pp. 808–814, 2004.
[25] J. Li and T. B. Brill, "Kinetics of solid polymorphic phase transitions of CL-20," *Propellants, Explosives, Pyrotechnics*, vol. 32, no. 4, pp. 326–330, 2007.
[26] T. P. Russell, P. J. Miller, G. J. Piermarini, and S. Block, "Pressure/temperature phase diagram of hexanitrohexaazaisowurtzitane," *Journal of Physical Chemistry*, vol. 97, no. 9, pp. 1993–1997, 1993.
[27] S. Zeman and M. Krupka, "New aspects of impact reactivity of polynitro compounds, Part III. Impact sensitivity as a function of the imtermolecular interactions," *Propellants, Explosives, Pyrotechnics*, vol. 28, no. 6, pp. 301–307, 2003.
[28] X. W. Fan, X. H. Ju, Q. Y. Xia, and H. M. Xiao, "Strain energies of cubane derivatives with different substituent groups," *Journal of Hazardous Materials*, vol. 151, no. 1, pp. 255–260, 2008.

Characterization of the Chemical Kinetics in an O_2/HMDSO RF Plasma for Material Processing

Ruggero Barni, Stefano Zanini, and Claudia Riccardi

Dipartimento di Fisica Occhialini, Università degli Studi di Milano-Bicocca, Piazza della Scienza 3, 20126 Milano, Italy

Correspondence should be addressed to Ruggero Barni, ruggero.barni@mib.infn.it

Academic Editor: Hiroshi Onishi

Experimental study of the plasma gas phase in low-pressure radiofrequency discharges of oxygen and hexamethyldisiloxane is presented. The plasma phase has been studied by means of optical emission spectroscopy. Mass spectroscopy of the neutral and of the charged species has been performed too, directly sampling the plasma gas phase, by a dedicated spectrometer. We also measured the ion energy distribution. We have studied the influence of the operating conditions on the plasma gas-phase composition which plays a primary role in the formation process of SiO_2 films, which are known for their important applicative uses.

1. Introduction

Plasma-enhanced chemical vapour deposition (PECVD) of silicon oxides is one of the most promising techniques allowing industrial-scale deposition of high-quality, transparent, and thin coatings [1, 2]. SiO_2-like films deposited by plasma polymerisation was first developed for microelectronics. Then, they have been applied also as protective coatings in optoelectronics and packaging [3]. The most favourable option is to use organosilicon compounds as precursor, like hexamethyldisiloxane (HMDSO) mixed with O_2 and/or noble gases. It was found that composition and structure of films change with the deposition conditions [1]. A systematic control of plasma is then needed to obtain films with the desired properties. Better understanding of the chemical kinetics processes happening in the discharge is then mandatory and requires suitable diagnostics to probe directly the plasma gas phase. Previous investigations employed mass spectroscopy (MS), optical emission spectroscopy (OES), and infrared absorption spectroscopy (FT-IR) [1, 4–7]. However, many points are still open, for instance, concerning the role of ions in the gas phase, the deposition and the film growth process [1]. We have investigated O_2/HMDSO plasmas produced by a resonant inductive radiofrequency (RF) plasma source and used to deposit SiO_2-like films on polyethylene terephthalate (PET).

Diagnostics include OES and on-line MS of neutrals and ions from the discharged gas phase. Our mass spectrometer allows to change the electron energy in the ionization source, in order to better identify the parent radical sampled from the plasma. We were able to measure also ion energy spectra. The properties of the film have been studied by attenuated total reflectance infrared (ATR) spectroscopy, electron microscopy, and by contact angle measurements.

2. Experimental Setup and Diagnostics

A scheme of the setup is shown in Figure 1. The plasma source was developed by CCR Technologies Gmbh [8, 9]. It is mounted on the top of a cubic vacuum chamber and consists of a single-turn coil coupled to a RF supply. The coupling of the discharge is enhanced by the presence of a static magnetic field B which allows resonant energy absorption at the electron cyclotron frequency when its value matches the 13.56 MHz of RF supply. A constant flow of O_2 and Ar (or He) is provided, while HMDSO in a sealed bottle is evaporated into the chamber through a micrometer needle valve. Total pressure is measured just before entering into the plasma source and inside the chamber. To keep the desired pressure there, the effective pumping speed is varied by partially closing the gate valve. OES of the discharge is used to gain insight into the excited species produced.

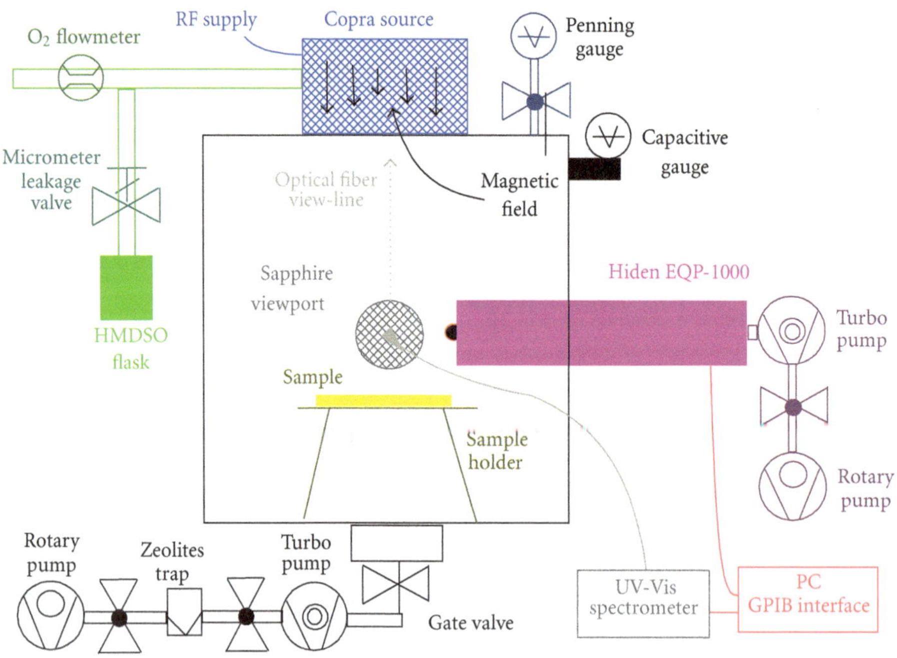

FIGURE 1: The experimental setup with diagnostics.

Measurements of the emission spectra are obtained with a wideband, low-resolution spectrometer. Spectra cover the UV-Vis-NIR range from 200 to 850 nm with a resolution of about 1 nm. Emission line intensity was measured, and atoms and molecules have been identified [10]. An EQP-1000 spectrometer by Hiden [9] allows MS of neutral species. Abundances and energy distribution of ions can be measured too. Ions and molecules enter through an orifice in a hollow tube mounted on a movable feed-through to probe plasma well inside the chamber. A sheath develops between the plasma and the grounded surface surrounding the orifice. Energy spectra are, thus, shifted by a value corresponding to the plasma potential V_p. Besides measuring the shift, the energy spectra were also fitted with a general form of the energy distribution function (EDF) [11, 12]. The following formula is general enough to describe the energy spectra recorded in the experiments

$$f(E) = N \cdot \sqrt{E} \cdot e^{-(1/\alpha)\cdot((E-qV_p)/\varepsilon)^{\alpha}}, \quad (1)$$

where ε is related to the mean energy of the ions, and α is a parameter describing how steep is the high energy tail of the distribution ($\alpha = 1$ is the Maxwellian thermal distribution). From the fitted parameters values, the mean energy could be then easily calculated and compared.

3. Plasma Characterization

Results are shown in Figures 2 to 5. In particular, data refers to discharges at a total pressure $P = 0.42$ Pa, an O_2 flowrate $Q = 10$ sccm and an O_2 to HMDSO partial pressure ratio of 3.8 : 1. This oxygen level is lower than the stechiometric ratio needed to have complete oxidation of HMDSO to SiO_2, CO_2, and water vapour (13 : 1). Data concerning these

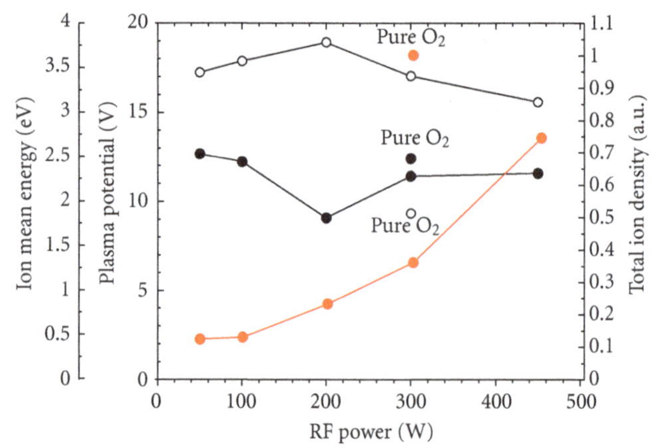

FIGURE 2: Mean ion energy (•) and plasma potential (∘) as a function of the RF power. On the right, in red, is the total plasma density normalized to that measured in a pure oxygen discharge.

experiments are plotted as a function of the RF power. At first we have analysed the charged particle produced in the plasma state. In the whole RF power range, the ion energy spectra are strongly peaked, with exponent α ranging between 3 and 5. The EDFs have high energy tails, nearly Maxwellian but with low population. Results derived from the EDF measurements are shown in Figure 2. Mean ion energies are about 2–2.5 eV, slightly decreasing with power and similar to those measured in pure O_2 discharges. Ion density increases strongly with RF power, but its values are much smaller than in pure O_2. Then HMDSO appears as a strong quenching agent of the discharge, probably because most of the electron energy is released into dissociation of the monomer, rather than contributing to the ionization

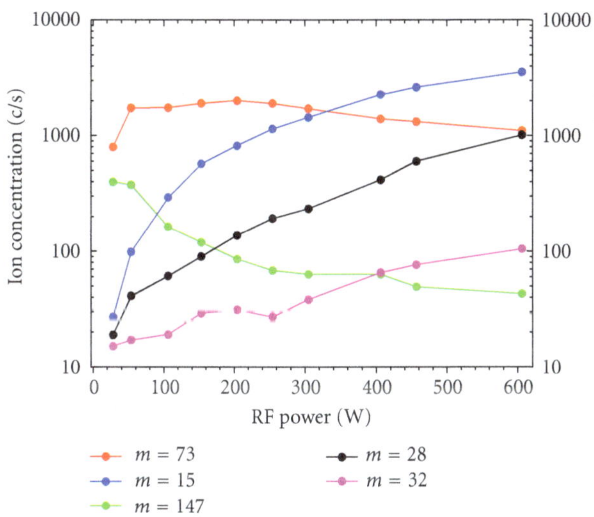

FIGURE 3: Concentration of a few major ionic species as a function of the RF power.

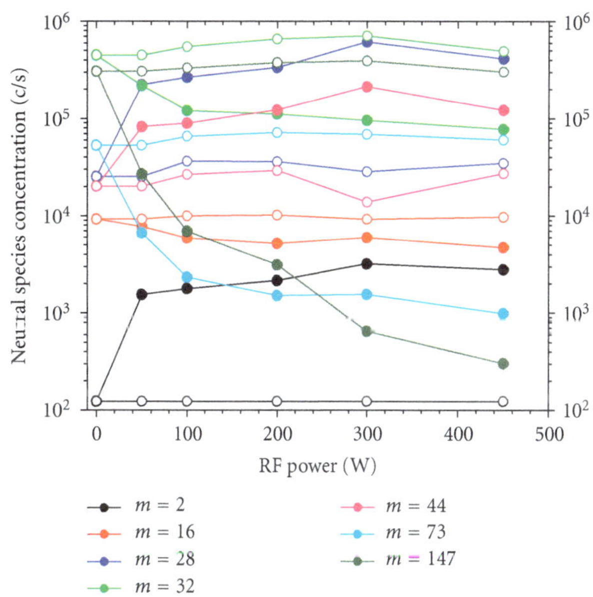

FIGURE 4: Neutral species concentration in the gas phase as a function of the RF power. Data could be compared when the discharge is ignited (• plasma on) or not (∘ plasma off).

processes sustaining plasma. The plasma potential is about 18 V, independent from the power. This is almost double than in pure O_2 and in Ar/HMDSO plasmas. Thus, if the film is grown on a grounded sample holder, ions are accelerated in the sheath up to energies of tens of eV, high enough to break all kinds of chemical bonds and induce radical formation at the surface of the growing film.

The measured ion mean energies correspond to the same range of electron temperatures, since ion energy distribution builds up in the presheath region, according to the Bohm criterion [13]. Electron impact ionization of HMDSO at such low energies mainly produces ions with the removal of a methyl group, that is, at a mass of 147 amu [6, 7]. From data shown in Figure 3, it could be seen that their concentration steadily declines. This is consistent with a strong dissociation level of HMDSO at high power, removing the source of this ion. The trend of the $Si(CH_3)_3^+$ ion (73 amu), which is formed from HMDSO after breaking of a Si–O bond, is similar. It reaches quickly a plateau and then declines, indicating a decrease of heavy fragments in the gas phase. Light ion density keeps increasing with power instead. This happens for CH_3^+ (15 amu) and for 28 amu ion. The latter grows almost exponentially in the whole RF power range. Its identity is not sure, since it could be CO^+ or Si^+, after complete dissociation/oxidation of HMDSO but also it could come from ionization of N_2, C_2H_4, or other hydrocarbons. The detection of strong emission lines from CO, with a similar trend with power, and the weak signals of CH, C_2, and Si from OES, favours the first hypothesis. O_2^+ increases too but it is always minor. Other peaks at odd masses like 27, 29, 31, 43, 45, and 59 can be assigned to (oxidated) carbon compound ions, while no peak was observed for masses larger than 150.

This indicates that polymerization and clusterization of HMDSO fragments proceeds slowly, if not at all, in the plasma gas phase.

Neutral species have been detected too, and we have compared the signal with and without plasma in Figure 4 [1, 6]. HMDSO (147 amu peak) disappears almost completely from the plasma, reducing by three orders of magnitude. In the actual operating conditions of the experiments the device works under the so called well-mixed reactor regime [14, 15]. Then in the main chamber the gas phase is mixed by diffusion before being evacuated by the pumping flow. This was also checked by varying the sampling position of our MS orifice. So our results confirm that already at the of 50 W more than 90% of HMDSO flow is actually dissociated before reaching the substrate holder where deposition occurs.

At higher-power level dissociation is virtually complete.

A less pronounced trend is observed by the 73 amu peak, a typical fragment of HMDSO, which could also be produced by heavier fragment of that monomer. Molecular and atomic oxygen decrease in the plasma reaching a constant concentration at power larger than 100 W. They are likely to be consumed in the oxidation of carbon compounds and of silicon ones. In particular, the absence of a significant increase in the atomic oxygen level warrants the high reactivity of plasma discharges, since it proves that O atoms react quickly in the gas phase without accumulation. Indeed in pure oxygen discharges atomic oxygen level increases with the power level, thanks to the dissociation by plasma electrons. The oxygen trend is related to the large increase observed for peaks at 2, 28, and 44 amu. The first is likely due to the formation of hydrogen. The other two can be assigned, respectively, to CO or Si and to CO_2 or SiO. They could receive contribution also from (oxidated) carbon compounds. Again OES results support prevalence of carbon monoxide over silicon. As happens for ions, no peak was observed for masses larger than 165 up to 1000 amu (at the sensitivity level of the spectrometer, 10 ppm). This confirms that the polymerization of HMDSO does not occur in gas phase, at least under our conditions [1].

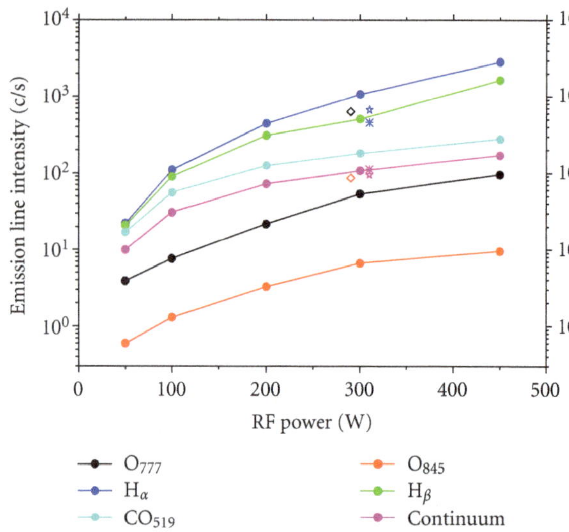

FIGURE 5: Emission line intensity as a function of the RF power. Data from pure O_2 (diamond) and HMDSO mixed with Ar (asterisk) and He (star) plasmas are shown for comparison.

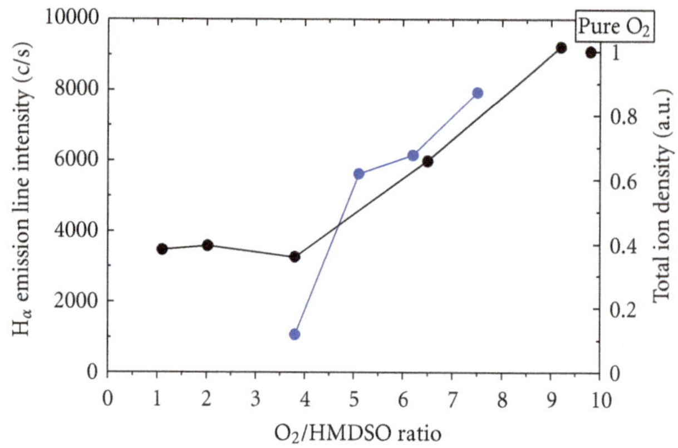

FIGURE 6: On the left scale is the H_α emission line intensity (blue) in the discharge as a function of the O_2/HMDSO ratio. On the right scale is the total plasma density normalized to that measured in a pure oxygen discharge.

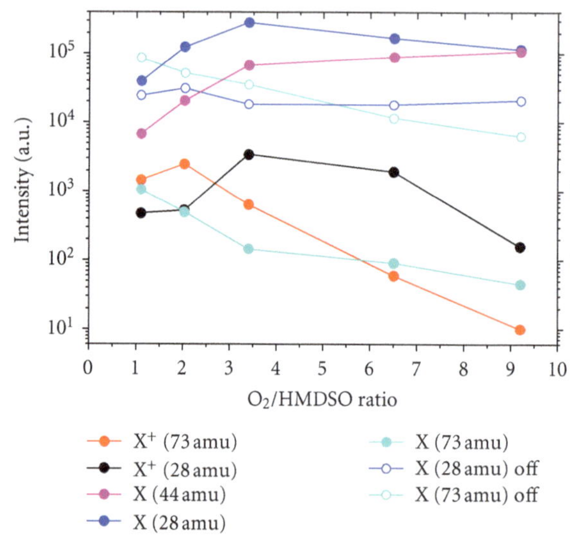

FIGURE 7: Major ion and neutral species concentration in the gas phase as a function of the O_2/HMDSO ratio (• plasma on, ∘ plasma off).

Spectra of the light emitted from the discharge show a broad continuum peaking at about 500 nm, with a rich structure of lines, mainly identified as due to the vibrational excitation levels of CO. However, the most intense line is usually the H_α line at 656.3 nm. Emission from O_2 and O_2^+ is not noticeable apart in pure oxygen, while atomic oxygen lines in the near infrared are detected (see Figure 5). Their intensity is, however, greatly reduced in the presence of HMDSO, again indicating that O is actively involved in the oxidation process. Search for atomic emission lines from C, Si, and their ions was unsuccessful. Also absent are the characteristic features due to CH and C_2 emission typically of methane plasmas [14]. This indicates that oxidation of carbon compounds proceeds quickly in our discharged gas phase. OES of HMDSO diluted in Ar or He plasmas shows that H_α intensity is greatly reduced, while the continuum is almost constant, as could be grasped from Figure 5. This indicates a reduced capability in HMDSO dissociation by plasmas containing noble gases compared to oxygen. The expected result is then a larger retention of organic groups in the film deposited from such plasmas [1].

Then we have performed experiments at a fixed-power level where oxygen flow was varied. In this way, we are effectively changing the O_2/HMDSO ratio, in order to ensure more complete oxidation level. Gate valve opening was regulated to have approximately the same pressure in the vacuum chamber, about 0.5 Pa. Results from the plasma diagnostics are shown in Figures 6 and 7. Both plasma density and H_α emission increase as HMDSO gets more diluted in oxygen. As for the former, at $O_2/\mathrm{HMDSO} = 9.2$ plasma density reaches almost the same value of pure oxygen discharges. At such level of dilution HMDSO no longer affects the ionization degree of the discharge. Atomic hydrogen emission could be detected only when O_2/HMDSO is larger than 3, despite that total ion density is almost the same. This is true for the whole light emission spectra, indicating a generalized decrease of electron density in pure or HMDSO dominated discharges. At higher dilution, H_α, intensity rises steadily. The increase appears even bigger than the ion density one, and it largely overcompensates for the decrease in the total hydrogen abundance at higher dilution.

The concentration of a few ionic and neutral species is displayed in Figure 7. The trend of both the ion and the neutral species at 28 and 73 amu is opposite. The former grows reaching a maximum at $O_2/\mathrm{HMDSO} = 3.8$ and then decreases, while the latter steadily declines with dilution. The 28 amu peak is more sensible to oxygen content, indicating that it controls the oxidation level of the HMDSO. At higher dilution, its concentration declines. The low level of the $Si(CH_3)_3$ radical (73 amu) in the discharge gas phase, two orders of magnitude lower, shows the high degree of

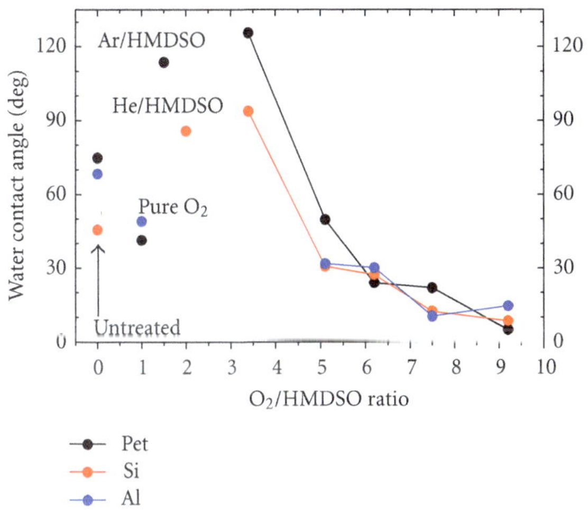

FIGURE 8: Water contact angle measured on different plasma-treated substrate samples as a function of the O_2/HMDSO ratio ($W = 300$ W, $P = 0.4$ Pa, and $T = 15'$).

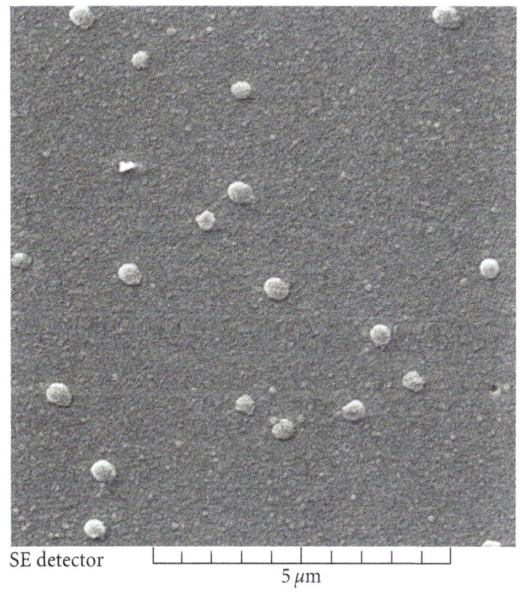

FIGURE 9: SEM image of the film deposited on silicon ($W = 300$ W, $P = 0.4$ Pa, O_2/HMDSO = 7.5, and $T = 15'$).

dissociation of HMDSO reached at such RF power even at the lowest dilution, while the 44 amu peak increases. All these observations are consistent with a picture where more complete dissociation and oxidation is favoured by higher HMDSO dilution in oxygen. Then a marked increase towards a pure SiO_2 composition of the deposited film can be anticipated at high dilution.

4. Deposited Film Characterization

Film deposition has been performed on 12 μm PET sheets exposed to plasma. Small samples of aluminium foils, Si wafers, and KBr tablets have been exposed too.

They have been employed mainly for evaluation of film properties purposes. Some analysis and characterization of the deposited film was already published [16–18]. Treatments of 15 minutes have been done at a fixed RF power of 300 W, high enough to completely dissociate HMDSO, as discussed above. Oxidation of the film precursors is the plasma state depends on the O_2/HMDSO ratio as discussed above. Although not complete, as it could be deduced from Figure 7, it appears to be sufficient to obtain a low carbon contamination in the film already at ratios above 5. The low carbon content of the film at high dilution is confirmed by ATR spectra recorded for the KBr samples [16]. They show only the typical Si–O–Si bands: the rocking, the bending, and the asymmetric stretching modes (450, 800, and 1000–1150 cm^{-1}). The CH_x stretching bands at 2900–2960 cm^{-1}, the Si–CH_3 rocking vibration band at 840 cm^{-1}, and the CH_3 symmetric bending in Si–CH_3 band at 1260 cm^{-1} do not appear in ATR spectra at high HMDSO dilution, indicating a high SiO_2-like character of the deposit. This correlates fairly well with the increasing hydrophilic nature of the deposited film, displayed in Figure 8.

On the contrary, when carbon compounds oxidation is not complete, many organic groups are embedded in the film, giving a hydrophobic character to the surface [6]. A similar result is obtained when deposition is performed from a plasma of HMDSO mixed with noble gases, like argon and helium. The deposited film is generally more hydrophilic than the original surface as well as the surface oxidated by a pure oxygen plasma treatment.

A SEM image of the film deposited at high dilution is shown in Figure 9. Here, the substrate was a silicon wafer just to have a nicely flat surface. A rather uniform film is obtained with a few micrometer dust particles scattered on the surface.

5. Conclusions

An experimental study of the plasma gas phase produced in a O_2/HMDSO mixture by a resonant inductively coupled radiofrequency plasma source has been performed. Strong dissociation of HMDSO is obtained at large RF power, exceeding 100 W, and high dilution. Ion mean energy was measured to be about 2 eV, while the plasma potential reached tens of volts. Application of the source to the treatment of PET results in the deposition of a highly inorganic SiO_2 film, with very low water contact angle of a few tens of degrees, when oxygen-to-HMDSO ratio is larger than 7. Preliminary measurements of the water vapour transmission rate show a decrease from 23 of untreated PET to 5.5 g/m^2/day at an O_2/HMDSO ratio of 7.5, which appears encouraging even if not yet comparable with results reported in the literature [17, 19]. Chemical kinetics in the gas phase of O_2/HMDSO plasmas largely controls the deposited film composition and characteristics. Suitable diagnostics of the gas phase then allows to have a clue linking the operating conditions with the characteristics of the deposition process. In particular, on-line mass spectroscopy of the plasma gas phase allows to identify all the radicals involved in the film growth and to monitor their concentration as a function of the operating parameters.

References

[1] H. Biederman, *Plasma Polymer Films*, Imperial College Press, London UK, 2004.

[2] P. Esena, S. Zanini, and C. Riccardi, "Plasma processing for surface optical modifications of PET films," *Vacuum*, vol. 82, no. 2, pp. 232–235, 2007.

[3] R. Lamendola and R. D'Agostino, "Process control of organosilicon plasmas for barrier film preparations," *Pure and Applied Chemistry*, vol. 70, no. 6, pp. 1203–1208, 1998.

[4] D. S. Wavhal, J. Zhang, M. L. Steen, and E. R. Fisher, "Investigation of gas phase species and deposition of SiO_2 films from HMDSO/O_2 plasmas," *Plasma Processes and Polymers*, vol. 3, no. 3, pp. 276–287, 2006.

[5] M. Creatore, Y. Barrell, J. Benedikt, and M. C. M. Van De Sanden, "On the hexamethyldisiloxane dissociation paths in a remote Ar-fed expanding thermal plasma," *Plasma Sources Science and Technology*, vol. 15, no. 3, pp. 421–431, 2006.

[6] D. Magni, C. Deschenaux, C. Hollenstein, A. Creatore, and P. Fayet, "Oxygen diluted hexamethyldisiloxane plasmas investigated by means of in situ infrared absorption spectroscopy and mass spectrometry," *Journal of Physics D*, vol. 34, no. 1, pp. 87–94, 2001.

[7] M. R. Alexander, F. R. Jones, and R. D. Short, "Mass spectral investigation of the radio-frequency plasma deposition of hexamethyldisiloxane," *Journal of Physical Chemistry B*, vol. 101, no. 18, pp. 3614–3619, 1997.

[8] M. Weiler, K. Lang, E. Li, and J. Robertson, "Deposition of tetrahedral hydrogenated amorphous carbon using a novel electron cyclotron wave resonance reactor," *Applied Physics Letters*, vol. 72, no. 11, pp. 1314–1316, 1998.

[9] R. Barni, S. Zanini, and C. Riccardi, "Diagnostics of reactive RF plasmas," *Vacuum*, vol. 82, no. 2, pp. 217–219, 2007.

[10] R. W. B. Pearse and A. G. Gaydon, *The Identification of Molecular Spectra*, John Wiley & Sons, New York, NY, USA, 1976.

[11] A. Schwabedissen, E. C. Benck, and J. R. Roberts, "Langmuir probe measurements in an inductively coupled plasma source," *Physical Review E*, vol. 55, no. 3, pp. 3450–3459, 1997.

[12] C. Riccardi, R. Barni, and M. Fontanesi, "Experimental study and simulations of electronegative discharges at low pressure," *Journal of Applied Physics*, vol. 90, no. 8, pp. 3735–3742, 2001.

[13] F. F. Chen, *Introduction to Plasma Physics*, Plenum, New York, NY, USA, 1984.

[14] J. M. Smith, *Introduction to Chemical Engineering Thermodynamics*, McGraw-Hill, New York, NY, USA, 1987.

[15] C. Riccardi, R. Barni, F. De Colle, and M. Fontanesi, "Modeling and diagnostic of an SF6 RF plasma at low pressure," *IEEE Transactions on Plasma Science*, vol. 28, no. 1, pp. 278–287, 2000.

[16] S. Zanini, C. Riccardi, M. Orlandi et al., "Surface properties of HMDSO plasma treated PET," *Surface and Coatings Technology*, vol. 200, no. 1–4, pp. 953–957, 2005.

[17] S. Zanini, C. Riccardi, M. Orlandi, and E. Grimoldi, "Characterisation of $SiO_xC_yH_z$ thin films deposited by low-temperature PECVD," *Vacuum*, vol. 82, no. 2, pp. 290–293, 2007.

[18] R. A. Siliprandi, S. Zanini, E. Grimoldi, F. S. Fumagalli, R. Barni, and C. Riccardi, "Atmospheric pressure plasma discharge for polysiloxane thin films deposition and comparison with low pressure process," *Plasma Chemistry and Plasma Processing*, vol. 31, no. 2, pp. 353–372, 2011.

[19] A. Sonnenfeld, A. Bieder, and P. R. von Rohr, "Influence of the gas phase on the water vapor barrier properties of SiO_x films deposited from RF and dual-mode plasmas," *Plasma Processes and Polymers*, vol. 3, no. 8, pp. 606–617, 2006.

8

Analytical Expressions Pertaining to the Concentration of Substrates and Product in Phenol-Polyphenol Oxidase System Immobilized in Laponite Hydrogels: A Reciprocal Competitive Inhibition Process

K. Indira and L. Rajendran

Department of Mathematics, The Madura College, Madurai 625011, India

Correspondence should be addressed to L. Rajendran, raj_sms@rediffmail.com

Academic Editor: Marc Koper

Theoretical analysis corresponding to the diffusion and kinetics of substrate and product in an amperometric biosensor is developed and reported in this paper. The nonlinear coupled system of diffusion equations was analytically solved by Homotopy perturbation method. Herein, we report the approximate analytical expressions pertaining to substrate concentration, product concentration, and current response for all possible values of diffusion and kinetic parameters. The numerical solution of this problem is also reported using Scilab/Matlab program. Also, we found excellent agreement between the analytical results and numerical results upon comparison.

1. Introduction

Theoretical modeling of biosensors usually provides some important insight into understanding the functioning of a biosensor. Usually, with the aid of an analytical device, it is not possible to measure the concentration of substrates inside the enzyme membranes. As a result, theoretical model in biosensors has been developed and employed as an important tool to study the analytical characteristics of biosensors. Initially, Goldman et al. [1] reported the theoretical modeling on biosensor. In that report [1], they have published an extensive theoretical treatment corresponding to substrate and product distribution in membranes containing enzymes. Later, Sundaram and Laidler [2] reported equations that describe the kinetics of reaction in an enzyme membrane immersed in a substrate solution. Kasche and coworkers [3] presented a model and equations that describe steady-state catalysis by an enzyme immobilized in spherical gel particles. Furthermore, they have demonstrated that the catalysis by a bounded enzyme at low substrate concentrations differs from the catalysis brought about by an unbound enzyme. Blaedel et al. [4] derived equations for steady-state fluxes of substrates and product through a membrane in simple systems.

Catalytic biosensors are sensors that use enzymes which catalyse a specific conversion of analyte [5]. The analysis of the action of the biosensors containing one and two enzymes was performed under internal diffusion limitation in [6]. Gough et al. [7] have simulated the performance of a cylindrical biosensor employed for glucose monitoring at steady state. Jobst et al. [8] reported a finite difference scheme for the discretization of the model equation. Bacha et al. [9, 10] have developed a model that takes into account a variety of configuration designs to describe the behaviour of an amperometric biosensor for glucose monitoring. Recently, Baronas et al. [11–14] have developed a theoretical model to examine the dynamic response of amperometric biosensors under stirred and nonstirred solutions. Coche-Guerente et al. [15] analyzed the amperometric response of PPO-based rotating disk bioelectrodes based on the kinetic model by considering the internal and external mass transport effects and a CEC' electroenzymatic mechanism.

Amperometric biosensors are rapid, sensitive, and highly stable candidates for environmental, clinical, and industrial applications. The general features of amperometric biosensors have been studied and analyzed extensively in the literature [16, 17]. However, due to the limited calculation possibilities and lack of model, the calculations were restricted by two critical concentrations of substrate (when enzyme acted at the first- and zero-order reaction conditions). Later, there are few reports that reported steady-state and nonstationary kinetics of amperometric biosensor response [17–21]. Baronas et al. [22] developed the model based on nonstationary diffusion equations containing a nonlinear term related to the enzymatic reaction. Ames [23] performed the digital simulation of the biosensor response with the implicit difference scheme.

Recently, Coche-Guerente et al. [24, 25] presented a theoretical analysis of the transport and kinetics of substrate and product in an immobilized enzyme layer involving an electrochemical substrate-recycling scheme. Baronas et al. [26] developed the mathematical model of the biosensor action to investigate the dynamics of the response of biosensors utilizing a triggered enzymatic reaction followed by the electrochemical and enzymatic product cyclic conversion scheme (CEC scheme). To the best of our knowledge, there are no analytical expressions corresponding to the steady-state substrate, product concentrations, and current for all values of parameters μ_1, μ_2, α_1, and α_2 at the rotating disk electrode for reciprocal competitive inhibition process [15]. These parameters are explained below (8). The purpose of this communication is to derive an analytical expression of steady-state concentration and current at electrodes for all values of parameters μ_1, μ_2, α_1, and α_2.

2. Mathematical Formulation of the Problem and Analysis

The complete mathematical formulation of this problem is described in Coche-Guerente et al. [15]. Figure 1 shows the schematic representation of a rotating disk electrode modified by a PPO enzymatic layer of thickness L [15]. The enzymatic reaction of substrates S_1, S_2 and the product P_2 follows reciprocal competitive inhibition process. Therefore, enzymatic rate of formation of *o*-quinone from phenol and catechol substrates, V_1 and V_2 can be expressed as follows:

$$V_1 = \frac{k_1[E_T][S_1]}{[S_1] + K_1(1 + [S_2]/K_2)},$$
$$V_2 = \frac{k_2[E_T][S_2]}{[S_2] + K_2(1 + [S_1]/K_1)}, \tag{1}$$

where $[E_T]$ represent the total concentration of active enzyme. K_1, K_2, k_1, and k_2 are the kinetic parameters. The nonlinear differential equations corresponding to the reciprocal competitive inhibition process with concentrations S_1, S_2 and P_2 within the enzymatic layer [15, 26] at steady state condition can be written as follows:

$$\frac{\partial^2[S_1]}{\partial X^2} - \frac{[S_1]}{\Lambda_1^2(1 + [S_1]/K_1 + [S_2]/K_2)} = 0,$$
$$\frac{\partial^2[S_2]}{\partial X^2} - \frac{[S_2]}{\Lambda_2^2(1 + [S_1]/K_1 + [S_2]/K_2)} = 0,$$
$$\frac{\partial^2[P_2]}{\partial X^2} + \frac{[S_1]}{\Lambda_1^2(1 + [S_1]/K_1 + [S_2]/K_2)} + \frac{[S_2]}{\Lambda_2^2(1 + [S_1]/K_1 + [S_2]/K_2)} = 0, \tag{2}$$

where X represents the distance from the electrode surface, Λ_1 and Λ_2 are the reaction lengths related to phenol substrate S_1, and catechol substrate S_2. Further, $\Lambda_1 = (D_f K_1/k_1[E_T])^{1/2}$, $\Lambda_2 = (D_f K_2/k_2[E_T])^{1/2}$, where D_f is the diffusion coefficient in the enzymatic layer. These equations are solved for the following initial and boundary conditions as described in the literature [15]:

$$\frac{\partial[S_1]}{\partial X} = 0, \quad [S_1] + [S_2] = [S_1]_\infty, \quad [P_2] = 0 \quad \text{when } X = 0,$$
$$[S_1] = [S_1]_\infty, \quad [S_2] = 0, \quad [P_2] = 0 \quad \text{when } X = L + \delta. \tag{3}$$

In addition, the mass conservation implies that $[S_1] + [S_2] + [P_2] = [S_1]_\infty$ when $X \geq L^+$, where L is the thickness of the enzymatic layer and δ is the thickness of the diffusion convection. At $X = L$, the fluxes J_{S_1}, J_{S_2}, and J_{P_1} of S_1, S_2, and P_2 on each side of the interface are equal [15]. Next, we introduce the following set of dimensionless variables:

$$u = \frac{[S_1]}{[S_1]_\infty}, \quad v = \frac{[S_2]}{[S_1]_\infty}, \quad w = \frac{[P_2]}{[S_1]_\infty}, \quad x = \frac{X}{L}, \tag{4}$$

where u, v, and w represent dimensionless concentration and x is the distance parameter. The dimensionless nonlinear equations for rotating disk electrode are as follows:

$$\frac{\partial^2 u}{\partial x^2} - \frac{\mu_1 u}{1 + \alpha_1 u + \alpha_2 v} = 0, \tag{5}$$

$$\frac{\partial^2 v}{\partial x^2} - \frac{\mu_2 v}{1 + \alpha_1 u + \alpha_2 v} = 0, \tag{6}$$

$$\frac{\partial^2 w}{\partial x^2} + \frac{\mu_1 u}{1 + \alpha_1 u + \alpha_2 v} + \frac{\mu_2 v}{1 + \alpha_1 u + \alpha_2 v} = 0, \tag{7}$$

where

$$\mu_1 = \frac{L^2}{\Lambda_1^2}, \quad \mu_2 = \frac{L^2}{\Lambda_2^2}, \quad \alpha_1 = \frac{[S_1]_\infty}{K_1}, \quad \alpha_2 = \frac{[S_1]_\infty}{K_2}, \tag{8}$$

where μ_1 is the ratio of the square of the thickness L and the reaction length Λ_1; μ_2 is the ratio of the square of the thickness L and the reaction length Λ_2; α_1 is the ratio of $[S_1]_\infty$ and the kinetic parameter K_1; α_2 is the ratio of $[S_1]_\infty$ and the kinetic parameter K_2. Now the initial and boundary conditions are given by the following equation:

$$\frac{\partial u}{\partial x} = 0, \quad u + v = 1, \quad w = 0 \quad \text{when } x = 0,$$
$$u = 1, \quad v = 0, \quad w = 0 \quad \text{when } x = 1 + \frac{\delta}{L} = m. \tag{9}$$

The dimensionless current is given by [15]

$$I = \frac{j_f}{2F} = -\left(\frac{\partial v}{\partial x}\right)_{x=0}, \quad (10)$$

where F is the Faraday constant.

3. Solution of Boundary Value Problem Using the Homotopy Perturbation Method

The HPM [27] uses the embedded parameter p as small, and only a few iterations are essential for an asymptotic solution. Using this method (see Appendix A), the solutions to (5)–(7) can be deduced as follows:

$$\begin{aligned} u(x) = {} & a\cosh\left(\sqrt{\mu_1}x\right) \\ & + \alpha_2 A_0\Big[2a\cosh\left(\sqrt{\mu_2}m\right)(\mu_2 - 2\mu_1)\sqrt{\mu_2} \\ & \quad \times \sinh\left(\sqrt{\mu_1}(m-x)\right) \\ & \quad + 4a\mu_1\sqrt{\mu_2}\sinh\left(\sqrt{\mu_1}m\right)\cosh\left(\sqrt{\mu_1}x\right) \\ & \quad - \left(\mu_2 + 2\sqrt{\mu_1\mu_2}\right)\sqrt{\mu_1}\sinh\left(\sqrt{\mu_1}x + \sqrt{\mu_2}(m-x)\right) \\ & \quad - \left(\mu_2 - 2\sqrt{\mu_1\mu_2}\right)\sqrt{\mu_1} \\ & \quad \times \sinh\left(-\sqrt{\mu_1}x + \sqrt{\mu_2}(m-x)\right)\Big] \\ & + \frac{\alpha_1 a^2}{6}\Big[a\cosh\left(\sqrt{\mu_1}x\right)\cosh\left(2\sqrt{\mu_1}m - 3\right) \\ & \quad - \cosh\left(2\sqrt{\mu_1}x\right) + 3\Big], \end{aligned} \quad (11)$$

$$\begin{aligned} v(x) = {} & (1-a)b\sinh\left(\sqrt{\mu_2}(m-x)\right) - \alpha_2 A_0\sinh\left(\sqrt{\mu_2}(m-x)\right) \\ & \times \left(2ab\sqrt{\mu_2}\sinh\left(\sqrt{\mu_1}m\right)\right. \\ & \quad \left.\times\left(\cosh\left(\sqrt{\mu_2}m\right)(\mu_2 - 2\mu_1) + 2\mu_1\right) - 2\sqrt{\mu_1\mu_2}\right) \\ & + A_1\Big[2\mu_1\sinh\left(\sqrt{\mu_2}(m-x)\right) + 4b\sqrt{\mu_1}\sqrt{\mu_2}\sinh\left(\sqrt{\mu_1}m\right) \\ & \quad \times \sinh\left(\sqrt{\mu_2}x\right) - \left(\mu_1 + 2\sqrt{\mu_1}\sqrt{\mu_2}\right) \\ & \quad \times \sinh\left(\sqrt{\mu_1}x + \sqrt{\mu_2}(m-x)\right) \\ & \quad - \left(\mu_1 - 2\sqrt{\mu_1}\sqrt{\mu_2}\right)\sinh\left(-\sqrt{\mu_1}x + \sqrt{\mu_2}(m-x)\right)\Big] \\ & - \alpha_1 a^2 b\sinh\left(\sqrt{\mu_2}(m-x)\right) \\ & \times \left(\frac{a\left(\cosh\left(2\sqrt{\mu_1}m\right) - 3\right)}{6} + \frac{1}{3}\right) \\ & + A_2\Big[\left(\cosh\left(2\sqrt{\mu_2}m\right) + 3\right)\sinh\left(\sqrt{\mu_2}(m-x)\right) \\ & \quad + 4\sinh\left(\sqrt{\mu_2}x\right) - \sinh\left(\sqrt{\mu_2}m\right) \\ & \quad \times \left(\cosh\left(2\sqrt{\mu_2}(m-x)\right) + 3\right)\Big], \end{aligned} \quad (12)$$

where $a = 1/(\cosh(\sqrt{\mu_1}m))$; $b = 1/(\sinh(\sqrt{\mu_2}m))$; $A_0 = (a(1-a)b\sqrt{\mu_1})/(2(\mu_2^2 - 4\mu_1\mu_2))$; $A_1 = (\alpha_1\mu_2 ab(1-a))/(2(\mu_1^2 - 4\mu_1\mu_2))$; $A_2 = (\alpha_2(1-a)^2 b^3)/6$. Adding (5) and (6) using the boundary conditions (9), we can obtain

$$w(x) = 1 - u(x) - v(x). \quad (13)$$

Equations (11)–(13) are the analytical expressions corresponding to the dimensionless concentrations as a function of dimensionless distance x. These equations satisfy the boundary conditions (9). When α_1 and $\alpha_2 \to 0$ (K_1 and $K_2 \to \infty$), the concentrations u, v, and w become

$$\begin{aligned} u(x) &= a\cosh\left(\sqrt{\mu_1}x\right), \\ v(x) &= (1-a)b\sinh\left(\sqrt{\mu_2}(m-x)\right), \\ w(x) &= 1 - u(x) - v(x). \end{aligned} \quad (14)$$

Using (10) the current density is

$$\begin{aligned} I = {} & (1-a)b\sqrt{\mu_2}\cosh\left(\sqrt{\mu_2}m\right) - \alpha_2 A_0\sqrt{\mu_2}\cosh\left(\sqrt{\mu_2}m\right) \\ & \times \left(2ab\sqrt{\mu_2}\sinh\left(\sqrt{\mu_1}m\right)\right. \\ & \quad \left.\times\left(\cosh\left(\sqrt{\mu_2}m\right)(\mu_2 - 2\mu_1) + 2\mu_1\right) - 2\sqrt{\mu_1\mu_2}\right) \\ & + A_1\Big[2\mu_1\sqrt{\mu_2}\cosh\left(\sqrt{\mu_2}m\right) - 4b\sqrt{\mu_1\mu_2}\sinh\left(\sqrt{\mu_1}m\right) \\ & \quad + \left(\mu_1 + 2\sqrt{\mu_1}\sqrt{\mu_2}\right)\cosh\left(\sqrt{\mu_2}m\right)\left(\sqrt{\mu_1} - \sqrt{\mu_2}\right) \\ & \quad - \left(\mu_1 - 2\sqrt{\mu_1}\sqrt{\mu_2}\right)\cosh\left(\sqrt{\mu_2}m\right)\left(\sqrt{\mu_1} + \sqrt{\mu_2}\right)\Big] \\ & - \alpha_1 a^2 b\sqrt{\mu_2}\cosh\left(\sqrt{\mu_2}m\right) \\ & \times \left(\frac{a\left(\cosh\left(2\sqrt{\mu_1}m\right) - 3\right)}{6} + \frac{1}{3}\right) \\ & + A_2\Big[\left(\cosh\left(2\sqrt{\mu_2}m\right) + 3\right)\sqrt{\mu_2}\cosh\left(\sqrt{\mu_2}m\right) - 4\sqrt{\mu_2} \\ & \quad - 2\sqrt{\mu_2}\sinh\left(\sqrt{\mu_2}m\right)\sinh\left(2\sqrt{\mu_2}m\right)\Big]. \end{aligned} \quad (15)$$

Equation (15) represents the dimensionless current as a function of dimensionless parameters $\mu_1, \mu_2, \alpha_1, \alpha_2$, and m.

4. Numerical Simulation

The system of steady-state nonlinear differential equations (5)–(7) are also solved in numerical methods. The function pdex4 in Scilab/Matlab software which is a function of solving partial differential equations is used to solve these equations. Its numerical solution is compared with the analytical solution obtained by HPM. Satisfactory agreement is noted. The Scilab/Matlab program is also given in Appendix B.

5. Results and Discussion

5.1. Concentration Profiles. Equations (11)–(13) are the new and simple analytical expressions of concentrations of phenol (substrate $u(\mu_1, \mu_2, \alpha_1, \alpha_2, \delta/L)$), catechol (substrate $v(\mu_1, \mu_2, \alpha_1, \alpha_2, \delta/L)$), and *o*-quinone (product $w(\mu_1, \mu_2, \alpha_1, \alpha_2, \delta/L)$), respectively. All substrate molecules are enzymatically oxidized within the enzymatic layer before reaching the electrode surface. Thus, the high sensitivity observed with a PPO bioelectrode working at a cathodic potential can be explained by an efficient electrochemical recycling of substrate which can enter a new enzymatic step in the proximity of the enzymatic layer interface. This interpretation agrees well with the substrates and product concentration profiles in the enzymatic layer calculated on the basis of (11)–(13) and reproduced in Figures 2–4. In these figures, the profiles of substrates (phenol and catechol) as well as product (*o*-quinone) concentration in the enzyme layer are presented for biosensor acting in CEC mechanism. The thickness of the enzymatic layer constitutes an important parameter that governs the biosensor sensitivity. The normalized concentration of phenol substrate u is represented in Figures 2(a)–2(c). From this figure, it is evident that the value of phenol substrate concentration u decreases when the thickness of the enzymatic layer increases and enzyme activity increases (μ_1, μ_2 increase) for all values of α_1 and δ/L. The concentration of phenol substrate u is equal to 1 when $\mu_1 \leq 0.1$ and u becomes zero when $\mu_1 \geq 30$ for all values α_1 and μ_2. Figures 3(a)–3(c) stand for the normalized concentration profile of catechol v. It is clear that the catechol concentration v increases when the thickness of the enzymatic layer increases. The concentration of catechol substrate v becomes zero when $\mu_1 \leq 0.1$ for all values of parameters. Figures 4(a)–4(c) represent the normalized *o*-quinone concentration w. It is clear that the concentration of *o*-quinone w increases when the thickness of the enzymatic layer increases. The concentration of *o*-quinone product w reaches the maximum value when $x = 0.3$ and then decreases to the minimum value zero when the diffusion-convection layer is equal to zero.

5.2. Current Response. The analytical expression of steady-state current $I(\mu_1, \mu_2, \alpha_1, \alpha_2, \delta/L)$ is given in (15). The steady-state current of membrane biosensors significantly depends on the thickness of the enzymatic layer. Figures 5(a) and 5(b) show the dimensionless current I for all values of parameters μ_1, μ_2, α_1 and α_2. From these figures it is inferred that the current increases gradually when μ_1 and μ_2 increase. When $\mu_2 \to 0$ at any $\mu_1 > 0$ current $I \to 0$ (refer to Figure 5(b)). Figures 5(c) and 5(d) represent the dimensionless current I for all values of kinetic parameters. From these figures it is evident that the current decreases when α_1 and α_2 increase. The current I is monotonously increasing function of both arguments μ_1 and μ_2. In the case of CEC mode an application of an active enzyme $[E_T]$ ($\mu_2 > 0$) stimulates an increase of the biosensor current. The steady-state current of membrane biosensor significantly depends on the thickness of the enzyme layer (L). In view of biosensor optimizations, it can be seen that the current I (refer to Figure 5(e)) is higher for low electrode rotation rate and small thickness of enzymatic layer.

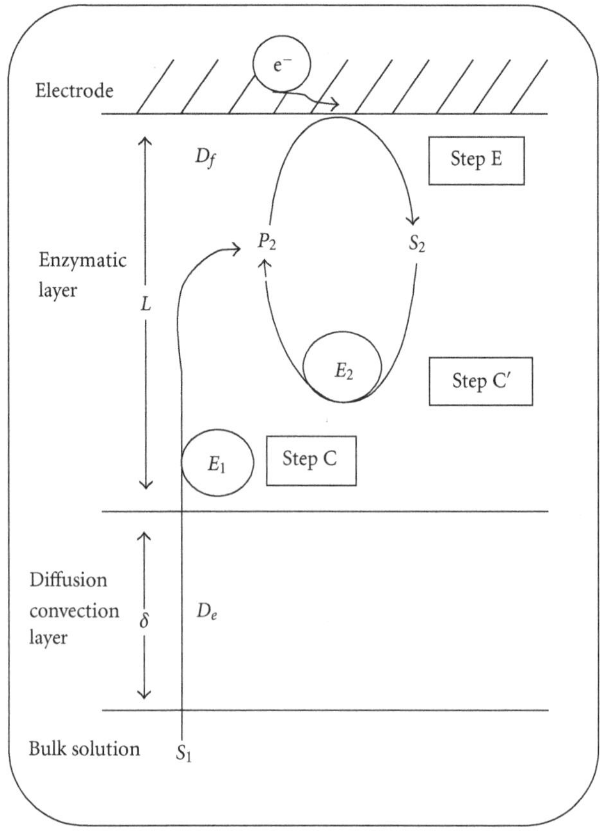

FIGURE 1: Schematic representation of rotating disk electrode modified by a PPO enzymatic layer and principle of the bioelectrode functioning in the presence of phenol substrates S_1, S_2 and product P_2 [15].

6. Conclusions

This paper reports a mathematical treatment for analyzing amperometric enzymatic reactions. In this paper, we have evaluated a theoretical model for a rotating disk electrode based on an immobilized enzyme layer. The novelty of this paper is an application of approximate method to the second-order nonlinear partial differential equations. The approximate analytical expressions for the steady state substrate concentrations and product concentration profiles for all values of μ_1, μ_2, α_1 and α_2 at the rotating disk electrode were obtained using Homotopy perturbation method. Furthermore, an analytical expression corresponding to the steady state current response is also presented. A satisfactory agreement with the existing limiting cases results is noted. The results presented in the paper extend to many applications related to biosensors modeling.

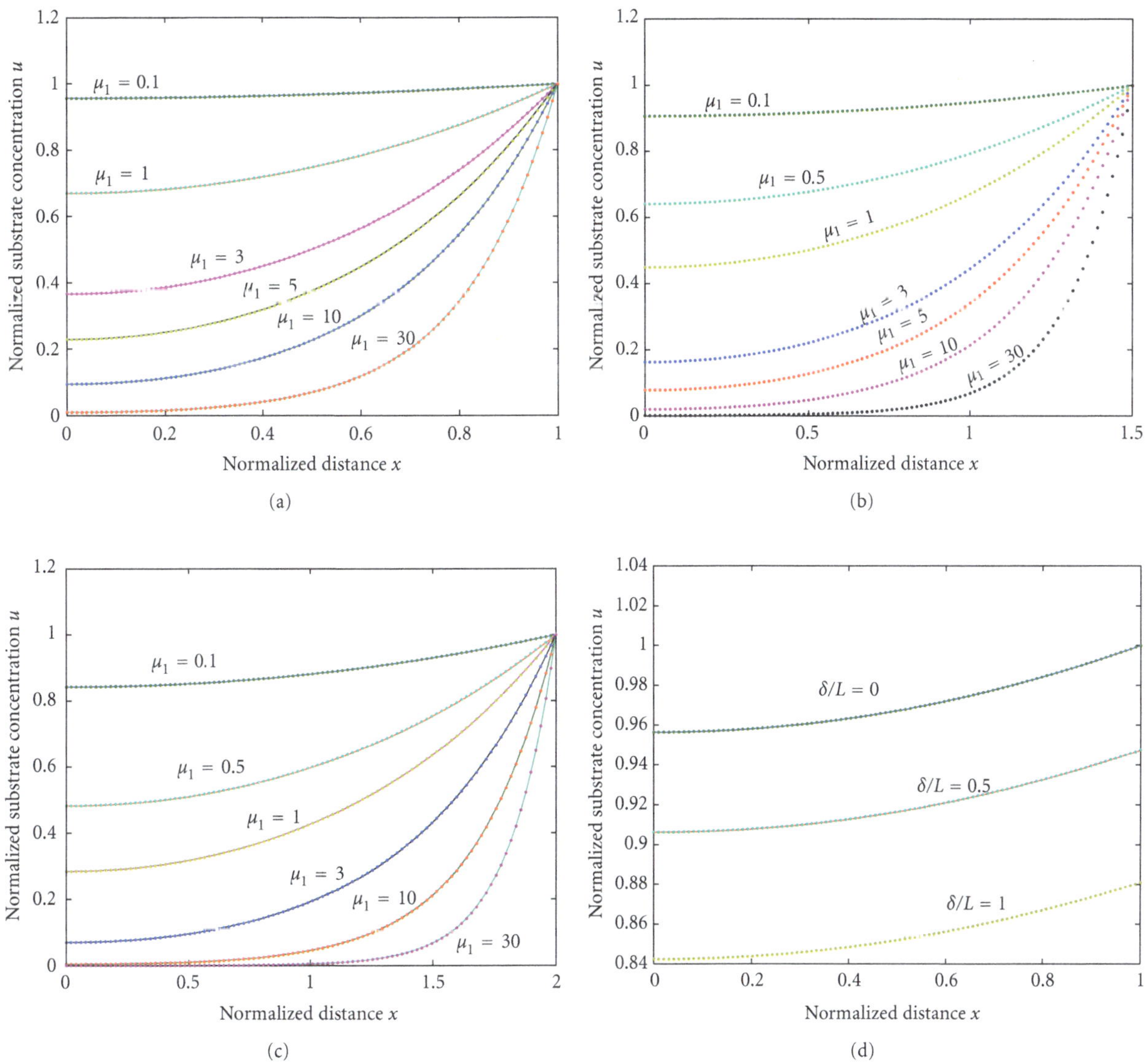

FIGURE 2: The profile of the normalized concentration of phenol substrate u versus the normalized distance x. The concentrations were computed using (11) when the parameters $\mu_2 = 1, \alpha_1 = 0.1, \alpha_2 = 0.1$ and for various values of μ_1 and δ/L. The key to the graph (stacked line) represents (11) and (dotted line) represents the numerical solution. (a) $\delta/L = 0$, (b) $\delta/L = 0.5$, (c) $\delta/L = 1$, (d) $\mu_1 = 0.1$.

Appendices

A. Solution of (5)–(7) Using Homotopy Perturbation Method

In this appendix, we indicate how (11)–(13) in this paper are deduced. In order to obtain the solution of (5)–(7), we first construct a Homotopy as follows:

$$(1-p)\left[\frac{\partial^2 u}{\partial x^2} - \mu_1 u\right] + p\left[\frac{\partial^2 u}{\partial x^2} - \mu_1 u + \alpha_1 u \frac{\partial^2 u}{\partial x^2} + \alpha_2 v \frac{\partial^2 u}{\partial x^2}\right] = 0,$$

$$(1-p)\left[\frac{\partial^2 v}{\partial x^2} - \mu_2 v\right] + p\left[\frac{\partial^2 v}{\partial x^2} - \mu_2 v + \alpha_1 u \frac{\partial^2 v}{\partial x^2} + \alpha_2 v \frac{\partial^2 v}{\partial x^2}\right] = 0. \quad \text{(A.1)}$$

The approximate solutions of (A.1) are

$$u = u_0 + p u_1 + p^2 u_2 + p^3 u_3 + \cdots,$$
$$v = v_0 + p v_1 + p^2 v_2 + p^3 v_3 + \cdots. \quad \text{(A.2)}$$

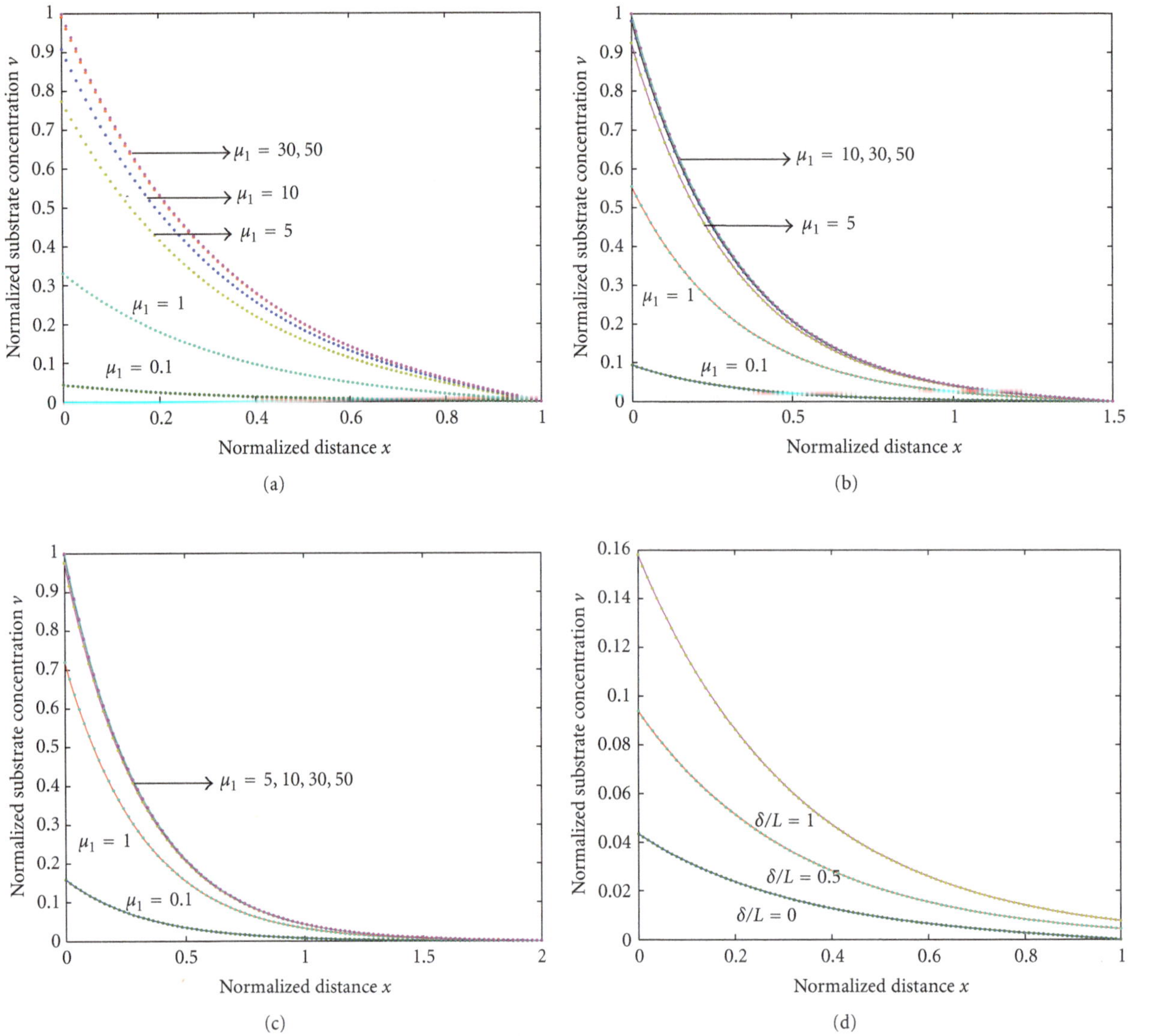

FIGURE 3: The profile of the normalized concentration of catechol substrate v versus the normalized distance x. The concentrations were computed using (12) when the parameters $\mu_2 = 10$, $\alpha_1 = 0.1$, $\alpha_2 = 0.1$ and for various values of μ_1 and δ/L. The key to the graph (stacked line) represents (12) and (dotted line) represents the numerical solution. (a) $\delta/L = 0$, (b) $\delta/L = 0.5$, (c) $\delta/L = 1$, (d) $\mu_1 = 0.1$.

Substituting (A.2) into (A.1) and comparing the coefficients of like powers of p

$$
\begin{aligned}
p^0 &: \frac{\partial^2 u_0}{\partial x^2} - \mu_1 u_0 = 0, \\
p^1 &: \frac{\partial^2 u_1}{\partial x^2} - \mu_1 u_1 + \alpha_1 u_0 \frac{\partial^2 u_0}{\partial x^2} + \alpha_2 v_0 \frac{\partial^2 u_0}{\partial x^2} = 0, \\
p^0 &: \frac{\partial^2 v_0}{\partial x^2} - \mu_2 v_0 = 0, \\
p^1 &: \frac{\partial^2 v_1}{\partial x^2} - \mu_2 v_1 + \alpha_1 u_0 \frac{\partial^2 v_0}{\partial x^2} + \alpha_2 v_0 \frac{\partial^2 v_0}{\partial x^2} = 0.
\end{aligned}
\tag{A.3}
$$

The boundary conditions are

$$
\begin{aligned}
&\frac{\partial u_0}{\partial x} = 0, \quad u_0 + v_0 = 1 \quad \text{when } x = 0, \\
&u_0 = 1, \quad v_0 = 0 \quad \text{when } x = 1 + \frac{\delta}{L} = m, \\
&\frac{\partial u_1}{\partial x} = 0, \quad u_1 + v_1 = 0 \quad \text{when } x = 0, \\
&u_1 = 0, \quad v_1 = 0 \quad \text{when } x = 1 + \frac{\delta}{L} = m.
\end{aligned}
\tag{A.4}
$$

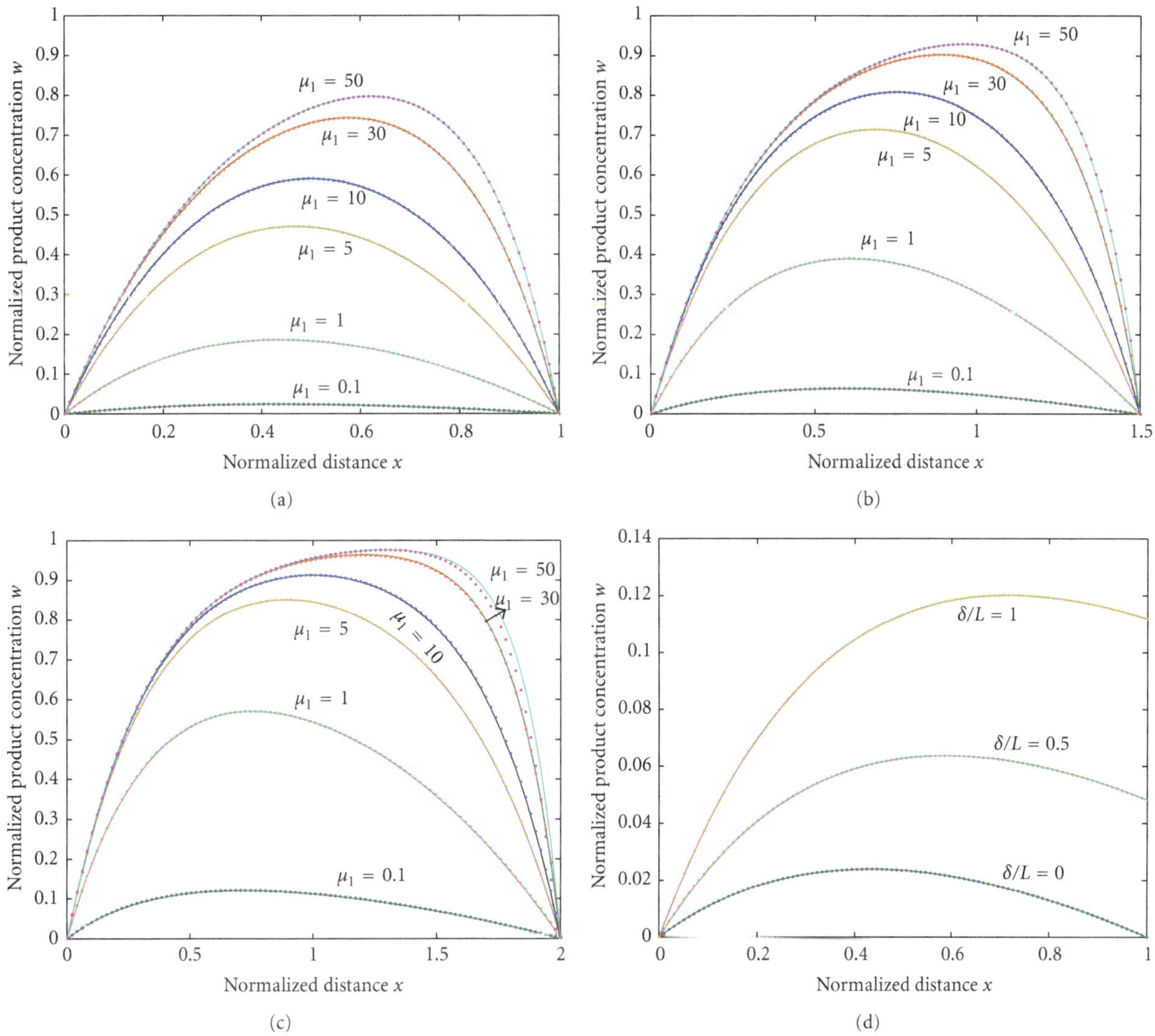

FIGURE 4: The profile of the normalized concentration of *o*-quinone product w versus the normalized distance x. The concentrations were computed using (13) when the parameters $\mu_2 = 10$, $\alpha_1 = 0.1$, $\alpha_2 = 0.1$ and for various values of μ_1 and δ/L. The key to the graph (stacked line) represents (13) and (dotted line) represents the numerical solution. (a) $\delta/L = 0$, (b) $\delta/L = 0.5$, (c) $\delta/L = 1$, (d) $\mu_1 = 0.1$.

Solving (A.3) using the boundary conditions (A.4), we can obtain

$$\begin{aligned}
u_0 &= a\cosh\left(\sqrt{\mu_1}x\right),\\
u_1 &= \alpha_2 A_0\Big[2a\cosh\left(\sqrt{\mu_2}m\right)(\mu_2 - 2\mu_1)\\
&\quad \times \sqrt{\mu_2}\sinh\left(\sqrt{\mu_1}(m-x)\right)\\
&\quad + 4a\mu_1\sqrt{\mu_2}\sinh\left(\sqrt{\mu_1}m\right)\cosh\left(\sqrt{\mu_1}x\right)\\
&\quad - \left(\mu_2 + 2\sqrt{\mu_1\mu_2}\right)\sqrt{\mu_1}\sinh\left(\sqrt{\mu_1}x + \sqrt{\mu_2}(m-x)\right)\\
&\quad - \left(\mu_2 - 2\sqrt{\mu_1\mu_2}\right)\sqrt{\mu_1}\\
&\quad \times \sinh\left(-\sqrt{\mu_1}x + \sqrt{\mu_2}(m-x)\right)\Big]\\
&\quad + \frac{\alpha_1 a^2}{6}\Big[a\cosh\left(\sqrt{\mu_1}x\right)\cosh\left(2\sqrt{\mu_1}m - 3\right)\\
&\quad - \cosh\left(2\sqrt{\mu_1}x\right) + 3\Big],
\end{aligned} \tag{A.5}$$

$$\begin{aligned}
v_0 &= (1-a)b\sinh\left(\sqrt{\mu_2}(m-x)\right),\\
v_1 &= -\alpha_2 A_0\sinh\left(\sqrt{\mu_2}(m-x)\right)\\
&\quad \times \Big(2ab\sqrt{\mu_2}\sinh\left(\sqrt{\mu_1}m\right)\\
&\quad \times \left(\cosh\left(\sqrt{\mu_2}m\right)(\mu_2 - 2\mu_1) + 2\mu_1\right)\\
&\quad - 2\sqrt{\mu_1\mu_2}\Big)
\end{aligned}$$

$$+A_1\Big[2\mu_1\sinh\big(\sqrt{\mu_2}(m-x)\big)$$
$$+4b\sqrt{\mu_1}\sqrt{\mu_2}\sinh\big(\sqrt{\mu_1}m\big)\sinh\big(\sqrt{\mu_2}x\big)$$
$$-\big(\mu_1+2\sqrt{\mu_1}\sqrt{\mu_2}\big)\sinh\big(\sqrt{\mu_1}x+\sqrt{\mu_2}(m-x)\big)$$
$$-\big(\mu_1-2\sqrt{\mu_1}\sqrt{\mu_2}\big)\sinh\big(-\sqrt{\mu_1}x+\sqrt{\mu_2}(m-x)\big)\Big]$$
$$-\alpha_1 a^2 b\sinh\big(\sqrt{\mu_2}(m-x)\big)$$
$$\times\left(\frac{a\big(\cosh\big(2\sqrt{\mu_1}m\big)-3\big)}{6}+\frac{1}{3}\right)$$
$$+A_2\Big[\big(\cosh\big(2\sqrt{\mu_2}m\big)+3\big)\sinh\big(\sqrt{\mu_2}(m-x)\big)$$
$$+4\sinh\big(\sqrt{\mu_2}x\big)-\sinh\big(\sqrt{\mu_2}m\big)$$
$$\times\big(\cosh\big(2\sqrt{\mu_2}(m-x)\big)+3\big)\Big]. \tag{A.6}$$

According to the HPM, we can conclude that

$$u=\lim_{p\to 1}u(x)\approx u_0+u_1, \tag{A.7}$$

$$v=\lim_{p\to 1}v(x)\approx v_0+v_1. \tag{A.8}$$

After putting (A.5) into (A.7) and (A.6) into (A.8), the final results can be described in (11) and (12) in the text.

B. Scilab/Matlab Program to Find the Numerical Solutions for (5), (6), and (7)

```
function pdex4
    m = 0;
    x = linspace(0,1);
    t = linspace(0,100000);
    sol = pdepe(m,@pdex4pde,@pdex4ic,@pdex4bc,x,t);
    u1 = sol(:,:,1);
    u2 = sol(:,:,2);
    u3 = sol(:,:,3);
    figure
    plot(x,u1(end,:))
    title("u1(x,t)")
    xlabel("Distance x")
    ylabel("u1(x,2)")
    figure
    plot(x,u2(end,:))
    title("u2(x,t)")
    xlabel("Distance x")
    ylabel("u2(x,2)")
    figure
    plot(x,u3(end,:))
    title("u3(x,t)")
    xlabel("Distance x")
    ylabel("u3(x,2)")
function [c,f,s] = pdex4pde(x,t,u,DuDx)
    c = [1; 1; 1];
    f = [1; 1; 1].* DuDx;
    M = 100;
    N = 0.1;
    P = 1;
    Q = 0.1;
    F = −M*u(1)/(1 + N*u(1) + Q*u(2));
    F1 = −P*u(2)/(1 + N*u(1) + Q*u(2));
    F2 = −F − F1;
    s = [F; F1; F2];
function u0 = pdex4ic(x);
    u0 = [0; 1; 0];
function [pl,ql,pr,qr] = pdex4bc(xl,ul,xr,ur,t)
    pl = [0; ul(2) – 1 + ul(1); ul(3)];
    ql = [1; 0; 0];
    pr = [ur(1)– 1; ur(2); ur(3)];
    qr = [0; 0; 0].
```

Nomenclature and Units

$[S_1]$: Concentration of the phenol substrate (mole cm^{-3})
$[S_2]$: Concentration of the catechol substrate (mole cm^{-3})
$[P_2]$: Concentration of the *o*-quinone product (mole cm^{-3})
$[S_1]_\infty$: Bulk concentration of the phenol substrate (mole cm^{-3})
$[S_2]_\infty$: Bulk concentration of the catechol substrate (mole cm^{-3})
$[P_2]_\infty$: Bulk concentration of the *o*-quinone product (mole cm^{-3})
$[E_T]$: Enzyme concentration (mole cm^{-3})
L: Thickness of the enzymatic layer (cm)

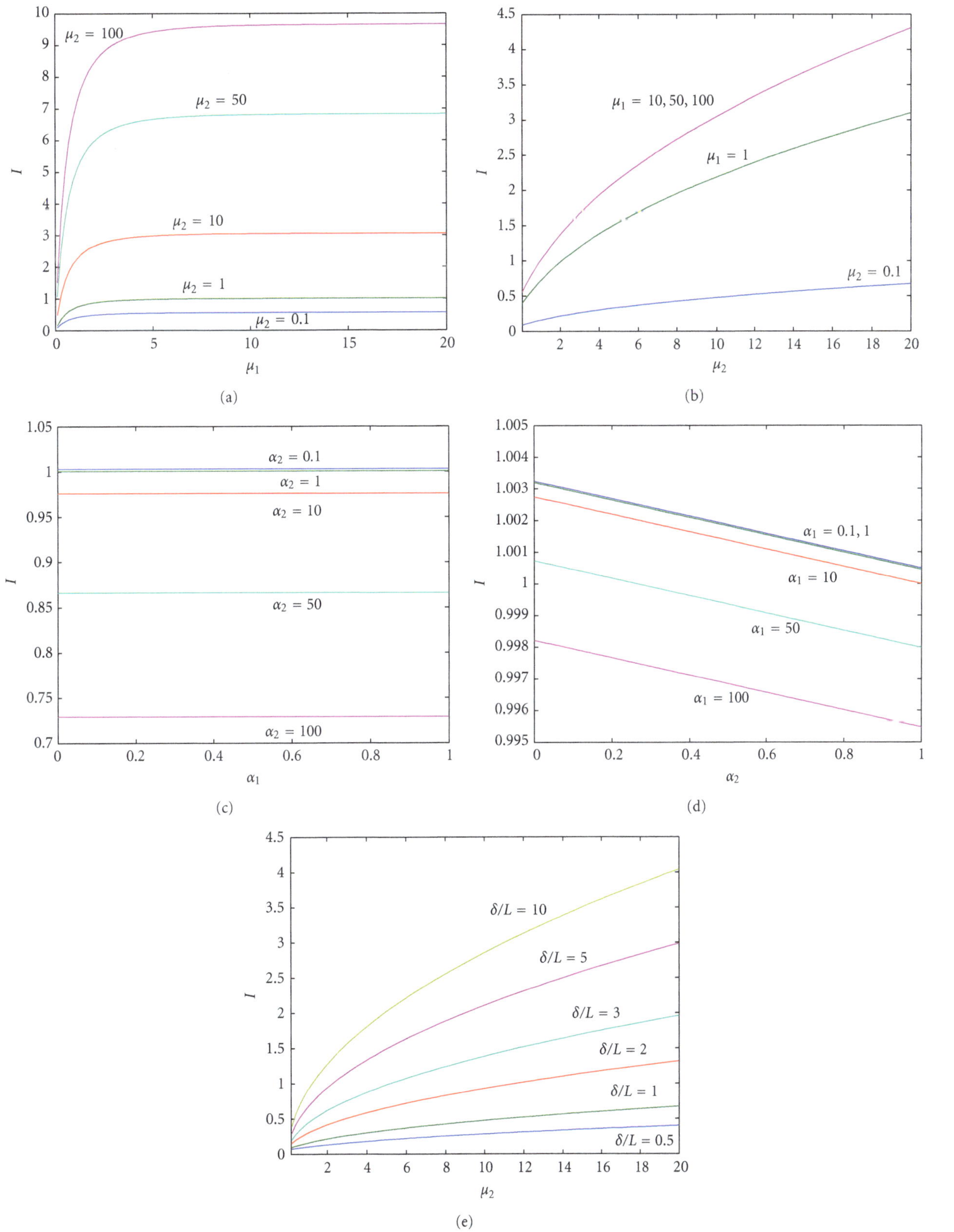

FIGURE 5: The steady state-current I versus dimensionless parameters. The current is computed using (15). (a) $\alpha_1 = 0.1$, $\alpha_2 = 0.1$, $\delta/L = 1$, (b) $\alpha_1 = 0.1$, $\alpha_2 = 0.1$, $\delta/L = 1$, (c) $\mu_1 = 100$, $\mu_2 = 0.01$, $\delta/L = 0$, (d) $\mu_1 = 100$, $\mu_2 = 0.01$, $\delta/L = 0$, (e) $\mu_1 = 0.1$, $\alpha_1 = 0.1$, $\alpha_2 = 0.1$.

δ: Thickness of the diffusion-convection layer (cm)
D_f: Diffusion coefficient in the enzymatic layer ($cm^2\, s^{-1}$)
D_e: Diffusion coefficient in the diffusion convection layer ($cm^2\, s^{-1}$)
Λ_1, Λ_2: Reaction lengths (cm)
K_1, K_2: Kinetic parameters (mole cm^{-3})
k_1, k_2: Kinetic parameters (s^{-1})
μ_1: Ratio of the square of thickness of the enzymatic layer and the reaction length (None)
μ_2: Ratio of the square of thickness of the enzymatic layer and the reaction length (None)
α_1: Dimensionless parameter (None)
α_2: Dimensionless parameter (None)
δ/L: Ratio of the thickness of the enzymatic layer and the thickness of the diffusion-convection layer (None)
F: Faraday constant (None)
u: Normalized concentration of the phenol substrate (None)
v: Normalized concentration of the catechol substrate (None)
w: Normalized concentration of the *o*-quinone product (None)
I: Dimensionless current (None).

Acknowledgments

This work was supported by the University Grants Commission (F. no. 39-58/2010 (SR)), New Delhi, India. The authors are thankful to Dr. R. Murali, The Principal, The Madura College, Madurai, and Mr. M. S. Meenakshisundaram, The Secretary, Madura College Board, Madurai, for their encouragement. They also thank the reviewers for their valuable comments to improve the quality of the paper. The author K. Indira is very thankful to the Manonmaniam Sundaranar University, Tirunelveli, for allowing to do the research work.

References

[1] R. Goldman, O. Kedem, and E. Katchalski, "Papain-Collodion membranes. II. Analysis of the kinetic behavior of enzymes immobilized in artificial membranes," *Biochemistry*, vol. 7, no. 12, pp. 4518–4532, 1968.

[2] P. V. Sundaram and K. J. Laidler, "Kinetic laws for solid-supported enzymes," *Canadian Journal of Biochemistry*, vol. 48, pp. 1498–1504, 1970.

[3] V. Kasche, H. Lundqvist, R. Bergman, and R. Axen, "A theoretical model describing steady-state catalysis by enzymes immobilized in spherical gel particles. Experimental study of α-chymotrypsin-sepharose," *Biochemical and Biophysical Research Communications*, vol. 45, no. 3, pp. 615–621, 1971.

[4] W. J. Blaedel, T. R. Kissel, and R. C. Boguslaski, "Kinetic behavior of enzymes immobilized in artificial membranes," *Analytical Chemistry*, vol. 44, no. 12, pp. 2030–2037, 1972.

[5] J. J. Kulys, "The development of new analytical systems based on biocatalysts," *Analytical Letters*, vol. 14, no. B6, pp. 377–397, 1981.

[6] J. J. Kulys, "Development of new analytical systems based on biocatalysers," *Enzyme and Microbial Technology*, vol. 3, no. 4, pp. 344–352, 1981.

[7] D. A. Gough, J. Y. Lucisano, and P. H. S. Tse, "Two-dimensional enzyme electrode sensor for glucose," *Analytical Chemistry*, vol. 57, no. 12, pp. 2351–2357, 1985.

[8] G. Jobst, I. Moser, and G. Urban, "Numerical simulation of multi-layered enzymatic sensors," *Biosensors and Bioelectronics*, vol. 11, no. 1-2, pp. 111–117, 1996.

[9] S. Bacha, A. Bergel, and M. Comtat, "Transient response of multilayer electroenzymatic biosensors," *Analytical Chemistry*, vol. 67, no. 10, pp. 1669–1678, 1995.

[10] S. Bacha, M. Montagné, and A. Bergel, "Modeling mass transfer with enzymatic reaction in electrochemical multilayer microreactors," *AIChE Journal*, vol. 42, no. 10, pp. 2967–2976, 1996.

[11] R. Baronas, F. Ivanauskas, J. Kulys, and M. Sapagovas, "Hypothesis on the solvability of parabolic equations with nonlocal conditions," *Nonlinear Analysis: Modelling and Control*, vol. 7, pp. 93–104, 2002.

[12] R. Baronas, F. Ivanauskas, J. Kulys, and M. Sapagovas, "Modelling of amperometric biosensors with rough surface of the enzyme membrane," *Journal of Mathematical Chemistry*, vol. 34, no. 3-4, pp. 227–242, 2003.

[13] R. Baronas, F. Ivanauskas, and J. Kulys, "The influence of the enzyme membrane thickness on the response of amperometric biosensors," *Sensors*, vol. 3, no. 7, pp. 248–262, 2003.

[14] R. Baronas, F. Ivanauskas, J. Kulys, and M. Sapagovas, "Modelling of amperometric biosensors with rough surface of the enzyme membrane," *Journal of Mathematical Chemistry*, vol. 34, no. 3-4, pp. 227–242, 2003.

[15] L. Coche-Guerente, P. Labbé, and V. Mengeaud, "Amplification of amperometric biosensor responses by electrochemical substrate recycling. 3. Theoretical and experimental study of the phenol-polyphenol oxidase system immobilized in laponite hydrogels and layer-by-layer self-assembled structures," *Analytical Chemistry*, vol. 73, no. 14, pp. 3206–3218, 2001.

[16] C. D. Mell and J. T. Maloy, "A model for the amperometric enzyme electrode obtained through digital simulation and applied to the immobilized glucose oxidase system," *Analytical Chemistry*, vol. 47, no. 2, pp. 299–307, 1975.

[17] C. D. Mell and J. T. Maloy, "Amperometric response enhancement of the immobilized glucose oxidase enzyme electrode," *Analytical Chemistry*, vol. 48, no. 11, pp. 1597–1601, 1976.

[18] J. J. Kulys, V. V. Sorochinskii, and R. A. Vidziunaite, "Transient response of bienzyme electrodes," *Biosensors*, vol. 2, no. 3, pp. 135–146, 1986.

[19] T. Schulmeister, "Mathematical modelling of the dynamic behaviour of amperometric enzyme electrodes," *Selective Electrode Reviews*, vol. 12, no. 2, pp. 203–260, 1990.

[20] V. V. Sorochinskii and B. I. Kurganov, "Steady-state kinetics of cyclic conversions of substrate in amperometric bienzyme sensors," *Biosensors and Bioelectronics*, vol. 11, no. 3, pp. 225–238, 1996.

[21] K. Yokoyama and Y. Kayanuma, "Cyclic voltammetric simulation for electrochemically mediated enzyme reaction and determination of enzyme kinetic constants," *Analytical Chemistry*, vol. 70, no. 16, pp. 3368–3376, 1998.

[22] R. Baronas, F. Ivanauskas, and J. Kulys, "Modelling a biosensor based on the heterogeneous microreactor," *Journal of Mathematical Chemistry*, vol. 25, no. 2-3, pp. 245–252, 1999.

[23] W. E. Ames, *Numerical Methods for Partial Differential Equations*, Academic Press, New York, NY, USA, 2nd edition, 1977.

[24] L. Coche-Guérente, V. Desprez, J. P. Diard, and P. Labbé, "Amplification of amperometric biosensor responses by electrochemical substrate recycling. Part I. Theoretical treatment of the catechol-polyphenol oxidase system," *Journal of Electroanalytical Chemistry*, vol. 470, no. 1, pp. 53–60, 1999.

[25] L. Coche Guérente, V. Desprez, P. Labbé, and S. Therias, "Amplification of amperometric biosensor responses by electrochemical substrate recycling. Part II. Experimental study of the catechol-polyphenol oxidase system immobilized in a laponite clay matrix," *Journal of Electroanalytical Chemistry*, vol. 470, no. 1, pp. 61–69, 1999.

[26] R. Baronas, J. Kulys, and F. Ivanauskas, "Mathematical model of the biosensors acting in a trigger mode," *Sensors*, vol. 4, no. 4, pp. 20–36, 2004.

[27] J. H. He, "Some asymptotic methods for strongly nonlinear equations," *International Journal of Modern Physics B*, vol. 20, no. 10, pp. 1141–1199, 2006.

New *ab Initio* Potential Energy Surfaces for the Renner-Teller Coupled $1^1A'$ and $1^1A''$ States of CH_2

Haitao Ma,[1] Chunfang Zhang,[1] Zhijun Zhang,[1] Xiaojun Liu,[2] and Wensheng Bian[1]

[1] *Beijing National Laboratory for Molecular Sciences, Institute of Chemistry, Chinese Academy of Sciences, Beijing 100190, China*
[2] *Key Laboratory of Luminescence and Optical Information, Ministry of Education, Institute of Optoelectronic Technology, Beijing Jiaotong University, Beijing 100044, China*

Correspondence should be addressed to Wensheng Bian, bian@iccas.ac.cn

Academic Editor: António Varandas

New *ab initio* potential energy surfaces (PESs) for the two lowest-lying singlet $1^1A'$ and $1^1A''$ electronic states of CH_2, coupled by the Renner-Teller (RT) effect and meant for the spectroscopic study, are presented. The surfaces are constructed using a dual-level strategy. The internally contracted multireference configuration interaction calculations with the Davidson correction, using the *aug-cc-pVQZ* basis set, are employed to obtain 3042 points at the lower level. The core and core-valence correlation effects are taken into account in the *ab initio* calculations with a modified optimized *aug-cc-pCVQZ* basis set for the higher-level points. The analytical representations of these PESs, with the inclusion of the nonadiabatic RT terms, are obtained by the nonlinear least-squares fit of the calculated points to three-body expansion. Quantum dynamical calculations are performed on these PESs, and the computed vibronic energy levels for the two singlet electronic states are in excellent agreement with experiment.

1. Introduction

The CH_2 biradical has been the subject of many theoretical and experimental studies, due to its distinct electronic characteristics and chemical and physical properties. It is the direct chemical precursor of the widely observed CH radical [1]. As a crucial link in the photodissociation sequence of cometary methane, CH_2 plays a significant role in the chemistry of hydrocarbon combustion and the astrophysics of interstellar medium [2–5].

Since Herzberg and Johns [2] carried out a detailed analysis of the high-resolution absorption spectrum of the singlet CH_2 in the near ultraviolet region half a century ago, several groups [6–11] have observed the direct absorption spectra or subsequent monitoring of fluorescence via laser-induced fluorescence, stimulated emission pumping, dispersed fluorescence, and so forth. Experimental investigations of the spectroscopy provided abundant and precise rovibronic energy levels [12–16]. Recently, bands in the CH_2 $\tilde{b}^1B_1$-$\tilde{a}^1A_1$ transition between 12500 and 13000 cm^{-1} were recorded at Doppler-limited resolution utilizing a transient frequency-modulation (FM) laser absorption spectrometer by Chang and coworkers [16]. An unexpected and particularly complicated rovibronic structure was detected for $b(1,1,0)^1$ and $b(0,3,0)^1$ vibronic levels. Comparison of the calculated spectra on the available potential energy surfaces (PESs) [17–19] with the experimental spectra of Herzberg and Johns [2] indicates that the labeling of some vibrational levels is quite inconsistent, which can be ascribed to the local perturbations from the vibrational resonances and Renner-Teller (RT) effect.

The study of the RT effect on the vibronic levels for nonlinear three-atom molecules has been an active area for several decades [8, 17, 20–24]. The $\tilde{b}^1B_1$-$\tilde{a}^1A_1$ transition of CH_2, which is seen in the red and yellow parts of the spectrum, is one of the best examples in which the RT coupling should be observed [5]. One of the consequences of the RT coupling is the inversion of the K_a rotational structure in the lower component state ($\tilde{a}^1A_1$) near the barrier to linearity. As explained by Jungen et al. [25], this reordering of the K_a energy levels results from the transformation at high bending levels from bent to linear configurations [8]. The two lowest-lying singlet electronic states of CH_2, which become a degenerate Δ pair at linearity, interact strongly with each other and are coupled dynamically [17, 24].

Theoretically, some researchers [3, 4, 26] examined a few electronic states of CH_2 with state-of-the-art *ab initio* methods and provided significant clues to the understanding of the electronic structures. Bussery-Honvault and coworkers [27] computed an *ab initio* global PES for the first singlet state of CH_2, where a mixed numerical and analytical method was employed in the PES construction. This surface shows no barrier for the C_{2V} insertion, while a barrier of 4319 cm^{-1} (12.35 kcal/mol) is present for the collinear approach. Unfortunately, visible discrepancies were found between the theoretical calculations on this surface and the experimental results [28]. Later, using the same *ab initio* methodology and similar fitting process, Bussery-Honvault and coworkers [29] constructed a PES for the second singlet state of CH_2; it should be mentioned that their two lowest-lying singlet PESs are not degenerate at linearity. Joseph and Varandas [30] constructed a more accurate PES for the lowest singlet state of CH_2 with the DMBE scaled-external-correlations method [31] and obtained very good agreement with the experimental rate constants, which recommends it for future dynamics studies [30, 32]. Furthermore, Dawes et al. [33] constructed the lowest singlet PES using the local interpolative moving least squares method and performed further spectroscopic calculations, which yielded $J = 0$ vibrational frequencies with a root-mean-square error of a few wavenumbers relative to available measurements.

Liu and co-workers [34] revealed various PES intersection seams among the $1^1A'$, $2^1A'$, $3^1A'$, $1^1A''$ and $2^1A''$, states in the $C(^1D)H_2$ reactive system systematically and determined the minimum energy crossing points (MECPs) accurately. The nonadiabatic interaction near MECPs may play an important role in spectroscopy and dynamics [35–40]. The lowest MECP [34] in $C(^1D)H_2$, which is only 8797 cm^{-1} above the CH_2 $1^1A'$ minimum and much lower in energy than all the other MECPs, is between the two low-lying $\tilde{a}^1A_1$ and $\tilde{b}^1B_1$ states, which become the degenerate Δ_g pair at linearity and hence are strongly coupled by the RT effect. All the other MECPs [34] are above or somewhat below the $C(^1D)+H_2$ asymptote in energy and thus are not expected to intervene most of the vibronic spectra of the two lowest-lying singlet electronic states. However, it is clear that the RT coupling must be taken into account in the vibronic energy level calculations of the two lowest-lying states.

A few PESs for the RT coupled $\tilde{a}^1A_1$ and $\tilde{b}^1B_1$ states have been developed to simulate the vibronic structure and electronic spectrum with the vibronic coupling included [17, 19, 24, 41]. However, the *ab initio* PESs of Green Jr. et al. [17] were empirically adjusted or shifted to fit the experimental data; other PESs [19, 24, 41] were constructed by fitting experimental data and a few *ab initio* points [13]. However, there are therefore accurate and fully *ab initio* PESs for the two lowest-lying singlet electronic states $1^1A'$ and $1^1A''$ of CH_2 with the RT terms required.

So far most of the *ab initio* work concerning CH_2 has been based on the traditional correlated *ab initio* electronic structure calculations, that is, the so-called frozen core approximation, in which correlation effects involving the electrons in *1s* core orbital of carbon are neglected. However, as noted by Peterson and Dunning [42], if the goals of a calculation are to obtain chemical accuracy of thermochemical properties, the effects of correlating the electrons in the core orbital generally must be addressed in the calculations. Of course, *ab initio* calculations including correlation effects of core electrons are very time-consuming. In this work, not only appropriate active space but also an optimized basis set with additional functions for describing core and core-valence correlation effects (called CV) is employed in our *ab initio* calculations, which can also guarantee that the two lowest-lying singlet PESs are degenerate at linearity. We further construct fully *ab initio* PESs for the two lowest-lying singlet electronic states $1^1A'$ and $1^1A''$ of CH_2 with the inclusion of the nonadiabatic RT terms.

The organization of the present article is as follows. Section 2 describes the *ab initio* electronic structure calculations. The fitting of the *ab initio* energy points is presented in Section 3. The fitted PESs of CH_2 (called MZB) and vibronic energy level calculations are discussed in Section 4. Finally, a summary is given in Section 5.

2. Electronic Structure and *ab Initio* Calculations

2.1. Electronic Structure. For computational convenience, the molecule is placed in the *yz* plane, and the electronic configuration of ground state is $(1a_1)^{\uparrow\downarrow}(2a_1)^{\uparrow\downarrow}(1b_2)^{\uparrow\downarrow}(3a_1)^{\uparrow}(1b_1)^{\uparrow}$. The $(1a_1)$ and $(2a_1)$ molecular orbitals (MOs) mostly have carbon *1s* and *2s* characters, respectively. The $(3a_1)$, $(1b_1)$, and $(1b_2)$ MOs mainly have the carbon *2p* character lying along the twofold *z* axis, perpendicular and parallel with respect to the *yz* plane, respectively. In the C_s symmetry, the a_1 and b_2 orbitals become the a' orbitals, and the b_1 and a_2 orbitals become the a'' orbitals.

The electronic configurations of CH_2 can be represented as shown in Table 1 in the C_{2V} or C_s symmetry.

The $\tilde{X}^3B_1$ and $\tilde{b}^1B_1$ states, which share the same spatial orbital configuration with two open-shell electrons parallelly or antiparallelly distributed in the outer $3a_1$ and $1b_1$ orbitals, correspond to the triplet and singlet configurations, respectively. While the $\tilde{a}^1A_1$ and $\tilde{c}^1A_1$ states correspond primarily to configurations with double occupation of the $3a_1$ or $1b_1$ orbital, they could be appropriately described by the two main configuration wave functions. The two CI coefficients C_1 and C_2 for $\tilde{a}^1A_1$ have opposite signs with $|C_1| \geq |C_2|$, while those for the $\tilde{c}^1A_1$ state have the same sign with $|C_3| \leq |C_4|$. Hence, multireference methods are required for an accurate *ab initio* description of the above excited states. The $\tilde{a}^1A_1$ and $\tilde{b}^1B_1$ states become the degenerate $1^1\Delta_g$ pair at linearity and hence are strongly mixed by the RT coupling, which is important to the vibronic calculations. The $\tilde{c}^1A_1$ state correlates with the $1^1\Sigma_g^+$ state at linearity. The next two states, 1^1A_2 and 1^1B_2, become the degenerate $1^1\Pi_u$ pair

Table 1

$\tilde{X}^3B_1:(1a_1)^{\uparrow\downarrow}(2a_1)^{\uparrow\downarrow}(1b_2)^{\uparrow\downarrow}(3a_1)^{\uparrow}(1b_1)^{\uparrow}$	$X^3A'':(1a')^{\uparrow\downarrow}(2a')^{\uparrow\downarrow}(3a')^{\uparrow\downarrow}(4a')^{\uparrow}(1a'')^{\uparrow}$
$\tilde{a}^1A_1:C_1(1a_1)^{\uparrow\downarrow}(2a_1)^{\uparrow\downarrow}(1b_2)^{\uparrow\downarrow}(3a_1)^{\uparrow\downarrow}(1b_1)^{0}+$	$1^1A':C_1(1a')^{\uparrow\downarrow}(2a')^{\uparrow\downarrow}(3a')^{\uparrow\downarrow}(4a')^{\uparrow\downarrow}(1a'')^{0}+$
$C_2(1a_1)^{\uparrow\downarrow}(2a_1)^{\uparrow\downarrow}(1b_2)^{\uparrow\downarrow}(3a_1)^{0}(1b_1)^{\uparrow\downarrow}$	$C_2(1a')^{\uparrow\downarrow}(2a')^{\uparrow\downarrow}(3a')^{\uparrow\downarrow}(4a')^{0}(1a'')^{\uparrow\downarrow}$
$\tilde{b}^1B_1:(1a_1)^{\uparrow\downarrow}(2a_1)^{\uparrow\downarrow}(1b_2)^{\uparrow\downarrow}(3a_1)^{\uparrow}(1b_1)^{\downarrow}$	$1^1A'':(1a')^{\uparrow\downarrow}(2a')^{\uparrow\downarrow}(3a')^{\uparrow\downarrow}(4a')^{\uparrow}(1a'')^{\downarrow}$
$\tilde{c}^1A_1:C_3(1a_1)^{\uparrow\downarrow}(2a_1)^{\uparrow\downarrow}(1b_2)^{\uparrow\downarrow}(3a_1)^{\uparrow\downarrow}(1b_1)^{0}+$	$2^1A':C_3(1a')^{\uparrow\downarrow}(2a')^{\uparrow\downarrow}(3a')^{\uparrow\downarrow}(4a')^{\uparrow\downarrow}(1a'')^{0}+$
$C_4(1a_1)^{\uparrow\downarrow}(2a_1)^{\uparrow\downarrow}(1b_2)^{\uparrow\downarrow}(3a_1)^{0}(1b_1)^{\uparrow\downarrow}$	$C_4(1a')^{\uparrow\downarrow}(2a')^{\uparrow\downarrow}(3a')^{\uparrow\downarrow}(4a')^{0}(1a'')^{\uparrow\downarrow}$
$1^1A_2:(1a_1)^{\uparrow\downarrow}(2a_1)^{\uparrow\downarrow}(1b_2)^{\uparrow}(3a_1)^{\uparrow\downarrow}(1b_1)^{\downarrow}$	$2^1A'':(1a')^{\uparrow\downarrow}(2a')^{\uparrow\downarrow}(3a')^{\uparrow}(4a')^{\uparrow\downarrow}(1a'')^{\downarrow}$
$1^1B_2:(1a_1)^{\uparrow\downarrow}(2a_1)^{\uparrow\downarrow}(1b_2)^{\uparrow}(3a_1)^{\downarrow}(1b_1)^{\uparrow\downarrow}$	$3^1A':(1a')^{\uparrow\downarrow}(2a')^{\uparrow\downarrow}(3a')^{\uparrow}(4a')^{\downarrow}(1a'')^{\uparrow\downarrow}$

at linearity. The spin-orbit coupling between the $\tilde{X}^3B_1$ and singlet states is small in the well region and thus is neglected in this work.

2.2. ab Initio Calculations. Two levels of *ab initio* calculations were performed for the PES construction which involves a dual-level strategy [43]. The electronic energies in the lower-level calculations are calculated with the state-averaged complete active space self-consistent field (SA-CASSCF) and internally contracted multireference configuration interaction (*ic*MRCI) methods [44–47]. The active space consists of six electrons distributed among seven orbitals, which correspond to all valence electrons and valence orbitals and one additional *3s* orbital of carbon which is of *Rydberg* character. The carbon *1s* orbital is not correlated but optimized at the SA-CASSCF level. To get a good description of the two lowest-lying singlet states, especially in view of the importance of correct degeneracy between $1^1A'$ and $1^1A''$ at linear geometries and the influence of PES intersections, we simultaneously consider the five singlet states which correlate with the $C(^1D)+H_2$ asymptote in the present *ab initio* calculations. Consequently, the SA-CASSCF calculations including three and two roots in the A' and A'' symmetries are carried out to obtain the orbitals for further *ic*MRCI calculations. The Davidson correction (denoted as +*Q*) is employed to include the correlation energy due to higher excitations. Dunning's correlation-consistent polarized valence quadruple-zeta basis set augmented with diffuse functions (*aug-cc-pVQZ*) is used.

In the higher-level calculations, the methods and algorithm are the same as the lower-level, but the active space and basis set are different. The active space consists of all electrons distributed among eight orbitals, which include 1s, all valence, and 3s orbital of carbon. The carbon *1s* orbital is correlated and also optimized at the SA-CASSCF level. To obtain the CV contribution to the energies, the *aug-cc-pVQZ* basis set for C atom is modified. The details of this scheme could be found elsewhere [42, 48–51], and only a brief outline will be given here. The seven inner *1s* functions are contracted to two functions using the coefficients from the *aug-cc-pVQZ* basis set. The outer five *s* functions are uncontracted as the six *p* functions. Two tight *d* and *f* functions are added to the *(3d, 2f, 1g)* polarized set given by Peterson and Dunning [42] and Woon and Dunning [48]. The additional functions are even tempered extensions of the valence sets, and the exponents of the added functions are 5.262 and 14.984 for the *d* functions and 4.152 and 12.147 for the *f* functions which are obtained by successively multiplying the corresponding tightest functions by the ratio of the first two compact functions [50, 51]. This core basis set, developed to treat both core and valence correlations, is of the form *(12s6p5d4f1g)*/*[7s6p5d4f1g]* and is designated as *optACVQZ*, which is much smaller than Dunning's standard *aug-cc-pCVQZ* [42, 48, 49] and gives a good description of core electrons [50, 51].

For the *ab initio* calculation of nonadiabatic terms, the SA-CASSCF method is used. The active space is the same as that of the lower-level energy calculations, and the basis set employed here is the uncontracted *aug-cc-pVQZ* basis. The nonadiabatic terms, required for calculating for the RT coupling, are the matrix elements of electronic orbital angular momentum $\hat{L}$, and they are obtained as expectation values over the SA-CASSCF wave functions. The RT coupling takes effect when the molecule approaches linearity, that is, lies on the *z*-axis, hence the matrix elements of $\hat{L}_x$ and $\hat{L}_y$ can be neglected. The details of the RT Hamiltonian have been discussed in another publication [52]. So, three matrix elements of electronic angular momentum (designated as L_z^{ab}, L_{zz}^{aa}, and L_{zz}^{bb}) are calculated, where $\langle \tilde{a}^1A_1 \mid \hat{L}_z \mid \tilde{b}^1B_1 \rangle = -i\hbar L_z^{ab}$, $\langle \tilde{a}^1A_1 \mid \hat{L}_z^2 \mid \tilde{a}^1A_1 \rangle = \hbar^2 L_{zz}^{aa}$, $\langle \tilde{b}^1B_1 \mid \hat{L}_z^2 \mid \tilde{b}^1B_1 \rangle = \hbar^2 L_{zz}^{bb}$.

All *ab initio* calculations reported in the present work were carried out using the MOLPRO2006.1 package of *ab initio* programs [53].

2.3. PES Grid. In order to cover the region of spectroscopic interest with two deep potential wells, *ab initio* calculations at icMRCI(6e, 7o)+Q/AVQZ and *ic*MRCI(8e, 8o)+Q/*optACVQZ* levels were performed at 3042 and 273 symmetry unique geometries, respectively. These were chosen carefully to accurately represent the dynamically important regions, particularly the vicinities of the minimum and near linearity. As pointed out by Liu and co-workers [34], the $1^1\Pi$ pair ($3^1A'$ and $2^1A''$) states conically cross not only the $1^1\Delta$ pair ($1^1A'$ and $1^1A''$) states at around 3.4 bohr, but also the $1^1\Sigma_g^+$ state ($2^1A'$) at 3.1 bohr in the stretching potential curves of linear H–C–H (with one of the C–H bond lengths fixed at 2.1 bohr).

We concentrate on the geometries with the CH bond length smaller than 3.0 bohrs in the process of the present PES construction. Dozens of geometries are also selected for the description of the higher energy regions. In the important regions, points were computed with small increments of 0.1–0.5 bohrs for bond length and 2.0–5.0° for bond angle, while in other regions coarser grids of 1.0-2.0 bohrs and 10.0–20.0° were used. Geometries with energies higher than 100 kcal/mol above the global minimum of $\tilde{a}^1A_1$ were not totally neglected but assigned a very small weight. In the calculations of *ic*MRCI(8e, 8o)/*optACVQZ*, the grids were chosen as the following: 1.4, 1.8, 2.0, 2.05, 2.2, 2.6, and 3.0 bohrs for CH distance; 80, 100, 120, 143, 155, 175, and 180° for ∠HCH. In the calculations for the nonadiabatic coupling terms, almost the same grids were selected as those used in the higher-level calculations.

3. Fit of the Potential Energy Surfaces

To construct the PESs for the $1^1A'$ and $1^1A''$ states of CH_2, we choose three-body expansion functional forms for the analytical representation of the PESs with respect to the internal coordinates (R_1, R_2, θ) using a dual-level strategy, and nonadiabatic coupling terms are also taken into account. The idea of the dual-level strategy is to use two levels of *ab initio* calculations so as to reduce the number of higher-level points needed for fitting [43], and the basic scheme is as follows. First, a set of lower-level *ab initio* points without the CV effect, which are calculated at the *ic*MRCI(6e, 7o)+Q/*AVQZ* level, is generated to construct a zeroth-order PES, V_x^0 ($x = a, b$ stands for the $1^1A'$ and $1^1A''$ states of CH_2, resp.). Then, a set of higher-level points with the CV effect, calculated at the *ic*MRCI(8e, 8o)+Q/*optACVQZ* level, is generated, and the data set of the energy differences of the two levels is used to construct a surface, V_x^{core}. And it should be noted that the energies of the lower level are obtained from V_x^0, instead of the *ab initio* calculations. In addition, the nonadiabatic coupling terms, L_z^{ab}, L_{zz}^{aa}, and L_{zz}^{bb} ($\hbar = 1$ hereafter), are fitted into functions in the third stage for the future calculations. Each of these steps will now be described in more details as follows. The final adiabatic potential V is expressed as

$$V = V_x^0 + V_x^{\text{core}}. \tag{1}$$

By applying the Levenberg-Marquardt technique for the nonlinear optimization, it was found that there are numerical problems for $M \geq 11$ due to near linear dependence of the parameters, and thus quadruple precision arithmetic was needed to obtain convergence [54]. To improve the fit, energy points below 60 kcal/mol relative to the PES minimum are weighted by a factor of 100, and energy points over 100 kcal/mol are weighted by 0.1. To accurately fit the intersection seam of the two lowest-lying singlet PESs at the linear configurations, all the points with ∠HCH larger than 170° are weighted by a factor of 100.

3.1. Analytical Representation of V^0. For the analytical representation of V^0, the calculated energy points were fitted to a three-body expansion in curvilinear coordinates, which are believed to match the shape of the adiabatic PESs. For both electronic states, a Morse-type coordinate

$$\rho = 1 - e^{-\alpha(R-R_0)} \tag{2}$$

was used for the CH stretching mode, where $(R - R_0)$ is the displacement from the equilibrium geometry of the corresponding electronic state. For the angular coordinate, several types of polynomials were tested, including $\cos(\beta\theta)$, $\cos(\beta(\pi-\theta))$, $\cos(\beta(\theta-\theta_0))$, and $[(\theta-\theta_0)+\beta(\theta-\theta_0)^2+\beta(\theta-\theta_0)^3]$. Finally, $\cos(\beta(\pi-\theta))$ was employed to describe the ∠HCH bending:

$$\begin{aligned} V_x^0 &= \sum_{ijk} C_{ijk}^x \rho_1^i \rho_2^j \rho_3^k \\ &= \sum_{ijk} C_{ijk}^x \left[1 - e^{-\alpha_1(R_1-R_0^x)}\right]^i \left[1 - e^{-\alpha_2(R_2-R_0^x)}\right]^j \\ &\quad \times \left[\cos(\beta(\pi-\theta))\right]^k. \end{aligned} \tag{3}$$

The parameters α_1, α_2, β, and the C_{ijk} are determined by performing unequally weighted least squares fit to the *ab initio* data.

3.2. Analytical Representation of V^{core}. The core correlation surface V^{core} was constructed in a similar way to the construction of V^0, and the surfaces can be expressed as

$$\begin{aligned} V_x^{\text{core}} &= C_{000}^x + \exp\left\{\alpha\left[(R_1-R_0^x)^2 + (R_2-R_0^x)^2\right]\right\} \\ &\quad \times \sum_{ijk} C_{ijk}^x \rho_1^i \rho_2^j \rho_3^k \\ &= C_{000}^x + \exp\left\{\alpha\left[(R_1-R_0^x)^2 + (R_2-R_0^x)^2\right]\right\} \\ &\quad \times \sum_{ijk} C_{ijk}^x \left[\frac{(R_1-R_0^x)}{R_0^x}\right]^i \left[\frac{(R_1-R_0^x)}{R_0^x}\right]^j \\ &\quad \times \left[\cos(\theta) - \cos(\theta_0^x)\right]^k. \end{aligned} \tag{4}$$

3.3. Analytical Representation of the Nonadiabatic Coupling Terms. To fit the matrix elements of $\hat{L}_z$ and $\hat{L}_z^2$ into analytical representations, several types of polynomials have been tested, and the following one provides a good description of the nonadiabatic coupling terms, L_z^{ab}, L_{zz}^{aa}, and L_{zz}^{bb}:

$$\begin{aligned} &C_{000} + \cos^2\left(\frac{\theta}{2}\right) \\ &\quad \times \sum_{ijk} C_{ijk}^n \left[1 - e^{-\alpha_1 R_1}\right]^i \left[1 - e^{-\alpha_2 R_2}\right]^j \left[\beta(\pi-\theta)\right]^k. \end{aligned} \tag{5}$$

In the analytical representation above, the $\cos^2(\theta/2)$ guarantees that L_{zz}^{aa} and L_{zz}^{bb} have a correct behavior at linearity. Several types of cos functions instead of $\beta(\pi - \theta)$ were tested as angular coordinate, but very good results were not obtained. $\beta(\pi - \theta)$ is found to be suitable for the description of the bending mode.

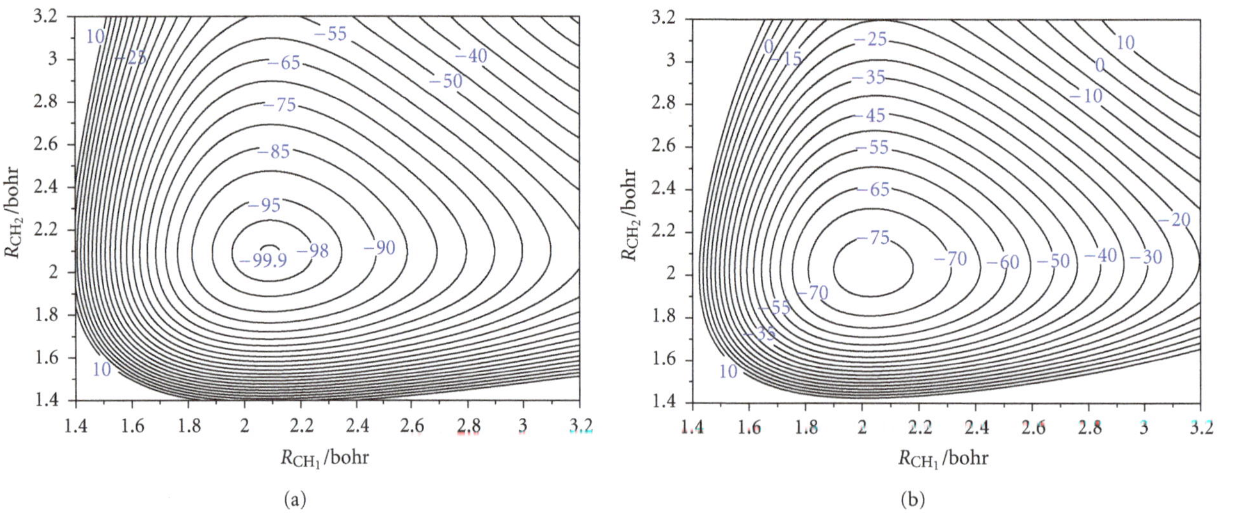

FIGURE 1: (a) Contour plot for the $1^1A'$ PES of CH_2 as a function of R_{CH_1} and R_{CH_2} (bohr) with ∠HCH fixed at the equilibrium value 102.45°. (b) Contour plot for the $1^1A''$ PES of CH_2 as a function of R_{CH_1} and R_{CH_2} (bohr) with ∠HCH fixed at the equilibrium value 144.36°. Energies (kcal/mol) are relative to the $C(^1D)+H_2$ asymptote.

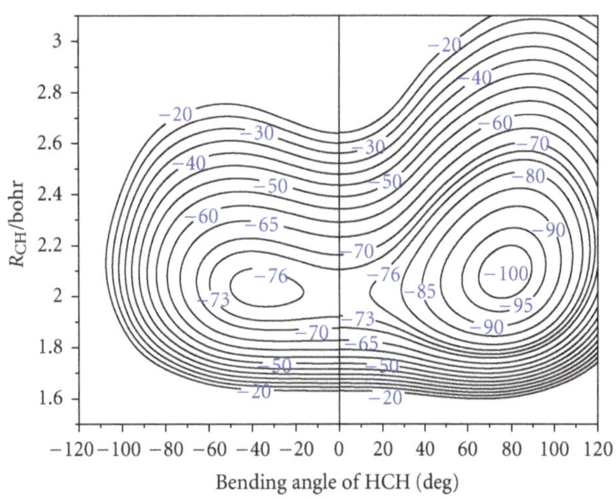

FIGURE 2: Contour plots for the $\tilde{a}^1A_1$ ($\rho \geq 0$) and $\tilde{b}^1B_1$ ($\rho \leq 0$) PESs as functions of R_{CH} and the bending angle of HCH. The intersection seam is shown as a bold line at $\rho = 0$, where ρ is the angle of bending. Energies (kcal/mol) are relative to the $C(^1D)+H_2$ asymptote.

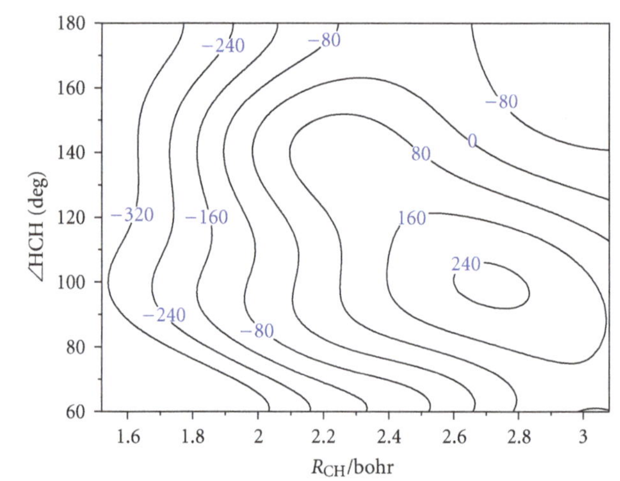

FIGURE 3: Contour plot (cm^{-1}) for V^{core} of the $\tilde{a}^1A_1$ state of CH_2 as a function of R_{CH} and ∠HCH.

Many test calculations were performed with different polynomial orders M, and the dependences are shown in Table S1 (see Table S1 in the Supplementary Material available online at doi: 10.1155/2012/236750). Although, the RMS errors in the fit become smaller as M goes higher, we choose $M = 10$ for the fit of V^0 due to numerical problems. The complete set of parameters amounts to a total of 286 linear coefficients and 3 nonlinear coefficients. The fit for V^0 has RMS errors of 31.88 and 142.88 cm^{-1} for the $1^1A'$ and $1^1A''$ states, respectively. Below 60 kcal/mol, the RMS errors are 6.64 and 5.94 cm^{-1}, respectively. In the fit of V^{core}, M is taken as 6. The complete set of parameters amounts to a total of 85 linear coefficients. The fit for V^{core} has RMS errors of 14.29 and 68.21 cm^{-1} for the $1^1A'$ and $1^1A''$ states, respectively. Below 60 kcal/mol, the RMS errors are 3.15 and 11.89 cm^{-1}, respectively. The numerical values of all parameters to generate the surfaces and coupling terms reported in the present study are presented in Tables S2, S3, and S4.

4. Features of the Fitted Surfaces and Vibronic Energy Level Calculations

4.1. Adiabatic PESs. Figure 1 presents the contour plots for CH bonds stretching of our PESs keeping ∠HCH fixed at 102.45° for the $\tilde{a}^1A_1$ state, 144.36° for the $\tilde{b}^1B_1$ state. We found the fitted potentials to be smooth and without any artificial oscillations. The degeneracy of the two lowest-lying singlet adiabatic potentials is illustrated in Figure 2.

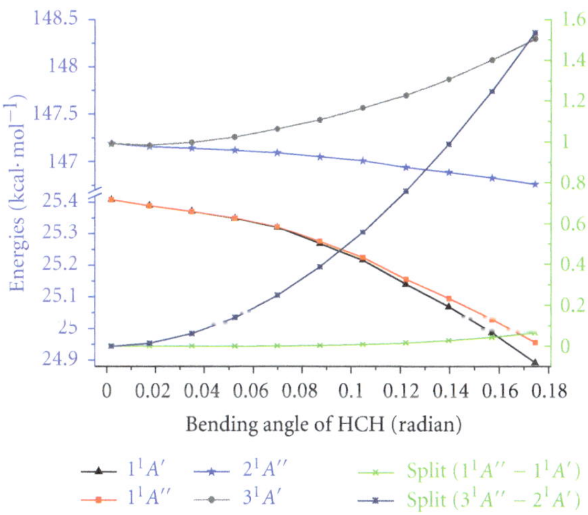

FIGURE 4: The bending potential energy curves for the four singlet states ($1^1A'$, $1^1A''$, $2^1A''$, and $3^1A'$) of CH_2 calculated at the *ic*MRCI(6e, 7o)+Q/*AVQZ* level. (The $2^1A'$ state, which lies between the $1^1A'$ and $3^1A'$ state, is not shown.) $R_{CH_1} = R_{CH_2} = 2.01$ bohrs. The zero energy is taken at the CH_2 ($1^1A'$) minimum.

They are contour plots as functions of R_{CH} and $\angle$HCH. Our fitted PESs for the two singlet states are degenerate at $R_{CH} = (1.6, 2.6)$ bohrs with $\angle\text{HCH} = 180°$. The PES difference at linear geometries in higher energy regions (with energies higher than 60 kcal/mol above the global minimum of $\tilde{a}^1A_1$) is due to the PES fitting error. To further improve the behavior of our PESs at linear geometries, we used an assumed angle dependence switching function to smoothly *connect* the two states above at linearity. Thus, this function is defined as follows:

$$f^{sw} = \frac{1}{2}\cos^4(\pi - \theta)\{1 - \tanh[200(1 + \cos\theta)]\}. \quad (6)$$

Two adiabatic potentials for the $\tilde{a}^1A_1$ and $\tilde{b}^1B_1$ states of CH_2, which are going to be degenerate at linearity, are represented as follows:

$$V_a = (V_a^0 + V_a^{\text{core}}) + \left[\left(V_b^0 + V_b^{\text{core}}\right) - (V_a^0 + V_a^{\text{core}})\right]f^{sw},$$
$$V_b = \left(V_b^0 + V_b^{\text{core}}\right) - \left[\left(V_b^0 + V_b^{\text{core}}\right) - (V_a^0 + V_a^{\text{core}})\right]f^{sw}. \quad (7)$$

From the experimental side, a wide number of studies have led to the determination of accurate equilibrium geometries for the $\tilde{a}^1A_1$ and $\tilde{b}^1B_1$ states of CH_2. Geometries and relative energies of minima obtained from our work along with the available experimental and other theoretical values are given in Table 2. We find the PES minimum of the $\tilde{a}^1A_1$ state on our PESs to be located at $R_{CH} = 2.092$ bohrs and $\angle\text{HCH} = 102.45°$. Obviously, these results are in very good agreement with the experimental value CH_2 available, $R_{CH} = 2.092 \pm 0.004$ bohrs and $\angle\text{HCH} = 102.38°$. Despite the ground singlet state being widely studied, there have been limited research on the first excited singlet state $\tilde{b}^1B_1$.

TABLE 2: Geometries and relative energies of the minima of the two lowest-lying singlet states of CH_2. R_e is the equilibrium interatomic distance, and θ_e is the equilibrium $\angle$HCH.

		Geometries R_e (bohr)	Geometries θ_e (degree)	Relative energies (kcal·mol⁻¹)*
$\tilde{a}^1A_1$	*Ab initio* without core[a]	2.0965	102.10	−100.54
	Ab initio with core[b]	2.0914	102.38	−100.70
	Our PESs[c]	2.0920	102.45	−100.70
	Liu et al.[d]	2.098	102.0	−100.3
	Flores and Gdanitz[e]	2.0917	102.31	
	Bussery-Honvault et al.[f]	2.09	102.5	−99.7
	DMBE[g]	2.09	102.4	−99.75
	Exp.[h]	2.099	102.38	
	Exp.[i]	2.092 ± 0.004	102.4 ± 0.4	
$\tilde{b}^1B_1$	*Ab initio* without core[a]	2.0316	143.12	−77.71
	Ab initio with core[b]	2.0300	144.60	−77.96
	Our PESs[c]	2.0300	144.36	−77.97
	Liu et al.[d]	2.032	143.2	−76.8
	Flores and Gdanitz[e]	2.0165	143.39	
	Bussery-Honvault et al.[f]	2.02	141	−79.9
	Exp.[h]	1.990	140 ± 15	
	Exp.[j]	2.052	139.30	

*Energies are relative to the C(^1D)+H_2 asymptote.
[a]Our *ab initio* values using *ic*MRCI(6e, 7o)+Q/*AVQZ*.
[b]Our *ab initio* values using *ic*MRCI(8e, 8o)+Q/*optACVQZ*.
[c]From our PESs.
[d]*Ab initio* values using *ic*MRCI(6e, 7o)/AVQZ from [34].
[e]*Ab initio* values using *ic*MR-ACPF(8e, 7o)/*ACVQZ* from [55].
[f]PES values from [27, 29].
[g]PES values from [30].
[h]Experimental values from [2].
[i]Experimental values from [11].
[j]Experimental values from [56].

As can be seen, this state is the open-shell singlet analog of the $\tilde{X}^3B_1$ state. As expected, its molecular orbitals and Mulliken atomic distributions are very similar to those of the ground state, which makes the *ab initio* calculation more difficult. The PES minimum for the $\tilde{b}^1B_1$ state on our PESs is located at $R_{CH} = 2.030$ bohrs and $\angle\text{HCH} = 144.36°$. In Figure 3, we have plotted the contours of the core corrections V^{core}, as functions of R_{CH} and $\angle$HCH of the $\tilde{a}^1A_1$ state. The CV effects have a geometry dependence, and the shifts can be positive or negative. The core correlations varies from -400 to $300\,\text{cm}^{-1}$. The inclusion of core and core-valence correlation decreases the bond lengths by 0.0051 and 0.0016 bohrs and increases the bond angle by 0.28 and 1.48° for the $\tilde{a}^1A_1$ and $\tilde{b}^1B_1$ states, respectively.

The $1^1\Delta_g$ state of HCH at linear configuration, splitting into the two lowest-lying singlet electronic excited states

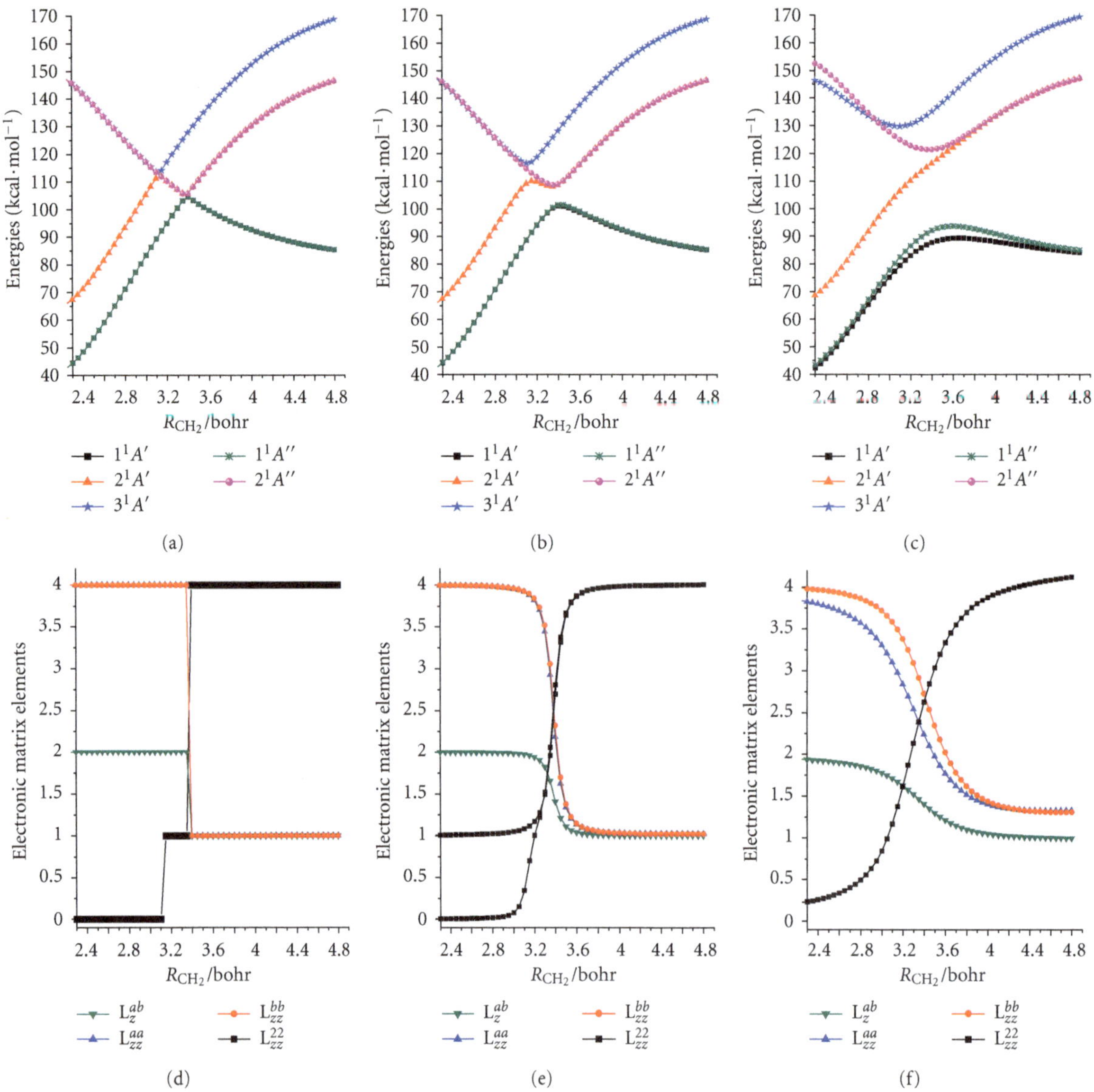

FIGURE 5: The stretching potential curves of the five singlet states ($1^1A'$, $2^1A'$, $3^1A'$, $1^1A''$, and $2^1A''$ state) of CH_2 as functions of the bond length R_{CH_2} with the ∠HCH fixed at (a) 180°, (b) 175°, and (c) 160°, respectively; The Renner-Teller terms L_z^{ab}, L_{zz}^{aa}, L_{zz}^{bb}, and L_{zz}^{22} ($\langle 2^1A' \mid \hat{L}_z^2 \mid 2^1A' \rangle$) as functions of the bond length R_{CH_2} with ∠HCH fixed at (d) 180°, (e) 175°, and (f) 160°, respectively. R_{CH_1} is fixed at 2.09 bohrs, and the zero energy is taken at the CH_2 ($1^1A'$) minimum.

$\tilde{a}^1A_1$ and $\tilde{b}^1B_1$ as the molecule is bent, is a prime example of the vibronic RT effect. In this case, the splitting will be proportional to ρ^n, where $n = 2\Lambda$ and Λ ($\Lambda = 1$ for Π, and 2 for Δ state) is the eigenvalue of $\hat{L}_z$, the axial component of electronic angular momentum at linearity. Although the other two higher excited states ($3^1A'$ and $2^1A''$) calculated with *ic*MRCI(6e, 7o)+Q/*AVQZ* are not involved in the present PESs, it is clearly seen that in this region the *ab initio* computed PECs show the ρ^2 behavior for the doubly degenerate Π pair ($3^1A'$ and $2^1A''$) in Figure 4. The energy splittings between $1^1\Delta_g$ pair and $1^1\Pi_g$ pair states go quartically and quadratically, respectively, when approaching linearity.

While the energy changes with the CH bond stretched, the degeneracy of the $1^1A'$ and $1^1A''$ (or $3^1A'$ and $2^1A''$) of CH_2 is not lifted so long as the molecule is linear. The barrier to linearity plays a very important role in quantum mechanical calculations of vibronic energy levels when the RT effect is considered [18]. The height of the barrier to linearity in $\tilde{a}^1A_1$ CH_2 has been a long standing source of controversy. The range of reported barrier heights for linearity in the $\tilde{a}^1A_1$ state of CH_2 is quite large, varying from 8000 to 10000 cm^{-1}, which is summarized in Table 3. Herzberg and Johns originally gave a value of 8000 cm^{-1}, estimated from the spacing of the $\tilde{b}^1B_1$ bending vibrational levels [2]. Duxbury and Jungen [56] obtained a barrier of

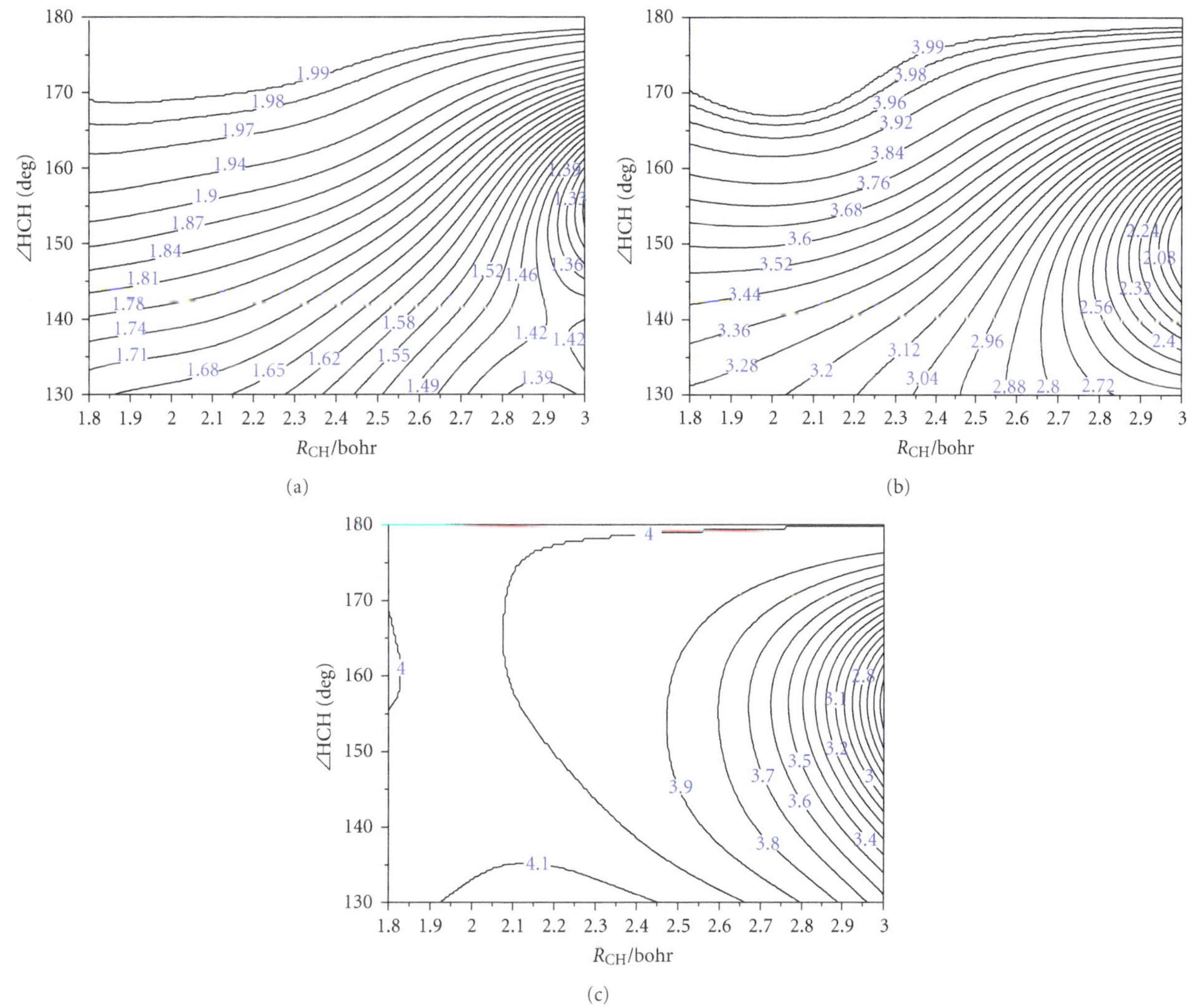

FIGURE 6: Contour plots for the Renner-Teller terms of CH_2 for the two lowest-lying singlet electronic states as functions of R_{CH} and $\angle$HCH: (a) L_z^{ab}, (b) L_{zz}^{aa}, and (c) L_{zz}^{bb}.

9800 cm^{-1} by fitting a bending potential function to the $(0, v_2, 0)$ levels. An earlier *ab initio* calculation gave a barrier height of 9600 cm^{-1} [57]. Kalemos et al. [3] calculated this value to be 9217.7 cm^{-1}. In 2009, the DMBE PES predicted 9644 cm^{-1} in agreement with the experimental determination of 9800 cm^{-1} [56].

Green Jr. et al. [17] calculated an *ab initio* value of 9125 cm^{-1} and empirically adjusted it to about 8800 cm^{-1} according to the visible spectra around 15000 cm^{-1}. And this value is in very good agreement with the derived value of 8600 ± 400 cm^{-1} from the experiment by Hartland et al. [8]. The value of 8666 cm^{-1} was obtained from the PESs constructed by Gu et al. [19] via fitting experimental data and a few *ab initio* points. Liu et al. gave the value as 8797 cm^{-1} based upon the *ic*MRCI(6e, 7o)+Q/*AVQZ* calculations with three reference states in the A' symmetry. Our *ab initio* calculation at the same level with five reference states gives the barrier to linearity as 8895.1 cm^{-1}, but when the core correlation is taken into account, we obtain the *ab initio* calculated barrier to linearity as 8735.8 cm^{-1}. The core correlations reduce this value by 160 cm^{-1}. It may be due to the fact that the *1s* electron is affected when the valence orbitals change from sp^2 hybridization at the minimum area to *sp* hybridization at linearity. Our PESs predict a height of 8715.02 and 760.2 cm^{-1} for the barriers to linearity in the $\tilde{a}^1A_1$ and $\tilde{b}^1B_1$ states of CH_2, respectively.

4.2. Renner-Teller Nonadiabatic Coupling Terms. The fit of the RT nonadiabatic coupling terms has an RMS error of 0.0060, 0.0225, and 0.0148 for L_z^{ab}, L_{zz}^{aa}, and L_{zz}^{bb}, respectively. The polynomial order M is taken as 9. Note that $\langle 1^1A' | \hat{L}_z | 1^1A'' \rangle$ is imaginary and its absolute value is thus used. Figures 5(a), 5(b), and 5(c) describe the stretching potential curves of the five singlet states ($1^1A'$, $2^1A'$, $3^1A'$, $1^1A''$ and $2^1A''$ state) of CH_2 as functions of bond length R_{CH_2}, with R_{CH_1} fixed at 2.09 bohrs and $\angle$HCH at 180°, 175° and 160°, respectively; at linearity, $1^1A'$ and $1^1A''$ states become a degenerate Δ pair. Due to the conical intersection between Π and Δ at R_{CH_2} = 3.38 bohrs, L_z^{ab}, L_{zz}^{aa}, and L_{zz}^{bb} in

TABLE 3: The barrier to linearity of the $\tilde{a}^1A_1$ and $\tilde{b}^1B_1$ states.

Barrier to linearity $\tilde{a}^1A_1$ (cm^{-1})	$\tilde{b}^1B_1$ (cm^{-1})	Note
9073.5	885.2	Our calculation at *ic*MRCI(6e, 7o)/*AVQZ*
8895.1	907.3	Our calculation at *ic*MRCI(6e, 7o)+Q/*AVQZ*
8735.8	706.2	Our calculation at *ic*MRCI(8e, 8o)+Q/*optACVQZ*
8715.1	760.2	Our PESs
8800		The empirically adjusted value based on the visible spectra[a]
8797		*Ab initio* value[b]
8666	725	The fit of empirically spectra and *ab initio* calculations[c]
8600 ± 400		The derived value from experiment[d]
8000		The derived value from experiment[e]
9217.7	953.2	*Ab initio* value[f]
9644		*Ab initio* value[g]
9451.1	1193.0	Others[h]
9750.0 ± 71.0	1616.6 ± 94	Others[i]
9600		Others[j]
9870		Others[k]
9144	879	Others[l]
9356	1049	Others[m]

[a]From Ref. [17]. [b]From [34]. [c]From [8]. [d]From [19]. [e]From [2]. [f]From [3]. [g]From [30]. [h]From [5]. [i]From [56]. [j]From [57]. [k]From [7]. [l]From [41]. [m]From [24].

TABLE 4: The calculated $J = 0$ vibronic energy levels of the $\tilde{a}^1A_1$ and $\tilde{b}^1B_1$ states on our *ab initio* PESs employing the MCTDH method, compared with other theoretical results and experimental values. The energy levels are in cm^{-1}, relative to the zero point energy of the $\tilde{a}^1A_1$ state.

$\tilde{a}/\tilde{b}$	ν_1	ν_2	ν_3	Green Jr. et al.[a]	Gu et al.[b]	Ours	Expt.
$\tilde{a}$	0	1	0	1356	1351.2	1350.9	1352.6[d]
$\tilde{a}$	0	2	0	2675	2664.1	2666.9	2667.7[d]
$\tilde{a}$	1	0	0	2808	2807.5	2808.9	2806.0[e]
$\tilde{a}$	0	0	1	2863	2864.5	2862.5	2865.0[e]
$\tilde{a}$	0	3	0	3962	3945.6	3950.6	3950.5[d]
$\tilde{a}$	1	1	0	4159	4156.5	4150.4	4152.8[f]
$\tilde{a}$	0	4	0	5216	5191.5	5199.2	5196.6[d]
$\tilde{a}$	1	2	0	5452	5437.6	5444.1	5444.9[f]
$\tilde{a}$	2	0	0	5538	5529.3	5529.3	5531.4[f]
$\tilde{a}$	0	5	0	6430	6397.9	6406.8	6403.0[d,f]
$\tilde{a}$	1	3	0		6706.4	6712.0	6714.1[f]
$\tilde{b}$	0	0	0	8383	8354	8349	8350[h]
$\tilde{b}$	0	1	0	9566	9537	9534	9537[h]
$\tilde{b}$	0	2	0	10848	10831	10828	10827[h,i]
$\tilde{b}$	0	3	0	12231	12226	12220	12220[g,i,j]
$\tilde{b}$	0	4	0	13681	13684	13673	13678[g]
$\tilde{b}$	1	2	0	13850	13840	13835	13834[g]
$\tilde{b}$	1	3	0	15116	15121	15106	15114[g]
$\tilde{b}$	0	5	0	15317	15326	15313	15319[g]
$\tilde{b}$	2	2	0	16749	16749	16738	16742[k]
$\tilde{b}$	0	6	0	16929	16948	16934	16941[c,g]
$\tilde{b}$	1	5	0	18186	18201	18182	18192[k]
$\tilde{b}$	0	7	0	18590	18617	18603	18610[c,g]

[a]From [17]. [b]From [19]. [c]From [58]. [d]From [8]. [e]From [11]. [f]From [59]. [g]From [2]. [h]From [10]. [i]From [12]. [j]From [9]. [k]From [60].

Figure 5(d), which shows the dependence of the electronic matrix elements with R_{CH_2} stretching, are almost constants (2, 4, and 4) for $R_{CH_2} < 3.38$ bohrs due to quantization of the electronic angular momenta ($L_z^{ab} = 2$, $L_{zz}^{aa,bb} = 4$ for the Δ state); however, they change rapidly to 1 at $R_{CH_2} \approx 3.38$ bohrs. Almost similar changing could be found in Figures 5(e) and 5(f) with $\angle$HCH = 175 and 160° due to the PESs intersections. And much more interesting rapid changing can also be found in the other three states ($2^1A'$, $3^1A'$, and $2^1A''$), for example, $\langle 2^1A' \mid \hat{L}_z^2 \mid 2^1A' \rangle$, but this is beyond the subject of the present work.

In Figure 6, the contour plots for L_z^{ab}, L_{zz}^{aa}, and L_{zz}^{bb} of the $\tilde{a}^1A_1$ and $\tilde{b}^1B_1$ states of CH_2 as function of CH stretching and $\angle$HCH bending are presented. It is well noticed that along the HCH axis, the values of $L_z^{ab} \simeq 2$ and $L_{zz}^{aa,bb} \simeq 4$ but then as the molecule deviates from linearity the values of L_z^{ab} and L_{zz}^{aa} decrease, and the values of L_{zz}^{bb} increase at the short R_{CH} region but decrease at the long R_{CH} region. The values of L_z^{ab} and $L_{zz}^{aa,bb}$ begin to drop rapidly and monotonically to 1 at $R_{CH} = 3.1$ bohrs due to the Δ/Π PES intersection at linearity. Figure 7 shows the variation of the electronic matrix elements for the two lowest-lying singlet electronic states as functions of the bending angle $\angle$HCH, with the CH-distance optimized for the $\tilde{a}^1A_1$ state. It is shown that, as the molecule bends, the curve of L_z^{ab} decreases monotonically from 2, that of L_{zz}^{bb} rises from 4 basically in a monotonic way, but that of L_{zz}^{aa} first goes down from 4 and then goes up.

4.3. Vibronic Energy Level Calculations. We have calculated the vibronic energy levels of the $\tilde{a}^1A_1$ and $\tilde{b}^1B_1$ states on our *ab initio* PESs employing the block improved relaxation scheme [61, 62] in the multiconfiguration time-dependent Hartree (MCTDH) method [63–65]. The $J = 0$ energy results are listed in Table 4 and are compared with experiments and calculations by other groups. The variation of the electronic matrix elements with geometry is not considered, and other groups also adopted this treatment in previous calculations. For consistency, the energy levels in Table 4 are labeled by the bent molecular notation (ν_1, ν_2, ν_3).

Our calculated results are in excellent agreement with the experimental values, reflecting the accuracy of the constructed *ab initio* PESs. For the energy level of $\tilde{b}(0, 2, 0)$, our result 10828 cm^{-1} is closer to the experimental value 10827 cm^{-1} from Sears et al. [10, 12] than 10823 cm^{-1} from Herzberg and Johns [2]. The experimental results from Sears et al. are more reliable [10, 12, 16]. Compared with the

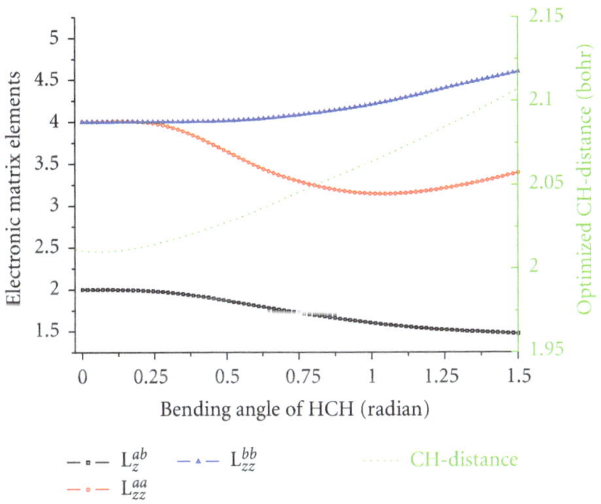

FIGURE 7: The Renner-Teller terms (L_z^{ab}, L_{zz}^{aa}, and L_{zz}^{bb}) of CH_2 for the two lowest-lying singlet electronic states as functions of the bending angle ∠HCH, with the CH-distance optimized for the $\tilde{a}^1A_1$ state.

calculated results from Green Jr. et al., our results are closer to experiment [17]. Generally speaking, our results are in slightly better agreement with the experimental values than the results of Gu et al. [19], and it should be noted that the semiempirical PESs used by Gu et al. were adjusted according to the experimental values while ours are fully *ab initio* ones.

5. Summary

In this work, we report fully *ab initio* PESs for the RT coupled $1^1A'$ and $1^1A''$ states of CH_2 suitable for the spectroscopic study, based on the *ic*MRCI+Q method using the *AVQZ* and a kind of optimized *ACVQZ* basis sets. The core and core-valence correlation effects are included, which are necessary for an accurate quantum chemical description of the CH_2 electronic states. The analytical representations of the two lowest-lying singlet PESs, with the inclusion of the matrix elements of electronic angular momentum L_z^{ab}, L_{zz}^{aa}, and L_{zz}^{bb}, are obtained by fitting. The obtained PESs are smooth, and the two adiabatic potentials are exactly degenerate at linearity. The minimal energy structures and the barriers to linearity predicted by our PESs are in excellent agreement with available experimental data. Furthermore, the MCTDH quantum dynamical calculations are carried out on these new PESs, and the calculated vibronic energy levels are in excellent agreement with the experimental values. Further work on the construction of the global PESs for the $1^1A'$ and $1^1A''$ states, suitable for the $C(^1D)+H_2$ reactive scattering studies, is in progress, and various PES intersections as revealed in our previous work [34], in particular conical intersections, will be included.

Acknowledgments

This work is supported by National Natural Science Foundation of China (nos. 20733005 and 21173232), Chinese Academy of Sciences, and Beijing National Laboratory for Molecular Sciences. The authors would like to thank Professor H. Partridge for useful discussions of the modified basis set *optACVQZ* for core-valence calculations.

References

[1] R. A. Beärda, M. C. Van Hemert, and E. F. Van Dishoeck, "Photodissociation of CH_2. I. Potential energy surfaces of the dissociation into CH and H," *The Journal of Chemical Physics*, vol. 97, no. 11, pp. 8240–8249, 1992.

[2] G. Herzberg and J. W. C. Johns, "The Spectrum and Structure of Singlet CH_2," *Proceedings of the Royal Society of London A*, vol. 295, no. 1441, pp. 107–128, 1966.

[3] A. Kalemos, T. H. Dunning, A. Mavridis, and J. F. Harrison, "CH_2 revisited," *Canadian Journal of Chemistry*, vol. 82, no. 6, pp. 684–693, 2004.

[4] C. D. Sherrill, M. L. Leininger, T. J. V. Huis, and H. F. Schaefer III, "Structures and vibrational frequencies in the full configuration interaction limit: predictions for four electronic states of methylene using a triple-zeta plus double polarization (TZ2P) basis," *The Journal of Chemical Physics*, vol. 108, no. 3, pp. 1040–1049, 1998.

[5] M. N. R. Ashfold, M. A. Fullstone, G. Hancock, and G. Duxbury, "Laser induced fluorescence spectroscopy of the CD_2 ($\tilde{a}^1A_1$) radical: Renner-teller effect in CH_2 and CD_2," *Molecular Physics* , vol. 45, no. 4, pp. 887–896, 1982.

[6] W. Xie, C. Harkin, H. L. Dai, W. H. Green, Q. K. Zheng, and A. J. Mahoney, "Transient vibrational spectroscopy of $\tilde{a}^1A_1$ CH_2 $v_2 = 2$," *Journal of Molecular Spectroscopy*, vol. 138, no. 2, pp. 596–601, 1989.

[7] W. Xie, C. Harkin, and H. L. Dai, "Bending overtones and barrier height of $\tilde{a}^1A_1$ CH_2 by flash photolysis stimulated emission pumping," *The Journal of Chemical Physics*, vol. 93, no. 7, pp. 4615–4623, 1990.

[8] G. V. Hartland, D. Qin, and H. L. Dai, "Renner-Teller effect on the highly excited bending levels of $\tilde{a}^1A_1$ CH_2 ," *The Journal of Chemical Physics*, vol. 102, no. 17, pp. 6641–6645, 1995.

[9] B. C. Chang, M. Wu, G. E. Hall, and T. J. Sears, "Near-infrared vibronic spectrum of the CH_2 $\tilde{b}^1B_1 \leftarrow \tilde{a}^1A_1$ transition," *The Journal of Chemical Physics*, vol. 101, no. 11, pp. 9236–9245, 1994.

[10] K. Kobayashi, L. D. Pride, and T. J. Sears, "Absorption spectroscopy of singlet CH_2 near 9500 cm^{-1}," *The Journal of Physical Chemistry A*, vol. 104, no. 45, pp. 10119–10124, 2000.

[11] H. Petek, D. J. Nesbitt, D. C. Darwin, P. R. Ogilby, C. B. Moore, and D. A. Ramsay, "Analysis of CH_2 $\tilde{a}^1A_1$ (1,0,0) and (0,0,1) Coriolis-coupled states, $\tilde{a}^1A_1$-$\tilde{X}^3B_1$ spin-orbit coupling, and the equilibrium structure of CH_2 $\tilde{a}^1A_1$," *The Journal of Chemical Physics*, vol. 91, no. 11, pp. 6566–6578, 1989.

[12] G. E. Hall, A. V. Komissarov, and T. J. Sears, "Doppler-resolved spectroscopy as an assignment tool in the spectrum of singlet methylene," *The Journal of Physical Chemistry A*, vol. 108, no. 39, pp. 7922–7927, 2004.

[13] A. J. Marr, T. J. Sears, and B. C. Chang, "Near-infrared spectroscopy of CH_2 by frequency modulated diode laser absorption," *The Journal of Chemical Physics*, vol. 109, no. 9, pp. 3431–3442, 1998.

[14] Z. Wang, Y. Kim, G. E. Hall, and T. J. Sears, "State mixing and predissociation in the $\tilde{c} \leftarrow \tilde{a}$ band system of singlet methylene studied by optical-optical double resonance," *The Journal of Physical Chemistry A*, vol. 112, no. 39, pp. 9248–9254, 2008.

[15] K. Kobayashi, G. E. Hall, and T. J. Sears, "The spectrum of CH_2 near 1.36 and 0.92 μm: reevaluation of rotational level structure and perturbations in $\tilde{a}(010)$," *The Journal of Chemical Physics*, vol. 124, no. 18, Article ID 184320, 2006.

[16] C. -H. Chang, Z. Wang, G. E. Hall, T. J. Sears, and J. Xin, "Transient laser absorption spectroscopy of CH_2 near 780 nm," *Journal of Molecular Spectroscopy*, vol. 267, no. 1-2, pp. 50–57, 2011.

[17] W. H. Green Jr., N. C. Handy, P. J. Knowles, and S. Carter, "Theoretical assignment of the visible spectrum of singlet methylene," *The Journal of Chemical Physics*, vol. 94, no. 1, pp. 118–132, 1991.

[18] G. Duxbury, A. Alijah, B. D. McDonald, and C. Jungen, "Stretch-bender calculations of the effects of orbital angular momentum and vibrational resonances in the spectrum of singlet methylene," *The Journal of Chemical Physics*, vol. 108, no. 6, pp. 2351–2360, 1998.

[19] J. P. Gu, G. Hirsch, R. J. Buenker et al., "A theoretical study of the absorption spectrum of singlet CH_2 ," *Journal of Molecular Structure*, vol. 517-518, pp. 247–264, 2000.

[20] R. Renner, "Zur Theorie der Wechselwirkung zwischen Elektronen- und Kernbewegung bei dreiatomigen, stabförmigen Molekülen," *Zeitschrift für Physik*, vol. 92, no. 3-4, pp. 172–193, 1934.

[21] G. J. Halasz, A. Vibok, R. Baer, and M. Baer, "Renner-Teller nonadiabatic coupling terms: an *ab-initio* study of the HNH molecule," *The Journal of Chemical Physics*, vol. 124, no. 8, Article ID 081106, 4 pages, 2006.

[22] G. J. Halász, Á. Vibók, R. Baer, and M. Baer, "D matrix analysis of the Renner-Teller effect: an accurate three-state diabatization for NH_2," *The Journal of Chemical Physics*, vol. 125, no. 9, Article ID 094102, 9 pages, 2006.

[23] S. Zhou, Z. Li, D. Xie, S. Y. Lin, and H. Guo, "An *ab initio* global potential-energy surface for NH_2 (A^2A') and vibrational spectrum of the Renner-TellerA^2A'-X^2A'' system," *The Journal of Chemical Physics*, vol. 130, no. 18, Article ID 184307, 10 pages, 2009.

[24] G. Duxbury, B. D. McDonald, M. Van Gogh, A. Alijah, C. Jungen, and H. Palivan, "The effects of vibrational resonances on Renner-Teller coupling in triatomic molecules: the stretch-bender approach," *The Journal of Chemical Physics*, vol. 108, no. 6, pp. 2336–2350, 1998.

[25] C. Jungen, D. N. Malm, and A. J. Merer, "Analysis of a ${}^1\Delta_u - {}^1\Sigma_g^+$ transition of CS_2 in the near ultraviolet," *Canadian Journal of Physics*, vol. 51, no. 14, pp. 1471–1490, 1973.

[26] B. Ostojić, "The dissociation of singlet methylene," *Journal of Molecular Spectroscopy*, vol. 212, no. 1, pp. 130–131, 2002.

[27] B. Bussery-Honvault, P. Honvault, and J. M. Launay, "A study of the $C(^1D) + H_2 \rightarrow CH + H$ reaction: global potential energy surface and quantum dynamics," *The Journal of Chemical Physics*, vol. 115, no. 23, pp. 10701–10708, 2001.

[28] L. Bañares, F. J. Aoiz, P. Honvault, B. Bussery-Honvault, and J. M. Launay, "Quantum mechanical and quasi-classical trajectory study of the $C(^1D) + H_2$ reaction dynamics," *The Journal of Chemical Physics*, vol. 118, no. 2, pp. 565–568, 2003.

[29] B. Bussery-Honvault, J. Julien, P. Honvault, and J.-M. Launay, "Global 1 1A″ potential energy surface of CH_2 and quantum dynamics of a sideways insertion mechanism for the $C(^1D) + H_2 \rightarrow CH(^2\Pi) + H$ reaction," *Physical Chemistry Chemical Physics*, vol. 7, no. 7, pp. 1476–1481, 2005.

[30] S. Joseph and J. C. Varandas, "Accurate double many-body expansion potential energy surface for the lowest singlet state of methylene," *The Journal of Physical Chemistry A*, vol. 113, no. 16, pp. 4175–4183, 2009.

[31] A. J. C. Varandas, "Intermolecular and intramolecular potentials: topographical aspects, calculation, and functional representation via a double many-body expansion method," *Advances in Chemical Physics*, vol. 74, pp. 255–338, 1988.

[32] S. Joseph, P. J. S. B. Caridade, and A. J. C. Varandas, "Quasiclassical trajectory study of the $C(^1D) + H_2$ reaction and isotopomeric variants: kinetic isotope effect and cd/ch branching ratio," *The Journal of Physical Chemistry A*, vol. 115, no. 27, pp. 7882–7890, 2011.

[33] R. Dawes, A. F. Wagner, and D. L. Thompson, "Ab initio wavenumber accurate spectroscopy: ${}^1CH_2$ and HCN vibrational levels on automatically generated IMLS potential energy surfaces," *The Journal of Physical Chemistry A*, vol. 113, no. 16, pp. 4709–4721, 2009.

[34] X. Liu, W. Bian, X. Zhao, and X. Tao, "Potential energy surface intersections in the $C(^1D)H_2$ reactive system," *The Journal of Chemical Physics*, vol. 125, Article ID 074306, 7 pages, 2006.

[35] H. Zhao, W. Bian, and K. Liu, "A theoretical study of the reaction of $O(^3P)$ with isobutene," *Journal of Physical Chemistry A*, vol. 110, no. 25, pp. 7858–7866, 2006.

[36] N. Matsunaga and D. R. Yarkony, "Energies and derivative couplings in the vicinity of a conical intersection. II. $CH_2(2^3A'',3^3A'')$ and $H_2S(1^1A''2^1A'')$, unexpected results in an ostensibly standard case," *The Journal of Chemical Physics*, vol. 107, no. 19, pp. 7825–7838, 1997.

[37] D. R. Yarkony, "Diabolical conical intersections," *Reviews of Modern Physics*, vol. 68, no. 4, pp. 985–1013, 1996.

[38] J. Ivanic, G. J. Atchity, and K. Ruedenberg, "Violation of the weak noncrossing rule between totally symmetric closed-shell states in the valence-isoelectronic series O_3, S_3, SO_2, and S_2O," *The Journal of Chemical Physics*, vol. 107, no. 11, pp. 4307–4317, 1997.

[39] G. J. Atchity and K. Ruedenberg, "Strong shifts in diabatic nondynamic electron correlations cause conical intersection between low-lying closed-shell adiabatic singlets of like symmetry in ozone," *The Journal of Chemical Physics*, vol. 99, no. 5, pp. 3790–3798, 1993.

[40] H. Ma, X. Liu, W. Bian, L. Meng, and S. Zheng, "A theoretical study of the mechanism and kinetics of $F + N_3$ reactions," *ChemPhysChem*, vol. 7, no. 8, pp. 1786–1794, 2006.

[41] P. Jensen, M. Brumm, W. P. Kraemer, and P. R. Bunker, "A treatment of the Renner effect using the MORBID hamiltonian," *Journal of Molecular Spectroscopy*, vol. 171, no. 1, pp. 31–57, 1995.

[42] K. A. Peterson and T. H. Dunning, "Accurate correlation consistent basis sets for molecular core-valence correlation effects: the second row atoms Al-Ar, and the first row atoms B-Ne revisited," *The Journal of Chemical Physics*, vol. 117, no. 23, pp. 10548–10560, 2002.

[43] J. Cao, Z. Zhang, C. Zhang, W. Bian, and Y. Guo, "Kinetic study on the $H+SiH_4$ abstraction reaction using an *ab initio* potential energy surface," *The Journal of Chemical Physics*, vol. 134, no. 2, 2011.

[44] H. J. Werner and P. J. Knowles, "A second order multiconfiguration SCF procedure with optimum convergence," *The Journal of Chemical Physics*, vol. 82, no. 11, pp. 5053–5063, 1985.

[45] H. J. Werner and W. Meyer, "A quadratically convergent multiconfiguration-self-consistent field method with simultaneous optimization of orbitals and Cl coefficients," *The Journal of Chemical Physics*, vol. 73, no. 5, pp. 2342–2356, 1980.

[46] H. J. Werner and W. Meyer, "A quadratically convergent MCSCF method for the simultaneous optimization of several

states," *The Journal of Chemical Physics*, vol. 74, no. 10, pp. 5794–5801, 1981.

[47] H. J. Werner and P. J. Knowles, "An efficient internally contracted multiconfiguration-reference configuration interaction method," *The Journal of Chemical Physics*, vol. 89, no. 9, pp. 5803–5814, 1988.

[48] D. E. Woon and T. H. Dunning, "Gaussian basis sets for use in correlated molecular calculations. V. Core-valence basis sets for boron through neon," *The Journal of Chemical Physics*, vol. 103, no. 11, pp. 4572–4585, 1995.

[49] A. K. Wilson, T. Van Mourik, and T. H. Dunning, "Gaussian basis sets for use in correlated molecular calculations. VI. Sextuple zeta correlation consistent basis sets for boron through neon," *Journal of Molecular Structure: THEOCHEM*, vol. 388, no. 1–3, pp. 339–349, 1996.

[50] H. Partridge and D. W. Schwenke, "The determination of an accurate isotope dependent potential energy surface for water from extensive *ab initio* calculations and experimental data," *The Journal of Chemical Physics*, vol. 106, no. 11, pp. 4618–4639, 1997.

[51] O. L. Polyansky, A. G. Császár, S. V. Shirin et al., "High-accuracy *ab initio* rotation-vibration transitions for water," *Science*, vol. 299, no. 5606, pp. 539–542, 2003.

[52] Z. Zhang, H. Ma, and W. Bian, "Accurate quantum mechanical study of the Renner-Teller effect in the singlet CH_2 ," *The Journal of Chemical Physics*, vol. 135, Article ID 154303, 10 pages, 2011.

[53] H.-J. Werner, P. J. Knowles, R. Lindh et al., MOLPRO, version 2006.1, a package of *ab initio* programs.

[54] W. Bian and H.-J. Werner, "Global *ab initio* potential energy surfaces for the ClH_2 reactive system," *The Journal of Chemical Physics*, vol. 112, no. 1, pp. 220–229, 2000.

[55] J. R. Flores and R. J. Gdanitz, "Accurately solving the electronic Schrödinger equation of small atoms and molecules using explicitly correlated (r_{12}-)MR-CI. VIII. Valence excited states of methylene (CH_2)," *The Journal of Chemical Physics*, vol. 123, no. 14, Article ID 144316, 8 pages, 2005.

[56] G. Duxbury and C. Jungen, "Effects of orbital angular momentum in CH_2 The Renner-Teller effect," *Journal of Polymer Science B*, vol. 63, no. 6, pp. 981–998, 1988.

[57] D. C. Comeau, I. Shavitt, P. Jensen, and P. R. Bunker, "An *ab initio* determination of the potential-energy surfaces and rotation-vibration energy levels of methylene in the lowest triplet and singlet states and the singlet—triplet splitting," *The Journal of Chemical Physics*, vol. 90, no. 11, pp. 6491–6500, 1989.

[58] H. Petek, D. J. Nesbitt, D. C. Darwin, and C. Bradley Moore, "Visible absorption and magnetic−rotation spectroscopy of $^1CH_2$: The analysis of the $\tilde{b}^1B_1$ state ," *The Journal of Chemical Physics*, vol. 86, no. 3, pp. 1172–1188, 1987.

[59] G. V. Hartland, D. Qin, and H. L. Dai, "Fourier transform dispersed fluorescence spectroscopy: observation of new vibrational levels in the 5000–8000 cm^{-1} region of $\tilde{a}^1A_1$ CH_2 ," *The Journal of Chemical Physics*, vol. 98, no. 3, pp. 2469–2472, 1993.

[60] W. H. Green, I. C. Chen, H. Bitto, D. R. Guyer, and C. B. Moore, "New vibrational bands of CH_2 ($\tilde{b}^1B_1$)," *Journal of Molecular Spectroscopy*, vol. 138, no. 2, pp. 614–629, 1989.

[61] L. J. Doriol, F. Gatti, C. Iung, and H. D. Meyer, "Computation of vibrational energy levels and eigenstates of fluoroform using the multiconfiguration time-dependent Hartree method," *The Journal of Chemical Physics*, vol. 129, no. 22, Article ID 224109, 9 pages, 2008.

[62] H. D. Meyer, F. L. Quéré, C. Léonard, and F. Gatti, "Calculation and selective population of vibrational levels with the Multiconfiguration Time-Dependent Hartree (MCTDH) algorithm," *Chemical Physics*, vol. 329, no. 1-3, pp. 179–192, 2006.

[63] H.-D. Meyer, U. Manthe, and L. S. Cederbaum, "The multiconfigurational time-dependent Hartree approach," *Chemical Physics Letters*, vol. 165, no. 1, pp. 73–78, 1990.

[64] M. H. Beck, A. Jäckle, G. A. Worth, and H. D. Meyer, "The multiconfiguration time-dependent Hartree (MCTDH) method: a highly efficient algorithm for propagating wavepackets," *Physics Report*, vol. 324, no. 1, pp. 1–105, 2000.

[65] G. A. Worth, M. H. Beck, A. Jäckle, and H.-D. Meyer, The MCTDH Package, Version 8.2, 2000, University of Heidelberg, Germany. H.-D. Meyer, Version 8.3, 2002, Version 8.4, 2007, http://mctdh.uni-hd.de.

10

Exploring Multiple Potential Energy Surfaces: Photochemistry of Small Carbonyl Compounds

Satoshi Maeda,[1,2] Koichi Ohno,[3] and Keiji Morokuma[2,4]

[1] The Hakubi Center, Kyoto University, Kyoto 606-8302, Japan
[2] Fukui Institute for Fundamental Chemistry, Kyoto University, Kyoto 606-8103, Japan
[3] Toyota Physical and Chemical Research Institute, Nagakute, Aichi 480-1192, Japan
[4] Department of Chemistry and The Cherry L. Emerson Center for Scientific Computation, Emory University, Atlanta, GA 30322, USA

Correspondence should be addressed to Satoshi Maeda, smaeda@fukui.kyoto-u.ac.jp
and Keiji Morokuma, keiji.morokuma@emory.edu

Academic Editor: Xinchuan Huang

In theoretical studies of chemical reactions involving multiple potential energy surfaces (PESs) such as photochemical reactions, seams of intersection among the PESs often complicate the analysis. In this paper, we review our recipe for exploring multiple PESs by using an automated reaction path search method which has previously been applied to single PESs. Although any such methods for single PESs can be employed in the recipe, the global reaction route mapping (GRRM) method was employed in this study. By combining GRRM with the proposed recipe, all critical regions, that is, transition states, conical intersections, intersection seams, and local minima, associated with multiple PESs, can be explored automatically. As illustrative examples, applications to photochemistry of formaldehyde and acetone are described. In these examples as well as in recent applications to other systems, the present approach led to discovery of many unexpected nonadiabatic pathways, by which some complicated experimental data have been explained very clearly.

1. Introduction

In order to theoretically unravel the entire photochemical reaction processes of a given system, one has to explore and characterize systematically several excited as well as the ground state potential energy surfaces (PESs). In the Franck-Condon (FC) approximation, a reaction starts from the FC point on an excited state PES. Then, the system undergoes either adiabatic or nonadiabatic pathways depending on a given excess energy and topographies on the PESs. In adiabatic paths, a bond reorganization occurs directly on the excited state PES through transition states (TSs). In nonadiabatic paths, on the other hand, the system makes nonadiabatic transition to a lower PES and then reacts on the lower PES. The system may cascade through several PESs via nonadiabatic transitions. Seams of intersection between two PESs often play a key role in various reactions [1–7]; near an intersection seam, nonadiabatic transitions can take place efficiently. When the two states have the same spin and space symmetry, the intersection hyperspace consists of f-2 dimensions and is called "conical intersection (CI)," where f is the number of vibrational degrees of freedom. If the spin multiplicity and/or the space symmetry are different, the PESs cross simply in f-1 dimensional hyperspace. In this paper, both of these two types of intersections, that is, conical intersection and simple intersection seam, are denoted as "seam-of-crossing." In many kinetic studies, minimum energy structures on seam-of-crossing hyperspaces (MSXs) have been searched as critical points of nonadiabatic transitions. In short, to find all possible photochemical reaction pathways of a given system, systematic search for TSs and MSXs is required for all PESs below a given photon energy.

There have been considerable efforts on development of geometry optimization methods for MSXs [8–16] as well as for TSs [17–26]. A variety of geometry optimization techniques, such as the gradient minimization [17, 18], the Berny optimization [19, 20], the eigenvector following [21], the geometry direct inversion in the iterative subspace [22], and the rational function optimization [23], have been a great help to locate the exact saddle point starting from a guessed TS structure. MSX optimization methods can be classified into three types: constrained optimization methods [8–10], the direct gradient (DG) method [11, 12], and penalty function (PF) methods [13, 14]. These methods were compared systematically, and the former two were shown to converge more quickly than a PF method [15]. On the other hand, PF methods have a significant advantage that nonadiabatic coupling derivative vector (CDV) calculations are not required in optimization of MSXs of the conical intersection type. Recently, CDV calculations can be avoided even in constrained optimization methods and in the DG method by using the branching plane updating method [16]. These methods have been successfully applied to many theoretical analyses of photochemical reactions.

In general, geometry optimization requires a good initial guess. However, guesses of MSX structures are sometimes very difficult, because their shapes are often highly deviated from those of stable molecular structures. A similar problem sometimes arises also in the search for TSs on a single PES. To avoid this problem, many automated TS (or minimum energy path) search methods have been developed [27–42]. Although most of these require guesses of a reaction mechanism such as key structures (product and intermediates) [27–34] or chemically relevant collective variables [35–37], some (relatively expensive) methods do not use any guess [38–42]. We also have developed two unique methods without needing any guess: the global reaction route mapping (GRRM) [43–45] method and the artificial force-induced reaction (AFIR) [46–48] method. GRRM can execute global mapping of reaction path networks of unimolecular isomerizations and dissociations on PESs of given atomic composition. AFIR can efficiently explore associative reaction paths among given reactants of multicomponent reactions. Although application of these automatic TS search methods to excited state PESs seems to be straightforward, it had not been successful probably because there are singular points in low-energy regions of excited state PESs due to conical intersections as discussed in Section 2.1. There has been no practical method for the automated MSX search either.

In this paper, we describe a recipe for exploring multiple PESs automatically. Two model functions, F^{AMF} and F^{SMF} [49, 50], are introduced for the TS search and the MSX search, respectively, in combination with the GRRM method. Although, in this study, it is applied to photochemical reactions, it can be applied to other nonadiabatic reactions such as ion-molecule reactions, organometallic reactions, and combustion reactions. Furthermore, the present recipe can adopt, instead of GRRM, any automated reaction path search method such as AFIR. In the present paper, we apply the proposed recipe with the GRRM method to photodissociation mechanisms of small carbonyl compounds such as formaldehyde H_2CO and acetone $(CH_3)_2CO$ [49, 51].

2. Theory

2.1. Avoiding Model Function (AMF) Approach. Figure 1(a) schematically illustrates a one-dimensional cut of two coupled adiabatic PESs. Along the curve for the upper PES, there is a local minimum in the left-hand side and a conical intersection in the right-hand side. In the conical intersection, the adiabatic PES is singular due to the coupling between two diabatic surfaces. To eliminate the singular point from the adiabatic PES, we use the following avoiding model function in automated search:

$$F^{\mathrm{AMF}}(\mathbf{Q}) = \frac{1}{2}\left(E^{\mathrm{State}\ 1}(\mathbf{Q}) + E^{\mathrm{State}\ 2}(\mathbf{Q})\right) + \frac{1}{2}\sqrt{\left(E^{\mathrm{State}\ 1}(\mathbf{Q}) - E^{\mathrm{State}\ 2}(\mathbf{Q})\right)^2 + 4U^2}, \quad (1)$$

$$U(\mathbf{Q}) = \frac{\alpha}{2}\exp\left[-\left(\frac{E^{\mathrm{State}\ 1}(\mathbf{Q}) - E^{\mathrm{State}\ 2}(\mathbf{Q})}{\alpha}\right)^2\right],$$

where $\mathbf{Q}$ is the atomic coordinates $\{Q_i\}$, $E^{\mathrm{State}\ 1}(\mathbf{Q})$ is an adiabatic PES of the target (upper) state, $E^{\mathrm{State}\ 2}(\mathbf{Q})$ is an adiabatic PES of the lower state, and α is a constant parameter. This F^{AMF} is very similar to the well-known equation for the diabatic/adiabatic transformation for two state systems, where $E^{\mathrm{State}\ n}$ should be diabatic energies in the equation. In this study, adiabatic energies were substituted for $E^{\mathrm{State}\ n}$. As seen in Figure 1(b), the model coupling function U in (1) changes the conical intersection region to an avoided crossing-like smooth curve. Thus, any automated search methods that require a smooth PES can be applied to F^{AMF}. TSs on F^{AMF} may be meaningless if the difference between F^{AMF} and $E^{\mathrm{State}\ 1}$ is significant in TS regions, and therefore U was designed to have effects only in limited regions with a small energy gap. Hence, in TS regions F^{AMF} is often very similar to $E^{\mathrm{State}\ 1}$. The accuracy (how well TSs on F^{AMF} reproduce TSs on $E^{\mathrm{State}\ 1}$) depends on the value of α. To our experience, α should be about 1/10 of the vertical excitation energy of target reactions. In this study, we use $\alpha = 30$ kJ/mol throughout.

An example is shown in Figure 1(c) for the direct dissociation channel of H_2CO on the S_1 PES. Starting from S_1-TS, the backward IRC (intrinsic reaction coordinate) reaches S_1-MIN, while the forward IRC crosses S_0/S_1-MSX (conical intersection) in a partially dissociated region. Behind these (true) critical structures, local minima and a saddle point on F^{AMF} (with $\alpha = 30$ kJ/mol) are shown in black. As seen in Figure 1(c), the saddle point structure on F^{AMF} perfectly overlaps with S_1-TS (identical within convergence criteria of geometry optimization: maximum absolute gradient $g_{\mathrm{MAX}} = 6.0 \times 10^{-5}$ hartree Å^{-1}, root mean square gradient $g_{\mathrm{RMS}} = 4.0 \times 10^{-5}$ hartree Å^{-1}, maximum absolute displacement $d_{\mathrm{MAX}} = 3.0 \times 10^{-4}$ Å, root mean square displacement $d_{\mathrm{RMS}} = 2.0 \times 10^{-4}$ Å). S_1-MIN is also reproduced perfectly by the minimum on F^{AMF}. Moreover, on F^{AMF}, S_0/S_1-MSX turned into a local minimum shown in black behind S_0/S_1-MSX. In other words, the conical intersection does not exist on F^{AMF}, which enables exploring the smooth F^{AMF} with an automated reaction path search method.

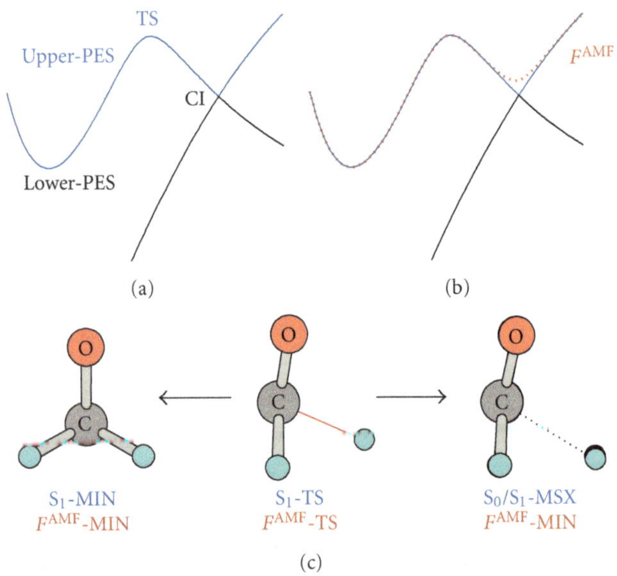

FIGURE 1: (a) A schematic one-dimensional curve of two coupled adiabatic PESs, (b) an example of F^{AMF} for the PESs in (a), and (c) comparisons between stationary points on F^{AMF} (with α = 30 kJ/mol) and those on the S_1-PES of H_2CO (see Section 2.4 for the ab initio calculation level). Each structure shown in (c) corresponds to S_1-MIN, S_1-TS, and S_0/S_1-MSX (conical intersection), respectively, and stationary structures on F^{AMF} are overlapped in black behind them. At S_1-MIN and S_1-TS, the stationary structures on F^{AMF} are identical to them within the convergence criteria (see text) of geometry optimization and cannot be seen. F^{AMF}-MIN can barely be seen behind S_0/S_1-MSX.

In short, TSs on an excited state (adiabatic) PES $E^{\text{State 1}}$ can be explored in two steps: a search for many TS-like structures as first-order saddles on F^{AMF} by an automated search method, optimization of true TSs on the PES $E^{\text{State 1}}$ using the TS-like structures as initial guesses. It should be noted that all TSs shown below are fully optimized first-order saddle points (true TSs) on adiabatic PESs.

2.2. Seam Model Function (SMF) Approach. In the automated search for MSXs, the following model function F^{SMF} is considered [49, 50]:

$$F^{SMF}(\mathbf{Q}) = \frac{1}{2}\left(E^{\text{State 1}}(\mathbf{Q}) + E^{\text{State 2}}(\mathbf{Q})\right) + \frac{\left(E^{\text{State 1}}(\mathbf{Q}) - E^{\text{State 2}}(\mathbf{Q})\right)^2}{\beta}, \quad (2)$$

where $\mathbf{Q}$ is the atomic coordinates $\{Q_i\}$, $E^{\text{State 1}}(\mathbf{Q})$ and $E^{\text{State 2}}(\mathbf{Q})$ are PESs of two target states, and β is a constant parameter. Minimization of F^{SMF} gives a geometry in which both the mean energy of the two states (the first term) and the energy gap between the two states (the second term) are small. Hence, minima on F^{SMF} can be good guesses of MSXs. Equation (2), which is also called penalty function (PF), or more complicated PFs have been employed in MSX geometry optimization [13, 14]. As discussed above, PF methods have a drawback concerned with convergence in MSX optimization [15]. However, in automated searches, use of the PF has a significant advantage that the PF is a single smooth function, which permits the application of automated search methods developed for single PESs without any modification. Hence, we proposed the SMF approach which consists of two steps: (a) a search for many MSX-like structures as local minima on the PF using an automated search method and (b) determination of (true) MSX structures by using one of MSX optimization methods and MSX-like structures obtained in (a) as initial guesses.

Although smaller β gives better (more accurate) candidates, minimizations of F^{SMF} with smaller β need more optimization steps since at $\beta = 0$ the function becomes singular. This is the cause of the slow convergence of a PF method observed in the systematic comparative study. To avoid this, use of relatively large β (typically $\beta \sim$ 30–50 kJ/mol) is recommended since the purpose of the first step (a) is to systematically collect many MSX-like structures as quick as possible. In this study, we use $\beta = 30$ kJ/mol throughout. It should be noted that all MSXs shown below are fully optimized true MSXs (the energy gap between two target states is smaller than 0.1 kJ/mol in all MSXs).

2.3. Global Reaction Route Mapping (GRRM) Method. In the GRRM method [43–45], global mapping of reaction path networks is executed by following the path network itself. To make such a search, one has to follow the paths in both uphill (minimum to saddle) and downhill (saddle to minimum) directions. The latter can be accomplished by following the IRC [52] using one of advanced steepest decent path methods [53–57]. The former, that is, uphill walk, had been difficult before 2004 until we proposed the anharmonic downward distortion following (ADDF) approach.

Many typical potential curves show a common feature that potential energy always become lower than the harmonic potential defined at the bottom of the curves in directions leading to TSs. Thus, reaction channels can be found by following ADD maxima starting from a local minimum on a PES. In many previous applications, the GRRM method based on the ADDF approach has found many unknown as well as nearly all known reaction channels automatically. The ADD maxima can be detected in multidimension by using the scaled hypersphere search (SHS) technique. Details of the SHS technique can be found in our previous papers [43–45].

Following all ADDs starting from all obtained local minima will provide an entire global map on PESs. Such a treatment called full-ADDF, denoted GRRM(f-ADDF), is very expensive, and its application is limited to small systems. An approach following only large ADDs, large-ADDF denoted GRRM(l-ADDF), is effective to survey low-energy regions of PESs in large systems [58]. Another useful option for speedup is double-ended ADDF, denoted GRRM(d-ADDF), which confines the search area between given reactant and product geometries [33, 34].

2.4. Computational Methods Used in Applications. In this section, we summarize quantum chemical methods used in the applications in the following sections. For details, see original articles [49, 51].

In an application to H_2CO, GRRM(f-ADDF) was applied to F^{SMF} at the 3SA-CAS(4e,3o)-SCF/6-31G and 3SA-CAS(2e,3o)-SCF/6-31G levels, where SA stands for state-average and the averaged three states are S_0, S_1, and T_1. All obtained MSX-like structures were fully optimized at the 2S-CAS(12e,10o)-PT2/aug-cc-pVDZ level for the singlet states and at the SS-CAS(12e,10o)-PT2/aug-cc-pVDZ level for the triplet state, where 2S and SS stand for two-state averaged and single-state, respectively. In the state averaging, equal weights were used. All (known) TSs and local minima on these PESs were optimized at the same (CASPT2) levels.

In an application to $(CH_3)_2CO$, GRRM(d-ADDF) was applied to F^{SMF} and F^{AMF} at the 2SA-CAS(8e,7o)-SCF/6-31G and SS-CAS(8e,7o)-SCF/6-31G levels for the S_0, S_1, and T_1 states, respectively, where SS here stands for state-specific. In addition to $(CH_3)_2CO$, three isomers, $CH_2{=}C(OH){-}CH_3$ (enol), $^{\bullet}CH_2{-}CH(O^{\bullet}){-}CH_3$ (biradical), and $CH_3{-}CO{-}CH_3$ (carbene), and a partially dissociated geometry $CH_3CO\cdots CH_3$ were considered as the end points for d-ADDF. All obtained MSX-like structures and TS-like structures were fully optimized at the 2S-CAS(8e,7o)-PT2/6-31+G* level for the singlet states and at the SS-CAS(8e,7o)-PT2/6-31+G* level for the triplet state. Local minima were optimized also at the same (CASPT2) levels.

Potential energies and gradients required for the searches as well as optimization were computed by the Molpro2006 program [59]. Using these quantities, all geometrical displacements were treated by the GRRM11 program [60].

3. Photodissociation of Formaldehyde

It was suggested experimentally that the photodissociation of H_2CO at low-photon energies occurs on the S_0 PES after an internal conversion (IC) from the S_1 PES [61–65]. Hence, dynamics as well as stationary points on the ground state S_0 PES have been studied extensively by theoretical calculations [39, 40, 43–45, 66–89]. The focuses in early studies had been on the following two channels:

$$H_2CO \longrightarrow CO + H_2, \tag{3}$$

$$H_2CO \longrightarrow HCO + H, \tag{4}$$

The TS for the channel (3), which is often called "tight-TS," was already determined in 1974 by ab initio calculations [67]. The radical dissociation (4) can occur both from the S_0 PES and from the T_1 PES, and hence the T_1 PES has also been studied [90, 91]. Knowledge of the higher singlet states was rather scarce until very recently [92, 93]. A new channel for the molecular products was experimentally proposed in 1993 [94]; the channel is often represented as

$$H_2CO \longrightarrow HCO\cdots H \longrightarrow CO + H_2, \tag{5}$$

where one of H atoms, once partially dissociated, roams around the HCO fragment and finally abstracts the other H atom to generate $CO + H_2$. This speculation was confirmed in 2004 by combined experimental and theoretical studies [95]. Since its discovery, this channel named "roaming" has been a hot topic as a new type of reaction mechanism [96–98], and it has been found in many reactions having implications to atmospheric chemistry [99–108] and combustion chemistry [88, 109, 110].

The IC was postulated to occur in the potential well of H_2CO in many previous studies. However, recently, several MSXs were found outside the potential well of H_2CO: two S_0/T_1-MSXs were located in the potential well of hydroxycarbene HCOH [111, 112], a S_0/S_1-MSX was located in a partially dissociated $H\cdots HCO$ region [113]. In 2009, we explored multiple PESs for the S_0, S_1, and T_1 states by the SMF-GRRM approach [49]. The potential energy profile we discovered is presented in Figure 2, where the high-energy regions for $H_2O\cdots C$ and $H_2C\cdots O$ are not shown. The two S_0/T_1-MSXs of HCOH [111, 112] are denoted as S_0/T_1-MSX2 and S_0/T_1-MSX3, and the partially dissociated S_0/S_1-MSX [113] is shown as S_0/S_1-MSX1. As seen in this figure, to isomerize on the T_1 PES or to partially dissociate on the S_1 PES, the system has to overcome a significant barrier. At low-photon energies ($<32\,000\,\mathrm{cm}^{-1} = 383\,\mathrm{kJ\,mol}^{-1}$), these barriers are not accessible. Hence, these known MSXs are not likely involved in the low-energy photodissociation process.

In addition to known MSXs, some new MSX structures were discovered in the automated search. One of those, S_0/T_1-MSX1, was discovered in the potential well of H_2CO. Adopting this new S_0/T_1-MSX1 and known excited state TSs [111, 114], we proposed the entire mechanism for the low-energy photolysis as follows. After the photoexcitation the system oscillates around the S_1-MIN for a long time since there is no S_0/S_1-MSX in the potential well and the lowest S_1/T_1-MSX1 is higher than the photon energies. The S_1/T_1 intersystem crossing (ISC) may be able to take place by trickling down from S_1 to T_1 at all the geometries, while the molecule in the S_1 state spends a long time oscillating around S_1-MIN, because PESs for the S_1 and T_1 states are very close and nearly parallel to each other in energy at any place in the basin of S_1-MIN. Although the probability at each geometry may be small because the spin-orbital coupling between the two states belonging to the same electronic configuration is small, the integrated probability over a long time could be substantial. Once the system comes down to T_1 from the S_1-MIN basin region, the T_1/S_0 ISC via the newly found S_0/T_1-MSX1 ($390\,\mathrm{kJ\,mol}^{-1}$) can take place in this H_2CO basin. This mechanism based on the CASPT2 energetics is consistent with the result obtained by the MRCISD(Q)/AV5Z calculations, where relative energy values of the three important barriers, that is, the S_1 (C–H bond scission) barrier, the T_1 barrier, and S_0/T_1-MSX1, are 456.4, 413.1 and 397.7 kJ/mol, respectively, at the MRCISD(Q)/AV5Z level [114]. After the transition to S_0, the dynamics on the S_0 PES should start in the H_2CO basin and propagate via the tight-TS or the roaming pathway to give the $CO + H_2$ product.

After the proposal of this mechanism, direct CASSCF dynamics studies have been conducted for the molecular dissociation at medium excitation energies [115, 116]. By these simulations, molecular dissociations involving an ISC or a direct IC in the partially dissociated $H\cdots HCO$ region were proposed. These paths undergo a C–H bond dissociation

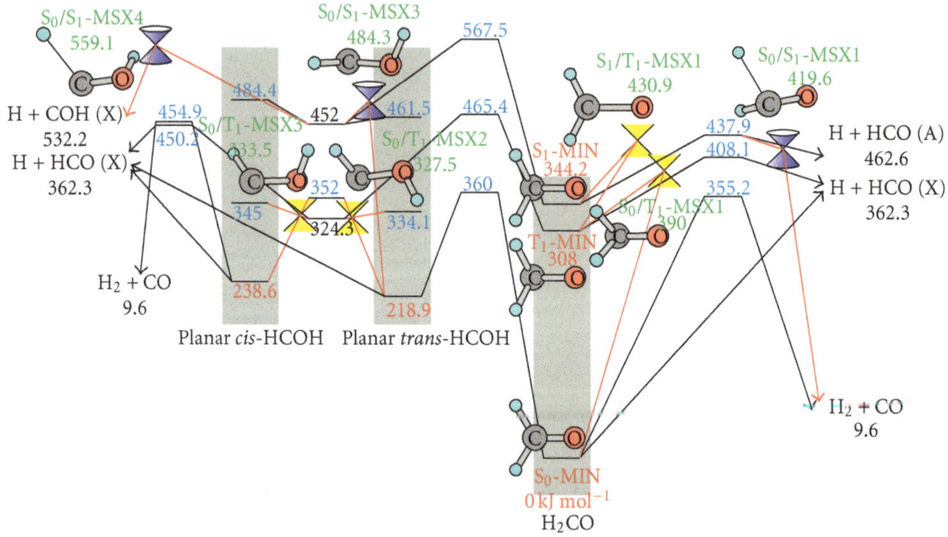

FIGURE 2: A CASPT2 potential energy profile (in kJ/mol) for the three low lying S_0, S_1, and T_1 states of H_2CO [49]. Reprinted with permission from [49]. Copyright 2009 American Chemical Society.

on either the T_1 or S_1 PESs. Obviously, the S_1 dissociation needs sufficient (medium) photon energies to overcome the S_1 barrier. The other path involving the partial dissociation on the T_1 PES may happen via tunneling or zero-point-energy effect even at the low photon energies. However, this (molecular dissociation involving the T_1 dissociation) was shown to be a very minor process in recent three-state trajectory surface hopping (3S-TSH) simulations including the S_0, S_1, and T_1 states [117]. In the 3S-TSH simulations, highly accurate (analytically fitted) PESs were employed and the hopping was treated by Tully's fewest switches algorithm. The 3S-TSH simulations demonstrated that the above decay mechanism involving the S_1-T_1 trickling down mechanism followed by the T_1-S_0 ISC through S_0/T_1-MSX1 in the potential well of H_2CO is the major process before the molecular dissociation. Furthermore, they discovered a new unexpected (but minor) dynamics in which the system once decays down to the S_0 PES and then isomerize to HCOH on the S_0 PES before the dissociation. Surprisingly, in such trajectories, the system hops up to the T_1 PES and then hops down to the S_0 PES through S_0/T_1-MSX1 or S_0/T_1-MSX2. This is energetically certainly allowed in Figure 2.

4. Photodissociation of Acetone

Another example is an application to $(CH_3)_2CO$ [51]. Photolysis of acetone is one of the most extensively studied photochemical reactions [118]. The CC bond dissociation, called "Norrish type I reaction" $(CH_3)_2CO \rightarrow CH_3CO + CH_3$, has been the main focus of many theoretical and experimental studies [119–133]. Roaming was also suggested experimentally [102], which was proposed to proceed on the S_0 PES as $(CH_3)_2CO \rightarrow CH_3CO \cdots CH_3 \rightarrow CO + CH_3\text{–}CH_3$.

In early experimental studies [119–121], it was proposed that the Norrish type I dissociation occurs on the T_1 PES (the T_1 path) after an ISC from the S_1 PES, as long as the photon energy exceeds the barrier of the CC bond breaking on the T_1 PES. A theoretical calculation at the CASSCF level indicated that the ISC happens around a S_1/T_1-MSX structure located below the T_1 barrier, and the T_1 path was suggested to be the most preferable channel (at least in terms of potential energy) [122]. However, in later combined femtosecond laser experimental and theoretical studies [123–125], it was suggested that the dissociation occurs through the S_1 barrier (the S_1 path) before the ISC takes place when photon energies are larger than the S_1 barrier. The S_1 path was then confirmed by another femtosecond laser study at 195 nm (613 kJ/mol) [126–128]. Relevant stationary structures for the S_1 path were examined by CASPT2 and MRCISD(Q) calculations [132]. We revisited [51] the PESs of $(CH_3)_2CO$ for two purposes: to find nonadiabatic paths to the S_0 PES systematically for the roaming channel and to explain the conflict between the fast S_1/T_1 ISC proposed by CASSCF calculations and the experimentally observed S_1 path which overcomes the S_1 barrier much higher than the CASSCF S_1/T_1-MSX.

Figure 3 presents a potential energy profile for the three low lying S_0, S_1, and T_1 states. For isomerizations on the S_1 and T_1 PESs, only the lowest channels for the $(CH_3)_2CO \leftrightarrow CH_2{=}C(OH)\text{–}CH_3$ reaction (S_1-TS2 and T_1-TS2) are shown in the profile. As seen in this figure, these excited state isomerizations have higher barriers and must be minor channels compared to the CC bond dissociations (S_1-TS1 and T_1-TS1). We should note that the S_1/T_1-MSX is much higher in energy than T_1-TS1 on the CASPT2 PESs in contrast to the case of the CASSCF calculations reported previously. Hence, if photon energy is higher than the S_1 barrier (S_1-TS1), the most favorable path is the direct dissociation. This explains the recent femtosecond laser experimental observation very well. Below the S_1 barrier, there is no exit channel from the S_1 PES. Hence and also because of very small spin-orbit

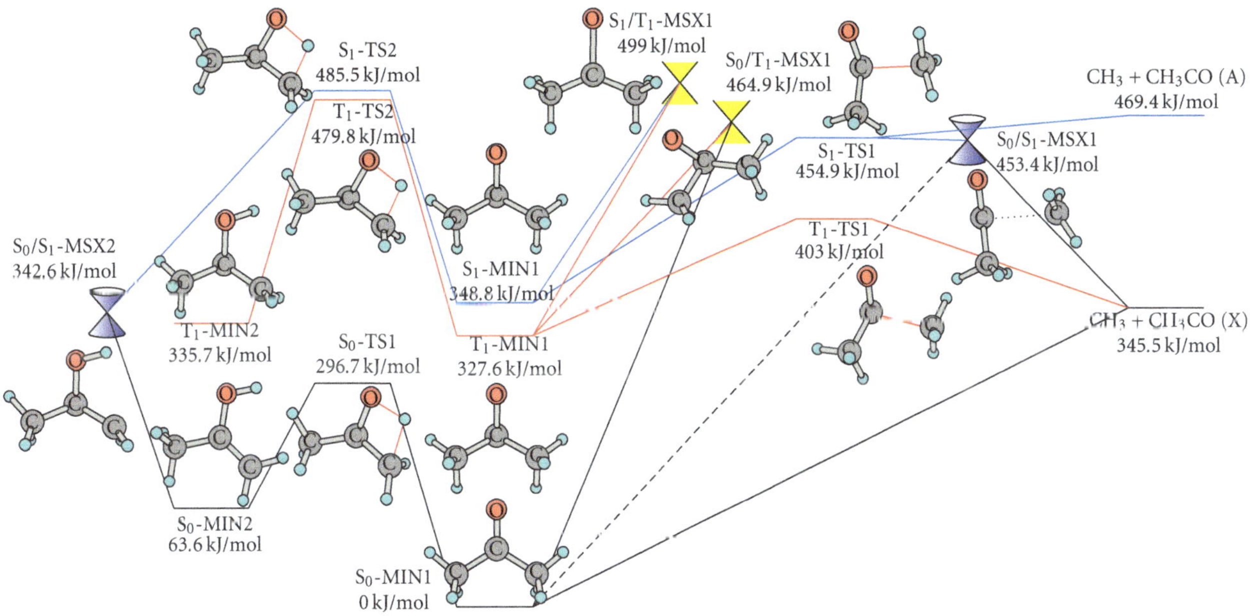

FIGURE 3: A CASPT2 potential profile for the three low lying states (S_0, S_1, and T_1) of $(CH_3)_2CO$ [51]. Reprinted with permission from [51]. Copyright 2010 American Chemical Society.

coupling between the S_1 and T_1 states, we proposed that the ISC occurs very slowly by trickling down from S_1 to T_1 while the molecule in the S_1 state spends a long time oscillating around S_1-MIN1. This may be able to happen because PESs for the S_1 and T_1 states are very close and nearly parallel to each other in energy at any place in the basin of S_1-MIN1, similarly to the case of H_2CO. There are some experimental evidences showing that the S_1/T_1 ISC is very slow [134]; recent time-resolved mass spectrometry and photoelectron spectroscopy experiments and direct dynamics simulations suggested that the system decays from the FC region to the S_1 minimum within 30 femtoseconds and then stays on the S_1 PES more than 100 picoseconds at 253–288 nm (473–415 kJ/mol) excitations [135, 136]. After coming down to the T_1 PES, the system can overcome the T_1 barrier to undergo the Norrish type I reaction.

In Figure 3, several decay paths can be found from the S_1 PES to the S_0 PES. Unlike H_2CO, S_0/T_1-MSX1 in the potential well of $(CH_3)_2CO$ is much higher in energy than T_1-TS1. The lowest path to the S_0 PES undergoes a dissociation on the S_1 PES through S_1-TS1 followed by a nonadiabatic transition in the nearly dissociated geometry through S_0/S_1-MSX1 and a recombination on the S_0 PES. The roaming dynamics was observed at 230 nm (520 kJ/mol) [102], at which the direct CC bond dissociation on the S_1 PES is the dominant path of the Norrish type I reaction. Hence, this is the most likely path to produce the ground state $(CH_3)_2CO$. At this photon energy, the excited state isomerization to the enol species followed by an IC through S_0/S_1-MSX2 and an enol-keto isomerization on the S_0 PES may be a minor channel to the ground state $(CH_3)_2CO$ before the roaming dynamics may take place.

5. S_0/T_1 Intersection Space

One can find candidates of saddle points on seam-of-crossing hyperspaces (SPSXs) as saddles on F^{SMF}. Minimum energy paths on seam-of-crossing hyperspaces (MEPSXs) starting from SPSXs were suggested to be useful for understanding dynamical trajectories crossing a high-energy region of a seam-of-crossing hyperspace [12]. In this section, SPSXs and MEPSXs for the S_0/T_1 intersection spaces of H_2CO and $(CH_3)_2CO$ are discussed because of their significance in the generation of the S_0 species. The optimization method for SPSXs from SMF (or guessed) structures and an algorithm to calculate an MEPSX from a SPSX are described in our previous paper [16].

Figure 4(a) shows a MEPSX for the S_0/T_1 intersection space of H_2CO at the CASPT2 level. The MEPSX was calculated from two SPSXs. Along the MEPSX, all points have the S_0/T_1 energy gap less than 1.0 kJ/mol. As seen in this figure, the S_0/T_1 intersection space was found to be connected from the HCOH area through the S_0/T_1-MSX1 geometry to the partially dissociated region. It follows that the S_0/T_1 ISC may take place not only at S_0/T_1-MSX1 but also with variety of geometries in this very wide intersection space.

Integration of an MEPSX with CASPT2 is expensive, and we also tested and used the UB3LYP method using Gaussian09 program [137]. At first, the MEPSX for H_2CO by the UB3LYP method is shown in Figure 4(b). For H_2CO the MEPSX by UB3LYP is similar to the one by the CASPT2 method. Therefore, we adopted the UB3LYP method for $(CH_3)_2CO$. Figure 5 shows two MEPSXs (a) for $H_2C{=}C(OH)CH_3 \leftrightarrow (CH_3)_2CO \leftrightarrow CH_3CO + CH_3$ and (b) for $CH_3{-}CO{-}CH_3 \leftrightarrow (CH_3)_2CO \leftrightarrow CH_3CO + CH_3$. Another MEPSX from an MSX of the $^{\bullet}CH_2{-}CH(O^{\bullet}){-}CH_3$ form (not shown) is directly connected to the partially dissociated

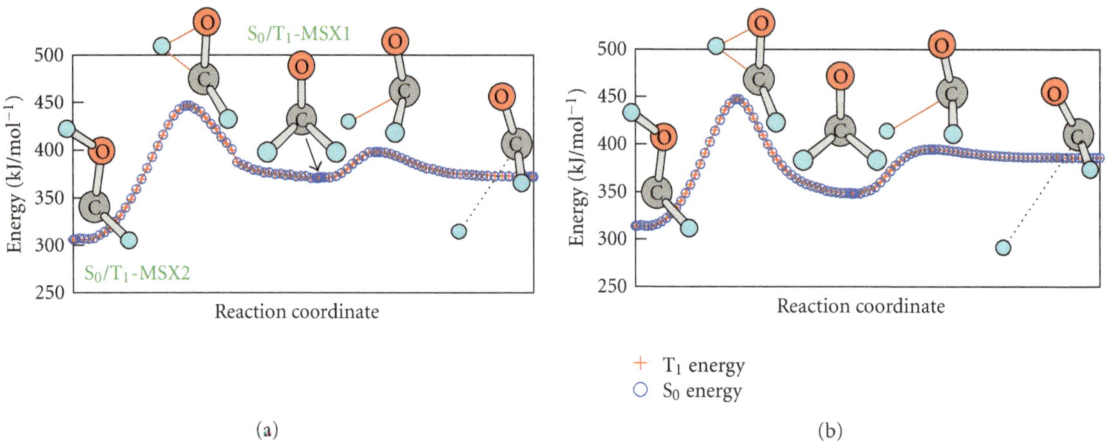

FIGURE 4: (a) A S_0/T_1-MEPSX curve for HCOH ↔ H_2CO ↔ HCO + H at the CASPT2 level (see Section 2.4 for details of the CASPT2 calculations) and (b) a S_0/T_1-MEPSX curve for the same reaction at the UB3LYP/aug-cc-pVDZ level.

$CH_3CO \cdots CH_3$ region through only one SPSX structure, that is, it does not enter the potential well of $(CH_3)_2CO$. In contrast to Figure 4, these intersection spaces lie in a very high-energy region in the potential well of $(CH_3)_2CO$. Only outside the potential well, the intersection space comes down to a low-energy region. Thus, the S_0/T_1 ISC is expected to be much slower in $(CH_3)_2CO$ than the case of H_2CO, although $(CH_3)_2CO$ also has a very wide S_0/T_1 intersection space.

6. A Merit of Automated Search

The most significant advantage of the present approach is that unexpected pathways can be discovered. Geometry optimization has been a very powerful mean to locate exact MSXs and TSs starting from guessed geometries. However, a guess structure leads only to a TS or a MSX anticipated from it. On excited state PESs, molecules frequently take unexpected geometries that are very different from those of stable ground state molecules. Hence, the advantage of the present method without guess is very helpful in analyses of photochemical reactions.

We already reported some unexpected paths discovered by the present recipe. The first is the S_0/T_1 ISC path for H_2CO as discussed above [49]. The second is an H-atom roundtrip path discovered in photolysis of methyl-ethyl-ketone $CH_3C(O)C_2H_5$ [138]. In this path, on the S_1 PES, an H atom in the ethylgroup at first transfers to the O atom to undergo an S_0/S_1 IC in a conical intersection of a diradical isomer, and then the H atom comes back to the original position on the S_0 PES, and finally the molecule dissociates into $CH_3CO + C_2H_5$ or $C_2H_5CO + CH_3$ on the S_0 PES. This path explained an experimental photodissociation quantum yields measurements very well. The third is the discovery for the first time of the excited state roaming pathway, which we found in photolysis of NO_3 [107, 108]. Here, one of the O atoms in NO_3 partially dissociates on the first excited doublet (D_1) PES and then roams around the NO_2 fragment before producing O_2 + NO on either the D_1 PES or the ground state D_0 PES. This mechanism also explained very well recent experimental results giving two channels, one with hot O_2 and the other with cold O_2 [105, 106]. Unexpected nonadiabatic ignition paths of unsaturated hydrocarbons were also discovered by the SMF-AFIR approach very recently [50].

7. Conclusion and Perspective

We reviewed our recipe to explore multiple PESs systematically by using an automated reaction path search method which has been used for the ground state (smooth) PESs [49, 50]. In this paper, it was applied together with the GRRM method [43–45] to small carbonyl compounds to study their photodissociation mechanisms. In an application to H_2CO [49], we discovered a new nonadiabatic (S_0/T_1 ISC) path to reproduce the ground state species in the potential well of H_2CO. This path was recently confirmed by extensive 3S-TSH simulations using very accurate PESs [117]. For $(CH_3)_2CO$ [51], many possible nonadiabatic decay paths from the S_1 PES were systematically located, and new insight into the slow S_1/T_1 ISC and new nonadiabatic paths to the ground state PES were obtained.

Three or more states may couple simultaneously in some reactions. An extension of the present approaches is straightforward since both (1) and (2) are similar to the equation for the diabatic/adiabatic transformation, although we have not yet made systematic tests for such cases.

Dynamic effects may play a significant role in many reactions. Although direct dynamics is very useful in simulations with moderate accuracies [139], it is still too expensive to find all the reaction events of a given system including slow events. There are many powerful approaches applied to ground state PESs to accelerate a dynamics for known mechanisms [140–142]. A prior mechanism search by the present approach followed by such accelerated dynamics may be a solution in the future for theoretical solving complicated reaction mechanisms. The present method can also be connected to the analytical PES fitting approach [143–146] which enabled the recent extensive 3S-TSH simulations of

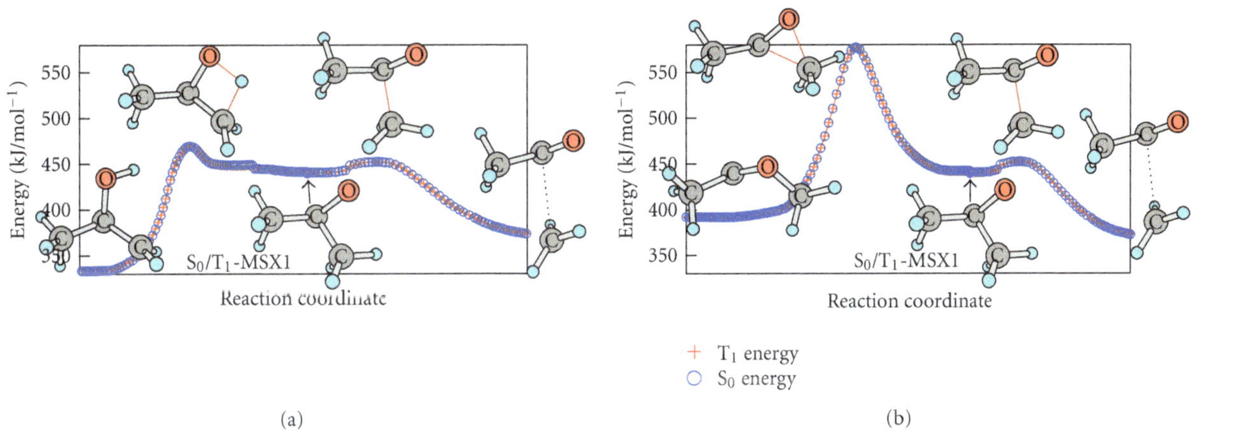

FIGURE 5: (a) A S_0/T_1-MEPSX curve for $H_2C{=}C(OH)CH_3 \leftrightarrow (CH_3)_2CO \leftrightarrow CH_3CO + CH_3$ and (b) a S_0/T_1-MEPSX curve for $CH_3{-}CO{-}CH_3 \leftrightarrow (CH_3)_2CO \leftrightarrow CH_3CO + CH_3$, at the UB3LYP/6-31+G* level.

H_2CO [117]. This approach permits the execution of highly accurate simulations, although applications are limited to small systems including less than 20 atoms. In constructing an analytical PES, prior knowledge about the mechanisms obtained by the present approach will be very useful for effective sampling of the PES data with the least number of ab initio calculations, avoiding exhaustive sampling on a perfect grid covering all configurations. One technical but significant issue in this approach is how to represent the square root topology of conical intersections, and several approaches have been proposed [147–150].

Our particular interest is to apply the present approach to multiple PESs described by combined quantum mechanical and molecular mechanical (QM/MM) calculations [151–154]. We have already reported automated reaction path search on the ground state PES of the QM/MM-ONIOM method [155] by combined microiteration [156, 157] and GRRM methods [158], where use of AFIR instead of GRRM has already been possible. This development is expected to greatly expand the applicability of the present approach to large systems.

Acknowledgements

This work is partly supported by a grant from Japan Science and Technology Agency with a Core Research for Evolutional Science and Technology (CREST) in the area of high-performance computing for multiscale and multiphysics phenomena at Kyoto University as well as a grant from US AFOSR (Grant no. FA9550-10-1-0304) at Emory University.

References

[1] F. Bernardi, M. Olivucci, and M. A. Robb, "Potential energy surface crossings in organic photochemistry," *Chemical Society Reviews*, vol. 25, no. 5, pp. 321–328, 1996.

[2] D. R. Yarkony, "Conical intersections: diabolical and often misunderstood," *Accounts of Chemical Research*, vol. 31, no. 8, pp. 511–518, 1998.

[3] D. Schröder, S. Shaik, and H. Schwarz, "Two-state reactivity as a new concept in organometallic chemistry," *Accounts of Chemical Research*, vol. 33, no. 3, pp. 139–145, 2000.

[4] A. L. Sobolewski, W. Domcke, C. Dedonder-Lardeux, and C. Jouvet, "Excited-state hydrogen detachment and hydrogen transfer driven by repulsive $1\pi\sigma^*$ states: a new paradigm for nonradiative decay in aromatic biomolecules," *Physical Chemistry Chemical Physics*, vol. 4, no. 7, pp. 1093–1100, 2002.

[5] R. Poli and J. N. Harvey, "Spin forbidden chemical reactions of transition metal compounds. New ideas and new computational challenges," *Chemical Society Reviews*, vol. 32, no. 1, pp. 1–8, 2003.

[6] B. G. Levine and T. J. Martínez, "Isomerization through conical intersections," *Annual Review of Physical Chemistry*, vol. 58, pp. 613–634, 2007.

[7] S. Nanbu, T. Ishida, and H. Nakamura, "Future perspectives of nonadiabatic chemical dynamics," *Chemical Science*, vol. 1, pp. 663–674, 2010.

[8] N. Koga and K. Morokuma, "Determination of the lowest energy point on the crossing seam between two potential surfaces using the energy gradient," *Chemical Physics Letters*, vol. 119, no. 5, pp. 371–374, 1985.

[9] M. R. Manaa and D. R. Yarkony, "On the intersection of two potential energy surfaces of the same symmetry. Systematic characterization using a Lagrange multiplier constrained procedure," *The Journal of Chemical Physics*, vol. 99, no. 7, pp. 5251–5256, 1993.

[10] J. M. Anglada and J. M. Bofill, "A reduced-restricted-quasi-Newton-Raphson method for locating and optimizing energy crossing points between two potential energy surfaces," *Journal of Computational Chemistry*, vol. 18, no. 8, pp. 992–1003, 1997.

[11] M. J. Bearpark, M. A. Robb, and H. Bernhard Schlegel, "A direct method for the location of the lowest energy point on a potential surface crossing," *Chemical Physics Letters*, vol. 223, no. 3, pp. 269–274, 1994.

[12] F. Sicilia, L. Blancafort, M. J. Bearpark, and M. A. Robb, "New algorithms for optimizing and linking conical intersection points," *Journal of Chemical Theory and Computation*, vol. 4, no. 2, pp. 257–266, 2008.

[13] C. Ciminelli, G. Granucci, and M. Persico, "The photoisomerization mechanism of azobenzene: a semiclassical simulation of nonadiabatic dynamics," *Chemistry*, vol. 10, no. 9, pp. 2327–2341, 2004.

[14] B. G. Levine, J. D. Coe, and T. J. Martínez, "Optimizing conical intersections without derivative coupling vectors: application to multistate multireference second-order perturbation theory (MS-CASPT2)," *Journal of Physical Chemistry B*, vol. 112, no. 2, pp. 405–413, 2008.

[15] T. W. Keal, A. Koslowski, and W. Thiel, "Comparison of algorithms for conical intersection optimisation using semiempirical methods," *Theoretical Chemistry Accounts*, vol. 118, no. 5-6, pp. 837–844, 2007.

[16] S. Maeda, K. Ohno, and K. Morokuma, "Updated branching plane for finding conical intersections without coupling derivative vectors," *Journal of Chemical Theory and Computation*, vol. 6, no. 5, pp. 1538–1545, 2010.

[17] J. W. McIver Jr. and A. Komornicki, "Structure of transition states in organic reactions. General theory and an application to the cyclobutene-butadiene isomerization using a semiempirical molecular orbital method," *Journal of the American Chemical Society*, vol. 94, no. 8, pp. 2625–2633, 1972.

[18] A. Komornicki, K. Ishida, K. Morokuma, R. Ditchfield, and M. Conrad, "Efficient determination and characterization of transition states using ab-initio methods," *Chemical Physics Letters*, vol. 45, no. 3, pp. 595–602, 1977.

[19] H. B. Schlegel, "Optimization of equilibrium geometries and transition structures," *Journal of Computational Chemistry*, vol. 3, no. 2, pp. 214–218, 1982.

[20] Ö Farkas and H. B. Schlegel, "Methods for optimizing large molecules. II. Quadratic search," *Journal of Chemical Physics*, vol. 111, no. 24, pp. 10806–10814, 1999.

[21] C. J. Cerjan and W. H. Miller, "On finding transition states," *The Journal of Chemical Physics*, vol. 75, no. 6, pp. 2800–2801, 1981.

[22] P. Császár and P. Pulay, "Geometry optimization by direct inversion in the iterative subspace," *Journal of Molecular Structure*, vol. 114, no. 1-2, pp. 31–34, 1984.

[23] A. Banerjee, N. Adams, J. Simons, and R. Shepard, "Search for stationary points on surfaces," *Journal of Physical Chemistry*, vol. 89, no. 1, pp. 52–57, 1985.

[24] H. B. Schlegel, "Exploring potential energy surfaces for chemical reactions: an overview of some practical methods," *Journal of Computational Chemistry*, vol. 24, no. 12, pp. 1514–1527, 2003.

[25] F. Jensen, *Introduction to Computational Chemistry*, John Wiley & Sons, Chichester, UK, 2nd edition, 2007.

[26] H. B. Schlegel, "Geometry optimization," *Wiley Interdisciplinary Reviews: Computational Molecular Science*, vol. 1, no. 5, pp. 790–809, 2011.

[27] T. A. Halgren and W. N. Lipscomb, "The synchronous-transit method for determining reaction pathways and locating molecular transition states," *Chemical Physics Letters*, vol. 49, no. 2, pp. 225–232, 1977.

[28] M. J. S. Dewar, E. F. Healy, and J. J. P. Stewart, "Location of transition states in reaction mechanisms," *Journal of the Chemical Society, Faraday Transactions 2: Molecular and Chemical Physics*, vol. 80, no. 3, pp. 227–233, 1984.

[29] R. Elber and M. Karplus, "A method for determining reaction paths in large molecules: application to myoglobin," *Chemical Physics Letters*, vol. 139, no. 5, pp. 375–380, 1987.

[30] G. Henkelman, B. P. Uberuaga, and H. Jónsson, "Climbing image nudged elastic band method for finding saddle points and minimum energy paths," *Journal of Chemical Physics*, vol. 113, no. 22, pp. 9901–9904, 2000.

[31] W. Ren and E. Vanden-Eijnden, "String method for the study of rare events," *Physical Review B*, vol. 66, no. 5, Article ID 052301, 2002.

[32] B. Peters, A. Heyden, A. T. Bell, and A. Chakraborty, "A growing string method for determining transition states: comparison to the nudged elastic band and string methods," *Journal of Chemical Physics*, vol. 120, no. 17, pp. 7877–7886, 2004.

[33] S. Maeda and K. Ohno, "A new approach for finding a transition state connecting a reactant and a product without initial guess: applications of the scaled hypersphere search method to isomerization reactions of HCN, $(H_2O)_2$, and alanine dipeptide," *Chemical Physics Letters*, vol. 404, no. 1–3, pp. 95–99, 2005.

[34] S. Maeda and K. Ohno, "Conversion pathways between a fullerene and a ring among C_{20} clusters by a sphere contracting walk method: remarkable difference in local potential energy landscapes around the fullerene and the ring," *Journal of Chemical Physics*, vol. 124, no. 17, Article ID 174306, 2006.

[35] M. Černohorský, S. Kettou, and J. Koča, "VADER : new software for exploring interconversions on potential energy surfaces," *Journal of Chemical Information and Computer Sciences*, vol. 39, no. 4, pp. 705–712, 1999.

[36] A. Laio and M. Parrinello, "Escaping free-energy minima," *Proceedings of the National Academy of Sciences of the United States of America*, vol. 99, no. 20, pp. 12562–12566, 2002.

[37] S. K. Burger and P. W. Ayers, "Dual grid methods for finding the reaction path on reduced potential energy surfaces," *Journal of Chemical Theory and Computation*, vol. 6, no. 5, pp. 1490–1497, 2010.

[38] J. Q. Sun and K. Ruedenberg, "Gradient extremals and steepest descent lines on potential energy surfaces," *The Journal of Chemical Physics*, vol. 98, no. 12, pp. 9707–9714, 1993.

[39] K. Bondensgård and F. Jensen, "Gradient extremal bifurcation and turning points: an application to the H_2CO potential energy surface," *Journal of Chemical Physics*, vol. 104, no. 20, pp. 8025–8031, 1996.

[40] W. Quapp, M. Hirsch, O. Imig, and D. Heidrich, "Searching for saddle points of potential energy surfaces by following a reduced gradient," *Journal of Computational Chemistry*, vol. 19, no. 9, pp. 1087–1100, 1998.

[41] K. K. Irikura and R. D. Johnson, "Predicting unexpected chemical reactions by isopotential searching," *Journal of Physical Chemistry A*, vol. 104, no. 11, pp. 2191–2194, 2000.

[42] E. M. Müller, A. de Meijere, and H. Grubmüller, "Predicting unimolecular chemical reactions: chemical flooding," *Journal of Chemical Physics*, vol. 116, no. 3, pp. 897–905, 2002.

[43] K. Ohno and S. Maeda, "A scaled hypersphere search method for the topography of reaction pathways on the potential energy surface," *Chemical Physics Letters*, vol. 384, no. 4–6, pp. 277–282, 2004.

[44] S. Maeda and K. Ohno, "Global mapping of equilibrium and transition structures on potential energy surfaces by the scaled hypersphere search method: applications to ab initio surfaces of formaldehyde and propyne molecules," *Journal of Physical Chemistry A*, vol. 109, no. 25, pp. 5742–5753, 2005.

[45] K. Ohno and S. Maeda, "Global reaction route mapping on potential energy surfaces of formaldehyde, formic acid, and their metal-substituted analogues," *Journal of Physical Chemistry A*, vol. 110, no. 28, pp. 8933–8941, 2006.

[46] S. Maeda and K. Morokuma, "Communications: a systematic method for locating transition structures of A+B → X type reactions," *Journal of Chemical Physics*, vol. 132, no. 24, Article ID 241102, 2010.
[47] S. Maeda, S. Komagawa, M. Uchiyama, and K. Morokuma, "Finding reaction pathways for multicomponent reactions: the passerini reaction is a four-component reaction," *Angewandte Chemie*, vol. 50, no. 3, pp. 644–649, 2011.
[48] S. Maeda and K. Morokuma, ""Finding reaction pathways of type A + B → X: toward systematic prediction of reaction mechanisms," *Journal of Chemical Theory and Computation*, vol. 7, no. 8, pp. 2335–2345, 2011.
[49] S. Maeda, K. Ohno, and K. Morokuma, "Automated global mapping of minimal energy points on seams of crossing by the anharmonic downward distortion following method: a case study of H_2CO," *Journal of Physical Chemistry A*, vol. 113, no. 9, pp. 1704–1710, 2009.
[50] S. Maeda, R. Saito, and K. Morokuma, "Finding minimum structures on the seam of crossing in reactions of type A + B → X: exploration of nonadiabatic ignition pathways of unsaturated hydrocarbons," *Journal of Physical Chemistry Letters*, vol. 2, no. 8, pp. 852–857, 2011.
[51] S. Maeda, K. Ohno, and K. Morokuma, "A theoretical study on the photodissociation of acetone: insight into the slow intersystem crossing and exploration of nonadiabatic pathways to the ground state," *Journal of Physical Chemistry Letters*, vol. 1, no. 12, pp. 1841–1845, 2010.
[52] K. Fukui, "A formulation of the reaction coordinate," *Journal of Physical Chemistry*, vol. 74, no. 23, p. 4161, 1970.
[53] K. Ishida, K. Morokuma, and A. Komornicki, "The intrinsic reaction coordinate. An ab initio calculation for HNC → HCN and H^- + CH4 → CH_4 + H^- ," *The Journal of Chemical Physics*, vol. 66, no. 5, pp. 2153–2156, 1976.
[54] K. Müller and L. D. Brown, "Location of saddle points and minimum energy paths by a constrained simplex optimization procedure," *Theoretica Chimica Acta*, vol. 53, no. 1, pp. 75–93, 1979.
[55] M. Page and J. W. McIver Jr., "On evaluating the reaction path Hamiltonian," *The Journal of Chemical Physics*, vol. 88, no. 2, pp. 922–935, 1988.
[56] C. Gonzalez and H. Bernhard Schlegel, "An improved algorithm for reaction path following," *The Journal of Chemical Physics*, vol. 90, no. 4, pp. 2154–2161, 1989.
[57] H. P. Hratchian and H. B. Schlegel, "Using Hessian updating to increase the efficiency of a Hessian based predictor-corrector reaction path following method," *Journal of Chemical Theory and Computation*, vol. 1, no. 1, pp. 61–69, 2005.
[58] S. Maeda and K. Ohno, "Structures of water octamers $(H_2O)_8$: exploration on ab initio potential energy surfaces by the scaled hypersphere search method," *Journal of Physical Chemistry A*, vol. 111, no. 20, pp. 4527–4534, 2007.
[59] H.-J. Werner, P. J. Knowles, F. R. Manby et al., MOLPRO, version 2006.1, a package of ab initio programs, http://www.molpro.net .
[60] S. Maeda, Y. Osada, K. Morokuma, and K. Ohno, GRRM11, http://kmweb.fukui.kyoto-u.ac.jp/~smaeda/GRRM_PR000 .html.
[61] W. M. Gelbart, M. L. Elert, and D. F. Heller, "Photodissociation of the formaldehyde molecule: does it or doesn't it?" *Chemical Reviews*, vol. 80, no. 5, pp. 403–416, 1980.
[62] C. B. Moore and J. C. Weisshaar, "Formaldehyde photochemistry," *Annual Review of Physical Chemistry*, vol. 34, pp. 525–555, 1983.
[63] D. J. Clouthier and D. A. Ramsay, "The spectroscopy of formaldehyde and thioformaldehyde," *Annual Review of Physical Chemistry*, vol. 34, pp. 31–58, 1983.
[64] W. H. Green Jr., C. B. Moore, and W. F. Polik, "Transition states and rate constants for unimolecular reactions," *Annual Review of Physical Chemistry*, vol. 43, no. 1, pp. 591–626, 1992.
[65] C. B. Moore, "A spectroscopist's view of energy states, energy transfers, and chemical reactions," *Annual Review of Physical Chemistry*, vol. 58, pp. 1–33, 2007.
[66] D. M. Hayes and K. Morokuma, "Theoretical studies of carbonyl photochemistry. I. ab initio potential energy surfaces for the photodissociation H_2CO^* → H + HCO," *Chemical Physics Letters*, vol. 12, no. 4, pp. 539–543, 1972.
[67] R. L. Jaffe, D. M. Hayes, and K. Morokuma, "Photodissociation of formaldehyde-4: potential energy surfaces for H_2CO^* → H_2 + CO," *The Journal of Chemical Physics*, vol. 60, no. 12, pp. 5108–5109, 1974.
[68] J. A. Hernandez and R. Carbo, "Theoretical-study of formaldehyde photodissociation," *Journal de Chimie Physique et de Physico-Chimie Biologique*, vol. 72, no. 9, pp. 959–960, 1975.
[69] R. L. Jaffe and K. Morokuma, "MCSCF potential energy surface for photodissociation of formaldehyde," *The Journal of Chemical Physics*, vol. 64, no. 12, pp. 4881–4886, 1976.
[70] J. D. Goddard and H. F. Schaefer, "The photodissociation of formaldehyde: potential energy surface features," *The Journal of Chemical Physics*, vol. 70, no. 11, pp. 5117–5134, 1979.
[71] N. C. Handy and S. Carter, "An improved potential surface for formaldehyde," *Chemical Physics Letters*, vol. 79, no. 1, pp. 118–124, 1981.
[72] R. Schinke, "Rotational state distributions of H_2 and CO following the photofragmentation of formaldehyde," *The Journal of Chemical Physics*, vol. 84, no. 3, pp. 1487–1491, 1986.
[73] G. E. Scuseria and H. F. Schaefer, "The photodissociation of formaldehyde: a coupled cluster study including connected triple excitations of the transition state barrier height for $H_2CO \rightarrow H_2$ + CO," *The Journal of Chemical Physics*, vol. 90, no. 7, pp. 3629–3636, 1989.
[74] Y. T. Chang, C. Minichino, and W. H. Miller, "Classical trajectory studies of the molecular dissociation dynamics of formaldehyde: $H_2CO \rightarrow H_2$+CO," *The Journal of Chemical Physics*, vol. 96, no. 6, pp. 4341–4355, 1992.
[75] L. Deng, T. Ziegler, and L. Fan, "A combined density functional and intrinsic reaction coordinate study on the ground state energy surface of H_2CO," *The Journal of Chemical Physics*, vol. 99, no. 5, pp. 3823–3835, 1993.
[76] W. Chen, W. L. Hase, and H. B. Schlegel, "Ab initio classical trajectory study of $H_2CO \rightarrow H_2$+CO dissociation," *Chemical Physics Letters*, vol. 228, no. 4-5, pp. 436–442, 1994.
[77] G. H. Peslherbe and W. L. Hase, "Semiempirical MNDO, AM1, and PM3 direct dynamics trajectory studies of formaldehyde unimolecular dissociation," *Journal of Chemical Physics*, vol. 104, no. 20, pp. 7882–7894, 1996.
[78] H. Nakano, K. Nakayama, K. Hirao, and M. Dupuis, "Transition state barrier height for the reaction $H_2CO \rightarrow H_2$+CO studied by multireference Møller-Plesset perturbation theory," *Journal of Chemical Physics*, vol. 106, no. 12, pp. 4912–4917, 1997.
[79] F. Jensen, "Stationary points on the H_2CO potential energy surface: dependence on theoretical level," *Theoretical Chemistry Accounts*, vol. 99, no. 5, pp. 295–300, 1998.
[80] D. Feller, M. Dupuis, and B. C. Garrett, "Barrier for the $H_2CO \rightarrow H_2$+CO reaction: a discrepancy between high-level

electronic structure calculations and experiment," *Journal of Chemical Physics*, vol. 113, no. 1, pp. 218–226, 2000.

[81] X. Li, J. M. Millam, and H. B. Schlegel, "Ab initio molecular dynamics studies of the photodissociation of formaldehyde, $H_2CO \rightarrow H_2 + CO$: direct classical trajectory calculations by MP2 and density functional theory," *Journal of Chemical Physics*, vol. 113, no. 22, pp. 10062–10067, 2000.

[82] H. B. Schlegel, S. S. Iyengar, X. Li et al., "Ab initio molecular dynamics: propagating the density matrix with Gaussian orbitals. III. Comparison with Born-Oppenheimer dynamics," *Journal of Chemical Physics*, vol. 117, no. 19, pp. 8694–8704, 2002.

[83] T. Yonehara and S. Kato, "Role of isomerization channel in unimolecular dissociation reaction $H_2CO \rightarrow H_2+CO$: Ab initio global potential energy surface and classical trajectory analysis," *Journal of Chemical Physics*, vol. 117, no. 24, pp. 11131–11138, 2003.

[84] X. Zhang, S. Zou, L. B. Harding, and J. M. Bowman, "A global ab initio potential energy surface for formaldehyde," *Journal of Physical Chemistry A*, vol. 108, no. 41, pp. 8980–8986, 2004.

[85] X. Zhang, J. L. Rheinecker, and J. M. Bowman, "Quasiclassical trajectory study of formaldehyde unimolecular dissociation: H_2 CO → H_2 + CO, H + HCO," *Journal of Chemical Physics*, vol. 122, no. 11, Article ID 114313, 8 pages, 2005.

[86] T. Yonehara and S. Kato, "Quantum dynamics study on multichannel dissociation and isomerization reactions of formaldehyde," *Journal of Chemical Physics*, vol. 125, no. 8, Article ID 084307, 2006.

[87] J. M. Bowman and X. Zhang, "New insights on reaction dynamics from formaldehyde photodissociation," *Physical Chemistry Chemical Physics*, vol. 8, no. 3, pp. 321–332, 2006.

[88] L. B. Harding, S. J. Klippenstein, and A. W. Jasper, "Ab initio methods for reactive potential surfaces," *Physical Chemistry Chemical Physics*, vol. 9, no. 31, pp. 4055–4070, 2007.

[89] S. Maeda and K. Ohno, "A new global reaction route map on the potential energy surface of H_2CO with unrestricted level," *Chemical Physics Letters*, vol. 460, no. 1–3, pp. 55–58, 2008.

[90] Y. Yamaguchi, S. S. Wesolowski, T. J. Van Huis, and H. F. Schaefer, "The unimolecular dissociation of H_2CO on the lowest triplet potential-energy surface," *Journal of Chemical Physics*, vol. 108, no. 13, pp. 5281–5288, 1998.

[91] H. M. Yin, S. H. Kable, X. Zhang, and J. M. Bowman, "Signatures of H_2CO photodissociation from two electronic states," *Science*, vol. 311, no. 5766, pp. 1443–1446, 2006.

[92] K. K. Baeck, "The analytic gradient for the equation-of-motion coupled-cluster energy with a reduced molecular orbital space: an application for the first excited state of formaldehyde," *Journal of Chemical Physics*, vol. 112, no. 1, pp. 1–4, 2000.

[93] J. B. Simonsen, N. Rusteika, M. S. Johnson, and T. I. Sølling, "Atmospheric photochemical loss of H and H_2 from formaldehyde: the relevance of ultrafast processes," *Physical Chemistry Chemical Physics*, vol. 10, no. 5, pp. 674–680, 2008.

[94] R. D. Van Zee, M. F. Foltz, and C. B. Moore, "Evidence for a second molecular channel in the fragmentation of formaldehyde," *Journal of Chemical Physics*, vol. 99, no. 3, pp. 1664–1673, 1993.

[95] D. Townsend, S. A. Lahankar, S. K. Lee et al., "The roaming atom: straying from the reaction path in formaldehyde decomposition," *Science*, vol. 306, no. 5699, pp. 1158–1161, 2004.

[96] A. G. Suits, "Roaming atoms and radicals: a new mechanism in molecular dissociation," *Accounts of Chemical Research*, vol. 41, no. 7, pp. 873–881, 2008.

[97] N. Herath and A. G. Suits, "Roaming radical reactions," *Journal of Physical Chemistry Letters*, vol. 2, no. 6, pp. 642–647, 2011.

[98] J. M. Bowman and B. C. Shepler, "Roaming radicals," *Annual Review of Physical Chemistry*, vol. 62, pp. 531–553, 2011.

[99] P. L. Houston and S. H. Kable, "Photodissociation of acetaldehyde as a second example of the roaming mechanism," *Proceedings of the National Academy of Sciences of the United States of America*, vol. 103, no. 44, pp. 16079–16082, 2006.

[100] B. C. Shepler, B. J. Braams, and J. M. Bowman, "Quasiclassical trajectory calculations of acetaldehyde dissociation on a global potential energy surface indicate significant non-transition state dynamics," *Journal of Physical Chemistry A*, vol. 111, no. 34, pp. 8282–8285, 2007.

[101] B. R. Heazlewood, M. J. T. Jordan, S. H. Kable et al., "Roaming is the dominant mechanism for molecular products in acetaldehyde photodissociation," *Proceedings of the National Academy of Sciences of the United States of America*, vol. 105, no. 35, pp. 12719–12724, 2008.

[102] V. Goncharov, N. Herath, and A. G. Suits, "Roaming dynamics in acetone dissociation," *Journal of Physical Chemistry A*, vol. 112, no. 39, pp. 9423–9428, 2008.

[103] C. Chen, B. Braams, D. Y. Lee, J. M. Bowman, P. L. Houston, and D. Stranges, "Evidence for vinylidene production in the photodissociation of the allyl radical," *Journal of Physical Chemistry Letters*, vol. 1, no. 12, pp. 1875–1880, 2010.

[104] E. Kamarchik, L. Koziol, H. Reisler, J. M. Bowman, and A. I. Krylov, "Roaming pathway leading to unexpected water + vinyl products in C_2H_4OH dissociation," *Journal of Physical Chemistry Letters*, vol. 1, no. 20, pp. 3058–3065, 2010.

[105] M. P. Grubb, M. L. Warter, A. G. Suits, and S. W. North, "Evidence of roaming dynamics and multiple channels for molecular elimination in NO_3 photolysis," *The Journal of Physical Chemistry Letters*, vol. 1, pp. 2455–2458, 2010.

[106] M. P. Grubb, M. L. Warter, K. M. Johnson, and S. W. North, "Ion imaging study of NO_3 radical photodissociation dynamics: characterization of multiple reaction pathways," *Journal of Physical Chemistry*, vol. 115, no. 15, pp. 3218–3226, 2011.

[107] H. Xiao, S. Maeda, and K. Morokuma, "Excited-state roaming dynamics in photolysis of a nitrate radical," *Journal of Physical Chemistry Letters*, vol. 2, no. 9, pp. 934–938, 2011.

[108] S. W. North, "Roaming in the dark," *Nature Chemistry*, vol. 3, no. 7, pp. 504–505, 2011.

[109] A. Fernández-Ramos, J. A. Miller, S. J. Klippenstein, and D. G. Truhlar, "Modeling the kinetics of bimolecular reactions," *Chemical Reviews*, vol. 106, no. 11, pp. 4518–4584, 2006.

[110] L. B. Harding and S. J. Klippenstein, "Roaming radical pathways for the decomposition of alkanes," *Journal of Physical Chemistry Letters*, vol. 1, no. 20, pp. 3016–3020, 2010.

[111] P. Zhang, *Applications of electronic structure theory in the study of molecular processes*, Ph.D. thesis, Emory University, 2005.

[112] B. C. Shepler, E. Epifanovsky, P. Zhang, J. M. Bowman, A. I. Krylov, and K. Morokuma, "Photodissociation dynamics of formaldehyde initiated at the T_1/S_0 minimum energy crossing configurations," *Journal of Physical Chemistry A*, vol. 112, no. 51, pp. 13267–13270, 2008.

[113] M. Araujo, B. Lasorne, M. J. Bearpark, and M. A. Robb, "The photochemistry of formaldehyde: internal conversion and

molecular dissociation in a single step?" *Journal of Physical Chemistry A*, vol. 112, no. 33, pp. 7489–7491, 2008.

[114] P. Zhang, S. Maeda, K. Morokuma, and B. J. Braams, "Photochemical reactions of the low-lying excited states of formaldehyde: T_1/S_0 intersystem crossings, characteristics of the S_1 and T_1 potential energy surfaces, and a global T_1 potential energy surface," *Journal of Chemical Physics*, vol. 130, no. 11, Article ID 114304, 2009.

[115] M. Araújo, B. Lasorne, A. L. Magalhães, G. A. Worth, M. J. Bearpark, and M. A. Robb, "The molecular dissociation of formaldehyde at medium photoexcitation energies: a quantum chemistry and direct quantum dynamics study," *Journal of Chemical Physics*, vol. 131, no. 14, Article ID 144301, 2009.

[116] M. Arauéjo, B. Lasorne, A. L. Magalhaes, M. J. Bearpark, and M. A. Robb, "Controlling product selection in the photodissociation of formaldehyde: direct quantum dynamics from the S_1 barrier," *Journal of Physical Chemistry A*, vol. 114, no. 45, pp. 12016–12020, 2010.

[117] B. Fu, B. C. Shepler, and J. M. Bowman, "hree-state trajectory surface hopping studies of the photodissociation dynamics of formaldehyde on ab initio potential energy surfaces," *Journal of the American Chemical Society*, vol. 133, no. 20, pp. 7957–7968, 2011.

[118] Y. Haas, "Photochemical α-cleavage of ketones: revisiting acetone," *Photochemical and Photobiological Sciences*, vol. 3, no. 1, pp. 6–16, 2004.

[119] R. A. Copeland and D. R. Crosley, "Radiative, collisional and dissociative processes in triplet acetone," *Chemical Physics Letters*, vol. 115, no. 4-5, pp. 362–368, 1985.

[120] H. Zuckermann, B. Schmitz, and Y. Haas, "Dissociation energy of an isolated triplet acetone molecule," *Journal of Physical Chemistry*, vol. 92, no. 17, pp. 4835–4837, 1988.

[121] S. W. North, D. A. Blank, J. Daniel Gezelter, C. A. Longfellow, and Y. T. Lee, "Evidence for stepwise dissociation dynamics in acetone at 248 and 193 nm," *The Journal of Chemical Physics*, vol. 102, no. 11, pp. 4447–4460, 1995.

[122] D. Liu, W. H. Fang, and X. Y. Fu, "An ab initio study on photodissociation of acetone," *Chemical Physics Letters*, vol. 325, no. 1–3, pp. 86–92, 2000.

[123] E. W. G. Diau, G. Kötting, and A. H. Zewail, "Femtochemistry of norrish type-1 reactions. I. Experimental and theoretical studies of acetone and related ketones on the s1 surface," *ChemPhysChem*, vol. 2, no. 5, pp. 273–293, 2001.

[124] E. W. G. Diau, C. Kötting, T. I. Sølling, and A. H. Zewail, "Femtochemistry of Norrish type-I reactions. III. Highly excited ketones—theoretical," *ChemPhysChem*, vol. 3, no. 1, pp. 57–78, 2002.

[125] T. I. Sølling, E. W. G. Diau, C. Kötting, S. de Feyter, and A. H. Zewail, "Femtochemistry of Norrish type-I reactions. IV. Highly excited ketones—experimental," *ChemPhysChem*, vol. 3, no. 1, pp. 79–97, 2002.

[126] W. K. Chen, J. W. Ho, and P. Y. Cheng, "Ultrafast photodissociation dynamics of the acetone 3s Rydberg state at 195 nm: a new mechanism," *Chemical Physics Letters*, vol. 380, no. 3-4, pp. 411–418, 2003.

[127] W. K. Chen, J. W. Ho, and P. Y. Cheng, "Ultrafast photodissociation dynamics of acetone at 195 nm. I. Initial-state, intermediate, and product temporal evolutions by femtosecond mass-selected multiphoton ionization spectroscopy," *Journal of Physical Chemistry A*, vol. 109, no. 31, pp. 6805–6817, 2005.

[128] W. K. Chen and P. Y. Cheng, "Ultrafast photodissociation dynamics of acetone at 195 nm: II. Unraveling complex three-body dissociation dynamics by femtosecond time-resolved photofragment translational spectroscopy," *Journal of Physical Chemistry A*, vol. 109, no. 31, pp. 6818–6829, 2005.

[129] H. Somnitz, M. Fida, T. Ufer, and R. Zellner, "Pressure dependence for the CO quantum yield in the photolysis of acetone at 248 nm: a combined experimental and theoretical study," *Physical Chemistry Chemical Physics*, vol. 7, no. 18, pp. 3342–3352, 2005.

[130] V. Khamaganov, R. Karunanandan, A. Rodriguez, and J. N. Crowley, "Photolysis of $CH_3C(O)CH_3$ (248 nm, 266 nm), $CH_3C(O)C_2H_5$ (248 nm) and $CH_3C(O)Br$ (248 nm): pressure dependent quantum yields of CH_3 formation," *Physical Chemistry Chemical Physics*, vol. 9, no. 31, pp. 4098–4113, 2007.

[131] R. Nádasdi, G. Kovács, I. Szilágyi et al., "Exciplex laser photolysis study of acetone with relevance to tropospheric chemistry," *Chemical Physics Letters*, vol. 440, no. 1–3, pp. 31–35, 2007.

[132] I. Antol, M. Eckert-Maksić, M. Ončák, P. Slavíček, and H. Lischka, "Photodissociation pathways of acetone upon excitation into the 3s Rydberg state: adiabatic versus diabatic mechanism," *Collection of Czechoslovak Chemical Communications*, vol. 73, no. 11, pp. 1475–1494, 2008.

[133] H. Somnitz, T. Ufer, and R. Zellner, "Acetone photolysis at 248 nm revisited: pressure dependence of the CO and CO_2 quantum yields," *Physical Chemistry Chemical Physics*, vol. 11, no. 38, pp. 8522–8531, 2009.

[134] T. Shibata and T. Suzuki, "Photofragment ion imaging with femtosecond laser pulses," *Chemical Physics Letters*, vol. 262, no. 1-2, pp. 115–119, 1996.

[135] N. Rusteika, K. B. Møller, and T. I. Sølling, "New insights on the photodynamics of acetone excited with 253–288 nm femtosecond pulses," *Chemical Physics Letters*, vol. 461, no. 4–6, pp. 193–197, 2008.

[136] R. Y. Brogaard, T. I. Sølling, and K. B. Møller, "Initial dynamics of the Norrish Type I reaction in acetone: probing wave packet motion," *Journal of Physical Chemistry A*, vol. 115, no. 15, pp. 556–561, 2011.

[137] M. J. Frisch, G. W. Trucks, H. B. Schlegel et al., "Gaussian 09 Revision A.2," Wallingford CT, 2009.

[138] R. Nádasdi, G. L. Zügner, M. Farkas, S. Dõbé, S. Maeda, and K. Morokuma, "Photochemistry of methyl ethyl ketone: quantum yields and S_1/S_0-diradical mechanism of photodissociation," *ChemPhysChem*, vol. 11, no. 18, pp. 3883–3895, 2010.

[139] B. Lasorne, G. A. Worth, and M. A. Robb, "Excited-state dynamics," *Wiley Interdisciplinary Reviews: Computational Molecular Science*, vol. 1, no. 3, pp. 460–475, 2011.

[140] J. Kästner, "Umbrella sampling," *Wiley Interdisciplinary Reviews: Computational Molecular Science*, vol. 1, no. 6, pp. 932–942, 2011.

[141] H. Hu and W. Yang, "Free energies of chemical reactions in solution and in enzymes with ab initio quantum mechanics/molecular mechanics methods," *Annual Review of Physical Chemistry*, vol. 59, pp. 573–601, 2008.

[142] A. Barducci, M. Bonomi, and M. Parrinello, "Metadynamics," *Wiley Interdisciplinary Reviews: Computational Molecular Science*, vol. 1, no. 5, pp. 826–843, 2011.

[143] G. C. Schatz, "The analytical representation of electronic potential-energy surfaces," *Reviews of Modern Physics*, vol. 61, no. 3, pp. 669–688, 1989.

[144] M. A. Collins, "Molecular potential-energy surfaces for chemical reaction dynamics," *Theoretical Chemistry Accounts*, vol. 108, no. 6, pp. 313–324, 2002.

[145] J. M. Bowman, B. J. Braams, S. Carter et al., "Ab-initio-based potential energy surfaces for complex molecules and molecular complexes," *Journal of Physical Chemistry Letters*, vol. 1, no. 12, pp. 1866–1874, 2010.

[146] J. M. Bowman, G. Czakó, and B. Fu, "High-dimensional ab initio potential energy surfaces for reaction dynamics calculations," *Physical Chemistry Chemical Physics*, vol. 13, no. 18, pp. 8094–8111, 2011.

[147] C. R. Evenhuis, X. Lin, D. H. Zhang, D. Yarkony, and M. A. Collins, "Interpolation of diabatic potential-energy surfaces: quantum dynamics on ab initio surfaces," *Journal of Chemical Physics*, vol. 123, no. 13, Article ID 134110, 12 pages, 2005.

[148] Z. H. Li, R. Valero, and D. G. Truhlar, "Improved direct diabatization and coupled potential energy surfaces for the photodissociation of ammonia," *Theoretical Chemistry Accounts*, vol. 118, no. 1, pp. 9–24, 2007.

[149] Z. Wang, I. S. K. Kerkines, K. Morokuma, and P. Zhang, "Analytical potential energy surfaces for N_3 low-lying doublet states," *Journal of Chemical Physics*, vol. 130, no. 4, Article ID 044313, 2009.

[150] X. Zhu and D. R. Yarkony, "On the representation of coupled adiabatic potential energy surfaces using quasi-diabatic Hamiltonians: description of accidental seams of conical intersection," *Molecular Physics*, vol. 108, no. 19-20, pp. 2611–2619, 2010.

[151] A. Warshel and M. Levitt, "Theoretical studies of enzymatic reactions: dielectric, electrostatic and steric stabilization of the carbonium ion in the reaction of lysozyme," *Journal of Molecular Biology*, vol. 103, no. 2, pp. 227–249, 1976.

[152] U. C. Singh and P. A. Kollman, "A combined ab initio quantum mechanical and molecular mechanical method for carrying out simulations on complex molecular systems: applications to the $CH_3Cl + Cl^-$ exchange reaction and gas phase protonation of polyethers," *Journal of Computational Chemistry*, vol. 7, no. 6, pp. 718–730, 1986.

[153] M. J. Field, P. A. Bash, and M. Karplus, "A combined quantum mechanical and molecular mechanical potential for molecular dynamics simulations," *Journal of Computational Chemistry*, vol. 11, no. 6, pp. 700–733, 1990.

[154] F. Maseras and K. Morokuma, "IMOMM—a new integrated ab-initio plus molecular mechanics geometry optimization scheme of equilibrium structures and transition-states," *Journal of Computational Chemistry*, vol. 16, no. 9, pp. 1170–1179, 1995.

[155] M. Svensson, S. Humbel, R. D. J. Froese, T. Matsubara, S. Sieber, and K. Morokuma, "ONIOM: a multilayered integrated MO + MM method for geometry optimizations and single point energy predictions. A test for Diels-Alder reactions and $Pt(P(t\text{-}Bu)_3)_2 + H_2$ oxidative addition," *Journal of Physical Chemistry*, vol. 100, no. 50, pp. 19357–19363, 1996.

[156] T. Vreven, K. Morokuma, Ö Farkas, H. B. Schlegel, and M. J. Frisch, "Geometry optimization with QM/MM, ONIOM, and other combined methods. I. Microiterations and constraints," *Journal of Computational Chemistry*, vol. 24, no. 6, pp. 760–769, 2003.

[157] T. Vreven, M. J. Frisch, K. N. Kudin, H. B. Schlegel, and K. Morokuma, "Geometry optimization with QM/MM methods II: explicit quadratic coupling," *Molecular Physics*, vol. 104, pp. 701–714, 2006.

[158] S. Maeda, K. Ohno, and K. Morokuma, "An automated and systematic transition structure explorer in large flexible molecular systems based on combined global reaction route mapping and microiteration methods," *Journal of Chemical Theory and Computation*, vol. 5, no. 10, pp. 2734–2743, 2009.

11

Synergistic Effect of [Ru(III) + Ir(III)] in N-Bromosuccinimide Reaction with Certain Aliphatic Ketones: A Kinetic Study

P. Giridhar Reddy, K. Ramesh, S. Shylaja, K. C. Rajanna, and S. Kandlikar

Department of Chemistry, Osmania University, Hyderabad 500 001, India

Correspondence should be addressed to K. C. Rajanna, kcrajannaou@yahoo.com

Academic Editor: Benjaram M. Reddy

Dramatic rate enhancements have been observed in the oxidation of acetone (MMK), ethyl methyl ketone (EMK), and isobutyl methyl ketone (IBMK) by N-bromosuccinimide (NBS) in the presence of a mixture of metal ions [Ru(III) + Ir(III)]. The rate of oxidation in the the [Ru(III) + Ir(III)] system is much faster than either in the Ru(III) or in Ir(III) catalyzed system. The order in [NBS] was found to be zero both in catalyzed as well as uncatalyzed reactions. However, the order in [ketone] changed from unity to a fractional one in metal ion catalyzed systems. On the basis of kinetic features, the most plausible mechanism involving the oxidation of [Ir(III)-ketone] by Ru(III) to give [Ir(V)-ketone] and Ru(I) in the rate limiting step is proposed. Ru(I) thus formed is oxidized by NBS to Ru(III) in a fast step is discussed and individual rate parameters were evaluated.

1. Introduction

Bimetallic catalysts often exhibit improved properties than either of the single metal catalysts. This is generally called the synergistic effect. The presence of the synergistic activity of platinum and rhodium in bimetallic catalysts and possible reasons for the effects has been discussed in the literature [1–10]. The synergistic activity of certain mixed catalyst systems, namely, [Ag(I) + Mn(II)], [Ag(I) + Cu(II)], [Ru(III) + Mn(II)], [Ir(III) + Mn(II)] and [Ru(III) + Os(VIII)], and so forth, has been the subject of interest in the field of homogenous redox reactions by several workers [4–10] in recent years.

N-bromosuccinimide (NBS) has been used as a versatile oxidizing agent and analytical reagent in acid and alkaline media [11–24]. In recent past microconcentrations (10^{-6} to 10^{-5} M) of Ru(III) and Ir(III) have been employed as an efficient catalyst in N-bromosuccinimide (NBS) oxidations and other redox systems [25–31]. Recently we have reported that Ru(III) 34 catalyzed the oxidation of aliphatic ketones by NBS in aqueous acetic acid [27], wherein trace amount of Ru(III) is used as a catalyst. However, kinetic studies involving the mixture of metal ions as a catalyst also appeared to be scanty in NBS oxidations. Preliminary experiments involving the mixture of Ru(III) and Ir(III) as a catalyst in NBS-aliphatic ketone redox system in aqueous acetic acid medium containing H_2SO_4 indicated that the reaction rate is much greater than the sum of the reaction rates when metal ions such as Ru(III) and Ir(III) are separately employed as a catalyst. These striking features have become stimulus for us to take up a systematic kinetic study comprising the catalytic oxidation of certain aliphatic ketones by NBS by using Ir(III) alone and a mixture of [Ru(III) + Ir(III)] as homogeneous catalysts. Less than millimolar Ru(III) and Ir(III) have been used in the composition of mixed catalysts in this study.

2. Experimental

Reagents used in this study were analytical grade. Stock solutions of ketone and NBS were prepared by dissolving appropriate amounts of samples in doubly distilled water. The solution of NBS was standardized iodometrically against sodium thiosulphate by a known procedure [12]. Succinimide (NHS) was prepared by a known method and recrystallised from alcohol and the stock solution was prepared by dissolving the required quantity in distilled water. Solutions of all the other reagents were prepared

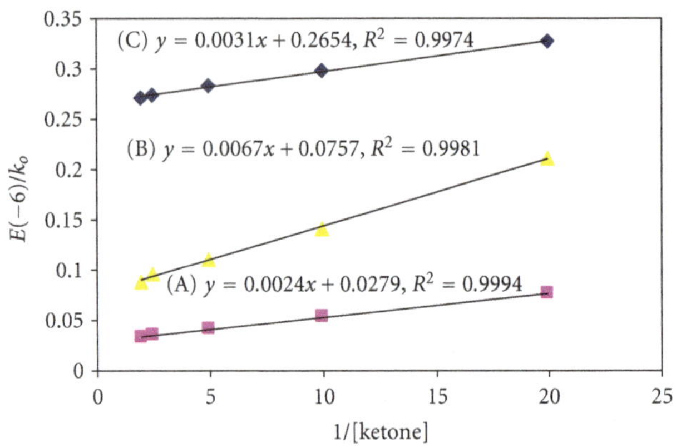

Figure 1: Plot of $[1/k_o]$ versus $(1/[\text{ketone}])$. (A) Acetone, (B) EMK, and (C) IBMK; 10^3 [NBS] = 5.00 mol dm^{-3}; 10^2 [H^+] = 5.00 mol dm^{-3}; HOAc = 10% (v/v); 10^5 [Ru(III)] = 2.09 mol dm^{-3}; 10^6 [Ir(III)] = 6.67 mol dm^{-3}; 10^2 [$Hg(OAC)_2$] = 2.00 mol dm^{-3}; Temp. = 300 K.

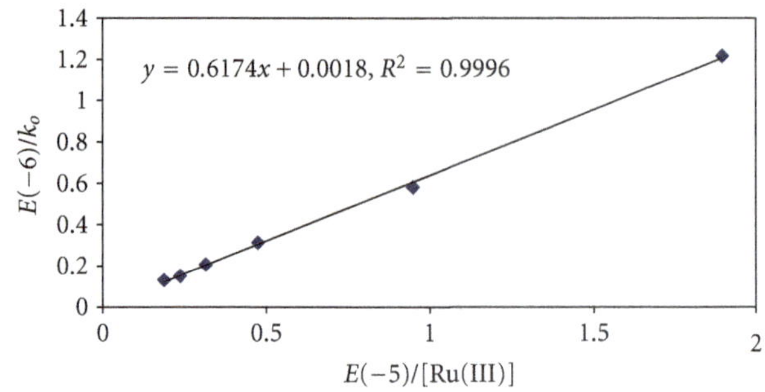

Figure 2: Plot of $1/k_o$ versus 1/[Ru(III)]. 10^3 [NBS] = 5.00 mol dm^{-3}; [EMK] = 0.100 mol dm^{-3}; HOAc = 10% (v/v); 10^2 [H^+] = 5.00 mol dm^{-3}; 10^7 [Ir(III)] = 5.05 mol dm^{-3}; 10^2 [$Hg(OAC)_2$] = 2.00 mol dm^{-3}; Temp. = 300 K.

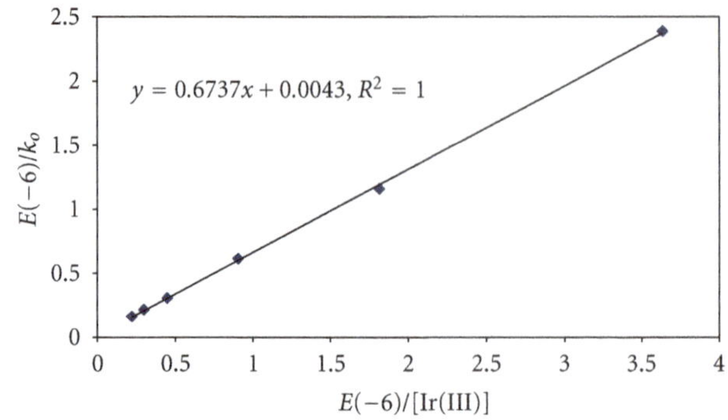

Figure 3: Plot of $1/k_o$ versus 1/[Ir(III)]. 10^3 [NBS] = 5.00 mol dm^{-3}; [EMK] = 0.100 mol dm^{-3}; HOAc = 10% (v/v); 10^2 [H^+] = 5.00 mol dm^{-3}; 10^5 [Ru(III)] = 2.09 mol dm^{-3}; Temp. = 300 K 10^2 [$Hg(OAC)_2$] = 2.00 mol dm^{-3}.

by dissolving the requisite amounts of samples in doubly distilled water. Sulphuric acid (H_2SO_4) and $KHSO_4$ or $NaClO_4$ were used to provide the required acidity and to maintain the ionic strength (μ), respectively.

3. Kinetic Measurements

All kinetic measurements were performed under pseudo-first-order conditions with [ketone] at least 10-fold in excess over [NBS] at a constant ionic strength (μ) and desired temperature. The reaction was initiated by mixing previously thermostated solutions of NBS and Ketone, which also contained the necessary quantities of acid and $NaClO_4$. The progress of the reaction was followed by the iodometric determination of the unreacted [NBS] in aliquots (5 mL each) of the reaction mixture withdrawn at regular time intervals. The initial rates (V) were evaluated from the tangential slopes of the plots of [NBS] versus time. The initial rate (V) values were reproducible within ±5%. The oxidation product was identified as a 1,2-dicarbonyl compound according to standard procedures cited in the literature.

The effect of dissolved oxygen on the rate of the reaction was studied by preparing the reaction mixture and following the reaction under nitrogen atmosphere. No significant difference between the results obtained under nitrogen atmosphere and those obtained in the presence of air were observed. However, fresh solutions were used during the experiments.

4. Results and Discussion

4.1. Salient Kinetic Features of the Study. (1) When [NBS] is taken for greater than [EMK], the stoichiometry of the reaction has been found to be 2[NBS] : 1[EMK]. The reaction kinetics indicated "zero"-order dependence on [NBS] and fractional in [Ketone] which could be seen from data presented in Table 1 and Figure 1. Further, the plots of 1/rate as a function of 1/[Ketone] were linear with positive slope and definite intercept on 1/rate axis (Figure 1) indicating a complex formation of ketone with the catalysts.

(2) At constant [H^+], ionic strength (μ), [EMK], [NBS], and [Ir(III)] order in [Ru(III)] was one as could be seen from the constancy of [k_o/[Ru(III)] values over a wide concentration range of Ru(III). Further, the plot of 1/rate versus 1/[Ru(III)] has been found linear passing through origin and with a positive slope (Table 2 and Figure 2).

(3) At constant [H^+], ionic strength (μ), [EMK], [NBS], [Ru(III)], and temperature order in [Ir(III)] was also unity, which could be evidenced from the constancy of [k_o/[Ir(III)] values over a wide variation of [Ir(III)]. The plot of 1/rate versus 1/[Ir(III)] was linear passing through origin and with a positive slope (Table 3 and Figure 3).

(4) A ten-fold concentration variation of additives such as [$KHSO_4$], [$NaClO_4$], or [H_2SO_4] did not alter the rate of oxidation to any significant extent in the present study.

(5) Rate data presented in Table 4 depicts that an increase in the [HOAc] increased the rate of oxidation. The increase in the rate of oxidation with increase in the polarity of the medium suggests that the transition state is more polar than the reactants. Laidler-Eyring's plot of $\log k_o$ against the inverse of the relative permittivity ($1/D$) indicated fairly good

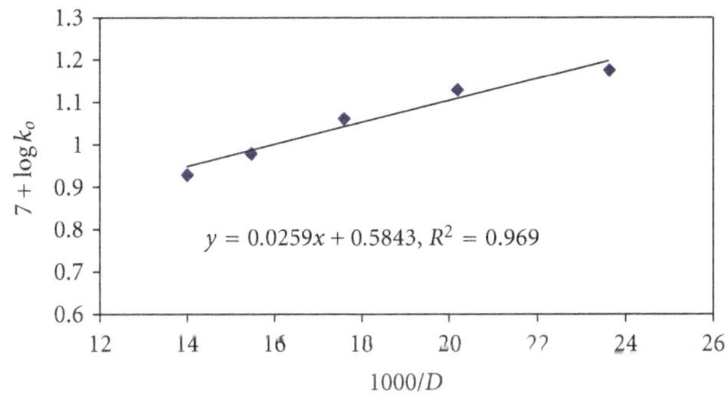

FIGURE 4: Effect of variation of solvent [% HOAc (v/v)] in NBS-EMK in [Ru(III) + Ir(III)] catalyzed reaction; Laidler-Eyring's plot of $\log k_o$ versus $1/D$. 10^3 [NBS] = 5.00 mol dm^{-3}; [EMK] = 0.100 mol dm^{-3}; 10^2 [$Hg(OAc)_2$] = 2.00 mol dm^{-4}; Temp. = 300 K; 10^6 [Ru(III)] = 5.25 mol dm^{-3}; 10^7 [Ir(III)] = 5.50 mol dm^{-3}.

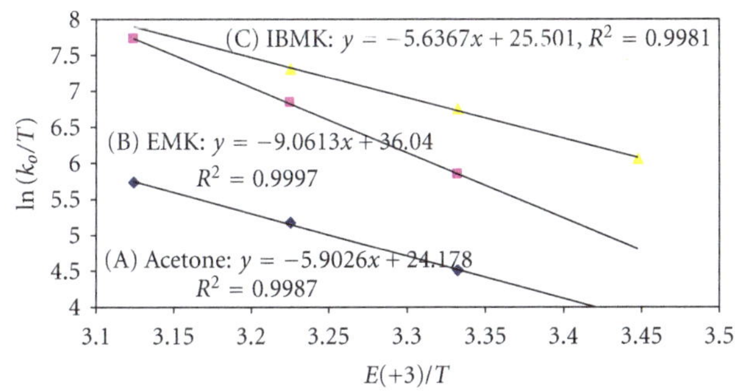

FIGURE 5: Eyring's plot of $\ln(k_o/T)$ versus $10^3/T$.

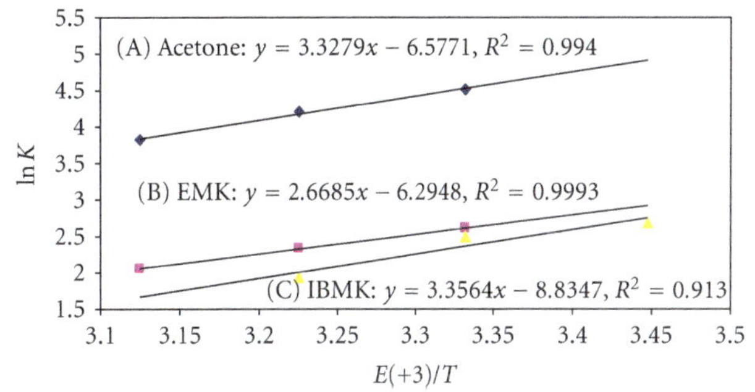

FIGURE 6: vant Hoff's plot of $\ln K$ versus $1/T$.

linearity ($R^2 = 0.969$) with a positive slope ($m = 0.025$). The value of m suggests that the transition state is more polar than the reactants. Thus considerable charge separation takes place in the transition state of the reaction.

(6) A four-fold change (0.02 to 0.08 M) in the concentration of mercuric acetate did not affect the rate of reaction to any considerable extent.

(7) Addition of olefin monomers such as acrylamide or acrylonitrile to the reaction system did not alter the reaction rate or indicate polymer formation ruling out the participation of free radical intermediates in the oxidation mechanism.

5. Mechanism of Oxidation

The rate of oxidation in [Ru(III) + Ir(III)] catalyzed system has been found to be greater than the sum of the reaction rates where Ru(III) and Ir(III) are employed separately as catalysts. Such a trend points out the synergistic activity of [Ru(III) + Ir(III)] on the rates of oxidation (Table 1). Further, the reaction kinetics in [Ru(III) + Ir(III)] catalyzed system is also similar to those when Ru(III) or Ir(III) is employed separately as catalysts. Observed reaction kinetics coupled with the linearity of reciprocal plot between 1/rate versus 1/[ketone] (Figure 1) suggests the formation of a precursor prior to the rate limiting oxidation step. At this point it is also pertinent to note that reaction rates are at least five to six times higher in Ir(III) catalyzed system as compared to Ru(III) catalyzed system. These observations probably indicate the formation of [EMK – Ir(III)] precursor prior to the slow decomposition step. Ru(III) being moderately good oxidizing agent may partake in the slow step to oxidize [Ir(III) – EMK] adduct and to give [Ir(V) – EMK] adduct as in Scheme 1 in ethyl methyl ketone (EMK) is as a typical example.

Scheme 1. Consider

$$\text{EMK} + \text{Ir(III)} \overset{K}{\rightleftharpoons} [\text{EMK} - \text{Ir(III)}] \quad (1)$$

$$[\text{EMK} - \text{Ir(III)}] + \text{Ru(III)} \xrightarrow[\text{slow}]{k} [\text{EMK} - \text{Ir(V)}] + \text{Ru(I)} \quad (2)$$

$$[\text{EMK} - \text{Ir(V)}] \xrightarrow{\text{fast}} \text{Ir(III)} + \text{P}_1 \quad (3)$$

$$\text{Ru(I)} + \text{NBS} \xrightarrow[\text{H}_2\text{O}]{\text{fast}} \text{Ru(III)} + \text{NHS} + \text{HBr} \quad (4)$$

$$\text{P}_1 + \text{NBS} \xrightarrow[\text{H}_2\text{O}]{\text{fast}} \text{NHS} + \text{final product} + \text{HBr}. \quad (5)$$

A rate law for the above mechanism could be written as

$$\text{Rate} = \frac{kK[\text{Ru(III)}][\text{Ir(III)}][\text{EMK}]}{1 + K[\text{EMK}]}. \quad (6)$$

Reciprocals of this equation yield

$$\frac{1}{\text{Rate}} = \frac{1}{kK[\text{Ru(III)}][\text{Ir(III)}][\text{EMK}]} + \frac{1}{k[\text{Ru(III)}][\text{Ir(III)}]}. \quad (7a)$$

This can be further simplified as

$$\frac{1}{\text{Rate}} = \frac{1}{k[\text{Ru(III)}][\text{Ir(III)}]}\left\{1 + \frac{1}{K[\text{EMK}]}\right\}. \quad (7b)$$

According to (7a) and (7b), plots of 1/rate versus 1/[Ru(III)], 1/rate versus 1/[Ir(III)] and 1/rate versus 1/[EMK] should be straight lines with positive slopes, the former two passing through origin and the later making an intercept on 1/rate axis. Such plots have been realized in the present study, keeping in with the proposed mechanism and validity of the rate law (Figures 2–4). The formation constant (K) value for

TABLE 1: The effect of variation of [NBS] and [EMK] in NBS-EMK system uncatalyzed and catalyzed by metal ions.

10^3 [NBS]	[EMK]	$10^6\, k_o$ mol dm^{-3} s^{-1}			
mol dm^{-3}		Uncatalyzed	[Ru(III)]	[Ir(III)]	[Ru(III)] + [Ir(III)]
2.00	0.400	0.099	1.28	7.51	11.5
4.00	0.040	0.098	1.27	7.53	11.4
5.00	0.040	0.098	1.29	7.50	11.6
8.00	0.040	0.098	1.28	7.52	11.5
5.00	0.080	0.195	2.22	11.40	17.8
5.00	0.100	0.245	2.63	12.8	20.0
5.00	0.200	0.486	4.00	16.7	27.0
5.00	0.400	0.975	5.85	19.6	32.2

10^2 [Hg(OAc)$_2$] = 2.00 mol dm^{-3}; HOAc = 10% (v/v); 10^2 [H$^+$] = 5.00 mol dm^{-3}; 10^5 [Ru(III)] = 2.09 mol dm^{-3}; 10^6 [Ir(III)] = 6.67 mol dm^{-3}; Temp. = 310 K.

TABLE 2: The effect of variation of Ru(III) in NBS-EMK system in the presence of Ir(III).

10^5 [Ru(III)] mol dm^{-3}	$10^6\, k_o$ mol dm^{-3} s^{-1}	k_o/[Ru(III)]
0.525	0.813	0.155
1.05	1.62	0.154
2.09	3.24	0.155
3.15	4.86	0.154
4.18	6.45	0.154
5.25	8.10	0.154

10^3 [NBS] = 5.00 mol dm^{-3}; [EMK] = 0.100 mol dm^{-3}; HOAc = 10% (v/v); 10^2 [H$^+$] = 5.00 mol dm^{-3}; 10^7 [Ir(III)] = 5.05 mol dm^{-3}; Temp. = 300 K. 10^2[Hg(OAC)$_2$] = 2.00 mol dm^{-3}.

TABLE 3: The effect of variation of [Ir(III)] in NBS-EMK system in the presence of Ru(III).

10^6 [Ir(III)] mol dm^{-3}	$10^6\, k_o$ mol dm^{-3} s^{-1}	k_o/[Ir(III)]
0.275	0.406	1.48
0.550	0.816	1.48
1.10	1.62	1.48
2.20	3.21	1.46
3.30	4.80	1.45
4.40	6.41	1.46

10^3 [NBS] = 5.00 mol dm^{-3}; [EMK] = 0.100 mol dm^{-3}; HOAc = 10% (v/v); 10^2 [H$^+$] = 5.00 mol dm^{-3}; 10^5 [Ru(III)] = 2.09 mol dm^{-3}; Temp. = 300 K. 10^2 [Hg(OAC)$_2$] = 2.00 mol dm^{-3}.

TABLE 4: The effect of variation of solvent [% HOAc (v/v)] in NBS-EMK in [Ru (III) + Ir (III)] catalyzed reaction.

HOAc % (v/v)	10	20	30	40	50
$10^{-6}\, k_o$ mol dm^{-3} s^{-1}	0.850	0.950	1.15	1.35	1.50

10^3 [NBS] = 5.00 mol dm^{-3}; [EMK] = 0.100 mol dm^{-3};
10^2 [Hg(OAc)$_2$] = 2.00 mol dm^{-4}; Temp. = 300 K;
10^6 [Ru(III)] = 5.25 mol dm^{-3}; 10^7 [Ir(III)] = 5.50 mol dm^{-3}.

[Ir(III) − EMK] adduct in mixed catalyzed system (Table 5) is by and large similar to that in Ir(III) catalyzed system. This observation also points out the validity of the proposed mechanism and rate law. Enthalpy ($\Delta H^{\#}$) and entropy of activation ($\Delta S^{\#}$) have been obtained from Eyring's plot (plot of $\ln(k_o/T)$ versus $10^3/T$, as shown in Figure 5), while the Gibbs-Helmholtz equation was used for the calculation of the free energy of activation ($\Delta G^{\#}$)

$$\Delta G^{\#} = \Delta H^{\#} - T\Delta S^{\#}. \tag{8}$$

Enthalpy (ΔH) and entropy (ΔS) of the precursor formation are computed from vant Hoff's plot (plot of $\ln(k/T)$ versus $10^3/T$, as shown in Figure 6). The free energy of the precursor formation (ΔG) was calculated using vant Hoff's reaction isotherm and entropy (ΔS) from the Gibbs-Helmholtz equation as usual

$$\Delta G = -RT \ln K. \tag{9}$$

Activation parameters involving (k) and thermodynamic parameters involving (K) are compiled in Table 5 accordingly.

5.1. Comparison of Catalytic Activity. In general, the reaction rates of catalyzed reactions are ten to twenty times greater than uncatalyzed systems under similar conditions. The rate of a mixed catalyzed system is much more than the sum of the rates when Ru(III) and Ir(III) are employed separately as catalysts. The catalytic activity is indicated by a trend [Ru(III) + Ir(III)] > [Ir(III)] > [Ru(III)]. This can be further seen from the magnitude of decomposition constant (K) which is 10^3 times greater over Ir(III) and 10^4 times to Ru(III) systems (Tables 1 and 5). These results certainly point out the synergistic activity of [Ru(III) + Ir(III)]. The decomposition constant values of mixed catalysis one far higher than those recorded in the presence of Ru(III) or Ir(III). The order of reactivity reveals the order IBMK > EMK > MMK. The formation of the precursor complex between metal ion and substrate can be supported from fairly a good magnitude of formation constants coupled with large negative ΔG and ΔS values (Table 5). Fairly good positive entropy of activation $\Delta S^{\#}$ values (Table 5) may suggest that the precursor is further undergoing a chemical change due to a reaction with another reagent.

Table 5: Kinetic and thermodynamic parameters.

Parameter	T (K)	MMK	EMK	IBMK
K (dm^3 mol^{-1} s^{-1})	290	—	—	1.23×10^5
	300	0.268×10^5	1.02×10^5	2.55×10^5
	310	0.541×10^5	2.86×10^5	4.59×10^5
	320	0.975×10^5	7.16×10^5	—
$\Delta H^{\#}$ kJ mol^{-1}		49.1	75.8	46.8
$\Delta G^{\#}$ kJ mol^{-1}		48.0	44.7	42.4
$\Delta S^{\#}$ j K^{-1} mol^{-1}		3.67	101	15.0
K dm^3 mol^{-1}	290	—	—	14.5
	300	90.0	13.4	12.0
	310	66.0	10.2	6.85
	320	45.0	7.69	—
$-\Delta H$ kJ mol^{-1}		27.7	22.4	28.1
$-\Delta G$ kJ mol^{-1}		11.2	6.00	6.20
$-\Delta S$ kJ mol^{-1}		55.0	53.0	73.0

Acknowledgments

The authors thank Professor P. K. Saiprakash for the constant encouragement and helpful discussions.

References

[1] K. C. Taylor, "Automobile catalytic converters," in *Catalysis: Science and Technology*, J. R. Anderson and M. Boudart, Eds., vol. 5, pp. 119–170, Springer, Berlin, Germany, 1984.

[2] K. C. Taylor, "Nitric oxide catalysis in automotive exhaust systems," *Catalysis Reviews: Science and Engineering*, vol. 35, no. 4, pp. 457–481, 1993.

[3] A. Gayen, T. Baidya, K. Biswas, S. Roy, and M. S. Hegde, "Synthesis, structure and three way catalytic activity of $Ce_{1-x}Pt_{x/2}Rh_{x/2}O_{2-\delta}$ (x = 0.01 and 0.02) nano-crystallites: synergistic effect in bimetal ionic catalysts," *Applied Catalysis A*, vol. 315, pp. 135–146, 2006.

[4] W. H. McCurdy Jr. and G. G. Guilbault, "Catalysts for cerium(IV) oxidimetry: determination of mixtures of mercury(I) and mercury(II)," *Analytical Chemistry*, vol. 32, no. 6, pp. 647–650, 1960.

[5] G. G. Guilbault and W. H. McCurdy Jr. Jr., "Catalysts for cerium(IV) oxidimetry: determination of polyhydric alcohols and metal chelates of 8-quinolinol," *Analytical Chemistry*, vol. 33, no. 4, pp. 580–582, 1961.

[6] G. G. Guilbault and W. McCurdy Jr., "Determination of phosphite, hypophosphite, tellurium and mercury," *Analytica Chimica Acta*, vol. 24, pp. 214–218, 1961.

[7] P. K. Saiprakash, B. Sethuram, and T. N. Rao, "($Ag^+ + Mn^{2+}$)–catalysed oxidation of isopropanol by ceric sulphate," *Indian Journal of Chemistry A*, vol. 14, p. 701, 1976.

[8] Y. Ramananda Sarma, *Kinetic study of redox reactions with one electron Oxidants [Ph.D. thesis]*, Osmania University, Andhra Pradesh, India, 1987.

[9] Y. R. Sarma, K. C. Rajanna, and P. K. Sai Prakash, "Kinetics of Ru(III) + Mn(III) nd Mn (III)catalysed oxidative decarboxylation of acetic and phenyl acetic acids by ceric sulphate," *Indian Journal of Chemistry A*, vol. 19, p. 825, 1980.

[10] Y. R. Sarma and P. K. Saiprakash, "Kinetics of Ir (III) # Ir(III) + Mn(II) catalysed oxidative decarboxylation and deamination of glycine, glutamic acid and serine by Ce(IV) in H_2SO_4media," *Indian Journal of Chemistry A*, vol. 19, p. 1139, 1980.

[11] R. Filler, "Oxidations and dehydrogenations with N-bromosuccinimide and related N-haloimides," *Chemical Reviews*, vol. 63, pp. 21–43, 1963.

[12] N. K. Mathur and C. K. Narang, *The Determination of Organic Compounds with NBS and Allied Reagents*, Acadamic Press, New York, NY, USA, 1975.

[13] D. L. Kamble, G. H. Hugar, and S. T. Nandibewoor, "Kinetics of oxidation of chromium(III) by N-bromosuccinimide in aqueous alkaline medium," *Indian Journal of Chemistry A*, vol. 35, no. 2, pp. 144–147, 1996.

[14] D. L. Kamble, R. B. Chougale, and S. T. Nandibewoor, "Kinetics and mechanism of uncatalyzed and ruthenium(III) catalyzed oxidation of allyl alcohol by N-bromosuccinimide in aqueous alkaline medium," *Indian Journal of Chemistry A*, vol. 35, no. 10, pp. 865–869, 1996.

[15] P. Saroja, B. K. Kumar, and S. Kandlikar, "Kinetic and mechanism of uncatalyzed and Ir (III) catalyzed oxidation of oxalate ion by N-bromosuccinimide in basic medium," *Indian Journal of Chemistry A*, vol. 28, p. 501, 1989.

[16] S. K. Mavalangi, M. R. Kembhavi, and S. T. Nandibewoor, "Oxidation of ethylenediaminetetraacetic acid by N-bromosuccinimide in aqueous alkaline medium—a kinetic study," *Turkish Journal of Chemistry*, vol. 25, no. 3, pp. 355–363, 2001.

[17] S. Gunasekaran and N. Venkatasubramanian, "Oxidation of diphenylmethanol by bromamine T. A kinetic and mechanistic study," *Journal of the Chemical Society, Perkin Transactions 2*, no. 7, pp. 949–953, 1983.

[18] J. P. Sharma, R. N. P. Singh, A. K. Singh, and B. Singh, "Kinetics and mechanism of Ru(III) catalysed oxidation of some polyhydric alcohols by N-bromosuccinimide in acidic media," *Tetrahedron*, vol. 42, no. 10, pp. 2739–2747, 1986.

[19] R. Ramachandrappa, Puttaswamy, S. M. Mayanna, and N. M. M. Gowda, "Kinetics and mechanism of oxidation of aspirin by bromamine-T, N-bromosuccinimide, and N-bromophthalimide," *International Journal of Chemical Kinetics*, vol. 30, no. 6, pp. 407–414, 1998.

[20] A. L. Harihar, M. R. Kembhavi, and S. T. Nandibewoor, "Kinetics and mechanism of N-bromosuccinimide oxidation of L-arginine in aqueous acidic medium," *Journal of the Indian Chemical Society*, vol. 76, no. 3, pp. 128–130, 1999.

[21] A. K. Singh, S. Rahmani, V. K. Singh, V. Gupta, D. Kesarwani, and B. Singh, "Iridium (III) catalysis of N-bromosuccinimide oxidation of reducing sugars in aqueous acid," *Indian Journal of Chemistry A*, vol. 40, no. 5, pp. 519–523, 2001.

[22] C. Karunakaran and K. Ganapathy, "Kinetic study on the reaction of N-bromosuccinimide with β, γ-unsaturated alcohols in aqueous acetic acid," *Indian Journal of Chemistry A*, vol. 29, p. 133, 1990.

[23] N. Venkatasubramanian and V. Thiagarajan, "Mechanism of oxidation of alcohols with N-bromo succinimide," *Canadian Journal of Chemistry*, vol. 47, no. 4, pp. 694–697, 1969.

[24] N. A. M. Farook, "Kinetics and mechanism of oxidation of 4-oxoacids by N-bromosuccinimide in aqueous acetic acid medium," *Journal of the Iranian Chemical Society*, vol. 3, no. 4, pp. 378–386, 2006.

[25] P. G. Reddy, Y. Ramesh, and S. Kandlikar, "Ru (III) catalyzed oxidation of ethanolamine, diethanolamine and tri ethanolamine by N-bromosuccinimide—a kinetic study," *Oxidation Communications*, vol. 7, p. 89, 1984.

[26] P. G. Reddy, A. V. Reddy, A. Ratnamala, S. V. Rao, and S. Kandlikar, "Ru(III) catalyzed oxidation of some aliphatic alcohols by N-bromosuccinimide (NBS). A kinetic study," *Reaction Kinetics and Catalysis Letters*, vol. 26, no. 1-2, pp. 209–213, 1984.

[27] P. G. Reddy, K. Ramesh, S. Shylaja, K. C. Rajanna, and S. Kandlikar, "Ru (III) catalyzed oxidation of aliphatic ketones by N-bromosuccinimide in aqueous acetic acid: a kinetic study," *The Scientific World Journal*, vol. 2012, Article ID 456516, 7 pages, 2012.

[28] K. N. Shivananda, B. Lakshmi, R. V. Jagadeesh, Puttaswamy, and K. N. Mahendra, "Mechanistic studies on the Ru(III)-catalyzed oxidation of some aromatic primary diamines by chloramine-T in hydrochloric acid medium: a kinetic approach," *Applied Catalysis A*, vol. 326, no. 2, pp. 202–212, 2007.

[29] K. N. Shivananda, R. V. Jagadeesh, Puttaswamy, and K. N. Mahendra, "Ru(III)-catalysed oxidation of some amines by chloramine-T in hydrochloric acid medium: mechanistic aspects and kinetic modeling," *Journal of Molecular Catalysis A*, vol. 255, no. 1-2, pp. 159–170, 2006.

[30] G. Caroling, J. Rajaram, and J. C. Kuriacose, "Kinetics and mechanism of Ru(III)-catalysed oxidation of organic sulphides and triphenylphosphine by N-methylmorpholine N-oxide," *Proceedings of the Indian Academy of Sciences*, vol. 100, no. 1, pp. 13–20, 1988.

[31] M. D. P. Rao, M. Ahmad, P. K. Pujapanda, and T. K. Kanungo, "Ru(III)-catalyzed oxidation of acetanilide, o-chloroacetanilide, p-chloroacetanilide and p-methylacetanilide by sodium perioda'te at pH4," *Reaction Kinetics and Catalysis Letters*, vol. 26, no. 3-4, pp. 375–379, 1984.

Predicting Heats of Explosion of Nitroaromatic Compounds through NBO Charges and ^{15}N NMR Chemical Shifts of Nitro Groups

Ricardo Infante-Castillo[1] and Samuel P. Hernández-Rivera[2]

[1] *Department of Physics and Chemistry, University of Puerto Rico, Arecibo Campus, P.O. Box 8152, Arecibo, PR 08152-00613, USA*
[2] *ALERT-DHS Center of Excellence/Center for Chemical Sensors Development, University of Puerto Rico, Mayagüez Campus, P.O. Box 9000, Mayaguez, PR 00681-9000, USA*

Correspondence should be addressed to Ricardo Infante-Castillo, ricinfante@gmail.com and Samuel P. Hernández-Rivera, samuel.hernandez3@upr.edu

Academic Editor: Michael D. Sevilla

This work presents a new quantitative model to predict the heat of explosion of nitroaromatic compounds using the natural bond orbital (NBO) charge and ^{15}N NMR chemical shifts of the nitro groups ($^{15}N_{Nitro}$) as structural parameters. The values of the heat of explosion predicted for 21 nitroaromatic compounds using the model described here were compared with experimental data. The prediction ability of the model was assessed by the leave-one-out cross-validation method. The cross-validation results show that the model is significant and stable and that the predicted accuracy is within 0.146 MJ kg^{-1}, with an overall root mean squared error of prediction (RMSEP) below 0.183 MJ kg^{-1}. Strong correlations were observed between the heat of explosion and the charges ($R^2 = 0.9533$) and ^{15}N NMR chemical shifts ($R^2 = 0.9531$) of the studied compounds. In addition, the dependence of the heat of explosion on the presence of activating or deactivating groups of nitroaromatic explosives was analyzed. All calculations, including optimizations, NBO charges, and $^{15}N_{Nitro}$ NMR chemical shifts analyses, were performed using density functional theory (DFT) and a 6-311+G(2d,p) basis set. Based on these results, this practical quantitative model can be used as a tool in the design and development of highly energetic materials (HEM) based on nitroaromatic compounds.

1. Introduction

Aromatic molecules with nitro groups are an important class of highly energetic materials (HEM). One of the most important thermodynamic properties that determine the performance of these explosives and propellants is the heat of explosion (HE). The HE is the quantity of heat liberated when an HEM undergoes detonation as a high explosive or deflagration as a low explosive. The processes of detonation and deflagration occur even in the absence of external oxygen or air because HEM contain oxygen themselves. The HE can be theoretically calculated [1] and experimentally determined [2]. The calculated value is determined by the difference between the energies of formation of the explosive components and the energies of formation of the explosion products. Experimentally, the HE is determined using a calorimetric bomb. The sample quantity is usually chosen to obtain a loading density of 0.1 g/cm^3. If a powder will not explode (compounds with heat of explosion lower than 800 cal/g), a "hot" powder with a known heat of explosion is added. Thus, the HE of the sample powder can be calculated from the mixture. The calorimetric values used in this work are based on liquid H_2O as a reaction product.

Predicting the performance and thermochemical properties of energetic materials from a given molecular structure with or without using experimental measurements is critical in the research and development of explosives. Several relationships have been found that relate explosive characteristics with measured and predicted molecular properties. The relationship between thermal [3], impact [4, 5], and electric spark [6] sensitivities and molecular structure of nitrocompounds and nitrates [7] has been the subject of many investigations. These studies concluded that the rate-determining step in the thermolysis was usually the homolytic cleavage of the $C–N_{Nitro}$, $N–N_{Nitro}$, and $O–N_{Nitro}$ bonds. In this context, the NO_2 (nitro) functional group corresponds to the reaction center of the HEM and should be correlated with the explosive characteristics of individual energetic materials. The electron configuration and the steric conditions within the reaction center of the molecule can be represented by NMR chemical shifts of the key atoms of the center. The shifts of these atoms should correlate with characteristics of initiation reactivity of individual energetic materials [6]. The dominant factor in the initiation by shock, electric spark, and in low-temperature thermolysis should be the electronic structure and close chemical ambient of the nitrogen atom of the primary leaving nitro group. In addition, for all nitro explosives in which the R-Nitro bond is the weakest, the charge in this group (Q_{Nitro}) can be regarded as another structural parameter to assess and predict explosive properties [8]. Recently, we established a quantitative model for predicting and calculating the heat of detonation and heat of explosion for a series of nitro paraffins, nitramines [9], and nitrate esters [10] by utilizing natural bond orbital (NBO) charge analysis and the ^{15}N NMR chemical shifts of the nitro groups. In addition, the calculated $^{15}N_{Nitro}$ NMR chemical shifts for the nitramines were obtained using a protocol established in our recent work [11], which uses the continuous set of gauge transformations (CSGT) method [12]. The *ab initio* calculations of the NMR chemical shifts for different nuclei have proven to be useful structural parameters in interpreting experimental data in several applications [13, 14]. There are many both empirical and theoretical efforts to predict detonation parameters from a given molecular structure. Rice and Hare [15] introduced a computational methodology that uses only quantum mechanical (QM) information about isolated molecules to predict the heats of detonation for pure and explosive formulations. In a series of studies, Keshavarz has established simple methods for predicting heats of detonation and other explosive properties [16–19]. In the application of these methods, there is no need to use any experimental and computed data of explosives. A simple approximation is introduced for calculating heats of detonation using chemical composition of explosives and their gas phase heat of formation that can be calculated by a group additivity rule [16]. Recently, a simple method to predict heats of detonation of important classes of energetic compounds including nitroaromatics, nitramines, nitroaliphatics, and nitrate esters was introduced [17]. The methodology is based on the ratios of oxygen to carbon and hydrogen to oxygen atoms as well as the contribution of some specific functional groups or structural parameters. Previously, Keshavarz's showed that the predicted values of the heats of detonation for aromatic compounds [18] are different from those of nonaromatic compounds [1] because aromaticity can affect the values of heats of detonation. The root mean squared (rms) deviation between predicted and experimentally determined heats of detonation was 0.32 and 0.54 kJ/g for aromatic and nonaromatic energetic compounds, respectively. Also, the heats of detonation for aromatic energetic materials were estimated assuming that the heat of detonation of an explosive compound of composition $C_aH_bN_cO_d$ can be approximated as the difference between the heat of formation of all H_2O-CO_2 arbitrary (H_2O, CO_2, N_2) detonation products and that of the explosive, divided by the formula weight of the explosive [19]. The rms deviation for this method was 0.34 kJ/g. Also, an empirical approach to calculate the heats of formation in condensed phase for different energetic compounds has been developed by Keshavarz [20]. It should be mentioned that the heat of explosion is not a constant of a high explosive and depends on the expansion ratio of the detonations products. The heat of explosion is reasonably included in the energetic characterization of the working capacity of commercial high explosives, but it has no relation to the power of brisance, which is determined by the propellant capacity. Even the detonation velocity is not a measure of the power of high explosives [21].

The purpose of this work was to establish a quantitative model to calculate and predict the heats of explosion of 35 nitroaromatic compounds from two structural parameters, nitro charges and ^{15}N NMR chemical shifts, obtained by theoretical methods. The predictive ability of the model was assessed by the leave-one-out (LOO) cross-validation method [22]. To show the reliability of the calculated heat of explosion using the quantitative model, the results for eight nitroaromatic compounds that were not used to build the model were compared with their experimental values. The model was applied to calculate heats of explosion of twenty energetic compounds and compared against results predicted using other empirical and quantum mechanical methods and against experimental values available. In addition, the model was employed to calculate the heat of explosion of fourteen nitroaromatic compounds whose experimental values were unavailable. Importantly, this work provides a simple and rapid method for estimating heats of explosion of HEM without experimental data for a systematic set of nitroaromatic compounds and should be of help in synthesizing and developing new high explosives.

2. Experimental Section

2.1. Computational Details. Gaussian 09 [23] software was used for all theoretical calculations. Molecular geometries were optimized using the B3LYP [24, 25] functional and 6-311+G(d,p) basis set. The absolute energies and $C–N_{Nitro}$ bond lengths for the thirty-five nitroaromatic compounds are shown in Tables 1, 2, and 3. The nitro group charges (Q_{Nitro}) were calculated according to an NBO scheme, which was implemented using Gaussian 09 software [26]. The $^{15}N_{Nitro}$ NMR calculations for all tested compounds were performed using the CSGT method and are reported in Tables 1–3. NMR shifts were computed using B3LYP/6-311+G(2d,p), and the values for the ^{15}N isotropic chemical shifts were referenced to the corresponding values for nitromethane. The effect of the solvent on the theoretical

TABLE 1: Absolute energies, code designations, $R_{C-N_{Nitro}}$ bond distances, nitro charges (Q_{Nitro}), $^{15}N_{Nitro}$ NMR calculated chemical shifts, and experimental and calculated heats of explosion for nitroaromatic compounds used to build a quantitative model.

Number	Chemical name	Code designation	Energy[a] (hartrees)	Distance $R_{C-N_{Nitro}}$ (Å)	NBO[b] charge $-Q_{Nitro}$	Calculated[c] $^{15}N_{Nitro}$ (ppm)	Experimental HE (H_2O liq.) (MJ kg^{-1})	Calculated[d] HE (MJ kg^{-1})
1	Tetryl	TETRYL	−1145	1.48604	0.191	−26.6	4.773	4.643
2	2,3,4,6-Tetranitroaniline	TeNA	−1105	1.47980	0.206	−25.6	4.378	4.336
3	2,4,6,2′,4′,6′-hexanitrodiphenylamine	HNDP	−1746	1.48079	0.210	22.8	4.075	4.099
4	Ethyl tetryl	EtTETRYL	−1185	1.48692	0.214	−23.1	4.058	4.052
5	2,4,6-Trinitrobenzoic acid	TNBA	−1035	1.48776	0.215	−23.6	3.964	4.066
6	2,4,6-Trinitrophenoxethyl nitrate	TNPHN	1280	1.48143	0.213	−21.3	3.911	3.959
7	Methyl picrate	MePi	−960	1.47516	0.222	−19.4	3.777	3.695
8	Ethyl picrate	EtPi	−999	1.47720	0.225	−19.1	3.515	3.628
9	2,4-Dinitrotoluene	24DNT	−681	1.48113	0.248	−15.7	3.192	3.043
10	2,4,6-Trinitrophenol (picric acid)	PiAc	−921	1.47425	0.228	−18.6	3.437	3.524
11	2,4,6-Trinitrocresol	TNC	−961	1.47516	0.231	−17.6	3.370	3.438
12	Dinitro-*ortho*-cresol	DNoC	−756	1.47521	0.244	14.4	3.027	3.029
13	*meta*-Dinitrobenzene	*m*DNB	−641	1.48414	0.250	9.5	2.666	2.632
							RMSEP	**0.0817**

[a]Calculated with B3LYP/6-311+G(d,p); [b]natural bond orbital analysis; [c]calculated with CSGT-B3LYP/6-311+G(2d,p); [d]predicted by (1).

Table 2: Absolute energies, code designations, $R_{C-N_{Nitro}}$ bond distances, nitro charges (Q_{Nitro}), $^{15}N_{Nitro}$ NMR calculated chemical shifts, and experimental and calculated heats of explosion for nitroaromatic compounds.

Number	Chemical name	Code designation	Energy[a] (hartrees)	Distance $R_{C-N_{Nitro}}$ (Å)	NBO[b] charge $-Q_{Nitro}$	Calculated[c] $^{15}N_{Nitro}$ (ppm)	Experimental HE (H_2O liq.) (MJ kg^{-1})	Calculated[d] HE (MJ kg^{-1})
14	1,3,5-Trinitrobenzene	TNB	−846	1.48024	0.219	−24.6	3.964	4.061
15	2,6-Dinitrotoluene	26DNT	−681	1.48266	0.240	−9.6	3.325	2.802
16	2,4,6-Trinitrotoluene	TNT	−885	1.48388	0.222	−18.9	3.766	3.665
17	2,4,6-Trinitroxylene	TNX	−925	1.48396	0.214	−12.4	3.533	3.400
18	2,4,6-Trinitroaniline	TNA	−901	1.47154	0.252	−21.3	3.589	3.318
19	2,4,6-Trinitro-1,3-dihydroxybenzene	TNR	−996	1.46312	0.243	−20.0	2.952	3.387
20	1,3,5-Trinitronaphtalene	135TNN	−999	1.48418	0.234	−15.3	3.521	3.248
21	Hexanitrostilbene	HNS	−1768	1.48668	0.205	−20.9	4.088	4.065
							RMSEP	**0.284**

[a]Calculated with B3LYP/6-311+G(d,p); [b]natural bond orbital analysis; [c]calculated with CSGT-B3LYP/6-311+G(2d,p); [d]predicted by (1).

TABLE 3: Absolute energies, code designations, $R_{C-N_{Nitro}}$ bond distances, nitro charges (Q_{Nitro}), $^{15}N_{Nitro}$ NMR calculated chemical shifts, and predicted heats of explosion for some nitroaromatic compounds.

Number	Chemical name	Code designation	Energy[a] (hartrees)	Distance $R_{C-N_{Nitro}}$ (Å)	NBO[b] charge $-Q_{Nitro}$	Calculated[c] $^{15}N_{Nitro}$ (ppm)	Calculated[d] HE ($MJ\,kg^{-1}$)
22	Nitrobenzene	NB	−437	1.48298	0.218	−8.7	3.109
23	*o*-Dinitrobenzene	*o*-DNB	−641	1.48063	0.207	−13.4	3.576
24	*p*-Dinitrobenzene	*p*-DNB	−641	1.48635	0.244	−19.4	3.344
25	3,5-Dinitrotoluene	35DNT	−631	1.48500	0.214	−19.3	3.393
26	1,3,5-Trimethyl-2,4,6-trinitrobenzene	TNMs	−963	1.48464	0.265	−77.0	2.276
27	1,3-Diamino-2,4,6,-trinitrobenzene	DATB	−957	1.45388	0.311	−21.1	2.337
28	1,3,5-Triamino-2,4,6-trinitrobenzene	TATB	−1012	1.43669	0.390	−20.0	1.301
29	1,3,5-Trihydroxy-2,4,6-trinitrobenzene	TNPg	−1071	1.45308	0.252	−20.9	3.294
30	2-amino-4,6-dinitrophenol (picramic acid)	PicAc	−772	1.47354	0.268	−16.3	2.751
31	Dinitrochlorobenzene	CDB	−1101	1.48101	0.225	−17.6	3.537
32	1-Chloro-2,4,6-trinitrobenzene	CTB	−1306	1.48454	0.203	−20.9	4.099
33	1,3-Dichloro-2,4,6-trinitrobenzene	DCTB	−1765	1.48489	0.169	−21.7	4.698
34	1,3,5-Trichloro-2,4,6-trinitrobenzene	TCTB	−2225	1.48645	0.143	−22.6	5.188
35	Tetranitronaphthalene	TeNN	−1204	1.48469	0.217	−19.5	3.783

[a]Calculated with B3LYP/6-311+G(d,p); [b]natural bond orbital analysis; [c]calculated with CSGT-B3LYP/6-311+G(2d,p); [d]predicted by (1).

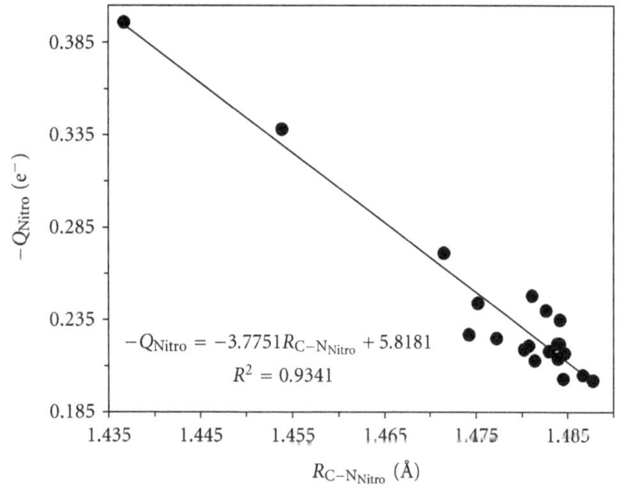

Figure 1: Dependence of the $R_{\text{C-N}_{\text{Nitro}}}$ bond lengths and the charges on the nitro group in nitroaromatic compounds.

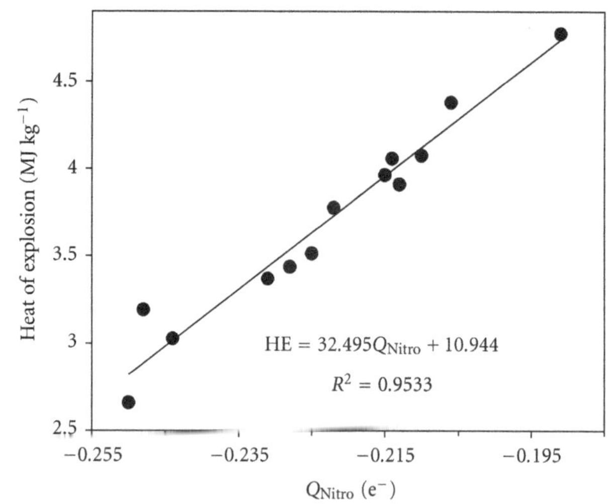

Figure 2: Relationship between heats of explosion and nitro charges (Q_{Nitro}) of nitroaromatic compounds.

NMR parameters was included using a default integral equation formalism polarized continuum model (IEF-PCM) [27] provided by the Gaussian 09 software suite. Dimethylsulfoxide (DMSO), which has a dielectric constant (ε) of 46.7, was used as the solvent. The experimental values of the heat of explosion used in the present paper were obtained from the literature [2].

3. Results and Discussions

3.1. Relationships between the Nitro Charges (Q_{Nitro}), the $R_{\text{C-N}_{\text{Nitro}}}$ Bond Lengths and Stability. The optimized C–N_{Nitro} bond lengths and nitro charges, with the corresponding total energies, for thirty-five nitroaromatic compounds are listed in Tables 1, 2 and 3. The charges on the nitro group (Q_{Nitro}) were calculated as the sum of the atomic charges on the nitrogen and oxygen atoms in the nitro group. It should be noted that the average values of $R_{\text{C-N}_{\text{Nitro}}}$ and Q_{Nitro} in polynitro compounds were used for all calculations, and these are listed in Tables 1–3. These results revealed that the compounds with higher stability (lower total energy) had higher nitro group charge values. Figure 1 shows a robust linear relationship ($R^2 = 0.9341$) between the $R_{\text{C-N}_{\text{Nitro}}}$ bond lengths and the nitro charges calculated with the NBO scheme.

Lower (more negative) nitro group charges were consistent with a decrease in the C–N_{Nitro} bond length. Therefore, the value of the charge on the nitro group reflects the strength of the corresponding bond and stability in these compounds. From the data shown in Tables 1–3, we concluded that the Q_{Nitro} was more sensitive to molecular changes than $R_{\text{C-N}_{\text{Nitro}}}$. For example, the $R_{\text{C-N}_{\text{Nitro}}}$ of all the isomers of dinitrobenzene (*o*, *m* and *p*) was 1.480 ± 0.006 Å; however, the Q_{Nitro} of these isomers was 0.207, 0.250, and 0.244 e, respectively. In addition, the average charge on the nitro groups was increased by substituents with σ- and π-donor ability (e.g., $-NH_2$, –OH, $-CH_3$) and decreased by substituents with σ- and π-electron acceptor ability (e.g., –COOH). The Q_{Nitro} for TNA, PiAc, TNT, TNB, and TNBA was 0.252, 0.228, 0.222, 0.219 and 0.215 e⁻, respectively.

All of the $R_{\text{C-N}_{\text{Nitro}}}$ values for the studied compounds were in the range of 1.4370 to 1.4880 Å. These values were higher than the calculated $R_{\text{C-N}_{\text{Nitro}}}$ bond lengths in nitramines (1.3640–1.4021 Å) and nitrate esters (1.4180–1.4370 Å) and lower than the $R_{\text{C-N}_{\text{Nitro}}}$ bond lengths in nitro paraffins (1.5029–1.5565 Å). Our calculations showed that the introduction of amino groups into multinitrobenzene decreased the NBO nitro group charges and shortened the C–N_{Nitro} bond length. This effect was strongest when one, two or three amino groups were introduced into 1,3,5-trinitrobenzene (TNB) to form TNA, DATB, and TATB, respectively. Global analysis of the results presented above indicated that the nitro group charges can be used to assess the stability and explosive properties of this class of compounds.

3.2. NBO Nitro Charges and Heat of Explosion. The experimental heats of explosion and calculated Q_{Nitro} values obtained by the NBO scheme for 35 nitroaromatic compounds are listed in Tables 1–3. As shown in Figure 2, the relationship between the calculated Q_{Nitro} values and the experimental heat of explosion is linear, and the correlation coefficient was 0.9533, indicating that the NBO charge calculated at the B3LYP/6-311+G(d,p) level efficiently reflected the molecular environment of these compounds. The nitroaromatic compounds with higher average nitro group charges had greater experimental heat of explosion.

3.3. $^{15}N_{\text{Nitro}}$ NMR Chemical Shifts and Heat of Explosion. To find the relationship between the $^{15}N_{\text{Nitro}}$ chemical shifts of the nitro group and the heat of explosion for the nitroaromatic compounds, the isotropic ^{15}N NMR shielding tensors were investigated theoretically at the B3LYP/6-311+G(2d,p) level within the CSGT model. In our recent work, a strong

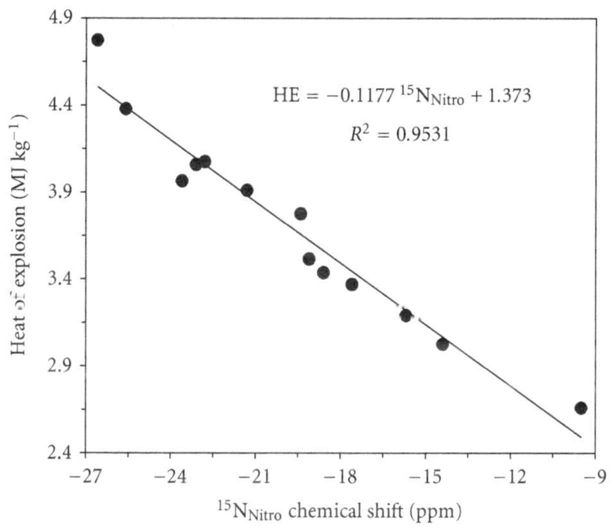

FIGURE 3: Relationship between heats of explosion and $^{15}N_{Nitro}$ NMR chemical shifts for nitroaromatic compounds.

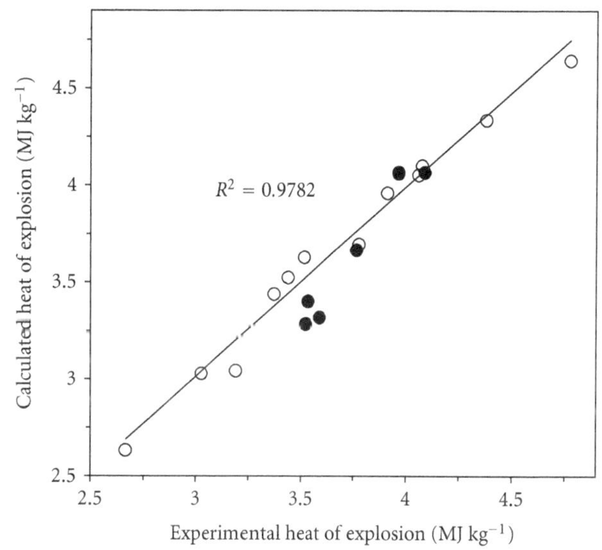

FIGURE 4: Correlation between calculated and experimental heats of explosion for nitroaromatic compounds. Molecules used to build a quantitative model (∘) and other nitroaromatic compounds (•).

correlation was found between the heat of detonation and heat of explosion for nitramines, nitro paraffins, and nitrate esters and the corresponding $^{15}N_{Nitro}$ chemical shifts. Zeman's work is noteworthy for its correlation between explosive properties (impact and electric sparks sensitivities, detonation, and thermal decomposition) and the ^{13}C and ^{15}N NMR chemical shifts of polynitro compounds [28]. The $^{15}N_{Nitro}$ NMR chemical shift results for the 35 nitroaromatic studied are listed in Tables 1–3. The correlation between heat of explosion and the calculated $^{15}N_{Nitro}$ chemical shifts is shown in Figure 3. As seen in this figure, the heat of explosion increased with a decreasing $^{15}N_{Nitro}$ chemical shift. The regression coefficient (R^2) obtained for the correlation between the heat of explosion and $^{15}N_{Nitro}$ chemical shifts was 0.9531. Although NMR studies in solution neglect important crystal-lattice interactions affecting some explosive properties, this particular correlation provided a consistent method for the assessment of the heat of detonation without the use of experimental data for a systematic set of nitroaromatic compounds.

3.4. Quantitative Relationships between the Heat of Explosion (HE), Q_{Nitro} and $^{15}N_{Nitro}$ NMR Chemical Shifts. To verify the relationship between the two parameters described in the previous section (Q_{Nitro} and $^{15}N_{Nitro}$) and the heat of explosion, 13 nitroaromatic compounds, each with a heat of explosion that had been measured experimentally, were correlated with our calculated values. These were obtained from the theoretical structural parameters (listed in Table 1). The following equation was determined from the linear regression fit:

$$\mathrm{HE} = 16.423 \times (Q_{Nitro}) - 0.06 \times (^{15}N_{Nitro}) - 6.15. \quad (1)$$

The significance of the results can be evaluated by the results of the statistical parameters R^2 = 0.9782, SE = 0.093, and RMSEP = 0.0817 MJ kg^{-1}, where R^2, SE, and RMSEP are the correlation coefficient, the standard error, and the root mean squared error prediction, respectively. As seen from (1), a nitroaromatic explosive with a large average Q_{Nitro} value and a low $^{15}N_{Nitro}$ chemical shift has a high heat of explosion. The stability and prediction capability of the quantitative model established by (1) was assessed using the leave-one-out (LOO) cross-validation method. In the LOO method procedure, one explosive is removed from the training set (the explosives are listed in Table 1), and a model is developed with remaining N-1 explosives. At each step, the heat of explosion value of the removed explosive is predicted. This process was repeated until each molecule of the training set is predicted. The prediction capability of the model was quantified in terms of r_{cv} and S_{cv}, which are defined as follows:

$$r_{cv}^2 = 1.0 - \frac{\sum_{i-1}^{n}(y_i - \hat{y}_i)^2}{\sum_{i=1}^{n}(y_i - \overline{y})^2},$$
$$S_{cv} = \sqrt{\frac{\sum_{i-1}^{n}(y_i - \hat{y}_i)^2}{N - D - 1}}, \quad (2)$$

where y_i and $\hat{y}_i$ are the experimental and predictive values, respectively, $\overline{y}$ is the mean value of y_i, N is the number of explosives to be used for building the model, and D is the number of descriptors in the model. The r_{cv} and S_{cv} values were 0.9727 and 0.146, respectively, which indicated that (1) was reasonably stable. The predicted accuracy of (1) for the heat of explosion of the nitroaromatic compounds was within 0.146 MJ kg^{-1}. Figure 4 shows a comparison between the experimental heat of explosion, as predicted by (1), and the corresponding regression coefficient for the nitroaromatic compounds. The calculated heats of explosion

Table 4: Comparison of calculated heats of explosion, for nitroaromatic compounds, according to (1), HE_{R-H} [15], HE_{MHK1} [18], and HE_{MHK2} [17] method with experimental values [2].

Code designation	Experimental HE (H_2O liq.) ($MJ\ kg^{-1}$)	HE_{R-H} (H_2O liq.) ($MJ\ kg^{-1}$)			HE_{MHK2} ($MJ\ kg^{-1}$)	HE_{MHK1} ($MJ\ kg^{-1}$)	Calculated HE ($MJ\ kg^{-1}$)
		(K-J, t)[a]	(K-J, e)[b]	(K-Jmod., e)[c]			
Tetryl	4.773	5.318	6.234	4.740	4.118	4.205	4.643
TeNA	4.378	5.485	6.138	4.912	4.245	4.594	4.336
TNB	3.964	5.067	5.845	4.711	3.617	4.109	4.061
24DNT	3.192	3.795	5.322	3.740	3.225	2.959	3.043
26DNT	3.325	3.908	5.435	3.845	3.225	2.959	2.802
TNT	3.766	4.694	5.795	4.259	3.546	3.725	3.665
TNA	3.589	4.656	5.565	4.368	3.822	4.004	3.318
PiAc	3.437	4.853	5.602	4.305	3.751	3.321	3.524
*m*DNB	2.666	4.289	5.443	3.870	3.228	3.388	2.632
HNS	4.088	5.017	5.749	4.284	4.027	3.934	4.065
	RMSEP	**1.046**	**2.023**	**0.671**	**0.332**	**0.362**	**0.204**

[a]Heat of detonation calculated using Kamlet-Jacobs method. The t denotes that the values used for the heats of formation for the products are predicted. [b]The e denotes that the values used for the heats of formation for the products are the experimental values. [c]Modified Kamlet and Jacobs method.

TABLE 5: Comparison of calculated heats of explosion for nitrate esters according to HE_{I-H} [10] and HE_{MHK2} method [17] with experimental values [2].

Chemical name	Experimental HE (H_2O liq.) ($MJ\,kg^{-1}$)	HE_{MHK2} ($MJ\,kg^{-1}$)	HE_{I-H} ($MJ\,kg^{-1}$)
Erythritol tetranitrate	6.356	6.555	6.362
Glycerol trinitrate	6.671	6.520	6.847
n-propyl nitrate	3.272	2.963	3.377
Isopropyl nitrate	3.126	2.963	3.106
Diethylene glycol dinitrate	4.566	4.624	4.389
Dioxiethylnitramine dinitrate	5.458	5.019	5.307
Dipentaerythritol hexanitrate	5.143	4.995	5.523
Butanetriol trinitrate	6.022	5.464	5.824
Metriol trinitrate	4.945	4.774	5.202
Pentaerythritol tetranitrate	6.322	5.717	6.316
	RMSEP	**0.333**	**0.186**

values were very similar to the values obtained in the experiments. These results validate our new method for calculating heat of explosion using the NBO charges on and ^{15}N NMR chemical shifts of nitro groups as structural parameters.

3.5. Predicting the Heat of Explosion for Nitroaromatic Explosives. The predicted detonation heat by (1) for 14 nitroaromatic explosives with unmeasured HEs is given in Table 3. Based on the HE predictions for the halogen derivatives of nitrobenzene, the HE had the following trend: CDB < CTB < DCTB < TCTB. Nitroaromatic compounds with a higher number of halogen atoms exhibit a higher heat of explosion. However, this theoretical study was carried out in solution neglect important crystal-lattice effects that are important in the determination of explosive properties. Zeman's work [29] established an inverse relationship between the volatility and Chapman-Jouguet (C-J) pressure. The increased volatility due to multiple substitution is linked with the deformation of the molecules that accompany the substitution. Therefore, it is of utmost importance the intermolecular nonbonding interaction. Global analysis of the results presented in Tables 1–3 indicated that the HE values of 2,4,6-trinitrobenzene derivatives decreased systematically with the number of activating groups in the molecular structure of the nitroaromatic explosive. The HE trend for the halogenated nitrobenzenes agreed with the σ- and π-donor ability of the compounds. In addition, the heat of explosion for 2,4,6-trinitrobenzene monosubstituted derivatives increased with substituents that had a larger σ- and π-electron acceptor ability, which was consistent with smaller nitro group charges and $^{15}N_{Nitro}$ chemical shifts.

3.6. Comparison of the Calculated Heats of Explosion with Other Empirical and Quantum Mechanical Methods. The heats of explosion for various nitro-aromatic and nitrate esters are given in Tables 4 and 5 and compared with experimental values, two empirical methods (KHM1 [18] and KHM2 [17]) and Rice-Hare methods [15]. Predicted heats of explosion for $H_2O_{(l)}$ using quantum mechanical approach including predicted and experimental values for the heat of formation have RMSEP from experiment of 1.046 and 2.023 $MJ\,kg^{-1}$, respectively, whereas the modified version of Kamlet Jacobs's method results has a substantially lower deviation from experiment (0.671 $MJ\,kg^{-1}$). As shown in Table 4, the RMSEP of empirical methods (0.332 and 0.362 $MJ\,kg^{-1}$) are lower than the quantum mechanical methods. As indicated in Tables 4 and 5, the predicted heats of explosion calculated by the new model deviated from experiment by less than 0.204 $MJ\,kg^{-1}$. As is evident, the proposed model based on the calculation of two structural parameters shows the best agreement with experimental data.

This method for predicting the heat of explosion of nitroaromatic explosives, that is, using NBO charges and ^{15}N NMR chemical shifts as references, can be generalized for other homologous systems if these parameters are calculated with the same computational approach. Hence, the computation of the heat of explosion equation will be useful in the design and development of strong nitroaromatic explosives when experimental data are not available. However, the heat of explosion depends on many factors besides nitro group charges or $^{15}N_{Nitro}$ NMR chemical shifts; a completely monotonic correlation between these factors was not found, which indicates that this quantitative model is only applicable to the nitroaromatic explosives whose stabilities are determined by the chemical bonds linked to the nitro group. Based on the previous discussion, it can be concluded that a consistent correlation between the heat of explosion and the Q_{Nitro} and $^{15}N_{Nitro}$ chemical shifts exists in such explosives.

4. Conclusion

NBO charge and ^{15}N NMR chemical shift analyses are two key parameters that can be used to determine the heat of explosion of aromatic explosives. Using these two structural parameters, a good quantitative model was established for predicting the heat of explosion of 35 nitroaromatic explosives. The model was significant and stable and displayed prediction accuracy within 0.146 $MJ\,kg^{-1}$, with an overall root mean squared error of prediction (RMSEP) below

$0.183\,\mathrm{MJ\,kg^{-1}}$. The heat of explosion of nitroaromatic compounds decreased systematically with the type and number of activating substituents (σ- and π-donors) and increased with the number of σ- and π-electron acceptors added to the corresponding molecular structure. Based on these results, the proposed model could be used to predict the heat of explosion without experimental data for a systematic set of nitroaromatic compounds and could aid in the design of these highly energetic materials. The calculated heats of explosion by this model give better results than other empirical and quantum mechanical methods available in the literature.

Acknowledgments

This work was supported by the Center for Chemical Sensors Development (CCSD) of the Chemistry Department of the University of Puerto Rico, Mayagüez and the Research and Creation Center (RCC) of the University of Puerto Rico, Arecibo. The authors also acknowledge the U.S. Department of Homeland Security for its support (Award number 2008-ST-061-ED0001). However, the views and conclusions contained in this document are those of the authors and should not be interpreted as necessarily representing the official policies, either expressed or implied, of the U.S. Department of Homeland Security.

References

[1] M. H. Keshavarz, "Determining heats of detonation of non-aromatic energetic compounds without considering their heats of formation," *Journal of Hazardous Materials*, vol. 142, no. 1-2, pp. 54–57, 2007.

[2] R.. Meyer, J. Köhler, and A. Homburg, *Explosives*, Wiley-VHC, Weinheim, Germany, 2007.

[3] A. Delpuech and J. Cherville, "Relation entre La structure electronique et la sensibilité au choc des explosifs seconaires nitrés," *Propellants Explosives, Pyrotechnics*, vol. 4, no. 6, pp. 121–128, 1979.

[4] C. Cao and S. Gao, "Two dominant factors influencing the impact sensitivities of nitrobenzenes and saturated nitro compounds," *Journal of Physical Chemistry B*, vol. 111, no. 43, pp. 12399–12402, 2007.

[5] B. M. Rice and J. J. Hare, "A quantum mechanical investigation of the relation between impact sensitivity and the charge distribution in energetic molecules," *Journal of Physical Chemistry A*, vol. 106, no. 9, pp. 1770–1783, 2002.

[6] S. Zeman, "New aspects of initiation reactivities of energetic materials demonstrated on nitramines," *Journal of Hazardous Materials*, vol. 132, no. 2-3, pp. 155–164, 2006.

[7] M. A. Hiskey, K. R. Brower, and J. C. Oxley, "Thermal decomposition of nitrate esters," *Journal of Physical Chemistry*, vol. 95, no. 10, pp. 3955–3960, 1991.

[8] C. Zhang, "Review of the establishment of nitro group charge method and its applications," *Journal of Hazardous Materials*, vol. 161, no. 1, pp. 21–28, 2009.

[9] S. P. Hernández-Rivera and R. Infante-Castillo, "A systematic theoretical investigation of the relationship between heats of detonation and NBO charges and ^{15}N NMR chemical shifts of nitro groups in nitramines and nitro paraffins," *Journal of Molecular Structure*, vol. 960, no. 1–3, pp. 57–62, 2010.

[10] S. P. Hernández-Rivera and R. Infante-Castillo, "Predicting heats of explosion of nitrate esters through their NBO charges and ^{15}N NMR chemical shifts on the nitro groups," *Computational and Theoretical Chemistry*, vol. 963, no. 2-3, pp. 279–283, 2011.

[11] R. Infante-Castillo and S. P. Hernández-Rivera, "On the choice of optimal protocol for calculation of ^{13}C and ^{15}N NMR isotropic chemical shifts in nitramine systems," *Journal of Molecular Structure*, vol. 940, no. 1–3, pp. 124–128, 2010.

[12] T. A. Keith and R. F. W. Bader, "Calculation of magnetic response properties using a continuous set of gauge transformations," *Chemical Physics Letters*, vol. 210, no. 1–3, pp. 223–231, 1993.

[13] R. Infante-Castillo and S. P. Hernández-Rivera, "Experimental and theoretical studies of the molecular structure of 1-(2-pyridinylmethyl)-2-methylbenzimidazole," *Journal of Molecular Structure*, vol. 968, no. 1–3, pp. 6–12, 2010.

[14] R. Infante-Castillo and S. P. Hernández-Rivera, "DFT studies of the structure and vibrational and NMR spectra of 1-(2-methylpropenyl)-2-methylbenzimidazole," *Journal of Molecular Structure*, vol. 917, no. 2-3, pp. 158–163, 2009.

[15] B. M. Rice and J. Hare, "Predicting heats of detonation using quantum mechanical calculations," *Thermochimica Acta*, vol. 384, no. 1-2, pp. 377–391, 2002.

[16] M. H. Keshavarz, "Simple procedure for determining heats of detonation," *Thermochimica Acta*, vol. 428, no. 1-2, pp. 95–99, 2005.

[17] M. H. Keshavarz, "A simple way to predict heats of detonation of energetic compounds only from their molecular structures," *Propellants, Explosives, Pyrotechnics*, vol. 37, no. 1, pp. 93–99, 2012.

[18] M. H. Keshavarz, "Quick estimation of heats of detonation of aromatic energetic compounds from structural parameters," *Journal of Hazardous Materials*, vol. 143, no. 1-2, pp. 549–554, 2007.

[19] M. H. Keshavarz, "Estimating heats of detonation and detonation velocities of aromatic energetic compounds," *Propellants, Explosives, Pyrotechnics*, vol. 33, no. 6, pp. 448–453, 2008.

[20] M. H. Keshavarz, "Theoretical prediction of condensed phase heat of formation of nitramines, nitrate esters, nitroaliphatics and related energetic compounds," *Journal of Hazardous Materials*, vol. 136, no. 2, pp. 145–150, 2006.

[21] V. I. Pepekin and S. A. Gubin, "Heat of explosion of commercial and brisant high explosives," *Combustion, Explosion and Shock Waves*, vol. 43, no. 2, pp. 212–218, 2007.

[22] L. M. Hall, L. H. Hall, and L. B. Kier, "QSAR modeling of β-lactam binding to human serum proteins," *Journal of Computer-Aided Molecular Design*, vol. 17, no. 2–4, pp. 103–118, 2003.

[23] M. J. Frisch, G. W. Trucks, H. B. Schlegel et al., *Gaussian 09 Rev. A.1.*, CT. Gaussian, Wallingford, UK, 2009.

[24] A. D. Becke, "Density-functional thermochemistry. III. The role of exact exchange," *Journal of Chemical Physics*, vol. 98, no. 7, pp. 5648–5652, 1993.

[25] C. Lee, W. Yang, and R. G. Parr, "Development of the Colle-Salvetti correlation-energy formula into a functional of the electron density," *Physical Review B*, vol. 37, no. 2, pp. 785–789, 1988.

[26] E. D. Glendening, A. E. Reed, J. E. Carpenter, and F. Weinhold, *NBO Version 3.1*, 2009.

[27] B. Mennucci and J. Tomasi, "Continuum solvation models: a new approach to the problem of solute's charge distribution and cavity boundaries," *Journal of Chemical Physics*, vol. 106, no. 12, pp. 5151–5158, 1997.

[28] S. Zeman, "Modified Evans-Polanyi-Semenov relationship in the study of chemical micromechanism governing detonation initiation of individual energetic materials," *Thermochimica Acta*, vol. 384, no. 1-2, pp. 137–154, 2002.
[29] S. Zeman, "The relationship between differential thermal analysis data and the detonation characteristics of polynitroaromatic compounds," *Thermochimica Acta*, vol. 41, no. 1, pp. 199–212, 1980.

Excess Transport Properties of Binary Mixtures of Quinoline with Xylenes at Different Temperatures

Sk. Fakruddin,[1] Ch. Srinivasu,[2] B. R. Venkateswara Rao,[1] and K. Narendra[1]

[1] *Department of Physics, Velagapudi Ramakrishna Siddhartha Engineering College, Andhra Pradesh, Vijayawada 520007, India*
[2] *Department of Physics, Andhra Loyola College, Vijayawada 520008, India*

Correspondence should be addressed to K. Narendra, narenk75@gmail.com

Academic Editor: Leonardo Palmisano

The ultrasonic velocity and density of binary liquid mixtures of quinoline with o-xylene, m-xylene, and p-xylene have been measured over the entire range of composition at T = 303.15, 308.15, 313.15, and 318.15 K. Using these data, various parameters like adiabatic compressibility (β), intermolecular free length (L_f), and acoustic impedance (Z) and some excess parameters like excess adiabatic compressibility (β^{E}), excess intermolecular free length (L_f^E), excess acoustic impedance (Z^E), and excess ultrasonic velocity (u^E) have been calculated for all the three mixtures. The calculated deviations and excess functions have been fitted to Redlich-Kister polynomial equation. The observed deviations have been explained on the basis of the intermolecular interactions present in these mixtures.

1. Introduction

In studying the molecular interactions and physicochemical behaviour of binary liquid mixtures [1, 2], the knowledge of thermodynamic and acoustical properties is of great importance. The results are frequently used in process design in many chemical and industrial processes. The studies of binary mixtures containing aromatic hydrocarbons are interesting because they find applications in preferential interactions of polymers in mixed solvents and the studies of polymer phase diagrams. Quinoline is widely used in the manufacturing of dyes, pesticides, and solvent for resins and terpenes. Xylenes are used in printing, rubber, and leather industries. Basically, a binary liquid mixture is formed by the replacement of like contacts in pure liquids by unlike contacts in the mixture. The quantitative and qualitative analyses of excess functions provide information about the nature of molecular interactions in the binary mixtures [3–5]. The literature survey reveals that there has been practically no study of the binary mixtures of these systems from the point of view of their ultrasonic behaviour.

In an attempt to explain the nature of interactions occurring between quinoline and xylenes, ultrasonic velocity and density of binary liquid mixtures of quinoline + o-xylene/m-xylene/p-xylene have been determined over the entire range of composition at T = 303.15, 308.15, 313.15, and 318.15 K. Using the experimental values of u and ρ, adiabatic compressibility (β), intermolecular free length (L_f), and acoustic impedance (Z) and some excess parameters like excess adiabatic compressibility (β^E), excess intermolecular free length (${L_f}^E$), excess acoustic impedance (Z^E), and excess ultrasonic velocity (u^E) have been calculated. The results were fitted to the Redlich-Kister polynomial equation. In the light of excess parameters, the intermolecular interactions have been estimated [8].

2. Experimental

2.1. Materials. The mass fraction of the liquids (obtained from Merck) is as follows: o-xylene (0.980), m-xylene (0.980), p-xylene (0.990), and quinoline (0.987). All the liquids used were further purified by standard procedure [9]. The purity of the samples was checked by comparing the experimental values of ultrasonic velocity and density with those available in the literature [6, 7] and these values are compiled in Table 1.

TABLE 1: Comparison of experimental values of ultrasonic velocities (u) and densities (ρ) of pure liquids with the values reported in the literature.

Liquid	T/K	$u/(\mathrm{m\cdot s^{-1}})$ Experimental	$u/(\mathrm{m\cdot s^{-1}})$ Literature	$\rho/(\mathrm{Kg\cdot m^{-3}})$ Experimental	$\rho/(\mathrm{Kg\cdot m^{-3}})$ Literature
Quinoline	303.15	1553.68	1547[a]	1085.45	1085.79[a]
o-xylene	308.15	1338.75	1328.30[b]	870.70	870.73[b]
m-xylene	313.15	1304.21	1300.34[b]	855.70	855.47[b]
p-xylene	318.15	1288.43	1289.68[b]	852.80	852.26[b]

[a]Nath [6].
[b]Al-Kandary et al. [7].

TABLE 2: Ultrasonic velocities (u) and densities (ρ) for binary mixtures at T = 303.15, 308.15, 313.15, and 318.15 K.

x_1	T = 303.15 K $u/(\mathrm{m\cdot s^{-1}})$	T = 303.15 K $\rho/(\mathrm{Kg\cdot m^{-3}})$	T = 308.15 K $u/(\mathrm{m\cdot s^{-1}})$	T = 308.15 K $\rho/(\mathrm{Kg\cdot m^{-3}})$	T = 313.15 K $u/(\mathrm{m\cdot s^{-1}})$	T = 313.15 K $\rho/(\mathrm{Kg\cdot m^{-3}})$	T = 318.15 K $u/(\mathrm{m\cdot s^{-1}})$	T = 318.15 K $\rho/(\mathrm{Kg\cdot m^{-3}})$
{Quinoline (1) + o-xylene (2)}								
0.0000	1338.75	870.70	1315.00	869.40	1297.50	867.70	1278.15	865.90
0.1022	1384.84	901.40	1368.21	896.79	1357.89	893.69	1348.42	891.31
0.2039	1404.42	922.99	1386.95	919.02	1376.84	915.51	1366.21	911.50
0.3051	1424.21	942.22	1408.89	936.59	1399.42	932.06	1387.95	929.86
0.4058	1444.32	961.85	1430.84	956.70	1422.53	952.25	1411.02	948.11
0.5060	1465.26	981.69	1453.95	977.92	1445.79	974.98	1436.16	967.28
0.6058	1484.84	1001.73	1475.37	998.64	1469.05	995.28	1460.42	990.71
0.7050	1504.47	1027.16	1496.32	1022.00	1492.00	1018.22	1484.05	1013.23
0.8038	1523.58	1047.92	1517.95	1042.71	1513.47	1039.33	1506.68	1036.86
0.9021	1541.21	1067.14	1538.42	1063.23	1534.95	1060.07	1528.63	1057.61
1.0000	1553.68	1085.45	1550.68	1082.11	1547.37	1078.60	1541.05	1074.99
{Quinoline (1) + m-xylene (2)}								
0.0000	1304.21	855.70	1285.27	848.70	1266.32	845.80	1244.21	840.50
0.1038	1375.05	887.18	1362.89	880.54	1354.74	875.63	1345.26	871.03
0.2067	1398.26	913.84	1384.79	910.65	1373.68	904.25	1361.05	900.75
0.3088	1420.05	931.94	1407.74	925.52	1395.26	921.60	1382.63	918.29
0.4100	1440.16	952.56	1428.68	945.83	1418.37	942.91	1406.89	940.30
0.5104	1461.95	978.74	1452.63	974.27	1443.47	969.91	1432.84	966.98
0.6099	1482.53	998.14	1474.35	996.30	1466.89	992.34	1457.26	989.29
0.7086	1502.32	1014.85	1495.16	1011.43	1488.84	1009.59	1479.89	1006.23
0.8066	1521.42	1043.74	1516.79	1038.14	1511.32	1034.56	1503.21	1031.79
0.9037	1532.89	1066.84	1529.26	1062.01	1525.79	1059.32	1519.47	1056.74
1.0000	1553.68	1085.45	1550.68	1082.11	1547.37	1078.60	1541.05	1074.99
{Quinoline (1) + p-xylene (2)}								
0.0000	1288.43	852.80	1275.79	846.00	1253.68	843.00	1234.73	836.90
0.1041	1371.89	885.76	1361.74	878.51	1351.57	874.51	1342.10	869.20
0.2073	1395.10	909.41	1384.63	903.39	1375.52	901.61	1360.89	898.74
0.3095	1415.89	929.81	1406.57	918.62	1402.10	917.24	1389.47	915.55
0.4108	1436.00	949.85	1428.02	942.34	1424.21	938.40	1412.73	938.07
0.5112	1457.79	976.74	1451.47	971.93	1446.31	969.00	1434.68	965.25
0.6107	1480.37	997.73	1473.89	995.26	1472.73	991.22	1463.10	988.58
0.7093	1500.15	1013.63	1494.50	1010.43	1493.68	1009.37	1484.73	1005.40
0.8071	1518.26	1043.03	1514.63	1037.13	1509.16	1033.54	1501.05	1030.18
0.9040	1530.73	1066.53	1527.10	1061.60	1522.63	1058.00	1515.31	1056.23
1.0000	1553.68	1085.45	1550.68	1082.11	1547.37	1078.60	1541.05	1074.99

TABLE 3: Coefficients of Redlich-Kister equation (A_i) and standard deviations (σ) for excess adiabatic compressibility (β^E), excess intermolecular free length ($L_f{}^E$), excess acoustic impedance (Z^E), and excess ultrasonic velocity (u^E) for binary liquid mixtures.

Properties	T/K	A_0	A_1	A_2	σ
		{Quinoline (1) + o-xylene (2)}			
$10^{-11}\,\beta^E/(\mathrm{m^2 \cdot N^{-1}})$	303.15	−14.2750	15.7470	−19.3987	0.6231
	308.15	−28.7822	50.6552	−42.1525	2.1127
	313.15	−17.6239	21.2062	−27.0003	0.9585
	318.15	−18.7916	26.9112	−35.7657	1.2192
$10^{-4}\,L_f{}^E/(\mathrm{m})$	303.15	−15.4753	18.3256	−25.1069	0.7669
	308.15	−17.0736	20.3448	−29.2354	0.9519
	313.15	−18.8797	24.2345	−34.6635	1.1721
	318.15	−19.8516	30.8846	−45.9429	1.4765
$10^4\,Z^E/(\mathrm{Kg \cdot m^2 \cdot s^{-1}})$	303.15	3.8618	−13.7678	27.2629	0.6353
	308.15	3.3493	−12.9783	28.4726	0.7100
	313.15	3.7404	−14.2036	31.6728	0.8641
	318.15	1.8696	−17.0587	42.6292	0.9230
$u^E/(\mathrm{m \cdot s^{-1}})$	303.15	72.2489	−105.2525	169.0334	5.2762
	308.15	80.2265	−123.8911	219.0926	6.9440
	313.15	89.2298	−154.6148	264.9103	8.2814
	318.15	102.3393	−204.3521	328.8972	11.0408
		{Quinoline (1) + m-xylene (2)}			
$10^{-11}\,\beta^E/(\mathrm{m^2 \cdot N^{-1}})$	303.15	−21.9058	31.4522	−26.7769	1.1162
	308.15	−24.3389	36.3768	−31.2803	1.3641
	313.15	−26.3980	42.3813	−39.2213	1.7172
	318.15	−30.0673	51.5898	−49.7464	2.1494
$10^{-4}\,L_f{}^E/(\mathrm{m})$	303.15	−24.3607	37.6519	−32.0435	1.3789
	308.15	−26.8359	43.2278	−37.1303	1.6805
	313.15	−28.8479	49.9906	−46.9235	2.0995
	318.15	−32.7902	60.3208	−59.4502	2.5997
$10^4\,Z^E/(\mathrm{Kg \cdot m^2 \cdot s^{-1}})$	303.15	10.0579	−27.656	25.6108	1.0966
	308.15	10.5394	−30.9755	26.8524	1.3238
	313.15	10.1966	−32.6801	33.4784	1.4769
	318.15	11.9598	−37.0021	41.4323	1.6993
$u^E/(\mathrm{m \cdot s^{-1}})$	303.15	127.5056	−281.9268	210.6969	11.5612
	308.15	134.0042	−310.4828	252.1364	13.3956
	313.15	142.3595	−363.6245	328.7575	16.7309
	318.15	157.3428	−427.7324	412.1425	20.2179
		{Quinoline (1) + p-xylene (2)}			
$10^{-11}\,\beta^E/(\mathrm{m^2 \cdot N^{-1}})$	303.15	−24.3371	40.1603	−34.8069	1.4993
	308.15	−26.1737	42.5118	−35.8559	1.7109
	313.15	−30.5317	50.6633	−42.1002	1.9289
	318.15	−33.3037	58.4176	−51.3519	2.2552
$10^{-4}\,L_f{}^E/(\mathrm{m})$	303.15	−26.8705	48.104	−41.4238	1.8424
	308.15	−28.7822	50.6552	−42.1525	2.1127
	313.15	−33.8045	59.9962	−48.8655	2.3611
	318.15	−36.5393	68.562	−59.8297	2.7239

TABLE 3: Continued.

Properties	T/K	A_0	A_1	A_2	σ
$10^4 Z^E/(\text{Kg}\cdot\text{m}^2\cdot\text{s}^{-1})$	303.15	10.6611	−35.6773	32.3543	1.4669
	308.15	10.9532	−36.633	29.7221	1.7861
	313.15	13.5849	−41.1183	31.3799	1.8255
	318.15	14.1107	−43.7213	39.128	1.8140
$u^E/(\text{m}\cdot\text{s}^{-1})$	303.15	143.0469	−360.5048	271.5399	14.6146
	308.15	148.7841	−368.4327	277.8589	15.3763
	313.15	179.5480	−431.3940	319.9555	17.7116
	318.15	184.2120	−485.6012	383.7827	21.1574

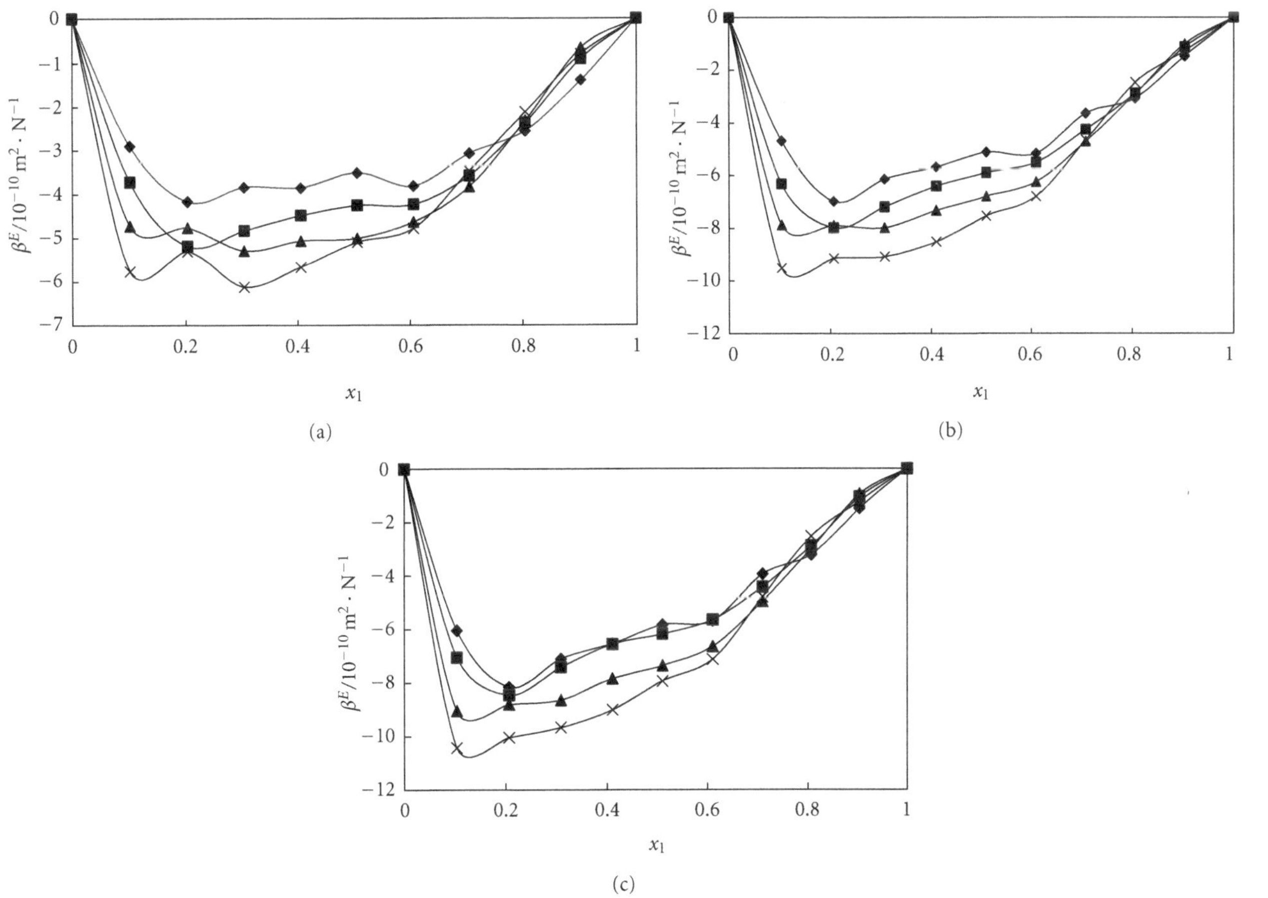

FIGURE 1: (a) Plot of excess adiabatic compressibility (β^E) versus mole fraction (x_1) for binary liquid mixtures of quinoline with o-xylene at temperatures: ♦, 303.15 K; □, 308.15 K; ▲, 313.15 K; ×, 318.15 K. (b) Plot of excess adiabatic compressibility (β^E) versus mole fraction (x_1) for binary liquid mixtures of quinoline with m-xylene at temperatures: ♦, 303.15 K; □, 308.15 K; ▲, 313.15 K; ×, 318.15 K. (c) Plot of excess compressibility (β^E) versus mole fraction (x_1) for binary liquid mixtures of quinoline with p-xylene at temperatures: ♦, 303.15 K; □, 308.15 K; ▲, 313.15 K; ×, 318.15 K.

2.2. Procedure. Job's method of continuous variation was used to prepare the mixtures of required proportions. The prepared mixtures were preserved in well-stoppard conical flasks. After mixing the liquids thoroughly, the flasks were left undisturbed to allow them to attain thermal equilibrium.

The ultrasonic velocities were measured by using single crystal ultrasonic pulse echo interferometer (Mittal enterprises, India; Model: F-80X). It consists of a high-frequency generator and a measuring cell. The measurements of ultrasonic velocities were made at a fixed frequency of 3 MHz. The capacity of the measuring cell is 12 mL. The calibration of the equipment was done by measuring the velocity in carbon tetrachloride and benzene. The results are in good agreement with those reported in the literature [10]. The ultrasonic velocity has an accuracy of $\pm 0.5\,\text{ms}^{-1}$. The temperature was controlled by circulating water around the liquid cell from thermostatically controlled constant temperature water bath (accuracy ± 0.01 K).

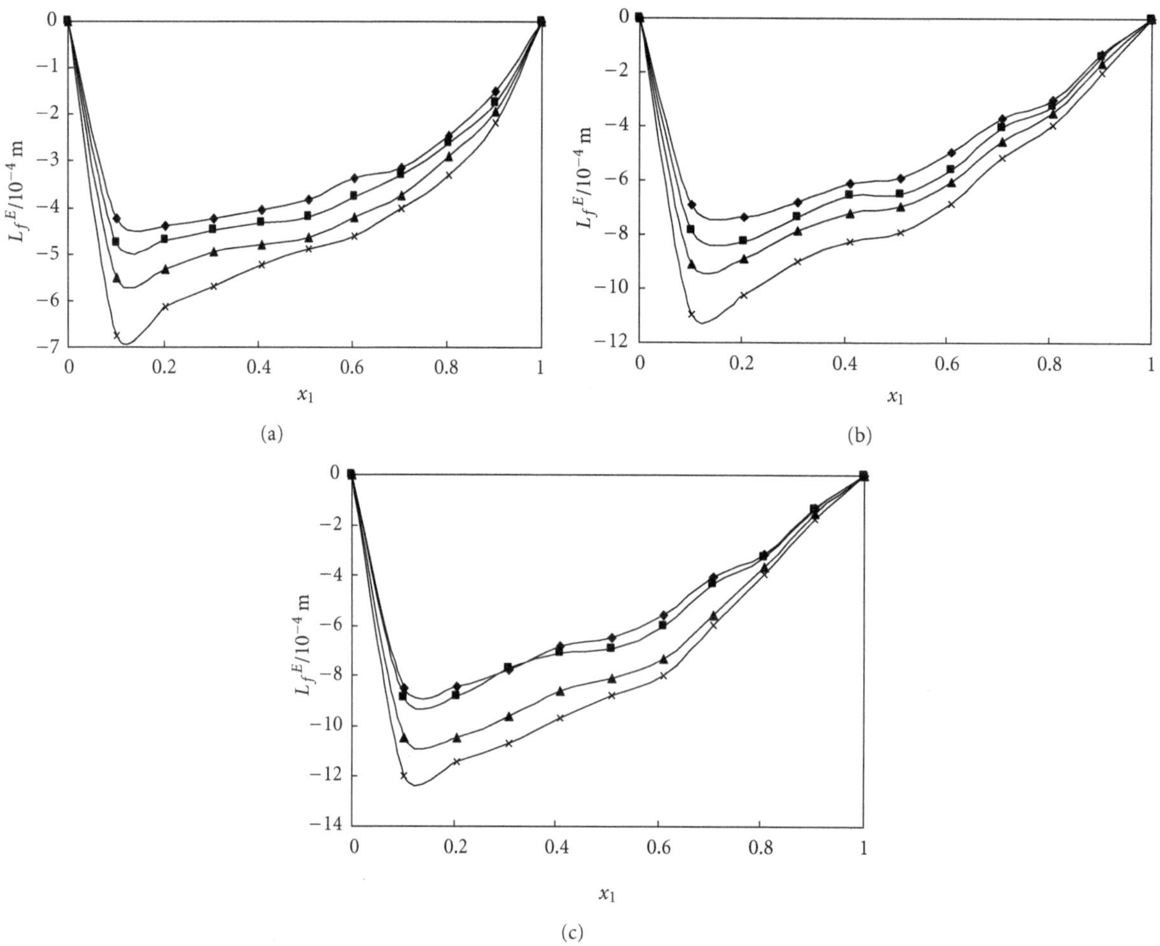

FIGURE 2: (a) Plot of excess free length (L_f^E) versus mole fraction (x_1) for binary liquid mixtures of quinoline with o-xylene at temperatures: ♦, 303.15 K; □, 308.15 K; ▲, 313.15 K; ×, 318.15 K. (b) Plot of excess free length (L_f^E) versus mole fraction (x_1) for binary liquid mixtures of quinoline with m-xylene at temperatures: ♦, 303.15 K; □, 308.15 K; ▲, 313.15 K; ×, 318.15 K. (c) Plot of excess free length (L_f^E) versus mole fraction (x_1) for binary liquid mixtures of quinoline with p-xylene at temperatures: ♦, 303.15 K; □, 308.15 K; ▲, 313.15 K; ×, 318.15 K.

The densities of pure liquids and liquid mixtures were measured by using a specific gravity bottle with an accuracy of ±0.5%. An electronic balance (Shimadzu AUY220, Japan), with a precision of ±0.1 mg, was used for the mass measurements. Averages of 4-5 measurements were taken for each sample.

3. Theory

From the experimental data of ultrasonic velocity and density various thermodynamic parameters are evaluated using standard equations.

Adiabatic compressibility:

$$\beta = \frac{1}{\rho u^2}, \tag{1}$$

where ρ and u are density and ultrasonic velocity, respectively.

Inter molecular free length:

$$L_f = K_T \cdot \beta^{1/2}, \tag{2}$$

where K_T is the temperature dependent constant [11].

Acoustic impedance:

$$Z = \rho u. \tag{3}$$

The strength of interaction between the component molecules of binary mixtures is well reflected in the deviation of the excess functions from ideality [12]. The excess properties such as β^E, L_f^E, Z^E, and u^E have been calculated using

$$Y^E = Y_{\text{mix}} - [x_1 Y_1 + x_2 Y_2], \tag{4}$$

where Y^E is β^E, L_f^E, Z^E, or u^E and x represents mole fraction of the component and subscript 1 and 2 for the components 1 and 2, respectively.

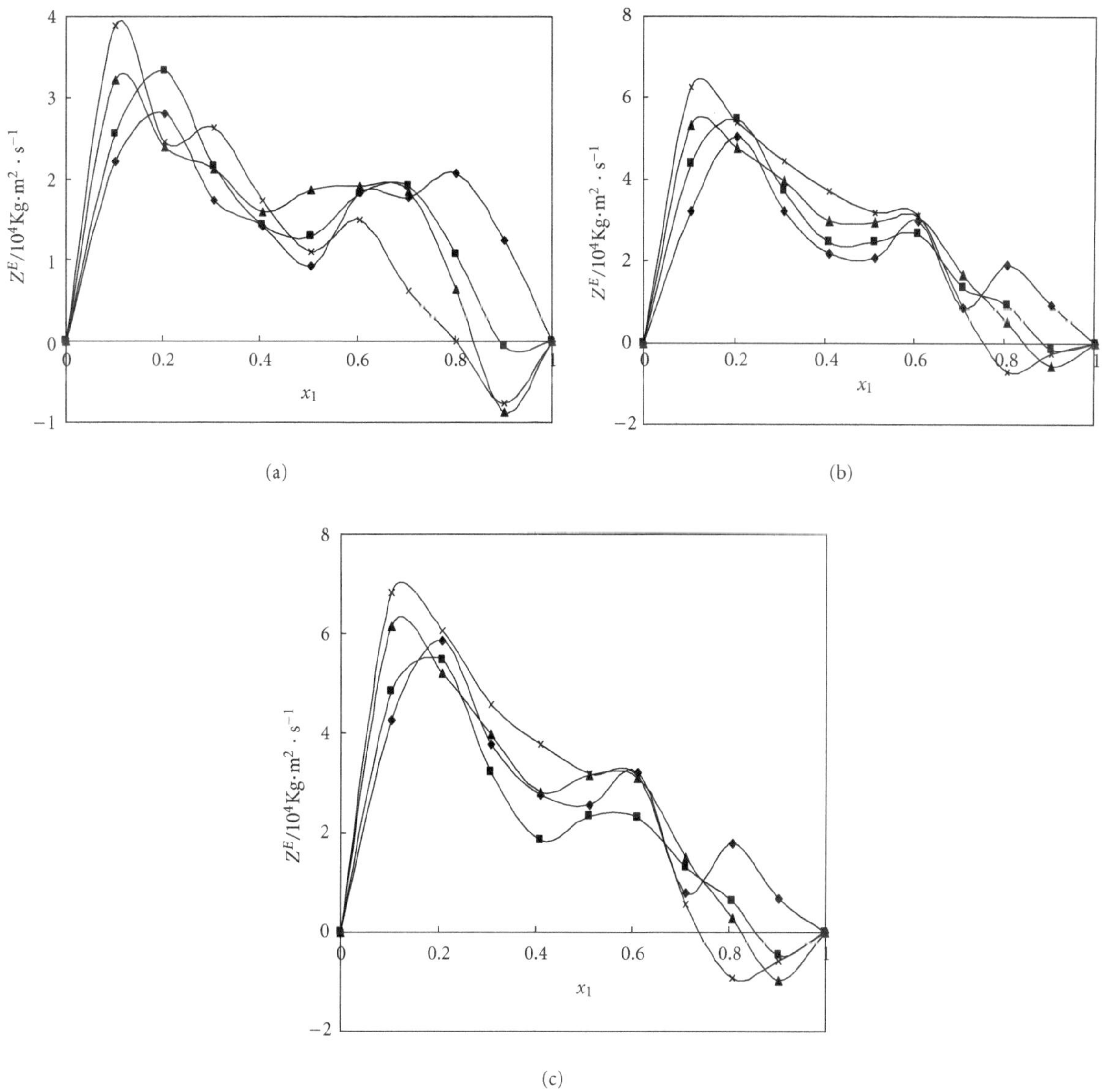

FIGURE 3: (a) Plot of excess acoustic impedance (Z^E) versus mole fraction (x_1) for binary liquid mixtures of quinoline with o-xylene at temperatures: ♦, 303.15 K; □, 308.15 K; ▲, 313.15 K; ×, 318.15 K. (b) Plot of excess acoustic impedance (Z^E) versus mole fraction (x_1) for binary liquid mixtures of quinoline with m-xylene at temperatures: ♦, 303.15 K; □, 308.15 K; ▲, 313.15 K; ×, 318.15 K. (c) Plot of excess acoustic impedance (Z^E) versus mole fraction (x_1) for binary liquid mixtures of quinoline with p-xylene at temperatures: ♦, 303.15 K; □, 308.15 K; ▲, 313.15 K; ×, 318.15 K.

The excess values of above parameters for each mixture have been fitted to Redlich-Kister polynomial equation [13]:

$$Y^E = X_1 \cdot X_2 \sum_{i=0}^{n} A_i (X_1 - X_2)^i. \quad (5)$$

The values of the coefficients A_i were calculated by a method of least squares along with the standard deviation $\sigma(Y^E)$. The coefficient is adjustable to the parameters for a better fit of the excess functions. The standard deviation values were obtained from

$$\sigma\left(Y^E\right) = \left[\frac{\sum_{i=1}^{n}\left(Y_{\text{expt}}^E - Y_{\text{cal}}^E\right)^2}{m-n}\right]^{1/2}, \quad (6)$$

where m is the number of experimental points, n is the number of parameters, and Y_{expt} and Y_{cal} are the experimental and calculated parameters, respectively.

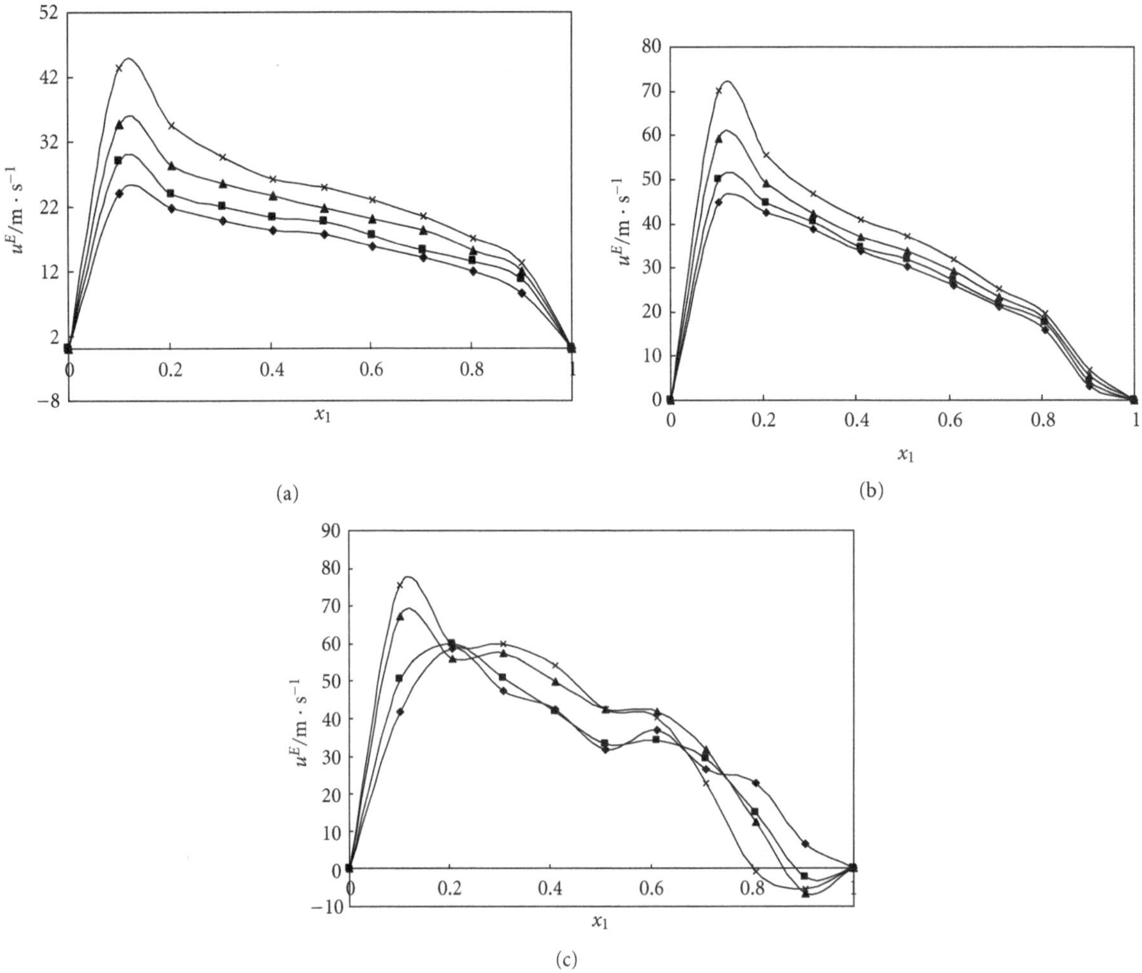

FIGURE 4: (a) Plot of excess ultrasonic velocity (u^E) versus mole fraction (x_1) for binary liquid mixtures of quinoline with o-xylene at temperatures: ♦, 303.15 K; □, 308.15 K; ▲, 313.15 K; ×, 318.15 K. (b) Plot of excess ultrasonic velocity (u^E) versus mole fraction (x_1) for binary liquid mixtures of quinoline with m-xylene at temperatures: ♦, 303.15 K; □, 308.15 K; ▲, 313.15 K; ×, 318.15 K. (c) Plot of excess ultrasonic velocity (u^E) versus mole fraction (x_1) for binary liquid mixtures of quinoline with p-xylene at temperatures: ♦, 303.15 K; □, 308.15 K; ▲, 313.15 K; ×, 318.15 K.

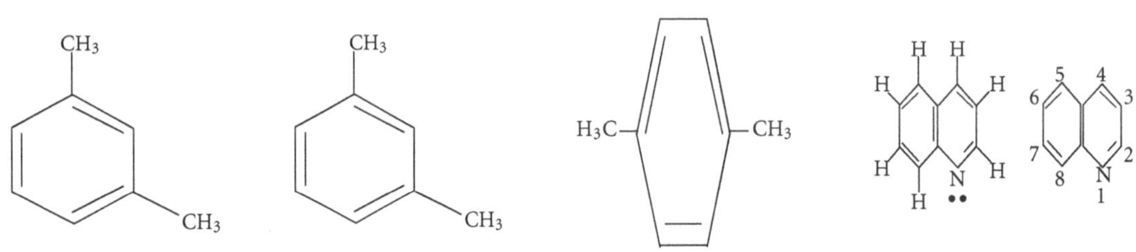

o-Xylene m-Xylene p-Xylene Quinoline

SCHEME 1

4. Results and Discussion

The structures of the components taken up for study are given in Scheme 1.

The experimental values of ultrasonic velocity (u) and density (ρ) for the three binary mixtures over the entire composition range and at temperatures T = 303.15, 308.15, 313.15, and 318.15 K are given in Table 2. The values of the Redlich-Kister polynomial coefficients A_i evaluated by the method of least squares along with standard deviation are given in Table 3. Plots of β^E, $L_f{}^E$, Z^E, and u^E against mole fraction of quinoline for all the three mixtures are given in Figures 1–4.

The excess transport properties of the mixtures are influenced by three main types of contributions, namely, (i) due to non specific Van der Waals type forces, (ii) due to hydrogen bonding, dipole-dipole, and donor-acceptor interaction between unlike molecules, and (iii) due to the fitting of smaller molecules into the voids created by the bigger molecules. The first effect leads to contraction in volume hence leads to negative contribution towards u^E and positive contribution towards β^E. However, the second effect leads to negative contribution towards β^E and positive contribution towards u^E.

For all the three mixtures, the results of the excess adiabatic compressibility β^E plotted in Figures 1(a)–1(c) are negative at all the temperatures studied. The negative values of β^E suggest that the mixtures are less compressible than the corresponding ideal mixture. According to Fort and Moore [14], the liquids of different molecular size usually mix with decrease in volume yielding negative β^E values. The strength of the interactions between component molecules increases when excess values tend to become increasingly negative. As the temperature increases, the β^E values also increase in all the three systems, suggesting that the thermal energy activates the molecules towards complex formation between unlike molecules [15]. It is also observed that the molecular interactions are stronger in quinoline + o-xylene mixture.

From Figures 2(a)–2(c) it is observed that ${L_f}^E$ values are negative for the entire mole fraction range, for all the four temperatures in case of all the three mixtures. The ${L_f}^E$ values are increasingly negative as the strength of interaction between component molecules increases [16].

The excess values of Z^E which are plotted in Figures 3(a)–3(c) are all positive in all the three liquid systems over the entire composition range and for all the four temperatures studied. The positive excess values of Z^E clearly suggest that there exist strong molecular interactions between the molecules of all the three mixtures [17].

The results for the excess ultrasonic velocity u^E plotted in Figures 4(a)–4(c) are positive for all the three mixtures at all the temperatures studied. The positive values of u^E increase with the increase in temperature which indicates the increase in strength of interaction with temperature in all the three mixtures. The higher positive values of u^E are observed in case of quinoline + o-xylene mixture, because the –OH group attached to the benzene ring in case of o-xylene is stabilized to a great extent through resonance as compared to other mixtures [18].

5. Conclusions

From the data of ultrasonic velocity and density, some thermoacoustical parameters and their excess parameters for the three binary liquid mixtures of quinoline with o-xylene, m-xylene, and p-xylene at T = 303.15, 308.15, 313.15, and 318.15 K are calculated for the entire composition range. From the results of these excess parameters, it is observed that there exists a strong molecular interaction between the unlike molecules in all the three mixtures. The interaction is stronger in case of quinoline + o-xylene mixture as compared to quinoline + m-xylene/p-xylene mixtures. These data will be useful in pharma and perfume industries for handling and mixing processes.

References

[1] K. Narendra, Ch. Srinivasu, Sk. Fakruddin, and P. N. Murthy, "Excess parameters of binary mixtures of anisaldehyde with o-cresol, m-cresol and p-cresol at T = (303.15, 308.15, 313.15, and 318.15) K," *The Journal of Chemical Thermodynamics*, vol. 43, no. 11, pp. 1604–1611, 2011.

[2] A. Ali, A. Yasmin, and A. K. Nain, "Study of intermolecular interactions in binary liquid mixtures through ultrasonic speed measurement," *Indian Journal of Pure and Applied Physics*, vol. 40, no. 5, pp. 315–322, 2002.

[3] S. I. Oswal and N. B. Patel, "Speed of sound, isentropic compressibility, viscosity, and excess volume of binary mixtures. 2. Alkanenitriles + dimethylformamide, + dimethylacetamide, and + dimethyl sulfoxide," *Journal of Chemical and Engineering Data*, vol. 40, no. 4, pp. 845–849, 1995.

[4] R. Mehra and R. Israni, "Effect of temperature on excess molar volumes of binary mixtures of hexadecane and butanol," *Indian Journal of Pure and Applied Physics*, vol. 38, no. 2, pp. 81–83, 2000.

[5] A. Ali and A. K. Nain, "Study of intermolecular interaction in binary mixtures of formamide with 2-propanol, 1,2-propanediol and 1,2,3-propanetriol through ultrasonic speed measurements," *Indian Journal of Pure and Applied Physics*, vol. 39, pp. 421–427, 2001.

[6] J. Nath, "Ultrasonic velocities, relative permittivities and refractive indices for binary liquid mixtures of trichloroethene with pyridine and quinoline," *Fluid Phase Equilibria*, vol. 109, no. 1, pp. 39–51, 1995.

[7] J. A. Al-Kandary, A. S. Al-Jimaz, and A. H. M. Abdul-Latif, "Viscosities, densities, and speeds of sound of binary mixtures of benzene, toluene, o-xylene, m-xylene, p-xylene, and mesitylene with anisole at (288.15, 293.15, 298.15, and 303.15) K," *Journal of Chemical and Engineering Data*, vol. 51, no. 6, pp. 2074–2082, 2006.

[8] S. Parveen, S. Singh, D. Shukla, K. P. Singh, M. Gupta, and J. P. Shukla, "Molecular interaction study of binary mixtures of THF with methanol and o-cresol—an optical and ultrasonic study," *Acta Physica Polonica A*, vol. 116, no. 6, pp. 1011–1017, 2009.

[9] D. D. Perrin and W. L. F. Armarego, *Purification of Laboratory Chemicals*, Pergamon Press, Oxford, UK, 3rd edition, 1980.

[10] D. Das and D. K. Hazra, "Molecular interaction study in binary mixtures of N,N-dimethyl acetamide with 2-ethoxyethanol at different temperatures," *Indian Journal of Physics B*, vol. 77, pp. 519–523, 2003.

[11] B. Jacobson, "Ultrasonic velocity in liquids and liquid mixtures," *Journal of Chemical Physics*, vol. 6, no. 5, pp. 927–928, 1952.

[12] D. R. Lide, Ed., *CRC Handbook of Chemistry and Physics*, CRC Press, Boca Raton, Fla, USA, 76th edition, 1995.

[13] O. Redlich and A. T. Kister, "Algebraic representation of thermodynamic properties and the classification of solutions," *Indian and Engineering Chemistry*, vol. 40, no. 2, pp. 345–348, 1948.

[14] R. J. Fort and W. R. Moore, "Adiabatic compressibilities of binary liquid mixtures," *Transactions of the Faraday Society*, vol. 61, pp. 2102–2111, 1965.

[15] S. N. Gour, J. S. Tomar, and R. P. Varma, "Study of molecular interactions in binary mixtures using excess thermodynamic parameters," *Indian Journal of Pure and Applied Physics*, vol. 24, p. 602, 1986.

[16] A. Ali and A. K. Nain, "Ultrasonic and volumetric study of binary mixtures of benzyl alcohol with amides," *Bulletin of the Chemical Society of Japan*, vol. 75, pp. 681–687, 2002.

[17] S. Thirumaran and D. George, "Ultrasonic study of intermolecular association through hydrogen bonding in ternary liquid mixtures," *Journal of Engineering and Applied Sciences*, vol. 4, no. 4, pp. 1–11, 2009.

[18] K. Tiwari, C. Patra, and V. Chakravortty, "Molecular interactions study on binary liquid mixtures of dimethylsulphoxide with benzene, carbontetrachloride and toluene from the excess properties of ultrasonic velocity, density and viscosity," *Acoustics Letters*, vol. 19, pp. 53–59, 1995.

14

Calculation of Thermal Pressure Coefficient of R11, R13, R14, R22, R23, R32, R41, and R113 Refrigerants by *pVT* Data

Vahid Moeini and Mahin Farzad

Department of Chemistry, Payame Noor University, P.O. Box 19395-3697, Tehran, Iran

Correspondence should be addressed to Vahid Moeini; v_moeini@yahoo.com

Academic Editor: Jan Skov Pedersen

For thermodynamic performance to be optimized particular attention must be paid to the fluid's thermal pressure coefficients and thermodynamic properties. A new analytical expression based on the statistical mechanics is derived for R11, R13, R14, R22, R23, R32, R41, and R113 refrigerants, using the intermolecular forces theory. In this paper, temperature dependency of the parameters of R11, R13, R14, R22, R23, R32, R41, and R113 refrigerants to calculate thermal pressure coefficients in the form of first order has been developed to second and third orders and their temperature derivatives of new parameters are used to calculate thermal pressure coefficients. These problems have led us to try to establish a function for the accurate calculation of the thermal pressure coefficients of R11, R13, R14, R22, R23, R32, R41, and R113 refrigerants based on statistical-mechanics theory for different refrigerants.

1. Introduction

Popular interest in the use of refrigerant blends started in the late 1950s. The emphasis was placed on energy savings through the reduction of irreversibility in the heat exchanger and on capacity variation during operation through the control of the fluid composition. Worldwide legislation has been enacted through the United Nations environmental program to reduce stratospheric ozone depletion. The Montreal Protocol was approved in 1987 to control production of the suspected ozone-depleting substances, among them chlorofluorocarbons and hydrochlorofluorocarbons, commonly used as refrigerants in the industry. For example, chlorofluorocarbons-(CFCs-) 11, 12 and 113 have been successfully used to determine groundwater recharge ages in the industry. Relatively good agreement exists between individual CFC ages and ages derived from other tracers [1–6].

The precise meaning of the internal pressure is contained in a generalized manner in the following well-known thermodynamic equations. United forces of external and internal pressure equalize the thermal pressure which tries to expand the matter. If the thermal pressure of a refrigerant is available, then the thermodynamics properties of refrigerant can be calculated easily. Liquids and dense fluids are usually considered to be complicated on a molecular scale, and a satisfactory theory of liquids only began to emerge in the 1960. However, they show a number of experimental regularities, some of which have been known by theoretical basis [7–10].

The first is the internal pressure regularity, in which $((\partial E/\partial V)_T/\rho RT)\ V^2$ is linear with respect to ρ^2 for each isotherm, where $\rho = 1/V$ is the molar density, E is the internal energy, and V is molar volume [9]. In the internal pressure regularity, which was originally devised for normal dense fluids, is based on the cell theory and considers only nearest adjacent interaction. Lennard-Jones potential function suitably describes the interactions between the molecules of a fluid under the condition that it behaves as a normal fluid. In the internal pressure regularity was attempted to calculate the internal pressure by modeling the average configurationally potential energy and then taking its derivative with respect to volume.

The second is an expression which is driven for as the thermal pressure coefficient of dense fluids (Ar, N_2, CO, CH_4, C_2H_6, n-C_4H_{10}, iso-C_4H_{10}, C_6H_6, and C_6H_5–CH_3) [11–17]. Only, *pVT* experimental data have been used for the calculation of thermal pressure coefficient [18].

The third is a regularity to predict metal-nonmetal transitions in cesium fluid. An accurate empirical potential has been found for dense cesium fluid and it is used to test the applicability of the theory. These theoretical predictions are in good agreement with experimental results [19–26].

The forth is the internal pressure of sodium, potassium, and rubidium attempt to predict X-ray diffraction and small-angle X-ray scattering to the range where the compressibility of the interacting electron gas has been theoretically predicted to become negative. Problems have led us to try to establish a function for the accurate calculation of the internal pressure and the prediction of metal-nonmetal transition alkali metals based on the internal pressure [27].

A property formulation is the set of equations used to calculate properties of a fluid at specified thermodynamic states defined by an appropriate number of independent variables. A typical thermodynamic property formulation is based on an equation of state which allows the correlation and computation of all thermodynamic properties of the fluid including properties such as entropy that cannot be measured directly. Modern equations of state at least with 17 terms for pure fluid properties are usually fundamental equations explicit in the Helmholtz energy as a function of density and temperature. The new class of equations of state for technical applications to dense fluids is formulated in the reduced Helmholtz energy. As usual, the reduced Helmholtz energy is split into one part which describes the behavior of the hypothetical ideal gas at given values of temperature and density and a second part which describes the residual behavior of the fluid. For some relevant properties the corresponding relations were given in [4, 9].

In 1993 a general regularity with 2 terms so-called the linear isotherm regularity has been reported for pure dense fluids, according to which $(Z-1)V^2$ is linear with respect to ρ^2, each isotherm as

$$(Z-1)V^2 = A + B\rho^2, \quad (1)$$

where $Z \equiv pV/RT$ is the compression factor, $\rho = 1/V$ is the molar density, and A and B are the temperature-dependent parameters. This equation of state works very well for all types of dense fluids, for densities greater than the Boyle density but for temperatures below twice the Boyle temperature. The regularity was originally suggested on the basis of a simple lattice-type model applied to a Lennard-Jones (12,6) fluid [28, 29]. At present work, the regularity has been used to calculate thermal pressure coefficient of dense Trichlorofluoromethane (R11), Chlorotrifluoromethane (R13), Tetrafluoromethane (R14), Chlorodifluoromethane (R22), Trifluoromethane (R23), Difluoromethane (R32), Fluoromethane (R41), and 1,1,2-Trichloro-1,2,2-trifluoroethane (R113) refrigerants [30]. In this paper, in Section 2.1, we present a simple method that keeps first order temperature dependency of parameters in the regularity versus inverse temperature. Then, the thermal pressure coefficient is calculated by this expression. In Section 2.2, temperature dependency of parameters in the regularity has been developed to second order. In Section 2.3, temperature dependency of parameters in the regularity has been developed to third order and then the thermal pressure coefficient is calculated in each state for R11, R13, R14, R22, R23, R32, R41, and R113 refrigerants.

2. Theory

We fist test the ability of the linear isotherm regularity [18]

$$\frac{p}{\rho RT} = 1 + A\rho^2 + B\rho^4. \quad (2)$$

2.1. First Order Temperature Dependency of Parameters. We first calculate pressure by the linear isotherm regularity and then use first order temperature dependency of parameters to get the thermal pressure coefficient for the dense fluid, where

$$A = A_2 - \frac{A_1}{RT}, \quad (3)$$

$$B = \frac{B_1}{RT}. \quad (4)$$

Here A_1 and B_1 are related to the intermolecular attractive and repulsive forces, respectively, while A_2 is related to the nonideal thermal pressure and RT has its usual meaning.

In the present work, the starting point in the derivation is (2). By substitution of (3) and (4) in (2) we obtain the pressure for R11, R13, R14, R22, R23, R32, R41, and R113 fluids:

$$p = \rho RT + A_2\rho^3 RT - A_1\rho^3 + B_1\rho^5. \quad (5)$$

We first drive an expression for thermal pressure coefficient using first order temperature dependency of parameters. The final result is TPC$^{(1)}$:

$$\left(\frac{\partial p}{\partial T}\right)_\rho = R\rho + A_2 R\rho^3. \quad (6)$$

According to (6), the experimental value of density and value of A_2 from Table 1 can be used to calculate the value of thermal pressure coefficient.

For this purpose we have plotted A versus $1/T$ whose intercept shows value of A_2. Table 1 shows the A_2 values for R11, R13, R14, R22, R23, R32, R41, and R113 fluids [1, 6, 30].

Then we obtain the thermal pressure coefficient of dense fluids. R13 serves as our primary test fluid because of the abundance of available thermal pressure coefficients data [6, 30]. For this purpose we have plotted A versus $1/T$ whose intercept shows value of A_2. Figures 1(a) and 1(b) show plots of A and B versus inverse temperature for R13, respectively. It is clear that A and B versus inverse temperature are not first order.

2.2. Second Order Temperature Dependency of Parameters. In order to solve this problem, the linear isotherm regularity equation of state in the form of truncated temperature series of A and B parameters has been developed to second order for dense fluids. Figures 1(a) and 1(b) show plots of A and B parameters versus inverse temperature for R13 fluid. It is clear that A and B versus inverse temperature are second order.

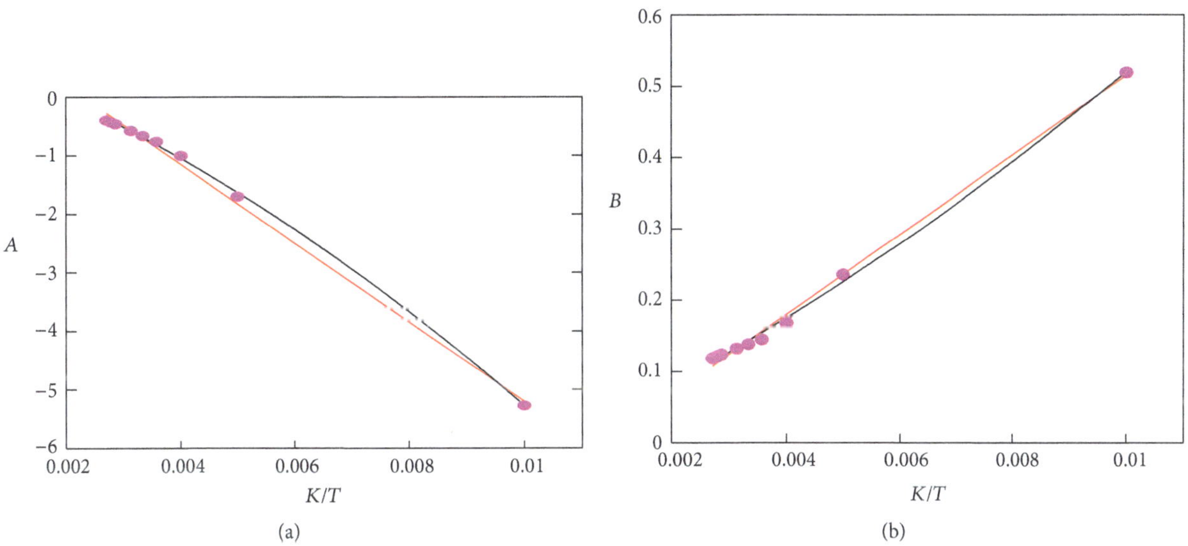

FIGURE 1: (a) Plot of A versus inverse temperature. Red and blue lines are the first and second order fit to the A data points, for R13, respectively. (b) Plot of B versus inverse temperature. Red and blue lines are the first and second order fit to the B data points, for R13, respectively.

TABLE 1: The calculated values of A_2 for different fluids using (3) and the coefficient of determination (R^2) [1, 6, 18, 30].

Fluid	A_2	$(T_{min}–T_{max})$/K	R^2
R11	2.2109	180–500	0.9849
R13	1.5748	100–370	0.9946
R14	1.1315	150–310	0.9894
R22	1.5616	120–500	0.9890
R23	1.0286	150–450	0.9905
R32	0.8650	150–430	0.9909
R41	0.6263	180–410	0.9927
R113	1.3863	300–525	0.9977

TABLE 2: The calculated values of A_1, A_3 using (7) and B_1, B_3 using (8) for different fluids and the coefficient of determination (R^2) [1, 6, 18, 30].

Fluid	A_1	A_3	R^2	B_1	B_3	R^2
R11	0.1354	-1.7517×10^5	0.9987	0.0625	1.4777×10^4	0.9997
R13	1.3594	−7165.5660	0.9898	−0.0650	−610.8607	0.9951
R14	0.1810	-4.2972×10^4	0.9992	0.1072	5461.0737	0.9997
R22	0.6210	-4.8725×10^4	0.9983	0.0196	3298.4237	0.9994
R23	0.4457	-3.4069×10^4	0.9993	0.0317	2449.0303	0.9997
R32	0.2404	-3.8638×10^4	0.9980	0.0528	3251.8493	0.9996
R41	0.0924	-3.7820×10^4	0.9982	0.0850	4669.1360	0.9996
R113	0.6053	-1.2415×10^5	0.9992	0.0316	7746.1673	0.9997

Thus, we obtain extending parameters A and B resulting in second order equation as follows:

$$A = A_1 + \frac{A_2}{T} + \frac{A_3}{T^2}, \qquad (7)$$

$$B = B_1 + \frac{B_2}{T} + \frac{B_3}{T^2}. \qquad (8)$$

The starting point in the derivation is (2) again. By substitution of (7)-(8) in equation (2) we obtain the pressure for R11, R13, R14, R22, R23, R32, R41, and R113 fluids [1, 6, 30]:

$$p = \rho RT + A_1\rho^3 RT + A_2\rho^3 R + \frac{A_3\rho^3 R}{T} + B_1\rho^5 RT + B_2\rho^5 R + \frac{B_3\rho^5 R}{T}. \qquad (9)$$

First, second, and third temperature coefficients and their temperature derivatives were calculated from this model and the final result is for the thermal pressure coefficient to form TPC$^{(2)}$:

$$\left(\frac{\partial p}{\partial T}\right)_\rho = \rho R + A_1\rho^3 R - \frac{A_3\rho^3 R}{T^2} + B_1\rho^5 R - \frac{B_3\rho^5 R}{T^2}. \qquad (10)$$

As (10) shows, it is possible to calculate the thermal pressure coefficient at each density and temperature by knowing A_1, A_3, B_1, B_3. For this purpose we have plotted extending parameters of A and B versus $1/T$ whose intercept and coefficients show the values of A_1, A_3, B_1, B_3 that are given in Table 2.

2.3. Third Order Temperature Dependency of Parameters. In another step, we test to form truncated temperature series of A and B parameters to third order:

$$A = A_1 + \frac{A_2}{T} + \frac{A_3}{T^2} + \frac{A_4}{T^3}, \qquad (11)$$

$$B = B_1 + \frac{B_2}{T} + \frac{B_3}{T^2} + \frac{B_4}{T^3}. \qquad (12)$$

TABLE 3: The calculated values of A_1, A_3, A_4 using (11) and B_1, B_3, B_4 using (12) for different fluids and the coefficient of determination (R^2) [1, 6, 18, 30].

Fluid	A_1	A_3	A_4	R^2	B_1	B_3	B_4	R^2
R11	−1.1003	-5.2851×10^5	3.2477×10^7	0.9999	0.3293	9.1064×10^4	-7.0120×10^5	0.9999
R13	−1.2784	-2.9997×10^5	1.5854×10^7	0.9999	0.2582	9.5263×10^4	-1.9424×10^6	0.9999
R14	0.8257	4.5033×10^4	-5.9456×10^6	0.9998	0.0642	−402.0298	3.6911×10^5	0.9993
R22	8.1×10^{-3}	-1.5322×10^5	6.9130×10^6	0.9999	0.0988	1.6789×10^4	-8.9251×10^5	0.9998
R23	0.2352	-7.5616×10^4	3.1780×10^6	0.9998	0.0618	8402.7408	-4.5542×10^5	0.9993
R32	−0.1924	-1.2103×10^5	6.2039×10^6	0.9999	0.1290	1.7760×10^4	-1.0925×10^6	0.9995
R41	0.6315	8.2985×10^4	-1.0303×10^7	0.9999	0.0298	−7705.9129	1.0555×10^6	0.9988
R113	−0.9445	-8.5187×10^5	9.2993×10^7	0.9998	0.2258	9.8942×10^4	-1.1654×10^7	0.9995

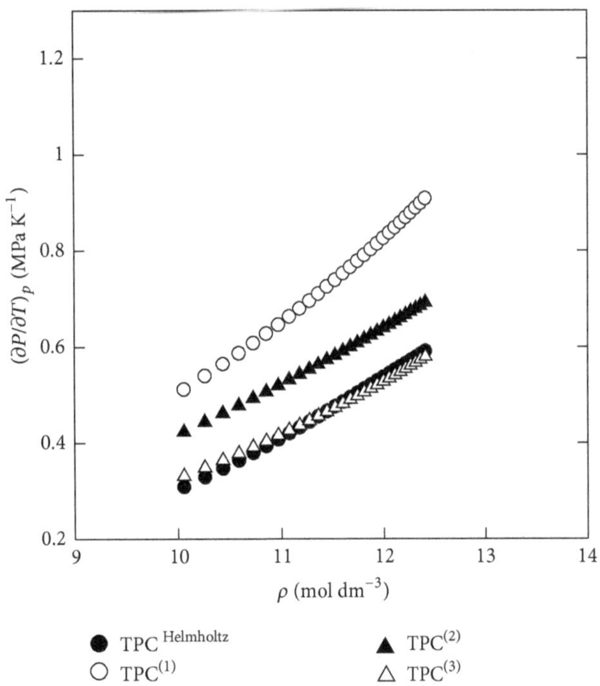

FIGURE 2: The experimental values of thermal pressure coefficient versus density for R13 fluid are compared with thermal pressure coefficient using the TPC$^{(1)}$, TPC$^{(2)}$, and TPC$^{(3)}$ at 300 K.

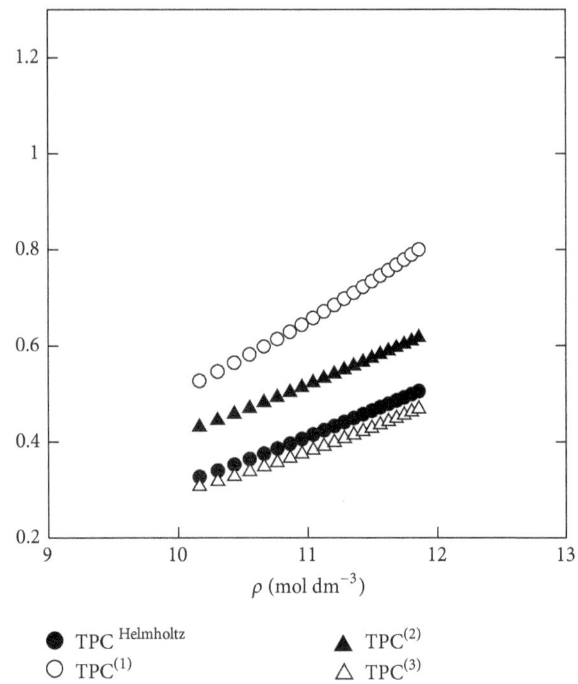

FIGURE 3: The experimental values of thermal pressure coefficient versus density for R13 fluid are compared with thermal pressure coefficient using the TPC$^{(1)}$, TPC$^{(2)}$, and TPC$^{(3)}$ at 320 K.

The starting point in the derivation is (2) again. By substitution of (11) and (12) in (2) we obtain the pressure equation for R11, R13, R14, R22, R23, R32, R41, and R113 fluids [1, 6, 30]:

$$p = \rho RT + A_1\rho^3 RT + A_2\rho^3 R + \frac{A_3\rho^3 R}{T} + \frac{A_4\rho^3 R}{T^2} + B_1\rho^5 RT + B_2\rho^5 R + \frac{B_3\rho^5 R}{T} + \frac{B_4\rho^5 R}{T^2}. \quad (13)$$

The final result is for the thermal pressure coefficient to form TPC$^{(3)}$:

$$\left(\frac{\partial p}{\partial T}\right)_\rho = \rho R + A_1\rho^3 R - \frac{A_3\rho^3 R}{T^2} - \frac{2A_4\rho^3 R}{T^3} + B_1\rho^5 R - \frac{B_3\rho^5 R}{T^2} - \frac{2B_4\rho^5 R}{T^3}. \quad (14)$$

Based on (14), to obtain the thermal pressure coefficient, it is necessary to determine values $A_1, A_3, A_4, B_1, B_3, B_4$ these values are given in Table 3. In contrast, Figures 2 and 3 show the experimental values of the thermal pressure coefficient versus density for R13 of liquid and supercritical fluids that are compared with the thermal pressure coefficient using the TPC$^{(1)}$, TPC$^{(2)}$, and TPC$^{(3)}$ at 300 and 320 K, respectively.

3. Experimental Tests and Discussion

A compromised ozone layer results in increased ultraviolet (UV) radiation reaching the earth's surface, which can have wide ranging health effects. Global climate change is believed to be caused by buildup of greenhouse gases in the atmosphere. The primary greenhouse gas is carbon dioxide (CO_2), created by fossil fuel-burning power plants.

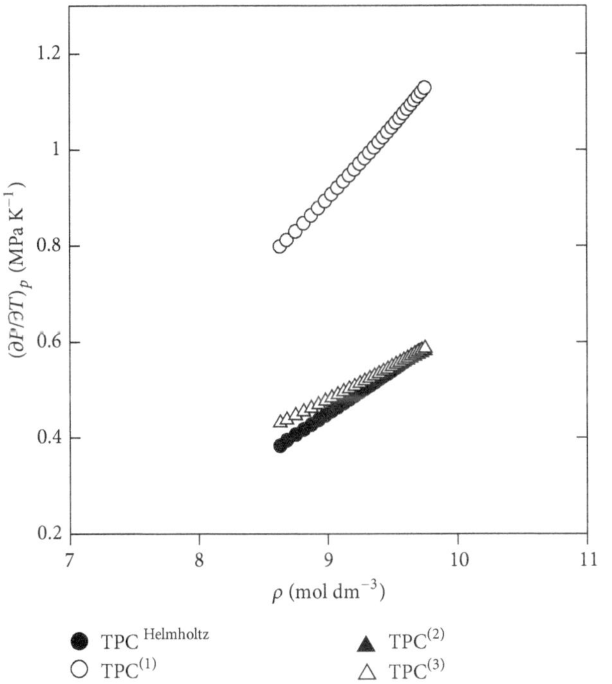

FIGURE 4: The experimental values of thermal pressure coefficient versus density for R11 fluid are compared with thermal pressure coefficient using the TPC$^{(1)}$, TPC$^{(2)}$, and TPC$^{(3)}$ at 400 K.

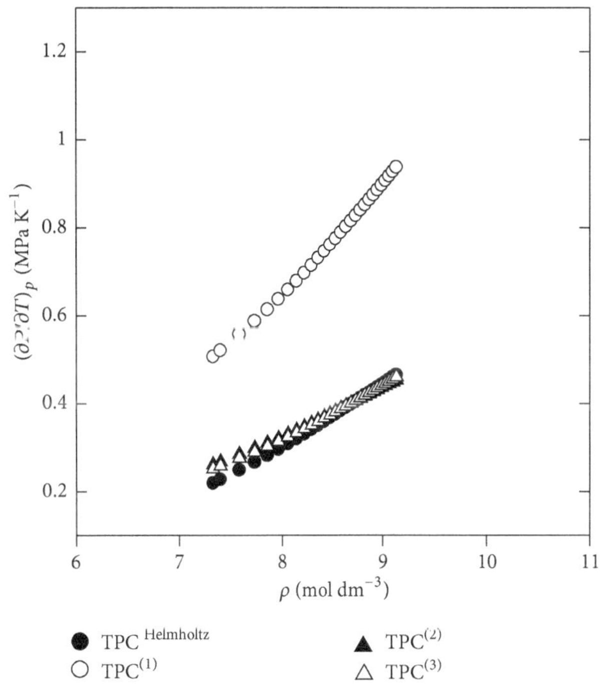

FIGURE 5: The experimental values of thermal pressure coefficient versus density for R11 fluid are compared with thermal pressure coefficient using the TPC$^{(1)}$, TPC$^{(2)}$, and TPC$^{(3)}$ at 420 K.

These gases trap the earth's heat, causing global warming. CFC, HCFC, and HFC refrigerants are considered greenhouse gases. The accurate description of thermodynamic properties of fluids over the large intervals of temperatures and densities with multiparameter equations of state has been a subject of active research, which has developed continuously during the last 30 years and will continue to do so. Generally, three categories of equations of state can be established according to their fundamentals: empirical, theoretical, and semiempirical. An empirical equation of state is usually needed to several experimental data or many adjustable parameters and therefore their applications are usually restricted to a very limited number of substances. A theoretical equation of state is also needed to the same number of molecular parameters, particularly to the intermolecular pair potential function [1–6, 30].

In this work, the thermal pressure coefficient is computed for refrigerant fluids of liquid and super critical fluids using three different models of the theoretical equation of state. R13 serves as our primary test fluid because of the abundance of available thermal pressure coefficients data. Such data are more limited for the other fluids examined. When we restricted temperature series of A and B parameters to first order it has been seen that the points from the low densities for TPC$^{(1)}$ deviate significantly from the experimental data. To decrease adequately deviation the thermal pressure coefficient from the experimental data, it was necessary to extend the temperature series of A and B parameters to second order. The present approach to

TABLE 4: Comparison of $(\partial p/\partial T)_\rho$ among the calculated and experimental values for R11 at 400 K [1, 18, 30].

p/MPa	ρ/mol·L^{-1}	$(\partial p/\partial T)_\rho$(MPa·K^{-1})			
		TPC^{Helmholtz}	TPC$^{(1)}$	TPC$^{(2)}$	TPC$^{(3)}$
1.4074	8.6318	0.3834	0.7986	0.4312	0.4310
2.1000	8.6827	0.3948	0.8120	0.4375	0.4375
3.1000	8.7509	0.4065	0.8301	0.4460	0.4462
4.1000	8.8142	0.4174	0.8472	0.4539	0.4544
5.1000	8.8732	0.4276	0.8633	0.4614	0.4621
6.1000	8.9286	0.4373	0.8787	0.4685	0.4694
7.1000	8.9810	0.4465	0.8933	0.4753	0.4764
8.1000	9.0306	0.4554	0.9074	0.4818	0.4831
9.1000	9.0780	0.4638	0.9210	0.4880	0.4895
10.1000	9.1232	0.4719	0.9340	0.4940	0.4957
11.1000	9.1665	0.4797	0.9467	0.4998	0.5017
12.1000	9.2081	0.4872	0.9589	0.5053	0.5075
13.1000	9.2481	0.4945	0.9708	0.5107	0.5131
14.1000	9.2868	0.5015	0.9824	0.5160	0.5185
15.1000	9.3241	0.5083	0.9937	0.5211	0.5238
16.1000	9.3602	0.5149	1.0046	0.5260	0.5289
17.1000	9.3952	0.5214	1.0154	0.5309	0.5339
18.1000	9.4292	0.5277	1.0259	0.5356	0.5388
19.1000	9.4623	0.5338	1.0361	0.5402	0.5436
20.1000	9.4944	0.5398	1.0462	0.5447	0.5483
21.1000	9.5257	0.5456	1.0560	0.5491	0.5529
22.1000	9.5562	0.5513	1.0657	0.5534	0.5574
23.1000	9.5859	0.5569	1.0752	0.5576	0.5618
24.1000	9.6150	0.5624	1.0845	0.5617	0.5661
25.1000	9.6434	0.5678	1.0937	0.5658	0.5703

TABLE 5: Comparison of $(\partial p/\partial T)_\rho$ among the calculated and experimental values for R11 at 440 K [1, 18, 30].

p/MPa	ρ/mol·L^{-1}	$(\partial p/\partial T)_\rho$(MPa·K^{-1})			
		TPC^{Helmholtz}	TPC$^{(1)}$	TPC$^{(2)}$	TPC$^{(3)}$
2.7745	7.3336	0.2189	0.5067	0.2614	0.2515
3.1000	7.4066	0.2271	0.5208	0.2679	0.2581
4.1000	7.5898	0.2486	0.5572	0.2846	0.2755
5.1000	7.7349	0.2664	0.5873	0.2983	0.2899
6.1000	7.8569	0.2818	0.6135	0.3102	0.3024
7.1000	7.9630	0.2955	0.6369	0.3207	0.3137
8.1000	8.0576	0.3080	0.6582	0.3304	0.3240
9.1000	8.1431	0.3195	0.6779	0.3392	0.3335
10.1000	8.2215	0.3302	0.6964	0.3475	0.3424
11.1000	8.2941	0.3403	0.7138	0.3552	0.3509
12.1000	8.3617	0.3498	0.7302	0.3626	0.3589
13.1000	8.4251	0.3587	0.7459	0.3695	0.3665
14.1000	8.4850	0.3673	0.7609	0.3762	0.3738
15.1000	8.5417	0.3755	0.7753	0.3826	0.3808
16.1000	8.5956	0.3833	0.7892	0.3887	0.3876
17.1000	8.6470	0.3909	0.8026	0.3946	0.3942
18.1000	8.6962	0.3981	0.8155	0.4003	0.4005
19.1000	8.7434	0.4052	0.8281	0.4058	0.4067
20.1000	8.7887	0.4120	0.8403	0.4112	0.4127
21.1000	8.8324	0.4185	0.8521	0.4164	0.4186
22.1000	8.8746	0.4249	0.8637	0.4215	0.4243
23.1000	8.9153	0.4312	0.8750	0.4264	0.4298
24.1000	8.9547	0.4372	0.8860	0.4312	0.4353
25.1000	8.9930	0.4431	0.8967	0.4359	0.4406
26.1000	9.0301	0.4489	0.9073	0.4404	0.4458

TABLE 6: Comparison of $(\partial p/\partial T)_\rho$ among the calculated and experimental values for R13 at 300 K [1, 6, 18, 30].

p/MPa	ρ/mol·L^{-1}	$(\partial p/\partial T)_\rho$(MPa·K^{-1})			
		TPC^{Helmholtz}	TPC$^{(1)}$	TPC$^{(2)}$	TPC$^{(3)}$
8.0000	10.0570	0.3095	0.5114	0.4231	0.3317
9.0000	10.2590	0.3295	0.5393	0.4434	0.3490
10.0000	10.4330	0.3476	0.5643	0.4614	0.3645
11.0000	10.5870	0.3642	0.5870	0.4776	0.3786
12.0000	10.7260	0.3796	0.6081	0.4925	0.3917
13.0000	10.8520	0.3940	0.6276	0.5061	0.4038
14.0000	10.9680	0.4075	0.6460	0.5189	0.4152
15.0000	11.0750	0.4203	0.6633	0.5309	0.4260
16.0000	11.1750	0.4324	0.6797	0.5421	0.4362
17.0000	11.2700	0.4440	0.6956	0.5530	0.4461
18.0000	11.3580	0.4550	0.7106	0.5631	0.4554
19.0000	11.4430	0.4656	0.7252	0.5729	0.4645
20.0000	11.5230	0.4757	0.7392	0.5823	0.4732
21.0000	11.5990	0.4854	0.7526	0.5912	0.4816
22.0000	11.6720	0.4948	0.7657	0.5999	0.4898
23.0000	11.7420	0.5038	0.7784	0.6082	0.4977
24.0000	11.8090	0.5125	0.7907	0.6163	0.5054
25.0000	11.8740	0.5208	0.8027	0.6242	0.5129
26.0000	11.9360	0.5290	0.8143	0.6317	0.5202
27.0000	11.9970	0.5368	0.8258	0.6391	0.5274
28.0000	12.0550	0.5444	0.8369	0.6463	0.5343
29.0000	12.1120	0.5517	0.8479	0.6533	0.5412
30.0000	12.1660	0.5587	0.8584	0.6600	0.5478
31.0000	12.2200	0.5657	0.8689	0.6667	0.5544
32.0000	12.2720	0.5723	0.8792	0.6732	0.5609

obtaining the thermal pressure coefficient from pVT data contrasts with the experimental data by extending the temperature series of A and B parameters to second order and its derivatives. That, the thermal pressure coefficient give to form TPC$^{(2)}$.

We also considered an even more accurate estimates namely, the extension of temperature series of A and B parameters to third order. The final result is for the thermal pressure coefficient to form TPC$^{(3)}$. In contrast, Figures 4 and 5 show the experimental values of the thermal pressure coefficient versus density for R11 of liquid fluid that are compared with the thermal pressure coefficient using the TPC$^{(1)}$, TPC$^{(2)}$, and TPC$^{(3)}$ at 400 and 440 K, respectively. Also, the experimental and calculated values of the thermal pressure coefficient using the TPC$^{(1)}$, TPC$^{(2)}$, and TPC$^{(3)}$ are compared in Tables 4, 5, 6, and 7 for R13 and R11 fluids. Although all three models capture the qualitative features for refrigerants, the calculated values of the thermal pressure coefficient using the TPC$^{(2)}$ model produce quantitative agreement, but Tables 4, 5, 6, and 7, which is a more test of these models, shows that the TPC$^{(3)}$ model is able to accurately predict both the thermal pressure coefficient of the liquid and the supercritical refrigerants.

4. Result

Refrigerants are the working fluids in refrigeration, air-conditioning, and heat-pumping systems. Accurate and comprehensive thermodynamic properties of refrigerants such as the thermal pressure coefficients are in demand by both producers and users of the materials. However, the database for the thermal pressure coefficients is small at present. Furthermore, the measurements of the thermal pressure coefficients made by different researchers often reveal systematic differences between their estimates. The researchers have led us to try to establish a correlation function for the accurate calculation of the thermal pressure coefficients for different fluids over a wide temperature and pressure ranges. The most straightforward way to derive the thermal pressure coefficient is the calculation of the thermal pressure coefficient with the use of the principle of corresponding states which covers wide temperature and pressure ranges. The principle of corresponding states calls for reducing the thermal pressure at a given reduced temperature and density to be the same for all fluids. The leading term of this correlation function is the thermal pressure coefficient of perfect gas, which each gas obeys in the low density range [31, 32].

TABLE 7: Comparison of $(\partial p/\partial T)_\rho$ among the calculated and experimental values for R13 at 320 K [1, 6, 18, 30].

p/MPa	ρ/mol·L^{-1}	$(\partial p/\partial T)_\rho$(MPa·K^{-1})			
		TPC^{Helmholtz}	TPC$^{(1)}$	TPC$^{(2)}$	TPC$^{(3)}$
15.0000	10.1650	0.3269	0.5262	0.4304	0.3063
16.0000	10.3040	0.3400	0.5457	0.4444	0.3175
17.0000	10.4310	0.3523	0.5640	0.4574	0.3281
18.0000	10.5490	0.3640	0.5813	0.4696	0.3381
19.0000	10.6600	0.3751	0.5980	0.4813	0.3479
20.0000	10.7630	0.3857	0.6138	0.4922	0.3571
21.0000	10.8610	0.3959	0.6290	0.5028	0.3660
22.0000	10.9530	0.4056	0.6436	0.5128	0.3746
23.0000	11.0410	0.4149	0.6578	0.5224	0.3829
24.0000	11.1240	0.4239	0.6713	0.5316	0.3909
25.0000	11.2040	0.4326	0.6846	0.5406	0.3988
26.0000	11.2810	0.4409	0.6975	0.5492	0.4064
27.0000	11.3540	0.4490	0.7099	0.5575	0.4138
28.0000	11.4250	0.4568	0.7221	0.5656	0.4211
29.0000	11.4930	0.4643	0.7339	0.5734	0.4282
30.0000	11.5590	0.4716	0.7455	0.5811	0.4352
31.0000	11.6220	0.4786	0.7567	0.5884	0.4419
32.0000	11.6840	0.4855	0.7679	0.5957	0.4486
33.0000	11.7440	0.4921	0.7788	0.6027	0.4552
34.0000	11.8020	0.4986	0.7894	0.6096	0.4616
35.0000	11.8580	0.5048	0.7997	0.6163	0.4679

In this paper, we drive an expression for the thermal pressure coefficient of R11, R13, R14, R22, R23, R32, R41, and R113 dense refrigerants using the linear isotherm regularity [18, 28]. Unlike previous models, it has been shown in this work that the thermal pressure coefficient can be obtained without employing any reduced Helmholtz energy [9]. Only *pVT* experimental data have been used for the calculation of the thermal pressure coefficient of R11, R13, R14, R22, R23, R32, R41, and R113 refrigerants [8]. Comparison of the calculated values of the thermal pressure coefficient using the linear isotherm regularity with the values obtained experimentally shows the validity of the use of the linear isotherm regularity for studying the thermal pressure coefficient of R11, R13, R14, R22, R23, R32, R41, and R113 refrigerants. In this work, it has been shown that the temperature dependences of the intercept and slope of using linear isotherm regularity are nonlinear. This problem has led us to try to obtain the expression for the thermal pressure coefficient of R11, R13, R14, R22, R23, R32, R41, and R113 refrigerants by extending the intercept and slope of the linearity parameters versus inversion of temperature to the third order. The thermal pressure coefficients predicted from this simple model are in good agreement with experimental data. The results show the accuracy of this method is generally quit good. The resulting model predicts accurately the thermal pressure coefficients from the lower density limit at the Boyle density from triple temperature up to about double the Boyle temperature. The upper density limit appears to be reached at 1.4 times the Boyle density. These problems have led us to try to establish a function for the accurate calculation of the thermal pressure coefficients based on equation of state theory for different refrigerants.

Acknowledgment

The authors thank the Payame Noor University for financial support.

References

[1] J. D. Happell, R. M. Price, Z. Top, and P. K. Swart, "Evidence for the removal of CFC-11, CFC-12, and CFC-113 at the groundwater-surface water interface in the Everglades," *Journal of Hydrology*, vol. 279, no. 1–4, pp. 94–105, 2003.

[2] R. Span and W. Wagner, "Equations of state for technical applications. III. Results for polar fluids," *International Journal of Thermophysics*, vol. 24, no. 1, pp. 111–162, 2003.

[3] K. A. Gillis, "Thermodynamic properties of seven gaseous halogenated hydrocarbons from acoustic measurements: $CHClFCF_3$, CHF_2CF_3, CF_3CH_3, CHF_2CH_3, $CF_3CHFCHF_2$, $CF_3CH_2CF_3$, and $CHF_2CF_2CH_2F$," *International Journal of Thermophysics*, vol. 18, no. 1, pp. 73–135, 1997.

[4] E. W. Lemmon and R. T. Jacobsen, "An international standard formulation for the thermodynamic properties of 1,1,1-trifluoroethane (HFC-143a) for temperatures from 161 to 450 K and pressures to 50 MPa," *Journal of Physical and Chemical Reference Data*, vol. 29, p. 521, 2000.

[5] M. J. Assael, J. Millat, V. Vesovic, and W. A. Wakeham, "The Thermal conductivity of methane and tetrafluoromethane in the limit of zero density," *Journal of Physical and Chemical Reference Data*, vol. 19, p. 1137, 1990.

[6] J. W. Magee, S. L. Outcalt, and J. F. Ely, "Molar heat capacity Cv, vapor pressure, and (p, ρ, T) measurements from 92 to 350 K at pressures to 35 MPa and a new equation of state for chlorotrifluoromethane (R13)," *International Journal of Thermophysics*, vol. 21, no. 5, pp. 1097–1121, 2000.

[7] R. Dohrn and J. M. Prausnitz, "A simple perturbation term for the Carnahan-Starling equation of state," *Fluid Phase Equilibria*, vol. 61, no. 1-2, pp. 53–69, 1990.

[8] L. Boltzmann, *Lectures on Gas Theory*, University of California Press, Berkeley, Calif, USA, 1964.

[9] V. Moeini, "New regularity for internal pressure of dense fluids," *Journal of Physical Chemistry B*, vol. 110, no. 7, pp. 3271–3275, 2006.

[10] V. Moeini and M. Deilam, "Determination of molecular diameter by *PVT*," *ISRN Physical Chemistry*, vol. 2012, Article ID 521827, 5 pages, 2012.

[11] R. B. Stewart and T. Jacobsen, "Thermodynamic properties of argon from the triple point to 1200 K with pressures to 1000 MPa," *Journal of Physical and Chemical Reference Data*, vol. 18, p. 639, 1989.

[12] R. D. Goodwin, "Carbon monoxide thermophysical properties from 68 to 1000 K at pressures to 100 MPa," *Journal of Physical and Chemical Reference Data*, vol. 14, p. 849, 1985.

[13] R. Span and W. Wagner, "A new equation of state for carbon dioxide covering the fluid region from the triple-point temperature to 1100 K at pressures up to 800 MPa," *Journal of Physical and Chemical Reference Data*, vol. 25, p. 1509, 1996.

[14] R. T. Jacobsen, R. B. Stewart, and M. Jahangiri, "Thermodynamic properties of nitrogen from the freezing line to 2000 K

at pressures to 1000 MPa," *Journal of Physical and Chemical Reference Data*, vol. 15, no. 2, pp. 735–909, 1986.

[15] B. A. Younglove and J. F. Ely, "Thermophysical properties of fluids. II. methane, ethane, propane, isobutane, and normal butane," *Journal of Physical and Chemical Reference Data*, vol. 16, p. 577, 1987.

[16] R. D. Goodwin, "Benzene thermophysical properties from 279 to 900 K at pressures to 1000 Bar," *Journal of Physical and Chemical Reference Data*, vol. 17, p. 1541, 1988.

[17] R. D. Goodwin, "Toluene thermophysical properties from 178 to 800 K at pressures to 1000 Bar," *Journal of Physical and Chemical Reference Data*, vol. 18, p. 1565, 1989.

[18] V. Moeini, F. Ashrafi, M. Karri, and H. Rahimi, "Calculation of thermal pressure coefficient of dense fluids using the linear isotherm regularity," *Journal of Physics Condensed Matter*, vol. 20, no. 7, Article ID 075102, 2008.

[19] V. Moeini, "Internal pressures of lithium and cesium fluids at different temperatures," *Journal of Chemical and Engineering Data*, vol. 55, no. 3, pp. 1093–1099, 2010.

[20] V. Moeini, "Calculation of thermal pressure coefficient of lithium fluid by *pVT* data," *ISRN Physical Chemistry*, vol. 2012, Article ID 724230, 11 pages, 2012.

[21] T. M. Reed and K. E. Gubbins, *Applied Statistical Mechanics*, McGraw-Hill, 1973.

[22] W. D. Weatherford, R. K. Johnston Jr., and M. L. Valtierra, "Kinematic Viscosity of Liquid Rubidium from 67° to 688°C," *Journal of Chemical and Engineering Data*, vol. 9, pp. 520–524, 1964.

[23] F. Roehlich, F. Tepper, and R. L. Rankin, "Surface tension of four alkali metals to 1000°C," *Journal of Chemical and Engineering Data*, vol. 13, no. 4, pp. 518–521, 1968.

[24] K. Mutsuda, K. Tamura, and M. Inui, "Instability of the electron gas in an expanding metal," *Physical Review Letters*, vol. 98, Article ID 096401, 2007.

[25] K. Matsuda, S. Naruse, K. Hayashi, K. Tamura, M. Inui, and Y. Kajihara, "Structural study of expanded fluid cesium," *Journal of Physics: Conference Series*, vol. 98, no. 1, Article ID 012003, 2008.

[26] K. Tamura, K. Matsuda, and M. Inui, "Structural and electronic properties of expanding fluid metals," *Journal of Physics: Condensed Matter*, vol. 20, Article ID 114102, 2008.

[27] V. Moeini, "Internal pressures of sodium, potassium, and rubidium fluids at different temperatures," *Journal of Chemical & Engineering Data*, vol. 55, no. 12, pp. 5673–5680, 2010.

[28] G. Parsafar and E. A. Mason, "Linear isotherms for dense fluids: a new regularity," *Journal of Physical Chemistry*, vol. 97, no. 35, pp. 9048–9053, 1993.

[29] G. Parsafar, V. Moeini, and G. Najafi, "Pressure dependency of liquid vapor pressure: gibbs prediction improvement," *Iranian Journal of Chemistry & Chemical Engineering*, vol. 20, pp. 37–43, 2001.

[30] http://webbook.nist.gov/chemistry/.

[31] J. O. Hirschfelder, C. F. Curtiss, and R. B. Bird, *Molecular Theory of Gases and Liquids*, John Wiley & Sons, New York, NY, USA, 2nd edition, 1964.

[32] Y. Ghayeb, B. Najafi, V. Moeini, and G. Parsafar, "Calculation of the viscosity of supercritical fluids based on the modified Enskog theory," *High Temperatures-High Pressures*, vol. 35-36, no. 2, pp. 217–226, 2003.

15

Study of Bovine Serum Albumin Solubility in Aqueous Solutions by Intrinsic Viscosity Measurements

Martin Alberto Masuelli

Instituto de Física Aplicada-CONICET, Facultad de Química, Bioquímica y Farmacia, Universidad Nacional de San Luis, Área de Química Física, Chacabuco 917, 5700 San Luis, Argentina

Correspondence should be addressed to Martin Alberto Masuelli; masuelli@unsl.edu.ar

Academic Editor: Sylvio Canuto

The behavior of bovine serum albumin (BSA) in water is scarcely studied, and the thermodynamic properties arising from the experimental measurements have not been reported. Intrinsic viscosity measurements are very useful in assessing the interaction between the solute and solvent. This work discussed in a simple determination of the enthalpy of BSA in aqueous solution when the concentration ranges from 0.2 to 36.71% wt. and the temperature from 35 to 40°C. The relationship between the concentration and intrinsic viscosity is determined according to the method of Huggins. The temperature increase reduces the ratio between inherent viscosity and concentration (η_i/c). This is reflected in the Van't Hoff curve. Furthermore, this work proposes hydrodynamic cohesion value as an indicator of the degree of affinity of protein with water and thermodynamic implications in conformational changes.

1. Introduction

Solubility is the ability of a substance to dissolve into another; it is given by the solubility constant which is in equilibrium with the solute excess or ions excess. Basic studies on proteins have focused on protein concentration, pH, ionic strength, polymeric additives, the dielectric properties of solvent and solvent mixtures, and effect temperature. In the case of proteins and polysaccharides, solubility studies are closely related to studies of gelation that try to determine the temperature and the concentration of gelation (T_{gel} and C_g) (Djabourov) [1].

The conformational stability of a protein is determined by intramolecular factors and solvent interactions (hydration). Solubility is determined primarily by intermolecular effects (protein/protein), but, as proteins are solvated, the hydration effects are also involved in changes in solubility. Conformational changes (changes in functional activity) can be induced by changes in temperature, pressure, and the solvent medium. The technological performance of proteins depends critically on conformation, hydration (water-holding capacity), and solubility. Like other polymers, proteins can be characterized by their chain conformations. However, methods of polymer statistics cannot be applied to most proteins because they adopt specific (native) conformations under different physiological conditions. Since proteins are polyelectrolytes, their solubility behavior is governed largely by electrostatic (ionic) interactions. In determining charge/charge interactions, pK_a and pK_b values of individual amino acids play an important role (Franks) [2].

Barton [3] worked with solubility parameters, trying to explain their nature and to extend solubility theory to liquid mixtures. Cohesion parameters (solubility parameters) provide one of the simplest methods for correlating and predicting the cohesive and adhesive properties of polymers and solvents based on the knowledge of the properties of the individual components isolated.

The nature of solubility may be understood in terms of molecular interactions broadly classified as either "reactive" (involving relatively strong "chemical" forces: complex formation, etc.) or "nonreactive" (involving relatively weak "physical" or "van der Waals" forces). Solution nonideality can of course be best explained when both "chemical" and "physical" forces are considered; the truth lies between these two extremes. The solubility parameter approach is basically "physical," but the introduction of specific interaction components has taken it some way toward a reasonably balanced position, except when "solvation" is considered. This restriction usually limits this approach to nonelectrolyte solutions, but an extension of it to ionic systems is possible. General theories of the liquid state and of solutions involve complex expressions linking molecular interaction potential energy, thermal energy, and volume; for many practical purposes it is convenient to use simple or semiempirical methods. It has been found that a good solvent for a certain (nonelectrolyte) solute such as a polymer has a "solubility parameter" (δ, defined below) value close to that of the solute. Often, a mixture of two solvents, one with a higher δ value and the other with a lower δ value than that of the solute, is a better solvent than each of the two components of the mixture (Barton [4]).

Hansen's [5] study of solubility parameter concepts can be used to interpret relations involved in liquid miscibility, polymer solubility, polymer compatibility, adsorption on solid surfaces, dispersion phenomena, solubility of inorganic and organic materials in organic liquids, and "salting in" phenomena in biological molecules. Mangarj [6] was the first author to work with intrinsic viscosities relating it the Hildebrand and Hansen solubility parameters.

Bozdogan [7] applied the intrinsic viscosity-temperature data of polystyrene (PS) fractions in decalin, cyclohexane, dioctyl phthalate and toluene solutions in theta temperatures, and obtained solubility parameters. This author proposed an equation using the critical volume fraction and segment number of the polymer (PS) for calculating the partial molar entropy of a polymer in dilute solution.

Shulgin and Ruckenstein [8] studied the local composition around protein molecules in aqueous mixtures containing polyethylene glycol (PEG) and solubility of proteins in water-PEG solutions. They concluded that their theory predicts that PEG acts as a salting-out agent for lysozyme, β-lactoglobulin, and bovine serum albumin.

Guner [9] determined the solubility parameters for dextrans/solvents systems using algorithms by group contribution and intrinsic viscosity measurements. Ravindra et al. [10] reported the solubility parameters for chitin and chitosan using group contribution methods. The solubility parameters of chitin and chitosan, as determined by these methods, are more or less equal to the experimental value of 41 $J^{1/2}$ $cm^{-3/2}$. They proposed a method for estimating the overall solubility parameter of chitosan with any deacetylation degree. Kong et al. [11] studied the basicity, water solubility, intrinsic viscosity, and molecular weight of carboxymethyl chitosan.

Naskar et al. [12] studied the viscosity and solubility behaviors of inulin in water solutions. This polysaccharide contains β $(2 \rightarrow 1)$ fructosyl fructose units and exhibits a compact (globular) in aqueous medium, and its solubility is a weak endothermic process.

Brodersen [13] studied the solubility of bilirubin in water and its interaction with phospholipids and albumin at pH 7.1–7.4 to 37°C, confirming that the logarithm of solubility changes linearly with pH. These authors conclude that the deviation of the slope from the theoretical value can be explained by the heterogeneity of bilirubin in the solid phase.

Minghetti et al. [14] investigated the solubility parameters using capillary viscometry (intrinsic viscosity), that may be useful in studies of patch formulation, as this parameter is predictive of the thermodynamic activity of a drug in a matrix.

Bustamante et al. [15] investigated partial solubility parameters from intrinsic viscosity measurements and suggested that this is a versatile method suitable to be used for the study of drugs and both nonpolymeric and polymeric excipients.

Adamska et al. [16] determined the solubility parameters from intrinsic viscosity measurements for a series of pharmaceutical excipients, this data can be used for predicting their behavior in a multicomponent system.

Cohn [17, 18] studied the ionic strength and the isoelectric point which are related to the solubility of globular proteins, concluding that the solubility of proteins in the neighborhood of their isoelectric points depends in a large part upon the degree of their dissociation, the more soluble proteins being the more dissociated.

Shaw et al. [19] studied the effect of net charge on the solubility of ribonuclease Sa, while Schmittschmitt and Scholtz [20] performed similar work in amyloid fibril.

Ferreira Machado et al. [21] performed studies of egg white protein solubility-density. Solubility increased with increases in pH, with the highest solubility at pH 9.0 and the lowest solubility at pH 4.6. This behavior was verified for all the salts analyzed. With an acid pH 3.0, a tendency of increasing solubility linked to the increase of the saline concentration was observed, due to the salting-in effect, and density increased with increases of the salt concentration.

Monkos [22, 23] in various works has made a detailed study of the viscosity and intrinsic viscosity of different globular proteins in terms of concentration and temperature in order to find a mathematical relationship that describes this phenomenon.

Olivares et al. [24] studied gelatin chain aggregation in dilute aqueous solutions at temperatures below the gel point ($T < T_g$) which were subject to different maturation temperatures T_m and maturation times t_m.

Curvale et al. [25] determined the intrinsic viscosity of BSA in aqueous solutions and found an abnormal behavior, according to measurements with the Huggins method; $[\eta]$ is an expression of the interaction between a biopolymer and solvent which reflects the solvent's ability to swell the macromolecule. Thus, we can see that BSA has a very low $[\eta]$ value at pH 7.4, which explains the assigned globular shape.

Sousa et al. [26] performed studies on the hydrodynamics of the Lupin protein, evaluating its intrinsic viscosity, and solubility. Shen [27] studied soy protein by solubility, intrinsic viscosity and optical rotation measurements.

Arakawa and Timasheff [28] worked on protein solubility. Their review said that protein concentration at equilibrium was a complex function of a number of factors such as the physical and chemical natures of the proteins themselves and of environmental parameters such as pH, temperature, nature of the salt, and the kind of organic solvent and its concentration.

Peters [29] wrote that the solubility of albumins was related to their high total electric charge, with corresponding strong hydrophilicity and attraction for water molecules; near neutrality, albumins were extremely soluble in water, 35% wt. in dilute salt solutions, and 50% wt. in water pure solutions. A similar analysis of protein solubility was proposed by Haworth [30], which highlights the hydrophobicity of the composition and the type of amino acids that compose it.

In concordance with the modified treatment of Bozdogan [7], in this work we studied the thermodynamic properties of BSA in aqueous solutions within a temperature range of 35 to 40°C; the increase in relative viscosity (η_i) was calculated for each case. In addition, this work proposes the hydrodynamic value of cohesion as an indicator of the degree of affinity of macromolecules with water and the thermodynamic implications of the conformational changes that water molecules present in association with BSA.

2. Materials and Methods

BSA (lyophilized and deionized powder, purity grade >98%) was obtained from Fedesa S.A.-UNSL; the BSA molecular weight is 66,500 g mol^{-1}, and $\nu_{(a/b)}$ universal shape function is 4.27 for physiological BSA (Curvale et al. [25]). Measurements were taken from fresh BSA solutions of 0.2 and 36.71% wt. in volume with pH 6.5. Solutions and dissolutions were prepared with deionized water. The different temperatures were maintained using a HAAKE C thermostatic bath (±0.1°C). Determinations were done using an Ubbelohde "suspended level" viscometer (IVA 1), with a water draining time of 34.25 s. Even though this viscometer works independently of the volume of the solution, it was here used for performing at least three measurements for each concentration and was later washed until the draining time of the solvent was obtained (Masuelli [31]). These solutions were previously prepared from a mother solution with a concentration determined by UV-absorbance at 278 nm with a Shimadzu UV-160A spectrophotometer. Density of each solution was measured using an Anton Paar DMA35N densimeter.

3. Theory

The solution equilibrium between the mixing solvent and the biopolymer, in concordance with the treatment of Bozdogan [7], is written as

$$S + B \overset{K}{\longleftrightarrow} S \ldots B. \quad (1)$$

The equilibrium constant is

$$K = \frac{[S \ldots B]}{[S]\,[B]}, \quad (2)$$

where $[S]$ is the solvent, and $[S \ldots B]$ and $[B]$ are the mixed and unmixed biopolymer concentrations, respectively. However, solvent concentration is greater than the concentration of biopolymer in solution. Therefore, (2) can be rewritten as

$$K_s = \frac{[S \ldots B]}{[B]}. \quad (3)$$

In the mixing process, the intrinsic viscosity is proportional to the mixed biopolymer concentration (Segarceanua and Leca [32]):

$$[\eta] = f\,[S \ldots B], \quad (4)$$

and unmixed biopolymer concentration is expressed by the following equation:

$$[\eta]_{\lim} - [\eta] = f\,[B], \quad (5)$$

where $[\eta]_{\lim}$ is the limiting intrinsic viscosity, which corresponds to the maximum swelling of the biopolymer, while f is the proportionality constant [7].

Equation (3) can be rewritten as

$$K_s^* = \frac{[\eta]}{[\eta]_{\lim} - [\eta]}. \quad (6)$$

In Huggins' method [25, 31], intrinsic viscosity $[\eta]$ is defined as the ratio of the increase in relative viscosity (η_i) to concentration (c in g cm^{-3}) when the latter tends to zero:

$$\frac{\eta_i}{c} = [\eta] + K_H[\eta]^2 c. \quad (7)$$

IUPAC recommends the term "increment of relative viscosity (η_i)," instead of "specific viscosity," because it has no attributions of specific quantity, meaning:

$$\eta_i = \eta_r - 1. \quad (8)$$

And let us remember that

$$\eta_r = \frac{\eta_s}{\eta_0} = \frac{\rho_s \cdot t_s}{\rho_0 \cdot t_0}, \quad (9)$$

where the subindex "s" indicates "solution" and "0" indicates "solvent."

Since intrinsic viscosity is relevant for dilute solutions of BSA in the range of 0.1 to 4% wt., when high concentrations are used, it is better to start with the first term of the Huggins equation "$\eta_i/c = \eta_{\text{red}}$."

The second law of thermodynamics is,

$$\overline{\Delta G_2} = -RT \ln K_{si}^*,$$
$$\overline{\Delta G_2} = \overline{\Delta H_2} - T\overline{\Delta S_2}. \quad (10)$$

Combining (6) and (7), according to a modified treatment of Bozdogan [7], the following equation is obtained:

$$\ln \frac{\eta_i/c}{(\eta_i/c)_{\lim} - (\eta_i/c)} = -\frac{\overline{\Delta H_2}}{RT} + \frac{\overline{\Delta S_2}}{R}, \quad (11)$$

where $\overline{\Delta G_2}$, $\overline{\Delta H_2}$, and $\overline{\Delta S_2}$ are Gibbs free energy, enthalpy, and entropy changes when mixing a biopolymer fraction, respectively. Plotting $\ln K_{si}^*$ as a function of T^{-1} will give a straight line with a slope of $-\overline{\Delta H_2}/R$ and intercept of $\overline{\Delta S_2}/R$.

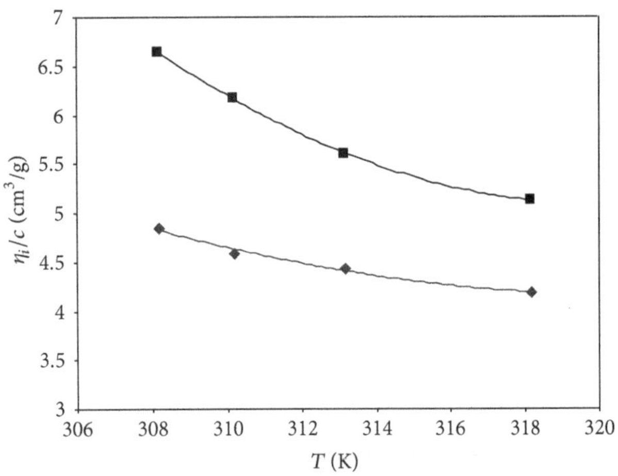

FIGURE 1: Plot of η_i/c versus T, where standard deviations are 0.9599 and 0.9729 for 0.2% and 36.71% wt.

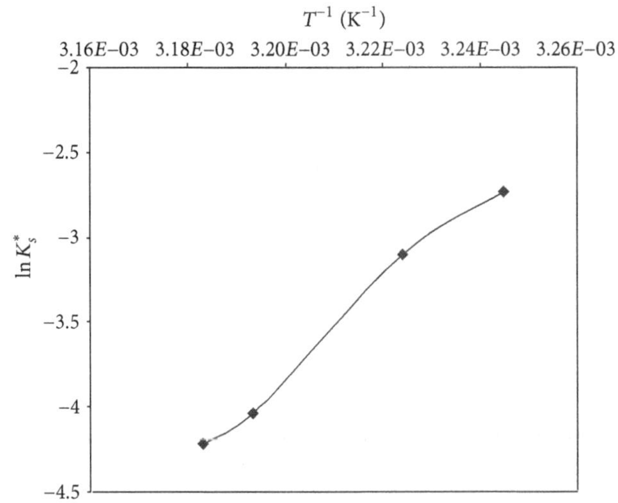

FIGURE 2: Plot $\ln K_{si}^*$ in function T^{-1} for water-BSA system.

The cohesion or Hildebrand solubility parameter "δ" of a polymer can be written as the following equation:

$$\delta = \left(\frac{\overline{\Delta H_2}}{V_M}\right)^{1/2}, \tag{12}$$

where V_M is the polymer molar volume (cm^3 mol^{-1}).

The hydrodynamic cohesion parameter "δ_{Hy}" for hydrosoluble biopolymers can be written as the following equation:

$$\delta_H = \left(\frac{\overline{\Delta H_2}}{V_H}\right)^{1/2}, \tag{13}$$

where V_H, is the hydrodynamic molar volume of biopolymer (cm^3 mol^{-1}), and can be written as the following equation:

$$V_H = \frac{M\,[\eta]}{\nu_{a/b}}. \tag{14}$$

Equation (13) gives an account of the thermodynamic changes and the degree of solvation of the macromolecule in a polar solvent and also shows the phase changes for different temperatures (Masuelli [31]).

4. Results and Discussion

The shape of the curve reduced viscosity ($\eta_i/c = \eta_{\mathrm{red}}$) as a function of temperature, shown in Figure 1, indicates a decrease of η_{red} with temperature, in agreement with other authors as Monkos [33, 34].

When studying the case of BSA in water it is very difficult to find works on solubility as a function of temperature and concentration determined by intrinsic viscosity measurements. Therefore in this work we propose a better alternative to working with ($\eta_i/c = \eta_{\mathrm{red}}$). We see in Figure 2 that for working concentrations the curve has a positive slope from 308 to 313 K.

Agostini et al. [35] investigated the relationship between the thermodynamics and the kinetics of protein aggregation, comparing the solubility of proteins with their aggregation rates. They found a significant correlation between these two quantities when considering a database of protein solubility values that were measured using an in vitro reconstituted translation system containing about 70% of *Escherichia coli* proteins. The existence of such correlation suggests that the thermodynamic stability of the native states of proteins with respect to the aggregate states is closely linked with the kinetic barriers that separate them. A similar study is released by Jiang et al. [36] in pH and thermal function.

Boye et al. [37] found that maximum thermal stability for denatured BSA was observed at pH 5 and 63°C. The denaturation of BSA, observed by IR, resulted in the loss of the 1,654 cm^{-1} band attributed to an α-helical structure and to the rise of two bands at 1,616 and 1,684 cm^{-1} attributed to the formation of an ordered nonnative β-sheet structure associated with aggregation.

Clark et al. [38] performed an extensive study on whey protein mixtures, where structure-property relationships are very difficult to establish. Furthermore, although presumably all globular proteins above the critical gelation concentration C_g can gelify given appropriate conditions, these authors worked on BSA and β-lactoglobulin gels. These authors failed to perform rheological studies of these proteins due to the difficulty of accurately measuring the thickness of the gel.

A similar study was realised by Pelegrine and Gasparetto [39] for whey proteins at a pH of 5.65, where protein solubility increased with temperature, indicating that there was neither coagulation or aggregation between protein molecules possibly because at this pH value the β-lactoglobulin is a dimmer that is dissociated in monomers at 50°C, and only above 60°C the proteins unfold and the hydrophobic groups react.

Analyzing Figure 2, the values of thermodynamic properties $\overline{\Delta H_2}$ and $\overline{\Delta S_2}$ are calculated for A 211 kJ mol^{-1}

and $-0.76\,\mathrm{kJ\,mol^{-1}\,K^{-1}}$ and for B $-31.3\,\mathrm{kJ\,mol^{-1}}$ and $0.078\,\mathrm{kJ\,mol^{-1}\,K^{-1}}$, respectively. The cutoff point between segments A and B is 308.16 K. This tells us that rank A is for endothermic, while rank B is for exothermic and spontaneous solubilization. This phenomenon of a change in the slope between segments A and B can be reasonably explained by taking into account the behavior of concentrate BSA in the water system because BSA-water is a complex solution that consumes energy in this system related with nonideality; read below.

The chemical potential for protein-water system can be written as

$$\mu - \mu_0 = -\frac{RT\nu_0 c}{M}\{1 + BMc + \cdots\}, \qquad (15)$$

where ν_0 is the molar volume of pure solvent (mol $\mathrm{cm^{-3}}$), and B is virial expansion of the solvent chemical potential. The quantity B, the second virial coefficient, serves as a convenient measure of solution nonideality (van Holde [40]).

The biopolymers in which there are strong side-chain interactions tend to coil up into dense globular conformation; these are called globular proteins (BSA). For a discussion of nonideal behavior of BSA solutions (because BSA interacts strongly with water) may be due to two effects: either $\Delta H_2 \neq 0$ or ΔS_2 may contain contributions from ordering or disordering of the solvent. A negative heat of mixing corresponds to favorable BSA-water interaction. This should lead to an even greater decrease in the solvent chemical potential than expected for ideal solutions and thus a positive value of B.

The derivation of that law involved the assumption that the solute molecules are of the same order of size as solvent molecules so that the solvent and solute might be interchanged at random in a hypothetical lattice. But this is by no means a solvent molecule; in fact, the monomer units of the protein are more comparable to the water molecules in size. In other words, a BSA solution more nearly resembles one in which the solute particles are required to move together in clumps. Put another way, this says that the distribution of solute molecules in a macromolecular solution can never be entirely random. The center of each molecule is excluded from a volume determined by the volumes occupied by all of the other molecules. It is not surprising, then, to find that the nonideality and hence the second virial coefficient depend on the excluded volume (ν_{exc}) in $\mathrm{cm^3\,g^{-1}}$ as a fellowship equation

$$B = \frac{N\nu_{\mathrm{exc}}}{2M^2}. \qquad (16)$$

The second virial coefficient, that defines nonideality, is determined by,

$$B = \frac{4\nu_s}{M}, \qquad (17)$$

where ν_s is the specific volume in $\mathrm{cm^3\,g^{-1}}$.

The conclusion to be drawn is that while this contribution to nonideality is small for spherical molecules, it can become large for very asymmetric rod or random coils (see Table 1). This is understandable for the number of ways in which such particles can be packed into the solution is quite limited. The macromolecular solutions can exhibit very nearly ideal behavior under some conditions. For example, for the random coil polymers there is often a particular temperature (called the θ temperature) at which $B = 0$. The excluded volume effect is always such as to lower the chemical potential of solvent. It is an entropic effect. On the other hand, in poor solvents the solute-solute interaction may be such as to increase the solvent chemical potential. This is an enthalpic term. There may exist a temperature (θ temperature) at which these enthalpic and entropic terms exactly are canceled, and the solution is ideal. Putting this mechanistic term, we regard the excluded volume effect as one that pushes molecules apart (they cannot interpenetrate). This may be compensated for by an interaction of the sort that makes solute molecules clump together. The work of a θ temperature for a particular macromolecule-solvent system is a prize indeed for it greatly simplifies all physical studies for ideal solutions. This θ temperature is very difficult to find in BSA since in our case this protein behaves globularly in the temperature range used, and the value of B is in the range between the spherical and rod-like form (van Holde [40] and Curvale [25]).

Table 1: Data of specific volume and second virial coefficient for different temperatures and concentrations.

T/K^{-1}	0.2% wt.		36.71% wt.	
	$\nu_s/\mathrm{cm^{-3}\,g}$	$B\times10^5$	$\nu_s/\mathrm{cm^{-3}\,g}$	$B\times10^5$
314.16	0.4151	2.4969	0.5089	3.0612
313.16	0.4396	2.6440	0.5562	3.3458
310.16	0.4558	2.7414	0.6135	3.6901
308.16	0.4809	2.8927	0.6596	3.9681

Furthermore, observing Table 1, it can be ensured that the effect of temperature on the estrangement of the ideality is higher at temperatures below 314 K, and the nonideality also increases with increasing the concentration of BSA in aqueous solution.

Fan et al. [41] studied BSA solutions (2.5%, pH 7.2) which under heat treatment can undergo structural changes. They found that α-helices are transformed to random coils at 67°C, resulting in an increased rotation angle. With subsequent heating, the transformed random coils may once again transform to nonnative β-sheets and restore the optical rotation angle. These two states are reversible. However, when the heating temperature goes up to 69°C, the denatured BSA starts to transform into a rigid network and becomes an irreversible state.

Wetzel et al. [42] studied the structural alterations of albumin, their dependence on concentration, and the role of free-SH groups in thermal denaturation, as well as the reversibility of thermally induced structural changes. The helix content changes with heat treatment, giving rise to beta structures which are amplified when cooled and which are correlated with the aggregation of albumin. With rising temperature and concentration the proportion of beta structures and aggregates increases. At denaturation degrees of

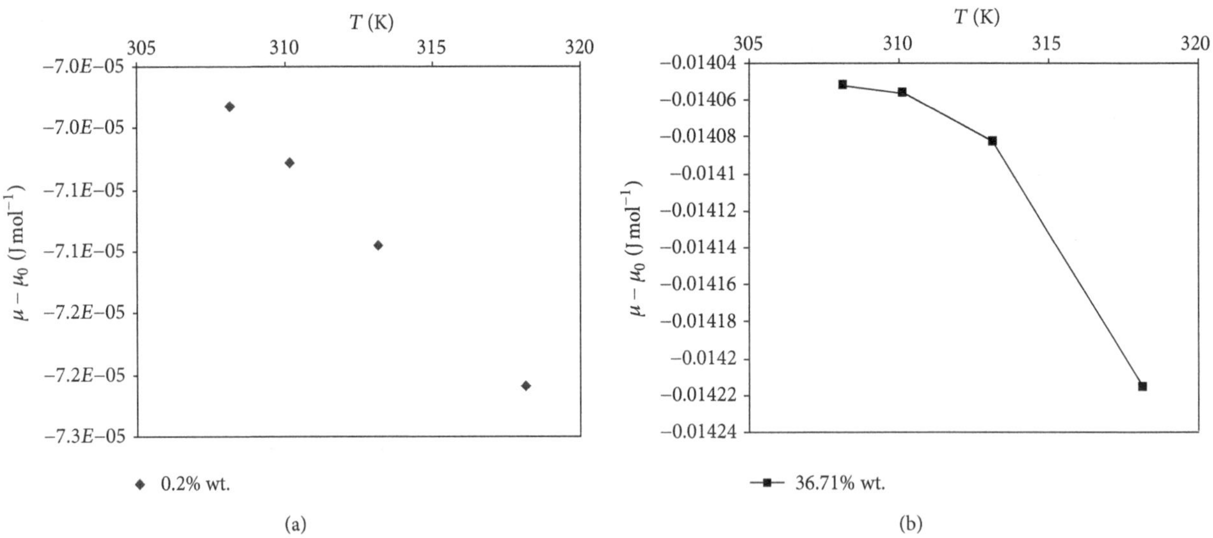

FIGURE 3: Plot of chemical potential in function of T: (a) 0.2% wt., (b) 36.71% wt.

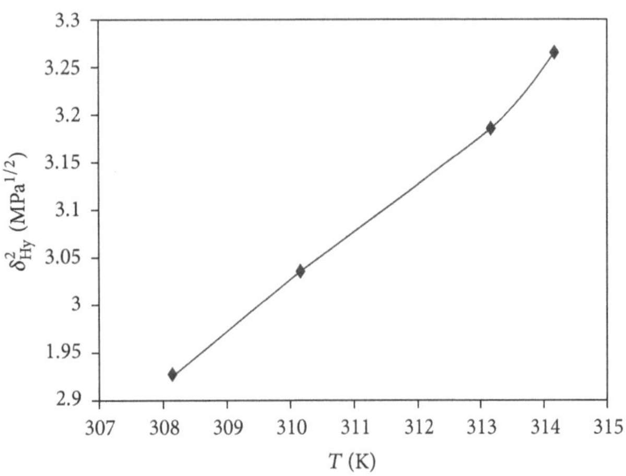

FIGURE 4: Plot of δ_{Hy} in function of T.

up to 20% complete renaturation is possibly in every case. The structure content is concentration dependent even at room temperature. It may be that intermolecular interactions induce additional alpha-helix structures, which are less stable than the ones stabilized by intramolecular interactions. Unfolding of the pocket containing the free -SH group of cysteine-34 enables disulphide bridges to be formed leading to stable aggregates and irreversible structural alterations. At temperatures below 65–70°C, oligomers are formed mainly via intermolecular beta structures.

Takeda et al. [43] studied the thermal denaturation of bovine serum albumin (BSA) at pH of 2.8 and 7.0 in the range of 2–65°C. They found, by the curve-fitting method of circular dichroism spectra, that the relative proportions of α-helix, β-structures, and disordered structures in the protein's conformation were determined as a function of temperature. With the rise of temperature at pH 7.0, the proportion of α-helix decreased above 30°C, and those of β-structures and disordered structures increased in the same temperature range. The structural change was reversible in the temperature range below 45°C. However, the structural change was partially reversible upon cooling to room temperature subsequent to heating to 65°C. On the other hand, the structural change of BSA at pH 2.3 was completely reversible in the temperature range of 2–65°C probably because the interactions between domains and subdomains disappear due to the acid expansion. The secondary structure of disulfide bridges-cleaved BSA remained unchanged during the heat treatment up to 65°C at pH of 2.8 and 7.0.

The Figure 3(a) is linear (i.e., an ideal system), and Figure 3(b) can be seen a nonideal system, which may be due to two effects: firstly a thermal effect and secondly a concentration effect, this is due to increase in temperature and increase in the BSA concentration in the system which causes departure from ideality.

As for the Hildebrand parameter, it remains constant from 309 to 313 K with a value of 1.37 $\mathrm{MPa}^{1/2}$ because the gyration radius of BSA does not change with temperature, and the denaturation of the protein begins at 314 K. The principal utility of the Hildebrand parameter is that it provides simple predictions of phase equilibrium based on a single parameter that is readily obtained for most materials. These predictions are often useful for nonpolar and slightly polar (dipole moment <2 debyes) systems without hydrogen bonding. It has found a particular use in predicting solubility and swelling of polymers by solvents. The principal limitation of the solubility parameter approach is that it applies only to associated solutions ("like dissolves like" or positive deviations from Raoult's law): it cannot account for negative deviations from Raoult's law that result from effects such as solvation or the formation of electron donor-acceptor complexes. Like any simple, predictive theory it can inspire overconfidence: it is best used for screening with data used to verify the predictions.

BSA is a water-soluble protein that is widely used in the biochemical, pharmaceutical, and food industries. The molecule possesses two amino acids, one between repeating units and the other in the ring besides the many hydroxyl groups. Due to this interesting structure, BSA has the ability to form hydrogen bonds both within its own structure and with polar solvents, providing BSA significant solubility or a tendency to form molecular associations. The fact that BSA has hydrogen bonding within its own structure, it interacts with the solvent through hydrogen bonds and is an important fact to be taken into account in the BSA-water system. The solubility parameter may easily be determined for liquids from their heats of vaporization; however, for polymers this process is inapplicable due to their nonvolatility. Yet, polymer solubility parameter values may be evaluated by indirect methods, such as by finding the liquids that cause maximum swelling of a slightly crosslinked network of the polymer or that yield a maximum limiting viscosity number from which the δ value of this solvent may be lower that of the polymer. The alternative method is calculating δ from the group molar attraction constants.

Analyzing the hydrodynamic cohesion parameter "δ_{Hy}" (Figure 4), a light increase can be observed from 309 to 313 K (2.93–3.19 $MPa^{1/2}$); this is because the hydrodynamic radius of BSA changes with temperature. The hydrodynamic cohesion parameter is used for polar molecules and macromolecules with dipole moment >2 debyes (BSA dipole moment is ≈400 debyes, obtained by Curvale, unpublished data); this system may have hydrogen bonding. It has been particularly useful for predicting solubility and swelling of biopolymers in aqueous solvents. It can account for negative deviations from Raoult's law that result from effects such as solvation or the formation of electron donor-acceptor complexes. The value of δ_{Hy} is about twice larger than that of δ due to the degree of solvation of BSA.

Solubility in water also increased significantly when the temperature increased (Pace et al. [44]). These excellent water solubility properties are due to the exposure of treated samples to the strong mechanical forces which cause particles to collide over short-time intervals, changing their size and shape. This behavior was previously confirmed by a significant decrease in the particle size of proteins accompanied by an increase in the number of charged groups $-NH_3^+$; $-COO^-$. Consequently, enhanced protein-water interactions occurred since electrostatic forces are higher, and more water interacts with protein molecules.

5. Conclusions

The viscosity of BSA solutions at temperatures of up to 308 K, in a wide range of concentrations at pH values near the isoelectric point, may be quantitatively described by the generalized Van't Hoff equation. The generalized Van't Hoff equation can be used for dilute concentrations of polymers (a condition for using intrinsic viscosity). Given the difficulty of measuring the intrinsic viscosity at high concentrations, this work proposes the use of the η_i/c ratio and molar concentration for solving this difficulty. The η_i/c decreases with increases in temperature. Both quantities depend on temperature, and at the high temperature limit, the η_i/c agrees well with the theoretical values obtained for the random coil conformation. In the hydrodynamic cohesion parameter "δ_{Hy}" (Figure 4), a slight increase can be observed from 308 to 313 K (2.93–3.19 $MPa^{1/2}$); this is because the hydrodynamic radius of BSA changes with temperature.

Acknowledgments

The author thanks Universidad Nacional de San Luis, Instituto de Física Aplicada (INFAP-CONICET) for their support and Dr. Rolando Curvale for their valuable contributions.

References

[1] M. Djabourov, "Gelation. A review," *Polymer International*, vol. 25, no. 3, pp. 135–143, 1991.

[2] F. Franks, *Solution Properties of Proteins*, Elsevier, 1988.

[3] A. F. M. Barton, "Applications of solubility parameters and other cohesion parameters in polymer science and technology," *Pure and Applied Chemistry*, vol. 57, no. 7, pp. 905–912, 1985.

[4] A. F. M. Barton, "Solubility parameters," *Chemical Reviews*, vol. 75, no. 6, pp. 731–753, 1975.

[5] C. M. Hansen, "The universality of the solubility parameter," *Industrial and Engineering Chemistry Product Research and Development*, vol. 8, no. 1, pp. 2–11, 1969.

[6] D. Mangarj, S. K. Bhatnagar, and S. B. Rath, "Cohesive-energy-densities of high polymers—part. III: estimation of C. E. D. by viscosity measurements," *Die Makromolekulare Chemie*, vol. 67, no. 1, pp. 75–83, 1963.

[7] A. E. Bozdogan, "A method for determination of thermodynamic and solubility parameters of polymers from temperature and molecular weight dependence of intrinsic viscosity," *Polymer*, vol. 45, no. 18, pp. 6415–6424, 2004.

[8] I. L. Shulgin and E. Ruckenstein, "Preferential hydration and solubility of proteins in aqueous solutions of polyethylene glycol," *Biophysical Chemistry*, vol. 120, no. 3, pp. 188–198, 2005.

[9] A. Guner, "The algorithmic calculations of solubility parameter for the determination of interactions in dextran/certain polar solvent systems," *European Polymer Journal*, vol. 40, no. 7, pp. 1587–1594, 2004.

[10] R. Ravindra, K. R. Krovvidi, and A. A. Khan, "Solubility parameter of chitin and chitosan," *Carbohydrate Polymers*, vol. 36, no. 2-3, pp. 121–127, 1998.

[11] X. P. Kong, J. Wang, C. J. Wang, and X. J. Wu, "Basicity, water solubility and intrinsic viscosity of carboxymethyl chitosan as a biofunctional material," *Advanced Materials Research*, vol. 531, pp. 507–510, 2012.

[12] B. Naskar, A. Dan, S. Ghosh, and S. P. Moulik, "Viscosity and solubility behavior of the polysaccharide inulin in water, water + dimethyl sulfoxide, and water + isopropanol media," *Journal of Chemical and Engineering Data*, vol. 55, no. 7, pp. 2424–2427, 2010.

[13] R. Brodersen, "Bilirubin. Solubility and interaction with albumin and phospholipid," *Journal of Biological Chemistry*, vol. 254, no. 7, pp. 2364–2369, 1979.

[14] P. Minghetti, F. Cilurzo, A. Casiraghi, and L. Montanari, "Application of viscometry and solubility parameters in miconazole

patches development," *International Journal of Pharmaceutics*, vol. 190, no. 1, pp. 91–101, 1999.

[15] P. Bustamante, J. Navarro-Lupión, and B. Escalera, "A new method to determine the partial solubility parameters of polymers from intrinsic viscosity," *European Journal of Pharmaceutical Sciences*, vol. 24, no. 2-3, pp. 229–237, 2005.

[16] K. Adamska, A. Voelkel, and K. Heberger, "Selection of solubility parameters for characterization of pharmaceutical excipients," *Journal of Chromatography A*, vol. 1171, no. 1-2, pp. 90–97, 2007.

[17] E. Cohn, "The relation between the iso-electric point of a globulin and its solubility and acid combining capacity in salt solution," *Proceedings of the National Academy of Sciences of the United States of America*, vol. 6, no. 5, pp. 256–263, 1920.

[18] E. J. Cohn and J. L. Hendry, "Studies in the physical chemistry of the proteins: II. The relation between the solubility of casein and its capacity to combine with base. The solubility of casein in systems containing the protein and sodium hydroxide," *Journal of General Physiology*, vol. 5, no. 5, pp. 521–554, 1923.

[19] K. L. Shaw, G. R. Grimsley, G. I. Yakovlev, A. A. Makarov, and C. N. Pace, "The effect of net charge on the solubility, activity, and stability of ribonuclease Sa," *Protein Science*, vol. 10, no. 6, pp. 1206–1215, 2001.

[20] J. P. Schmittschmitt and J. M. Scholtz, "The role of protein stability, solubility, and net charge in amyloid fibril formation," *Protein Science*, vol. 12, no. 10, pp. 2374–2378, 2003.

[21] F. Ferreira Machado, J. S. R. Coimbra, E. E. Garcia Rojas, L. A. Minim, F. C. Oliveira, and R. D. C. S. Sousa, "Solubility and density of egg white proteins: effect of pH and saline concentration," *LWT—Food Science and Technology*, vol. 40, no. 7, pp. 1304–1307, 2007.

[22] K. Monkos, "Concentration and temperature dependence of viscosity in lysozyme aqueous solutions," *Biochimica et Biophysica Acta*, vol. 1339, no. 2, pp. 304–310, 1997.

[23] K. Monkos, "Viscosity analysis of the temperature dependence of the solution conformation of ovalbumin," *Biophysical Chemistry*, vol. 85, no. 1, pp. 7–16, 2000.

[24] M. L. Olivares, M. B. Peirotti, and J. A. Deiber, "Analysis of gelatin chain aggregation in dilute aqueous solutions through viscosity data," *Food Hydrocolloids*, vol. 20, no. 7, pp. 1039–1049, 2006.

[25] R. Curvale, M. Masuelli, and A. Perez Padilla, "Intrinsic viscosity of bovine serum albumin conformers," *International Journal of Biological Macromolecules*, vol. 42, no. 2, pp. 133–137, 2008.

[26] I. M. N. Sousa, P. J. Morgan, J. R. Mitchell, S. E. Harding, and S. E. Hill, "Hydrodynamic characterization of Lupin proteins: solubility, intrinsic viscosity, and molar mass," *Journal of Agricultural and Food Chemistry*, vol. 44, no. 10, pp. 3018–3021, 1996.

[27] J. L. Shen, "Solubility profile, intrinsic viscosity, and optical rotation studies of acid precipitated soy protein and of commercial soy isolate," *Journal of Agricultural and Food Chemistry*, vol. 24, no. 4, pp. 784–788, 1976.

[28] T. Arakawa and S. N. Timasheff, "Theory of protein solubility," *Methods in Enzymology*, vol. 114, pp. 49–77, 1985.

[29] J. Peters, *All Albumin*, Elsevier, 1993.

[30] I. S. Haworth, *Protein Solubility. Chemistry: Foundations and Applications*, 2004, http://www.encyclopedia.com/doc/1G2-3400900422.html.

[31] M. A. Masuelli, "Viscometric study of pectin. Effect of temperature on the hydrodynamic properties," *International Journal of Biological Macromolecules*, vol. 48, no. 2, pp. 286–291, 2011.

[32] O. Segarceanua and M. Leca, "Improved method to calculate Hansen solubility parameters of a polymer," *Progress in Organic Coatings*, vol. 31, no. 4, pp. 307–310, 1997.

[33] K. Monkos, "Viscosity of bovine serum albumin aqueous solutions as a function of temperature and concentration," *International Journal of Biological Macromolecules*, vol. 18, pp. 61–68, 1996.

[34] K. Monkos, "On the hydrodynamics and temperature dependence of the solution conformation of human serum albumin from viscometry approach," *Biochimica et Biophysica Acta*, vol. 1700, no. 1, pp. 27–34, 2004.

[35] F. Agostini, M. Vendruscolo, and G. G. Tartaglia, "Sequence-based prediction of protein solubility," *Journal of Molecular Biology*, vol. 421, no. 2-3, pp. 237–241, 2012.

[36] J. Jiang, Y. L. Xiong, and J. Chen, "pH shifting alters solubility characteristics and thermal stability of soy protein isolate and its globulin fractions in different pH, salt concentration, and temperature conditions," *Journal of Agricultural and Food Chemistry*, vol. 58, no. 13, pp. 8035–8042, 2010.

[37] J. I. Boye, I. Alli, and A. A. Ismail, "Interactions involved in the gelation of bovine serum albumin," *Journal of Agricultural and Food Chemistry*, vol. 44, no. 4, pp. 996–1004, 1996.

[38] A. H. Clark, G. M. Kavanagh, and S. B. Ross-Murphy, "Globular protein gelation—theory and experiment," *Food Hydrocolloids*, vol. 15, no. 4–6, pp. 383–400, 2001.

[39] D. H. G. Pelegrine and C. A. Gasparetto, "Whey proteins solubility as function of temperature and pH," *LWT—Food Science and Technology*, vol. 38, no. 1, pp. 77–80, 2005.

[40] K. H. van Holde, *Physical Biochemistry*, Prentice Hall, Englewood Cliffs, NJ, USA, 1971.

[41] K. C. Fan, M. Song, and R. S. Lu, "Measurement system of tiny angle based on LED," in *Seventh International Symposium on Precision Engineering Measurements and Instrumentation*, vol. 8321, Yunnan, China, August 2011.

[42] R. Wetzel, M. Becker, J. Behlke et al., "Temperature behaviour of human serum albumin," *European Journal of Biochemistry*, vol. 104, no. 2, pp. 469–478, 1980.

[43] K. Takeda, A. Wada, K. Yamamoto, Y. Moriyama, and K. Aoki, "Conformational change of bovine serum albumin by heat treatment," *Journal of Protein Chemistry*, vol. 8, no. 5, pp. 653–659, 1989.

[44] C. N. Pace, S. D. Teviño, E. Prabhakaran, and J. M. Scholtz, "Protein structure, stability and solubility in water and other solvents," *Philosophical Transactions of the Royal Society B*, vol. 39, no. 1448, pp. 1225–1235, 2004.

16

The N· · · HF Interactions in the X-Pyridazine· · · $(HF)_2$ Complexes: Substituent Effects and Energy Components

Ali Ebrahimi, Mostafa Habibi-Khorassani, Farideh Badichi Akher, and Abdolkarim Farrokhzadeh

Department of Chemistry, University of Sistan and Baluchestan, P.O. Box 98135-674, Zahedan, Iran

Correspondence should be addressed to Ali Ebrahimi, ebrahimi@chem.usb.ac.ir

Academic Editor: Laimutis Bytautas

The effects of substituents on the N· · · HF interactions in the X-pyridazine· · · $(HF)_n$ (X = $N(CH_3)_2$, $NHCH_3$, NH_2, C_2H_5, CH_3, OCH_3, OH, CN, OF, NO_2, F, Br, Cl, and $n = 1, 2$) complexes have been studied at the B3LYP/6-311++G(d,p) level of theory. In all complexes, the binding energies increase for the electron-donating substituents and decrease for the electron-withdrawing substituents. A negative cooperativity is observed for two hydrogen bond interactions. There are meaningful relationships between the Hammett constants and the energy data and the results of population analysis in the binary and ternary complexes. Symmetryadapted perturbation theory (SAPT) analysis was also carried out to unveil the nature of hydrogen bond in the complexes **2** and **3**. The electron-donating substituents increase the magnitude of the SAPT interaction energy components and the electron-withdrawing substituents decrease those components. The highest/lowest change is observed for the E_{exch}/E_{disp} component. The effect of C_2H_5 (or CH_3) on different components is higher than OCH_3 in the complex **2** while the trend is reversed in the complex **3**. It is demonstrated that the electrostatic interaction plays a main role in the interaction, although induction and dispersion interactions are also important.

1. Introduction

The diazine rings are building blocks of many important natural and synthetic compounds [1]. They have been the subject of extensive research, particularly in the pharmaceutical and agrochemical areas due to their broad activities, such as antihypertensive and anti-inflammatory activity [2–5].

The hydrogen bond plays a crucial role in biology, chemistry and related disciplines [6–13]. Cooperativity is an important characteristic of hydrogen bond interactions. The role of hydrogen bond may be modified by the cooperativity of hydrogen bonds in many chemical and biological systems [14–16]. It plays an important role in controlling and regulating the processes occurring in living organisms. Many physical and chemical properties of materials are determined by hydrogen-bonding cooperativity [17–20]. For example, the hydrogen bond cooperativity is relevant for sustaining the stable conformers of biological molecules [21, 22] and constructing the crystal structures [22, 23].

The effects of substituents on the binding energies of the X-pyridazine· · · $(HF)_n$ ($n = 1, 2$) complexes (represented by **1–3** in Scheme 1) and the cooperativity of the H-bond interactions have been investigated in the present study. Substituent X is located at position 4 of pyridazine ring, which is meta relative to N_2 in **1** and **2** and para relative to N_1 in **1** and **3**. The relationship between binding energies and the Hammett constants have been studied for binary and ternary complexes. Also, the cooperativity of two hydrogen bond interactions and the strength of interactions have been investigated by the results of atoms in molecules (AIMs) [24], the natural bond orbital (NBO) [25], and molecular electrostatic potential (MEP) analysis. Symmetry-adapted perturbation theory (SAPT) [26] has been employed to determine the physically significant components of the total interaction energies for the complexes **2** and **3**.

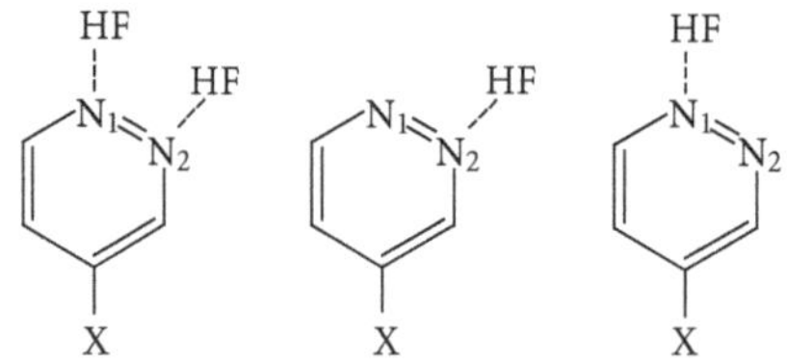

1 2 3

Scheme 1: The X-pyridazine$\cdots$(HF)$_n$ complexes considered in the present work (X = $N(CH_3)_2$, $NHCH_3$, NH_2, C_2H_5, CH_3, OCH_3, OH, CN, OF, NO_2, F, Br, Cl, and $n = 1, 2$).

2. Methods

The geometries of the complexes were fully optimized with the B3LYP [27] method using the 6-311++G(d,p) basis set by the Gaussian 03 program package [28]. The single-point calculations were carried out using the MP2 [29] and PBE1-KCIS [30] methods in conjunction with the 6-311++G(d,p) and aug-cc-pVDZ basis sets. The interaction energies were corrected with the basis set superposition error (BSSE) using the counterpoise method of Boys and Bernardi [31]. The frequency calculations were performed at the B3LYP/6-311++G(d,p) level of theory. The obtained wave functions at the B3LYP/6-311++G(d,p) computational level have been used to analyze the electron density within the AIM methodology by the AIM2000 program [32]. The NBO analysis has been performed using the HF method in conjunction with the 6-311++G(d,p) basis set using the NBO3.1 program in the Gaussian03 package. Also, the ChelpG [33] charges were calculated at the B3LYP/6-311++G(d,p) level of theory to investigate the charge transfer between two units. Cube files containing the MEP information have been generated for complexes at the B3LYP/6-311++G(d,p) level of theory. The freely available MOLEKEL program [34] has been used for the visualization of the MEP data. The most negative-valued MEP point ($V_{\min}$) can be obtained from visual inspection of MEP data for the lone-pair region of the nitrogen atoms in pyridazine. The B3LYP-optimized geometries with the 6-311++G(d,p) basis set were then used to perform interaction energy decomposition using the SAPT scheme. Molecular integrals were first obtained with the GAMESS package [35], and the SAPT partitioning was performed using the SAPT-2008 program [36].

3. Results and Discussion

The most important geometrical parameters of complexes optimized at the B3LYP/6-311++G(d,p) level of theory are gathered in Table 1. The maximum and minimum values of the N$\cdots$H bond length correspond to NO_2 and $N(CH_3)_2$ substituents, respectively, in all cases. Also, the N$\cdots$H bond length in the complex **1** is longer than that in the complexes **2** and **3**. On the other hand, the N$\cdots$H bond length in the complex **2** is longer than that in the complex **3**. These results show that the cooperativity of H-bond interactions leads to the elongation of the N$\cdots$H bond lengths.

The total binding energies of complexes (ΔE) calculated at the B3LYP/6-311++G(d,p) level of theory and corrected for BSSE are summarized in Table 2. The results show that the stabilization energies of the complex **3** are larger than those of complex **2**. As can be seen, the maximum and minimum values of ΔE calculated at the B3LYP/6-311++G(d,p) level of theory correspond to the $N(CH_3)_2$ and NO_2 substituents, respectively.

The results of single-point energy calculation at the MP2/6-311++G (d,p), MP2/aug-cc-pVDZ, PBE1KCIS/6-311++G(d,p), and PBE1KCIS/aug-cc-pVDZ levels of theory on the geometries optimized at the B3LYP/6-311++G(d,p) level of theory are also given in Table 2. The maximum and minimum ΔE values correspond to $N(CH_3)_2$ and NO_2 substituents, respectively, at all levels of theory.

The ΔE values calculated by the MP2 method increase by 13.8 to 7.6 percent going from 6-311++G(d,p) to aug-cc-pVDZ basis sets. The corresponding changes are in the range of -0.51 to -0.19 kcal mol^{-1} for the values calculated by the PBE1KCIS method.

The absolute values of ΔE calculated at the MP2/6-311++G(d,p) level are 2.07 to 4.22 kcal mol^{-1} smaller than those calculated at the B3LYP/6-311++G(d,p) level. The absolute values of ΔE calculated by the PBE1KCIS method are approximately 3.54 to 0.51 kcal mol^{-1} larger than the values calculated by the MP2 and B3LYP methods. The ΔE values calculated at the PBE1KCIS/aug-cc-pVDZ level are approximately identical with the values calculated at the B3LYP/6-311++G(d,p) level of theory.

The BSSE-corrected binding energies of the complex **1** are lower than the sum of binding energies of the complexes **2** and **3** (see Table 2). In addition, the stabilization energies of complexes become more negative by the electron-donating substituents (with the exception of the OH substituent in the complex **2**) relative to the pyridazine, while the behavior is reversed by the electron-withdrawing substituents.

The total substituent effect comprises inductive/field effects, which have electrostatic character, and resonance effects, which are not electrostatic in nature. Resonance effects are strongest at para position, while the electrostatic interactions inversely depend on distance [38, 39]. The Hammett substituent constants are presented in Table 1 [37]. For the $N(CH_3)_2$, $NHCH_3$, NH_2, CH_3, and C_2H_5 substituents, the σ_p values are more negative than the σ_m values, so, the electron donation of the aforementioned substituents in the complex **3** is higher than that in the complex **2**. The high

TABLE 1: The N· · · H bond lengths optimized at the B3LYP/6-311++G(d,p) level in (Å) and the Hammett constants.

	Bond length			Hammett constants				
	Complex **1**	Complex **2**	Complex **3**	σ_m	σ_p	σ_{total}	σ_I	σ_R
$N(CH_3)_2$	1.67 (1.69)	1.64	1.63	−0.15	−0.83	−0.98	0.15	−0.98
$NHCH_3$	1.68 (1.70)	1.65	1.63	−0.52	−0.74	−1.26	0.03	−0.73
NH_2	1.68 (1.71)	1.65	1.64	−0.16	−0.66	−0.82	0.08	−0.74
CH_3	1.71 (1.71)	1.66	1.66	−0.07	−0.17	−0.24	0.01	−0.18
C_2H_5	1.71 (1.71)	1.66	1.66	−0.07	−0.15	−0.22	0.00	−0.15
OCH_3	1.70 (1.72)	1.67	1.65	0.12	−0.27	−0.15	0.29	−0.56
OH	1.71 (1.73)	1.67	1.66	0.12	−0.37	−0.25	0.30	−0.70
H	1.72 (1.72)	1.67	1.67	0.00	0.00	0.00		
OF	1.73 (1.74)	1.70	1.68	0.47	0.36	0.83		
CN	1.75 (1.70)	1.71	1.70	0.56	0.66	1.22	0.51	0.15
NO_2	1.76 (1.76)	1.72	1.71	0.71	0.78	1.49	0.60	0.13
F	1.73 (1.74)	1.69	1.68	0.34	0.06	0.40	0.45	−0.39
Br	1.73 (1.74)	1.69	1.69	0.39	0.23	0.62	0.40	−0.22
Cl	1.73 (1.74)	1.69	1.68	0.37	0.23	0.60	0.42	−0.19

The data in the parentheses correspond to the $N_2 \cdots H$ bond length. The σ values are taken from [37].

TABLE 2: The binding energies ($-\Delta E$ in kcal mol^{-1}) corrected for BSSE calculated at different levels.

	Complex **1**			Complex **2**			Complex **3**		
	B3LYP	MP2	PBE1KCIS	B3LYP	MP2	PBE1KCIS	B3LYP	MP2	PBE1KCIS
$N(CH_3)_2$	25.75	21.90 (23.52)	26.22 (25.86)	13.92	11.90 (12.81)	14.19 (14.03)	14.48	11.72 (12.70)	14.71 (14.49)
$NHCH_3$	25.32	21.80 (23.23)	25.81 (25.42)	13.73	12.00 (12.74)	13.99 (13.82)	14.31	11.82 (12.60)	14.54 (14.31)
NH_2	24.51	20.30 (22.12)	25.06 (24.66)	13.27	11.30 (12.14)	13.55 (13.38)	13.88	11.14 (12.10)	14.15 (13.91)
CH_3	22.98	19.80 (21.10)	23.44 (23.07)	12.80	10.90 (11.63)	13.03 (12.85)	12.86	10.82 (11.50)	13.08 (12.88)
C_2H_5	23.34	19.80 (21.44)	23.88 (23.64)	13.01	10.90 (11.86)	13.32 (13.24)	13.09	10.78 (11.80)	13.40 (13.31)
OCH_3	23.19	19.90 (21.24)	23.65 (23.31)	12.61	10.90 (11.55)	12.85 (12.68)	13.20	10.99 (11.70)	13.43 (13.24)
OH	22.42	19.30 (20.69)	22.96 (22.67)	12.29	10.60 (11.34)	12.57 (12.43)	12.72	10.61 (11.30)	12.98 (12.81)
H	22.01	19.00 (20.28)	22.45 (22.07)	12.33	10.50 (11.17)	12.55 (12.35)	12.33	10.50 (11.20)	12.55 (12.35)
OF	20.10	17.60 (18.89)	20.70 (20.56)	10.99	9.59 (10.29)	11.29 (11.21)	11.56	9.82 (10.50)	11.86 (11.76)
CN	18.33	16.00 (17.10)	18.76 (18.42)	10.21	8.66 (9.26)	10.41 (10.22)	10.43	9.03 (9.62)	10.64 (10.46)
NO_2	17.59	15.50 (16.74)	18.14 (17.95)	9.85	8.40 (9.04)	10.10 (9.99)	10.03	8.88 (9.50)	10.30 (10.19)
F	20.30	17.60 (18.86)	20.84 (20.57)	11.20	9.65 (10.28)	11.46 (11.31)	11.63	9.86 (10.50)	11.89 (11.75)
Br	20.28	17.70 (18.86)	20.79 (20.42)	11.16	9.63 (10.23)	11.41 (11.23)	11.51	9.89 (10.50)	11.75 (11.58)
Cl	20.29	17.70 (18.81)	20.80 (20.44)	11.18	9.65 (10.22)	11.43 (11.24)	11.55	9.91 (10.50)	11.80 (11.61)

The data in the parenthesis correspond to the aug-cc-pVDZ basis set.

electronegativity of oxygen makes the OCH_3 and OH substituents electron-withdrawing by the inductive effect; this is reflected in the positive value for σ_m.

Recall that the meta substituents only affect the reaction center by the inductive effects, whereas the para substituents affect it by both the inductive and resonance effects. Thus, there is a satisfactory linear relationship between ΔE and $\sigma_I (R = 0.91)$, ΔE and $\sigma_I + \sigma_R (R = 0.98)$, respectively, in the complexes **2** and **3**. Comparing the σ values for the meta and para indicates that the electron-donating resonance effect (σ_R) dominates over the electron-withdrawing inductive effect (σ_I). Therefore, these substituents are more electron-donating in the complex **3** in comparison to the complex **2**. The σ_m and σ_p values are positive for the OF, CN, NO_2, F, Br, and Cl substituents. Thus, these substituents are electron-withdrawing in both the meta and para positions. Comparison between the σ values for the meta and para indicates that the σ_I dominates over the σ_R. Since the inductive effect is inversely related to distance, the electron-withdrawing inductive effects are stronger in the complex **2** in comparison with the complex **3**.

The σ_m and σ_p constants can be used as appropriate parameters for the description of the intermolecular interaction in the X-pyridazine$\cdots(HF)_2$ complex. The linear correlation coefficient between the ΔE values and the σ_m, σ_p, and $\sigma_{total}(\sigma_m + \sigma_p)$ constants is equal to 0.95, 0.97, and 0.98, respectively. This indicates that the total electrostatic effect of the substituents as well as induction and resonance vitally impacts the two intermolecular interactions (see Figure 1).

3.1. AIM and NBO Analysis. A way to characterize the hydrogen bond is AIM analysis that interprets these interactions in terms of critical points [40, 41]. The values of electron density (ρ) calculated at the $N_2 \cdots H$ and $N_1 \cdots H$ bond critical points (BCPs) of complexes **1–3** are gathered in Table 3.

TABLE 3: Electron densities ρ (in e/au^3) at the N$\cdots$H BCP in the X-pyridazine$\cdots$(HF)$_n$(n = 1, 2) complexes and individual H-bond energies E (in kcal mol^{-1}) calculated for the complex **1**.

	Complex 1		Complex 2	Complex 3
	$\rho_{BCP} \times 10^2$	E	$\rho_{BCP} \times 10^2$	$\rho_{BCP} \times 10^2$
$N(CH_3)_2$	**5.46**, *5.23*	**−14.41**, *−13.44*	6.03	6.16
$NHCH_3$	**5.40**, *5.15*	**−14.23**, *−13.18*	5.95	5.95
NH_2	**5.32**, *5.05*	**−13.85**, *−12.73*	5.82	5.97
CH_3	**4.99**, *4.95*	**−12.36**, *−12.21*	5.69	5.69
C_2H_5	**5.02**, *5.00*	**−12.76**, *−12.67*	5.73	5.74
OCH_3	**5.08**, *4.91*	**−12.98**, *−12.29*	5.60	5.79
OH	**5.02**, *4.79*	**−12.66**, *−11.75*	5.56	5.65
H	**4.83**, *4.83*	**−12.04**, *−12.04*	5.54	5.54
OF	**4.70**, *4.55*	**−11.37**, *−10.81*	5.19	5.35
CN	**4.45**, *4.43*	**−10.22**, *−10.15*	5.01	5.07
NO_2	**4.42**, *4.45*	**−10.08**, *−10.19*	4.90	4.90
F	**4.71**, *4.57*	**−11.46**, *−10.92*	5.23	5.36
Br	**4.69**, *4.61*	**−11.33**, *−11.02*	5.24	5.33
Cl	**4.69**, *4.60*	**−11.36**, *−11.02*	5.25	5.35

The bold and italicized values correspond to the $N_1\cdots H$ and $N_2\cdots H$ hydrogen bonds, respectively.

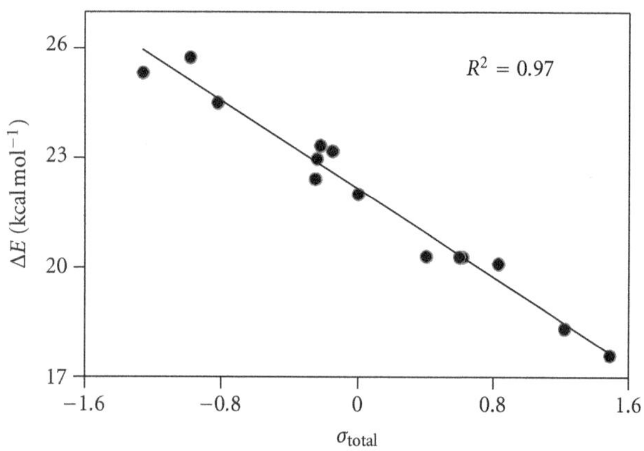

FIGURE 1: Correlation between the binding energies ($-\Delta E$ in kcal mol^{-1}) and Hammett constant σ_{total} for the complex **1**.

The maximum and minimum values correspond to the $N(CH_3)_2$ and NO_2 substituents, respectively, (with the exception of ρ_{N_2H} in the complex **1** where the minimum value corresponds to the CN substituent). The topological properties of ρ calculated at the BCP of the intermolecular hydrogen bonds may be treated as a measure of the hydrogen bond strength [42–44]. The electron-withdrawing substituents pull the lone pair of nitrogen atoms of pyridazine inside the ring and decrease the ρ values at the N$\cdots$H BCP, while the electron-donating substituents increase the ρ values and enhance the hydrogen bond strength. The ρ values calculated at the N$\cdots$H BCP in the complex **3** are higher than those in the complex **2**, and the complex **3** is more stable than the complex **2**. On the other hand, the ρ values of N$\cdots$H hydrogen bond of complex **1** are lower than the values calculated at the $N_2\cdots H$ and $N_1\cdots H$ hydrogen bonds in the complexes **2** and **3**. Thus, a negative cooperativity is predicted from the comparison between $\rho_{N\cdots H}$ values in the complex **1** and the complexes **2** and **3**. The N$\cdots$H bond length decreases linearly by the increase in the ρ values calculated at the N$\cdots$H BCP for all categories ($R^2 \simeq 0.99$). Also, there is a linear relationship between ρ and ΔE for all cases. A linear relationship ($R^2 = 0.99$) is observed between the ΔE values calculated at the B3LYP/6-311++G(d,p) level and the ρ values calculated at the N$\cdots$H BCPs in the complex **1**.

Because of linear relationship between ΔE and ρ_{NH} values, the $\Delta E = c\rho_{NH}$ can be used for the calculation of individual H-bond energies in the complex **1** (see Table 4). All individual H-bond energies calculated for the complex **1** are lower than the values calculated for the complexes **2** and **3**. Also, there is a linear relationship with high correlation coefficient ($R = 0.99$) between individual H-bond energies and the N$\cdots$H bond lengths.

For a better understanding of the hydrogen bond interaction in the complexes **1–3**, the NBO analysis has been carried out at the HF/6-311++G(d,p) level of theory. The $lp_N \rightarrow \sigma^*_{HF}$ interaction, which can be considered as a measure of charge transfer, has been evaluated in the complexes **1–3** (see Table 5). The donor-acceptor interaction energy (E^2) value of this interaction can be used to predict the strength of the N$\cdots$H hydrogen bond. In all cases, the maximum and minimum values correspond to the $N(CH_3)_2$ and NO_2 substituents, respectively. As can be seen in Table 4, the E^2 values of the $lp_{N_2} \rightarrow \sigma^*_{HF}$ and $lp_{N_1} \rightarrow \sigma^*_{HF}$ interactions in the complexes **2** and **3** are higher than those in the complex **1**. So, the negative cooperativity decreases the E^2 values of $lp_{N_2} \rightarrow \sigma^*_{HF}$ and $lp_{N_1} \rightarrow \sigma^*_{HF}$ donor-acceptor interactions in the complex **1** relative to the complexes **2** and **3**. The electron-donating substituents increase the electron density on the nitrogen atoms of pyridazine ring (with the exception of OH substituent in the complex **2**) and increase their inclination on polarization of HF, which increase the E^2 value of the $lp_N \rightarrow \sigma^*_{HF}$ interaction. Though OH is an electron-donating substituent, but it destabilizes the complex **2** relative to the pyridazine. This behavior is not observed for the OCH_3 functional group. The NBO atomic charge on the O (−0.71) of OH substituent is more negative than O (−0.61) of the OCH_3 substituent. Thus, electron-withdrawing induction effect is predominant over the electron-donating resonance effect by the high electronegativity of O atom in the OH substituent. Moreover, there are good linear relationships between the E^2 values and both the σ_{total} values and N$\cdots$H bond lengths.

The occupation numbers of lp_{N_2}, lp_{N_1}, and σ^*_{HF} are given in Table 4. The maximum and minimum occupancies for lp_{N_2} and lp_{N_1} correspond to NO_2 and $N(CH_3)_2$, respectively, the maximum and minimum occupancies of σ^*_{HF} correspond to $N(CH_3)_2$ and NO_2, respectively. As can be seen in Table 4, increasing the occupation numbers of lp_{N_2} and lp_{N_1} is accompanied with the decrease in the occupation number of σ^*_{HF} for all complexes. The occupancies of lp_{N_2} and lp_{N_1} in the complex **1** are larger than those in the complexes **2** and **3**, while the occupancy of σ^*_{HF} in the complexes **2** and **3** is larger than that in the complex **1**. Thus, the occupancy of the first lp_N/σ^*_{HF} increases/decreases in the presence of second

TABLE 4: The results of NBO analysis for the X-pyridazine· · · (HF)$_n$ (n = 1, 2) complexes at the HF/6-311++G(d,p) level of theory.

	Complex 1			Complex 2			Complex 3		
	E^2	lp_N	σ^*_{HF} ($\times 10^2$)	E^2	lp_N	σ^*_{HF} ($\times 10^2$)	E^2	lp_N	σ^*_{HF} ($\times 10^2$)
$N(CH_3)_2$	**34.16**, *30.47*	**1.911**, *1.914*	**4.509**, *5.030*	39.48	1.904	5.767	42.41	1.902	6.166
$NHCH_3$	**33.46**, *29.71*	**1.912**, *1.915*	**4.395**, *4.934*	38.66	1.906	5.651	41.79	1.903	6.087
NH_2	**32.52**, *28.72*	**1.913**, *1.917*	**4.236**, *4.793*	37.35	1.908	5.438	40.05	1.905	5.838
CH_3	**28.59**, *27.95*	**1.919**, *1.920*	**4.148**, *4.239*	36.07	1.911	5.289	36.44	1.910	5.333
C_2H_5	**28.94**, *28.42*	**1.919**, *1.919*	**4.213**, *4.288*	36.62	1.910	5.394	36.91	1.910	5.399
OCH_3	**29.60**, *27.51*	**1.917**, *1.920*	**5.604**, *5.946*	35.08	1.912	5.101	37.70	1.900	5.503
OH	**29.01**, *25.99*	**1.918**, *1.921*	**3.859**, *4.286*	34.39	1.912	5.031	36.12	1.910	5.273
H	**26.79**, *26.73*	**1.921**, *1.921*	**3.982**, *3.982*	34.69	1.912	5.093	34.69	1.912	5.093
OF	**25.44**, *23.61*	**1.922**, *1.924*	**4.931**, *5.238*	30.95	1.918	4.520	32.49	1.915	4.760
CN	**22.58**, *22.52*	**1.928**, *1.926*	**3.348**, *3.352*	28.57	1.919	4.200	29.04	1.921	4.259
NO_2	**21.48**, *21.68*	**1.929**, *1.928*	**4.566**, *4.556*	27.26	1.922	3.993	27.64	1.922	4.041
F	**25.50**, *23.92*	**1.922**, *1.925*	**3.533**, *3.785*	30.95	1.918	4.520	32.63	1.914	4.774
Br	**25.29**, *24.36*	**1.923**, *1.923*	**3.622**, *3.757*	31.14	1.915	4.575	32.23	1.916	4.723
Cl	**25.36**, *24.28*	**1.923**, *1.923*	**3.602**, *3.763*	31.16	1.916	4.570	32.40	1.915	4.745

The E^2 values are in kcal mol^{-1}. The bold and italicized values correspond to the $N_1 \cdots H$ and $N_2 \cdots H$ bonds, respectively.

TABLE 5: The results of MEP analysis ($V_{min} \times 10^3$) and charge transfer ($\Delta q \times 10^3$ in au) calculated from ChelpG charges.

	Complex 1		Complex 2		Complex 3	
	V_{min}	Δq	V_{min}	Δq	V_{min}	Δq
$N(CH_3)_2$	97	**205**, *194*	103	295	107	309
$NHCH_3$	96	**199**, *187*	101	282	106	293
NH_2	94	**193**, *189*	99	283	103	294
CH_3	87	**189**, *195*	96	294	97	291
C_2H_5	88	**190**, *200*	97	292	98	296
OCH_3	89	**181**, *204*	98	288	99	295
OH	86	**192**, *188*	92	293	96	295
H	84	**194**, *195*	94	300	94	300
OF	79	**182**, *182*	88	278	89	280
CN	73	**175**, *185*	83	277	82	279
NO_2	72	**178**, *182*	81	278	80	270
F	79	**186**, *181*	88	280	89	279
Br	80	**213**, *163*	89	325	90	328
Cl	80	**189**, *176*	89	275	90	285

The bold and italicized values correspond to the charge transfer from N_1 and N_2 atoms to HF unit, respectively.

N· · · HF interaction. This confirms the negative cooperativity in the complex **1**. There is a little difference between the occupancies of lp_{N_2} and lp_{N_1} in the complexes **2** and **3**. Also, the occupancy of σ^*_{HF} in the complex **3** is larger than that in the complex 2. Therefore, the changes of occupation numbers of lp_{N_2}, lp_{N_1}, and σ^*_{HF} are in agreement with the binding energies in all categories. A linear relationship (R^2 = 1.0) is observed between the ΔE values and the sum of E^2 values of $\text{lp}_{N_1} \rightarrow \sigma^*_{HF}$ and $\text{lp}_{N_1} \rightarrow \sigma^*_{HF}$ in the complex **1**. The high linear correlation (R^2 = 1.0) indicates the additivity of the E^2 values of two H-bond interactions.

There is also a linear relationship between the E^2 values and the occupancies of lp_{N_2}, lp_{N_1}, and σ^*_{HF}. In addition, there is a linear relationship between E^2 and N· · · H hydrogen bond lengths in three categories. Linear relationships are observed for the E^2 value of $\text{lp}_{N_2} \rightarrow \sigma^*_{HF}$ versus $r_{N_2\cdots H}$(R^2 = 0.98) and the E^2 values of $\text{lp}_{N_1} \rightarrow \sigma^*_{HF}$ versus $r_{N_1\cdots H}$(R^2 = 0.99) in the complex 1.

On the base of charges calculated by the ChelpG method (see Table 5), the charge transfer occurred from X-pyridazine to HF unit, The electron-donating substituents promote the charge transfer from X-pyridazine to HF unit, and enhance the basicity of the nitrogen atoms in the X-pyridazine. In the complex **3**, the charge transfer is larger and the basicity of N atom is higher than those in the complex **2**; this can be a reason for the stability of complex **3**. On the other hand, the charge transfer in the complex **1** is lower than that in complexes **2** and **3**; this is in agreement with the negative cooperativity of two hydrogen bond interactions.

3.2. MEP Analysis. The MEP is an important tool in exploring the nature of intermolecular interactions [45–51]. The capability of the N_2 and N_1 atoms to accept hydrogen bond was estimated through the minimum of the MEP (V_{min}) around the nitrogen atoms. As can be seen in Table 5, the MEP values become more negative with the electron-donating substituents. Thus, the electrostatic term depends on the electron donation or electron withdrawal character of the substituents. Also, a good linear relationship is observed between the V_{min} values and the σ_{total} values (see Figure 2). Therefore, those values can be used to predict the Hammett constants. The trend in the V_{min} values is **3** > **2** > **1**; so, those values become less negative around the nitrogen atoms when both interactions coexist.

The MEP, $V_{(r)}$, is directly related to electron density $\rho(r)$ through the Poisson's equation [52]:

$$\nabla^2 V(r) = 4\pi\rho(r). \tag{1}$$

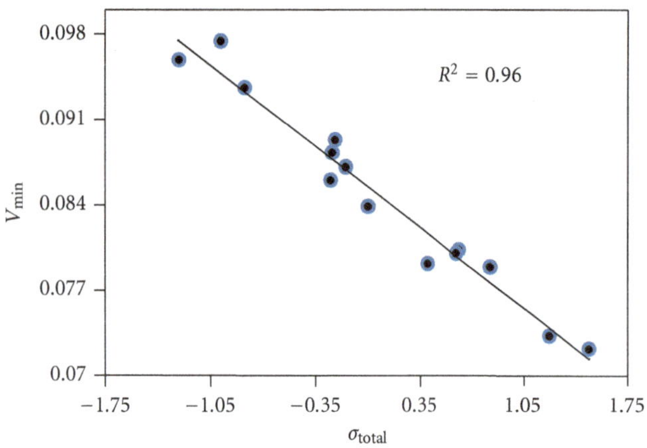

FIGURE 2: Linear correlation between molecular electrostatic potential minimum (V_{min}) around the nitrogen atoms of the pyridazine and the Hammett electronic parameters for the complex **1**.

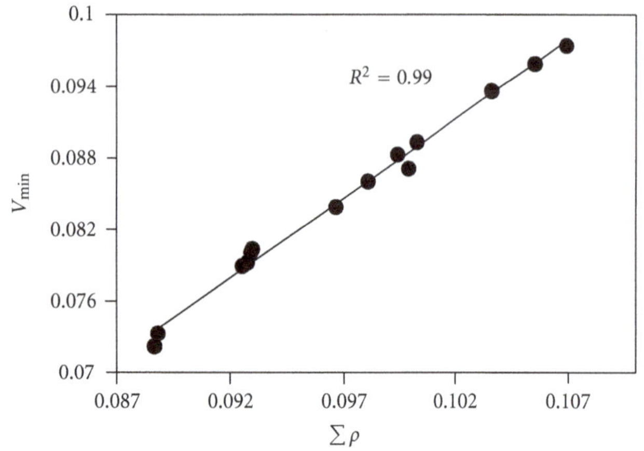

FIGURE 3: Interplay between the V_{min} value around the nitrogen of pyridazine versus the sum of ρ values calculated at the N$\cdots$H BCPs for complex **1**.

Is there a liner relationship between V_{min} and the ρ values calculated at the HBCPs? The V_{min} values become more negative with increasing ρ values calculated at HBCPs in X-pyridazine$\cdots(HF)_n$ complexes. The linear correlation coefficients between the V_{min} values and the ρ values are equal to 0.98 and 0.99, respectively, in the complexes **2** and **3**. As can be seen in Figure 3, there is a good correlation between V_{min} values and $\sum\rho$ values for the complex **1** ($R = 1.00$). Similarly, there is a linear relationship ($R = 0.99$) between V_{min} and E^2 of $lp_N \rightarrow \sigma^*_{HF}$ interaction, V_{min}, and the $N_2\cdots H$ and $N_1\cdots H$ bond lengths. Also, there is a good linear correlation between V_{min} values and $\sum E^2$ in the complex **1** ($R = 1.00$).

3.3. SAPT Energy Decomposition. The SAPT analysis is a method for investigation the nature of the intermolecular interactions [53–55]. The SAPT method provides detailed information on the intermolecular interaction, as this method directly calculates magnitude of each term (electrostatic, dispersion, etc.) of the intermolecular interactions [56]. A detailed description of SAPT and some of its applications can be found in some recent references [57–59].

To determine the nature of the hydrogen bond in the complexes **2** and **3**, the interaction energies were decomposed into physically meaningful components, including electrostatic, induction, dispersion, and exchange energies, using SAPT technique at the B3LYP/6-31G(d) level. It can be seen from Table 6 that the values of electrostatic energy (E_{els}), induction energy (E_{ind}), and dispersion energy (E_{disp}) are negative, whereas the exchange energy (E_{exch}) is positive for all complexes. Thus, the stabilization/destabilization of the complexes by substitution is represented simply by the sum from E_{els}, E_{ind}, and E_{disp} contributions. In both complexes, the absolute values for these three negative terms are largest for the $N(CH_3)_2$ substituent and are smallest for the NO_2 substituent. In the complex **3**, the absolute values of E_{els}, E_{ind}, and E_{disp} are larger than those values in the complex **2**. Thus, the complex **3** is more stable than the complex **2**. The electron-donating substituents have positive effect and increase the magnitude of the calculated SAPT interaction energy components (with the exception of OH substituent in the complex **2** that decreases the magnitude of electrostatic energy relative to pyridazine). The electron-withdrawing substituents have negative effect and decrease the magnitude of interaction energy components, with the exception of E_{disp} in the complex **2** that slightly increases in the presence of mentioned substituents. The dependence of the dispersion term to the substituent is smaller than other complexes. When position of HF changes from para to meta relative to the substituent, the electrostatic term changes by 2.8%, the induction term changes by 2.2%, and the dispersion term changes by 12.5%. Thus, the dispersion energy component of the SAPT analysis was found to be very sensitive to the position of HF relative to the substituent.

In the complex **2**, the electrostatic forces contribute about 52.0–54.6% to the total attractive interaction energy, the induction forces are about 35.6–43.6% of the total attractive energy, and the dispersion contribution is 9.7–11.9%. On the other hand, the electrostatic, induction, and dispersion contributions are 51.9–52.6%, 36.4–37.0%, and 10.5–11.7%, respectively, in the complex **3**. Thus, we believe that the electrostatic interactions are mainly responsible for the binding energies and formation of hydrogen bonds, although the induction and dispersion interactions are also important. Since the electrostatic plays an important role, the polarization correlates with the electrostatic energy. The electron-withdrawing substituents hinder the electron transfer-driven polarization to the HF unit (which results in destabilization), whereas the electron-donating substituents allow strong polarization to the HF unit (which results in strong stabilization). It is also interesting to note that the polarization for the electron-donating and electron-withdrawing substituents in the complex **2** is slightly weaker than that in the complex **3**. In the complex **3**, the electron density at the para position is an important stabilizing factor, and thus the stabilization/destabilization by substituent of a pyridazine is governed mostly by the electrostatic energy. In Figure 4, a linear

TABLE 6: SAPT interaction energy decomposition (in kcal mol^{-1}) for the complexes **2** and **3**.

	Complex 2					Complex 3				
	E_{els}	E_{ind}	E_{exch}	E_{disp}	E_{int}^{SAPT}	E_{els}	E_{ind}	E_{exch}	E_{disp}	E_{int}^{SAPT}
$N(CH_3)_2$	−25.53	−17.74	34.24	−5.16	−14.18	−26.21	−18.48	35.29	−5.27	−14.67
$NHCH_3$	−25.15	−17.41	33.69	−5.08	−13.94	−25.93	−18.29	35.00	−5.22	−14.45
NH_2	−24.34	−16.81	32.71	−4.95	−13.39	−25.16	−17.66	33.94	−5.09	−13.97
CH_3	−23.5	−16.35	31.87	−4.84	−12.83	−23.51	−16.40	31.90	−4.84	−12.84
C_2H_5	−23.73	−16.53	32.16	−4.88	−12.98	−23.75	−16.59	32.23	−4.88	−13.00
OCH_3	−22.95	−15.86	31.01	−4.75	−12.55	−23.99	−16.83	32.64	−4.94	−13.11
OH	−22.62	−15.72	30.90	−4.74	12.19	23.08	−16.16	31.54	−4.80	−12.51
H	−22.80	−14.89	29.63	−4.06	−12.12	−22.70	−15.73	30.81	−4.69	−12.31
OF	−20.06	−14.07	27.96	−4.41	−10.59	−21.04	−14.78	29.27	−4.54	−11.09
CN	−18.92	−13.48	26.85	−4.28	−9.84	−19.37	−13.65	27.22	−4.31	−10.11
NO_2	−18.05	−12.92	25.87	−4.17	−9.28	−18.64	−13.06	26.27	−4.20	−9.62
F	−20.35	−14.27	28.35	−4.43	−10.71	−21.13	−14.84	29.39	−4.54	−11.12
Cl	−20.52	−14.49	28.57	−4.48	−10.93	−21.13	−14.84	29.29	−4.55	−11.23

All energies are in kcal mol^{-1}. $E_{int}^{SAPT} = E_{els} + E_{ind} + E_{disp} + E_{exch}$.

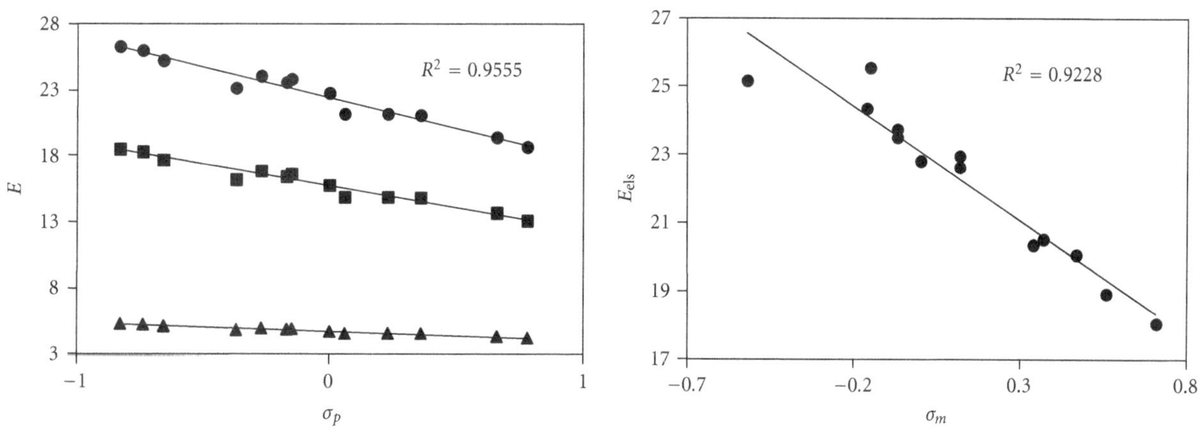

(a) (b)

FIGURE 4: Linear relationship between the calculated SAPT interaction energy components (•E_{els}, ■E_{ind}, and ▲E_{dis} in kcal mol^{-1}) and σ_p for the complex **3** (a), and the electrostatic energy and σ_m for the complex **2** (b).

relationship with high correlation coefficient ($R = 0.98$) is observed between the calculated SAPT interaction energy components and σ_p in the complex **3**. In the complex **2**, a linear relationship with correlation coefficient 0.96 is found between the E_{els} and σ_m.

4. Conclusions

The results of quantum mechanical calculations indicate that the binding energy in the complex **3** is larger than that in the complex **2**. The sum of binding energies of the complexes **2** and **3** is larger than the binding energy of the complex **1**. So, there is a negative cooperativity for two hydrogen bond interactions.

Very good linear correlations are observed between the binding energies and Hammett electronic parameters σ_{total} of the substituents. The electron-donating substituents stabilize and the electron-withdrawing substituents destabilize the complexes relative to pyridazine. According to the results of AIM analysis, the cooperativity effect decreases the electron density at the N···H BCPs in the X-pyridazine···$(HF)_2$ complex.

On the basis of the results of SAPT analysis, the portion of electrostatic force in the complex **3** is smaller than **2**, while the induction and dispersion portions are larger in the complex **3**. The E^2 values of $lp_{N_2} \rightarrow \sigma^*_{HF}$ and $lp_{N_1} \rightarrow \sigma^*_{HF}$ interactions in the complex **1** are smaller than those in the complexes **2** and **3**. Thus, the cooperativity effect decreases the E^2 values of $lp_{N_2} \rightarrow \sigma^*_{HF}$ and $lp_{N_1} \rightarrow \sigma^*_{HF}$ interactions in the complex **1**. The occupation numbers of lp_{N_2} and lp_{N_1} in the complex **1** are larger than those in the complexes **2** and **3**, respectively. The charge transfer, on the basis of ChelpG charges, is in agreement with the results of NBO and AIM analysis.

There are linear relationships between the V_{min} and ρ_{BCP} values, the V_{min} and the E^2 values of $lp_N \rightarrow \sigma^*_{HF}$ interaction, and between the V_{min} and the σ_{total} values.

The SAPT calculations show that the electrostatic is the dominating interaction component in the complexes **2** and **3**, although the induction and dispersion interactions are also important. The dispersion energy component of the SAPT analysis was found to be very sensitive to the position of HF relative to the substituent. There is a good correlation between σ_p and the calculated SAPT interaction energy components in the complex **3**, while the linear relationship is only found between the E_{els} and σ_m in the complex **2**.

References

[1] S. Breda, I. D. Reva, L. Lapinski, M. J. Nowak, and R. Fausto, "Infrared spectra of pyrazine, pyrimidine and pyridazine in solid argon," *Journal of Molecular Structure*, vol. 786, no. 2-3, pp. 193–206, 2006.

[2] X. J. Zou, G. Y. Jin, and Z. X. Zhang, "Synthesis, fungicidal activity, and QSAR of pyridazinonethiadiazoles," *Journal of Agricultural and Food Chemistry*, vol. 50, no. 6, pp. 1451–1454, 2002.

[3] S. G. Lee, J. J. Kim, H. K. Kim et al., "Recent progress in pyridazin-3(2H)-ones chemistry," *Current Organic Chemistry*, vol. 8, no. 15, pp. 1463–1480, 2004.

[4] R. V. A. Orru and M. de Greef, "Recent advances in solution-phase multicomponent methodology for the synthesis of heterocyclic compounds," *Synthesis*, no. 10, pp. 1471–1499, 2003.

[5] N. G. Kandile, M. I. Mohamed, H. Zaky, and H. M. Mohamed, "Novel pyridazine derivatives: synthesis and antimicrobial activity evaluation," *European Journal of Medicinal Chemistry*, vol. 44, no. 5, pp. 1989–1996, 2009.

[6] G. R. Desiraju and T. Steiner, *The Weak Hydrogen Bond in Structural Chemistry and Biology*, Oxford University Press, Oxford, UK, 1997.

[7] G. R. Desiraju and T. Steiner, *The Weak Hydrogen Bond*, Oxford University Press, Oxford, UK, 1999.

[8] M. Nishio, M. Hirota, and Y. Umezawa, *The CH/π Interaction*, Wiley-VCH, New York, NY, USA, 1998.

[9] S. Scheiner, Ed., *Molecular Interactions: from van der Waals to Strong Bound Complexes*, John Wiley & Sons, Chichester, UK, 1997.

[10] S. Scheiner, Ed., *Hydrogen Bonding: A Theoretical Perspective*, Oxford University Press, Oxford, UK, 1997.

[11] G. A. Jeffrey and W. Saenger, *Hydrogen Bonding in Biological Structures*, Springer, Berlin, Germany, 1991.

[12] G. A. Jeffrey, *An Introduction to Hydrogen Bonding*, Oxford University Press, Oxford, UK, 1997.

[13] W. Wang, Y. Zhang, and K. Huang, "Unconventional interaction in N(P)-related systems," *Chemical Physics Letters*, vol. 411, no. 4–6, pp. 439–444, 2005.

[14] W. Chen and M. S. Gordon, "Energy decomposition analyses for many-body interaction and applications to water complexes," *Journal of Physical Chemistry*, vol. 100, no. 34, pp. 14316–14328, 1996.

[15] A. D. Kulkarni, R. K. Pathak, and L. J. Bartolotti, "Effect of additional hydrogen peroxide to $H_2O_2 \cdots (H_2O)n$, $n = 1$ and 2 complexes: quantum chemical study," *Journal of Chemical Physics*, vol. 124, no. 21, Article ID 214309, 7 pages, 2006.

[16] Q. Li, X. An, F. Luan et al., "Cooperativity between two types of hydrogen bond in H_3 C-HCN-HCN and H_3 C-HNC-HNC complexes," *Journal of Chemical Physics*, vol. 128, no. 15, Article ID 154102, 6 pages, 2008.

[17] Y. L. Zhao and Y. D. Wu, "A theoretical study of β-sheet models: is the formation of hydrogen-bond networks cooperative?" *Journal of the American Chemical Society*, vol. 124, no. 8, pp. 1570–1571, 2002.

[18] R. Wieczorek and J. J. Dannenberg, "H-bonding cooperativity and energetics of α-helix formation of five 17-amino acid peptides," *Journal of the American Chemical Society*, vol. 125, no. 27, pp. 8124–8129, 2003.

[19] Q. Li, X. An, B. Gong, and J. Cheng, "Cooperativity between OH· · · O and CH· · · O hydrogen bonds involving dimethyl sulfoxide-H_2O-H_2O complex," *Journal of Physical Chemistry A*, vol. 111, no. 40, pp. 10166–10169, 2007.

[20] H. Guo, N. Gresh, B. P. Roques, and D. R. Salahub, "Many-body effects in systems of peptide hydrogen-bonded networks and their contributions to ligand binding: a comparison of the performances of DFT and polarizable molecular mechanics," *Journal of Physical Chemistry B*, vol. 104, no. 41, pp. 9746–9754, 2000.

[21] R. Viswanathan, A. Asensio, and J. J. Dannenberg, "Cooperative hydrogen-bonding in models of antiparallel β-sheets," *Journal of Physical Chemistry A*, vol. 108, no. 42, pp. 9205–9212, 2004.

[22] F. H. Allen, J. A. K. Howard, V. J. Hoy, G. R. Desiraju, D. Shekhar Reddy, and C. C. Wilson, "First neutron diffraction analysis of an O-H· · · π hydrogen bond: 2-ethynyladamantan-2-ol," *Journal of the American Chemical Society*, vol. 118, no. 17, pp. 4081–4084, 1996.

[23] B. S. Hudson, D. A. Braden, D. G. Allis et al., "The crystalline enol of 1,3-cyclohexanedione and its complex with benzene: vibrational spectra, simulation of structure and dynamics and evidence for cooperative hydrogen bonding," *Journal of Physical Chemistry A*, vol. 108, no. 36, pp. 7356–7363, 2004.

[24] R. F. W. Bader, *Atoms in Molecules: A Quantum Theory*, Oxford University Press, Oxford, UK, 1990.

[25] A. E. Reed, L. A. Curtiss, and F. Weinhold, "Intermolecular interactions from a natural bond orbital, donor-acceptor viewpoint," *Chemical Reviews*, vol. 88, no. 6, pp. 899–926, 1988.

[26] B. Jeziorski, R. Moszynski, and K. Szalewicz, "Perturbation theory approach to intermolecular potential energy surfaces of van der waals complexes," *Chemical Reviews*, vol. 94, no. 7, pp. 1887–1930, 1994.

[27] A. D. Becke, "Density-functional thermochemistry. III. The role of exact exchange," *Journal of Chemical Physics*, vol. 98, no. 7, pp. 5648–5652, 1993.

[28] M. J. Frisch, G. W. Trucks, H. B. Schlegel et al., *Gaussian 03 (Revision B.03)*, Gaussian, Pittsburgh, Pa, USA, 2003.

[29] C. Møller and M. S. Plesset, "Note on an approximation treatment for many-electron systems," *Physical Review*, vol. 46, no. 7, pp. 618–622, 1934.

[30] Y. Zhao and D. G. Truhlar, "Benchmark databases for non-bonded interactions and their use to test density functional theory," *Journal of Chemical Theory and Computation*, vol. 1, no. 3, pp. 415–432, 2005.

[31] S. F. Boys and F. Bernardi, "Calculation of small molecular interactions by differences of separate total energies—some procedures with reduced errors," *Molecular Physics*, vol. 19, no. 4, pp. 553–566, 1970.

[32] F. B. König, J. Schönbohm, and D. Bayles, "AIM2000—A program to analyze and visualize atoms in molecules," *Journal of Computational Chemistry*, vol. 22, no. 5, pp. 545–559, 2001.

[33] C. M. Breneman and K. B. Wiberg, "Determining atom-centered monopoles from molecular electrostatic potentials. The need for high sampling density in formamide conformational analysis," *Journal of Computational Chemistry*, vol. 11, no. 3, pp. 361–373, 1990.

[34] P. Flukiger, H. P. Luthi, S. Portmann, and J. Weber, *Molekel 4.0*, Swiss Center for Scientific Computing, Manno, Switzerland, 2000.

[35] M. W. Schmidt, K. K. Baldridge, J. A. Boatz et al., "General atomic and molecular electronic structure system," *Journal of Computational Chemistry*, vol. 14, no. 11, pp. 1347–1363, 1993.

[36] R. Bukowski, W. Cencek, P. Jankowski et al., *An Ab Initio Program for Many-Body Symmetry-Adapted Perturbation Theory Calculations of Intermolecular Interaction Energies. Sequential and Parallel Versions*, University of Delaware, Newark, Del, USA; University of Warsaw, Warsaw, Poland, 2008.

[37] C. Hansch, A. Leo, and R. W. Taft, "A survey of hammett substituent constants and resonance and field parameters," *Chemical Reviews*, vol. 91, no. 2, pp. 165–195, 1991.

[38] L. P. Hammett, "Some relations between reaction rates and equilibrium constants," *Chemical Reviews*, vol. 17, no. 1, pp. 125–136, 1935.

[39] F. Cozzi, F. Ponzini, R. Annunziata, M. Cinquini, and J. S. Siegel, "Polar interactions between stacked π systems in fluorinated 1,8-diarylnaphthalenes: Importance of quadrupole moments in molecular recognition," *Angewandte Chemie*, vol. 34, no. 9, pp. 1019–1020, 1995.

[40] M. Alcamí, O. Mó, and M. Yáñez, "Modelling intrinsic basicities: the use of the electrostatic potentials and the atoms-in-molecules theory," *Theoretical and Computational Chemistry*, vol. 3, pp. 407–456, 1996.

[41] C. H. Suresh and S. R. Gadre, "Electrostatic potential minimum on aromatic ring as a measure of substituent constant," *Inorganic Chemistry*, vol. 111, no. 4, pp. 710–714, 2007.

[42] A. Ebrahimi, M. Habibi, and H. R. Masoodi, "Theoretical study of the influence of para- and meta-substituents on X-pyridine· · · HF hydrogen bonding," *Chemical Physics*, vol. 340, no. 1–3, pp. 85–92, 2007.

[43] S. J. Grabowski and M. Małecka, "Intramolecular H-bonds: DFT and QTAIM studies on 3-(aminomethylene)pyran-2, 4-dione and its derivatives," *Journal of Physical Chemistry A*, vol. 110, no. 42, pp. 11847–11854, 2006.

[44] S. J. Grabowski, "Ab initio calculations on conventional and unconventional hydrogen bonds-study of the hydrogen bond strength," *Journal of Physical Chemistry A*, vol. 105, no. 47, pp. 10739–10746, 2001.

[45] E. Espinosa, C. Lecomte, N. E. Ghermani et al., "Hydrogen bonds: First quantitative agreement between electrostatic potential calculations from experimental X-(X + N) and theoretical ab initio SCF models," *Journal of the American Chemical Society*, vol. 118, no. 10, pp. 2501–2502, 1996.

[46] B. Galabov and P. Bobadova-Parvanova, "Molecular electrostatic potential as reactivity index in hydrogen bonding: ab initio molecular orbital study of complexes of nitrile and carbonyl compounds with hydrogen fluoride," *Journal of Physical Chemistry A*, vol. 103, no. 34, pp. 6793–6799, 1999.

[47] V. Dimitrova, S. Ilieva, and B. Galabov, "Electrostatic potential at atomic sites as a reactivity descriptor for hydrogen bonding. Complexes of monosubstituted acetylenes and ammonia," *Journal of Physical Chemistry A*, vol. 106, no. 48, pp. 11801–11805, 2002.

[48] I. Mata, E. Molins, I. Alkorta, and E. Espinosa, "Topological properties of the electrostatic potential in weak and moderate N· · · H hydrogen bonds," *Journal of Physical Chemistry A*, vol. 111, no. 28, pp. 6425–6433, 2007.

[49] P. Politzer, J. S. Murray, and M. C. Concha, "Halogen bonding and the design of new materials: organic bromides, chlorides and perhaps even fluorides as donors," *Journal of Molecular Modeling*, vol. 13, no. 6-7, pp. 643–650, 2007.

[50] P. W. Kenny, "Hydrogen bonding, electrostatic potential, and molecular design," *Journal of Chemical Information and Modeling*, vol. 49, no. 5, pp. 1234–1244, 2009.

[51] D. J. R. Duarte, M. M. de las Vallejos, and N. M. Peruchena, "Topological analysis of aromatic halogen/hydrogen bonds by electron charge density and electrostatic potentials," *Journal of Molecular Modeling*, vol. 16, no. 4, pp. 737–748, 2010.

[52] J. S. Murray, J. M. Seminario, M. C. Concha, and P. Politzer, "An analysis of molecular electrostatic potentials obtained by a local density functional approach," *International Journal of Quantum Chemistry*, vol. 44, no. 2, pp. 113–122, 1992.

[53] G. Chałasiński and M. M. Szczęśniak, "State of the art and challenges of the ab initio theory of intermolecular interactions," *Chemical Reviews*, vol. 100, no. 11, pp. 4227–4252, 2000.

[54] A. J. Stone, *The Theory of Intermolecular Forces*, Clarendon Press, Oxford, UK, 1996.

[55] S. Tsuzuki and A. Fujii, "Nature and physical origin of CH/π interaction: significant difference from conventional hydrogen bonds," *Physical Chemistry Chemical Physics*, vol. 10, no. 19, pp. 2584–2594, 2008.

[56] H. L. Williams and C. F. Chabalowski, "Using Kohn-Sham orbitals in symmetry-adapted perturbation theory to investigate intermolecular interactions," *Journal of Physical Chemistry A*, vol. 105, no. 3, pp. 646–659, 2001.

[57] R. Bukowski, K. Szalewicz, and C. F. Chabalowski, "Ab initio interaction potentials for simulations of dimethylnitramine solutions in supercritical carbon dioxide with cosolvents," *Journal of Physical Chemistry A*, vol. 103, no. 36, pp. 7322–7340, 1999.

[58] D. Kim, S. Hu, P. Tarakeshwar, K. S. Kim, and J. M. Lisy, "Cation-π interactions: a theoretical investigation of the interaction of metallic and organic cations with alkenes, arenes, and heteroarenes," *Journal of Physical Chemistry A*, vol. 107, no. 8, pp. 1228–1238, 2003.

[59] A. L. Ringer, M. S. Figgs, M. O. Sinnokrot, and C. D. Sherrill, "Aliphatic C-H/π interactions: methane-benzene, methane-phenol, and methane-indole complexes," *Journal of Physical Chemistry A*, vol. 110, no. 37, pp. 10822–10828, 2006.

17

Positronium in a Liquid Phase: Formation, Bubble State and Chemical Reactions

Sergey V. Stepanov,[1] Vsevolod M. Byakov,[1,2] Dmitrii S. Zvezhinskiy,[1] Gilles Duplâtre,[3] Roman R. Nurmukhametov,[1] and Petr S. Stepanov[1]

[1] *Institute of Theoretical and Experimental Physics, Moscow 117218, Russia*
[2] *D. Mendeleev University of Chemical Technology of Russia, Miusskaya Square 9, Moscow 125047, Russia*
[3] *Institut Pluridisciplinaire Hubert Curien, CNRS/IN2P3, BP 28, 67037 Strasbourg Cedex 2, France*

Correspondence should be addressed to Sergey V. Stepanov, stepanov@itep.ru

Academic Editor: Michael Deleuze

The present approach describes the e^+ fate since its injection into a liquid until its annihilation. Several stages of the e^+ evolution are discussed: (1) energy deposition and track structure of fast positrons: ionization slowing down, number of ion-electron pairs, typical sizes, thermalization, electrostatic interaction between e^+ and the constituents of its blob, and effect of local heating; (2) positronium formation in condensed media: the Ore model, quasifree Ps state, intratrack mechanism of Ps formation; (3) fast intratrack diffusion-controlled reactions: Ps oxidation and ortho-paraconversion by radiolytic products, reaction rate constants, and interpretation of the PAL spectra in water at different temperatures; (4) Ps bubble models. Inner structure of positronium (wave function, energy contributions, relationship between the pick-off annihilation rate and the bubble radius).

1. Introduction

Positrons (e^+) as well as positronium atoms (Ps) are recognized as nanoscale probes of the local structure in a condensed phase (liquid or solid) and of the early radiolytic physicochemical processes occurring therein. The parameters of positron annihilation spectra determined experimentally (e.g., positron and Ps lifetimes, angular and energetic widths of the spectra, and Ps formation probability) are highly sensitive to the chemical composition, the local molecular environment of Ps (free volume size), and the presence of structural defects. They are also sensitive to variation of temperature, pressure, external electric and magnetic fields, and phase transitions.

The informative potentiality of positron spectroscopy strongly depends on the reliability of any theory describing the behavior of positrons in matter, since it should help decipher the information coded in the annihilation spectra. So, realistic models are needed for e^+ track structure, e^+ energy losses, ionization slowing down and thermalization, intratrack reactions (ion-electron recombination, solvation, and interaction with scavengers), Ps formation process, Ps interaction with chemically active radiolytic species, and e^+/Ps trapping by structural defects.

Usually, treatment of the measured annihilation spectra is reduced to their resolution into a set of simple trial functions: sums of decaying time exponentials in the case of PALS (positron annihilation lifetime spectroscopy) and of Gaussians in the case of ACAR (angular correlation of annihilation radiation) and DBAR (doppler broadening of annihilation radiation). The outcome of such conventional analyses of positron annihilation data is the intensities of these components and the corresponding lifetimes/widths.

However, realistic theoretical models suggest more complex kinetics for describing physicochemical processes within the fast positron track, especially in its terminal part (which is called blob). Ps formation precisely occurs in the blob, and further Ps reactions with intratrack radiolytic products and solutes (oxidation, spin-conversion, and complex formation) also take place there. Because of inhomogeneous spatial distribution of the track species, their outdiffusion becomes an important factor: diffusion kinetics cannot be expressed

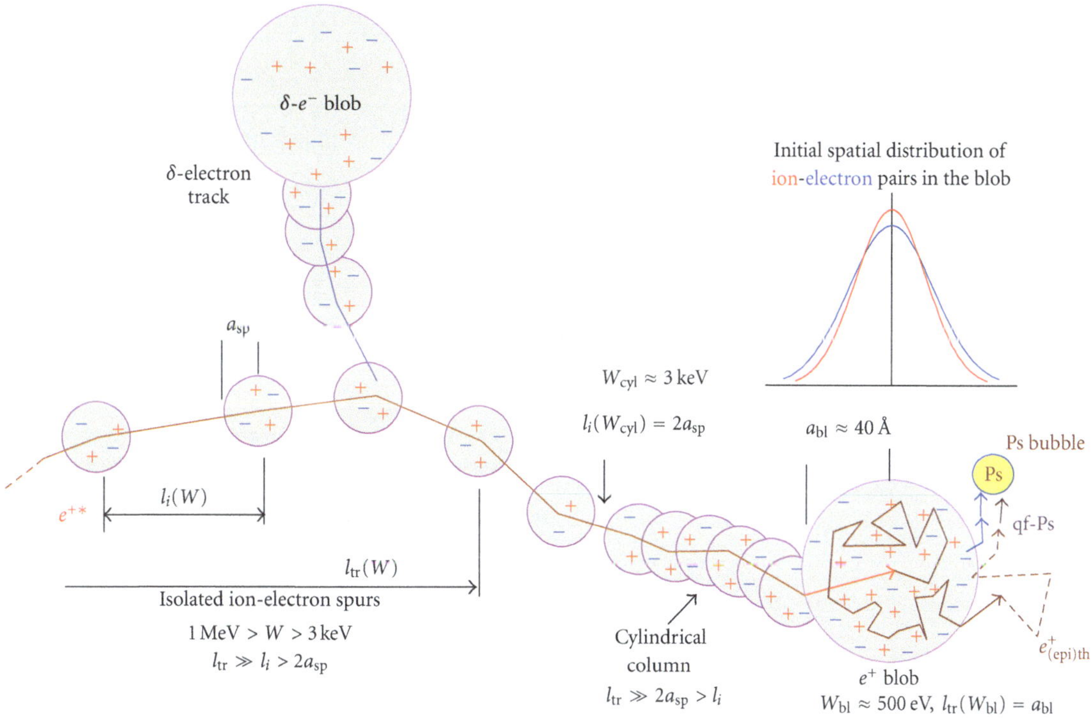

FIGURE 1: Scheme of the end part of the e^{+*} track and Ps formation.

in terms of mere exponentials or Gaussians. Obviously, more elaborated theoretical models should be used in the fitting procedure of the annihilation spectra. In this case, the adjustable parameters (reaction rate constants or reaction radii, diffusion coefficients, initial size of the terminal part of the e^+ track, and contact density in the Ps atom) would present a clear physical meaning instead of the above-mentioned "intensities" of some trial functions.

Here we describe a model that can be used for interpretation of the positron spectroscopy data in molecular liquids. The model describes the e^+ fate since its injection into a liquid until its annihilation.

2. Energy Deposition and Track Structure of the Fast Positron

2.1. Ionization Slowing Down. Positrons, produced in nuclear β^+-decay, have initial energies of about several hundreds of keV. Once injected into a medium they lose energy via molecular ionization. Within 10 ps the positron energy drops down to the ionization threshold. Further approach to thermal equilibrium proceeds primarily via excitations of intra- and intermolecular vibrations. This usually takes few tens of picoseconds [1, 2].

Roughly half of the positron kinetic energy is lost in rare head-on collisions, resulting in the knocking out of δ-electrons with kinetic energies of about several keV, the tracks of these electrons forming branches around the positron trail (Figure 1). The other half of the energy is spent in numerous glancing collisions with molecules. The average energy loss in such a collision is several tens of eV (up to 100 eV). A secondary electron knocked out in a glancing collision produces, in turn, a few ion-electron pairs inside a spherical nanovolume, called a "spur" in radiation chemistry. Its radius, a_{sp}, is determined by the thermalization length of the knocked-out electrons in the presence of the Coulombic attraction of the parent ions. With a large uncertainty a_{sp} in water may be estimated as 30–70 Å [3].

While the positron energy W is greater than $W_{cyl} \sim 3$ keV, the mean distance l_i between adjacent ionizations produced by the positron is greater than the spur size $2a_{sp}$. It ensues that at high positron energies, the spurs are well separated from each other. The trajectory of the fast positron is a quasistraight line because l_i is less than the positron transport path l_{tr}. The latter is the mean distance traveled by the positron before it changes its initial direction of motion by 90°.

When $l_i < 2a_{sp} < l_{tr}$ or $W_{bl} < W < W_{cyl}$ the spurs overlap, forming something like a cylindrical ionization column. When the e^+ energy becomes less than the blob formation energy, W_{bl} (<1 keV, see below), the positron is about to create a terminal blob. The diffusion motion of e^+ in the blob becomes more pronounced: the direction of its momentum changes frequently due to elastic scattering and the ionization of surrounding molecules. All intrablob ionizations are confined within a sphere of radius a_{bl}. The terminal positron blob contains a few tens of ion-electron pairs ($n_0 \approx W_{bl}/W_{iep} \approx 30$) because the average energy W_{iep} required to produce one ion-electron pair is 16–22 eV [4]. Finally, the positron becomes subionizing and its energy loss rate drops by almost 2 orders of magnitude [5].

The dependences of $l_i(W)$ and $l_{tr}(W)$ versus the e^+ energy in liquid water are shown in Figure 2 [6]. The calculation of the mean distance between adjacent ionizations $l_i(W) = W_{iep}/\text{LET}(W)$ is based on e^- slowing down data (http://physics.nist.gov/PhysRefData/Star/Text/ESTAR.html) and the estimation of the transport path has been done in the framework of the Born approximation (the Born amplitude was calculated by simulating an H_2O molecule as an isoelectronic atom).

As we have mentioned, e^+ moves in a diffusive way at the blob formation stage. So its spatial displacement squared should increase "linearly" with time, but because e^+ diffusion coefficient, $D_p(W)$, is energy dependent, this relationship becomes more complicated. When e^{+*} loses energy from W_i down to W_f, its mean square displacement is [7]

$$R^2_{\text{ion}}(W_i, W_f) = \int_{t(W=W_f)}^{t(W=W_i)} 6D_p(W)dt = \int_{W_f}^{W_i} l_{tr}(W)\frac{dW}{|-dW/dx|_{\text{ion}}}. \quad (1)$$

Here we used

$$D_p(W) = \frac{l_{tr}(W)v_p}{6}, \qquad dt = \frac{dx}{v_p} = \frac{1}{v_p}\cdot\frac{dW}{|-dW/dx|_{\text{ion}}} \quad (2)$$

and v_p is the positron velocity

The blob parameters, W_{bl} and a_{bl}, can be estimated from the following equations, Figure 3:

$$l_{tr}(W_{bl}) = a_{bl}, \qquad R^2_{\text{ion}}(W_{bl}, \text{Ry}) - a^2_{bl} = a^2_{bl}. \quad (3)$$

The first relationship implies that the positron with energy W_{bl} "jumps" to the center of the nascent blob from aside, passing the distance $l_{tr}(W_{bl})$ (the red arrow in Figure 3). After that, the positron is located at the center of the blob. The second equation corresponds to the rest of the e^+ diffusion motion (its slowing down till the subionizing energies; Ry = 13.6 eV stands here for a typical value of these energies). All this terminal part of the e^+ trajectory must be confined within the positron blob (i.e., within the sphere with the radius a_{bl}), that is, the e^+ displacement squared, $R^2_{\text{ion}}(W_{bl}, \text{Ry}) - a^2_{bl}$, equals the dispersion of the ion-electron pairs of the blob, a^2_{bl}.

In the case of liquid water one obtains $W_{bl} \approx 500$ eV and $a_{bl} \approx 40$ Å. It seems that values of a_{bl} and W_{bl} should not differ significantly from one liquid to another because the ionization slowing down parameters depend mostly on the ionization potential and the average electron density, parameters that are more or less the same in all molecular media.

2.2. Thermalization Stage. Interaction between the Positron and Its Blob. At the end of the slowing down by ionization and electronic excitation, the spatial distribution of the positron coincides with the distribution of the blob species (i.e., $\sim \exp(-r^2/a^2_{bl})$). Such a subionizing

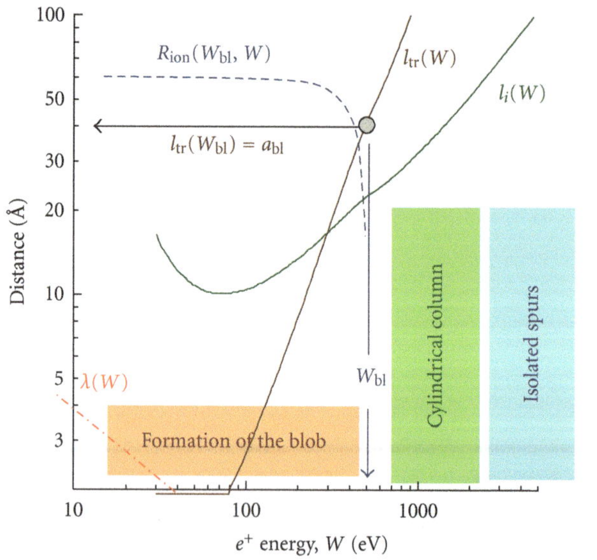

Figure 2: Dependences of $l_i(W)$ and $l_{tr}(W)$ versus the energy W of the positron in liquid water. This figure illustrates also the solution of (3) for the blob parameters W_{bl} and a_{bl}: $l_{tr}(W_{bl}) = a_{bl}$, $R_{\text{ion}}(W_{bl}, \text{Ry}) = \sqrt{2a^2_{bl}} = \sqrt{2}a_{bl}$. Red dash-and-dot line displays the de Broglie wavelength of the positron, $\lambda = h/mv \approx 12.3\ [\text{Å}]/\sqrt{W, \text{eV}}$.

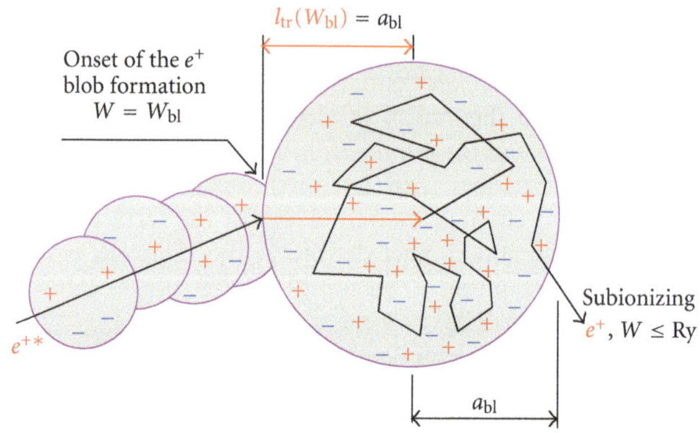

Figure 3: Scheme of the terminal positron blob.

positron having some eV of excess kinetic energy may easily escape from its blob. It is expected that by the end of thermalization, the e^+ distribution becomes broader with the dispersion:

$$a^2_p \approx a^2_{bl} + \langle R^2_{\text{vib}}(W_0, T)\rangle_{W_0}. \quad (4)$$

Here, $R^2_{\text{vib}}(T, W_0)$ is determined by analogy with R^2_{ion}, see (1), where $|dW/dx|_{\text{ion}}$ should be replaced by $|dW/dx|_{\text{vib}}$, because the stopping power of subionizing positrons relates to the excitation of vibrations, but not to ionizations (T is the temperature in energy units). The estimation of a_p requires quantitative data on $|dW/dx|_{\text{vib}}$, scattering properties of the subionizing positron, and the spectrum of its energies W_0 just after the last ionization event. In (4) $\langle\cdots\rangle_{W_0}$ denotes the average over W_0. In contrast to the parameters related to ionization slowing down, a_p depends on the properties of the investigated liquid and may reach hundreds of Å especially if the liquid is nonpolar.

Between the positively charged ions and the knocked-out intrablob electrons there exists a strong Coulombic attraction: outdiffusion of the electrons (even during their thermalization) is almost completely suppressed and their distribution is close enough to that of the ions. This case is known as ambipolar diffusion when ions and electrons expand with the same diffusion coefficient equal to the duplicated diffusion coefficient of the ions. The electrostatic potential in the blob comes out everywhere positive, that is, repulsive towards the positron (typical value of this repulsive energy is about several T) [7].

However, there is an opposite effect: while residing inside the blob, the thermalized e^+ rearranges the intrablob electrons so that the total energy of the system decreases because of the Debye screening. The corresponding energy drop may be estimated by using the Debye-Huckel theory. It is $\sim e^2/\varepsilon(r_D + a_{bl}/n_0^{1/3})$ where ε is the high frequency dielectric permittivity. However, the Debye radius $r_D \approx (4\pi r_c c_{iep})^{-1/2} \approx 4$ Å is quite small in comparison with $a_{bl}/n_0^{1/3}$ the average distance between intrablob electrons (here $r_c = e^2/\varepsilon T$ is the Onsager radius and $c_{iep} \approx n_0/(4/3)\pi a_{bl}^3$ is the concentration of ion-electron pairs within the blob). Thus, this screening energy $\sim T n_0^{1/3} r_c/a_{bl}$ becomes a dominant contribution, which is some tenths of eV. It may result in a trapping of the positron, which may finally thermalize close to the central part of its blob. Sometimes this version of the blob model describing Ps formation is called the "black blob model" [8] in contrast to the "white blob model," where this effect of in-blob trapping of thermalized and also slightly epithermal positrons is neglected [9, 10].

2.3. Effects of Local Heating and Premelting in the Terminal Part of the e^+ Track. We have estimated above that the ionization slowing down of energetic positrons (e^{+*}) and subsequent ion-electron recombination release an energy up to 1 keV in the terminal e^+ blob, this energy being finally converted into heat. Therefore, the temperature in the e^+ blob should be higher than the bulk temperature. This phenomenon may be called *the local heating effect*. This transient temperature regime may strongly affect, for example, the Ps bubble growth, by changing the viscosity of the medium. This effect may also influence the mobility of intratrack species and their reaction rates constants.

For quantitative estimations we simulate this process with the help of the macroscopic heat transfer equation [12]:

$$c_p\rho\frac{\partial T(r,t)}{\partial t} = \operatorname{div}(\lambda\nabla T)+q_+(r,t), \qquad T(r,t=0)=T_{bulk}. \tag{5}$$

Here, $T(r,t)$ is the local temperature, T_{bulk} is the bulk temperature of the medium, c_p is its specific heat capacity, ρ is the density, and λ is the thermal conductivity. The second term in the RHS quantifies the energy released by the positron when creating its blob: $q_+(r,t) \approx W_{bl}G(r,a)f(t,\tau)$, where $W_{bl} \approx 1$ keV is the blob formation energy, $G(r,a) \approx e^{-r^2/a_{bl}^2}/\pi^{3/2}a_{bl}^3$ describes the spatial distribution of the released energy, and $f(t,\tau) \approx \exp(-(t-1\text{ ps})^2/2\tau^2)/(\sqrt{2\pi}\tau)$ is its temporal distribution, where $\tau \approx 0.3$ ps is the typical time of ion-electron recombination.

In general, the system can be simultaneously solid and liquid. Then, the following method for obtaining a numerical solution of (5) can be used [12, 19]. To describe the deposition of the latent heat of melting, q_m, we added to $c_p(T)$ a term, which is nonzero only in a narrow temperature interval $T_m - \Delta T < T < T_m$, where T_m is the melting temperature and ΔT is the width of the phase transition (arbitrarily fixed to 0.25 K). Moreover, an integration of this term over temperature must give q_m. This additional overshot to $c_p(T)$ is simulated by a Gaussian function as follows:

$$\tilde{c}_p(T) = c_p(T) + \frac{q_m \exp\left(-(T-T_m)^2/2\Delta T^2\right)}{\sqrt{2\pi\Delta T^2}}. \tag{6}$$

Close to the phase transition region, the T-dependences of thermal conductivity and density were approximated by smooth functions:

$$\lambda(T) = \frac{\lambda^S}{\exp(-(T_m-T)/\Delta T)+1} + \frac{\lambda^L}{\exp((T_m-T)/\Delta T)+1},$$

$$\rho(T) = \frac{\rho^S}{\exp(-(T_m-T)/\Delta T)+1} + \frac{\rho^L}{\exp((T_m-T)/\Delta T)+1}. \tag{7}$$

This approach works well especially when it is hard to trace an interphase boundary.

Simulations were done for water, methanol, ethanol, and butanol. Some thermodynamic properties of these substances are given in Table 1. Temperature profiles $T(r,t)$ were calculated numerically within a spherically symmetric volume for $r < 200$ Å and for t up to 1 ns. Equation (5) was solved with the boundary condition $T(r = 200\text{ Å}) = T_{bulk}$ with the help of PDEPE subroutine from Matlab. Some temperature profiles in ice and in water close to the melting point are shown in Figure 4.

Close to the phase transition region, the state of the medium at a given spatial point may be deduced from its local temperature: if $T < T_m - \Delta T$, it is solid, and for $T > T_m + \Delta T$ we have a liquid; in between there is a "mixed" region. Figure 5 displays the maximum radius, R_{max}, of the molten region versus $T_{bulk} - T_m$ and the lifetime t_{ex} of the molten region. At $t > t_{ex}$ temperature of any point of the medium is below its melting point.

The local heating effect may result in a so-to-say "preliminary" formation of the so-called Ps bubble state. (It is a common stand point that in a liquid phase Ps atom forms a nanobubble and resides therein. Nature of this state is discussed below.) While the major fraction of the medium is solid, Ps bubble may be formed within the small premelted region close to the origin of the e^+ blob. This effect can

Table 1: Thermodynamical properties for water and some alcohols in solid and liquid phases near melting temperature [20–24].

Substance	T_m, K	λ^S/λ^L, W/m/K	q_m, J/g	ρ^S/ρ^L, g/cm^3
Methanol	175.6	0.32/0.21	99	0.98/0.79
Ethanol	159	0.27/0.17	109	1.06/0.79
Butanol	184.5	0.47/0.16	125	1.05/0.81
Water	273	2.38/0.56	334	0.92/1.00

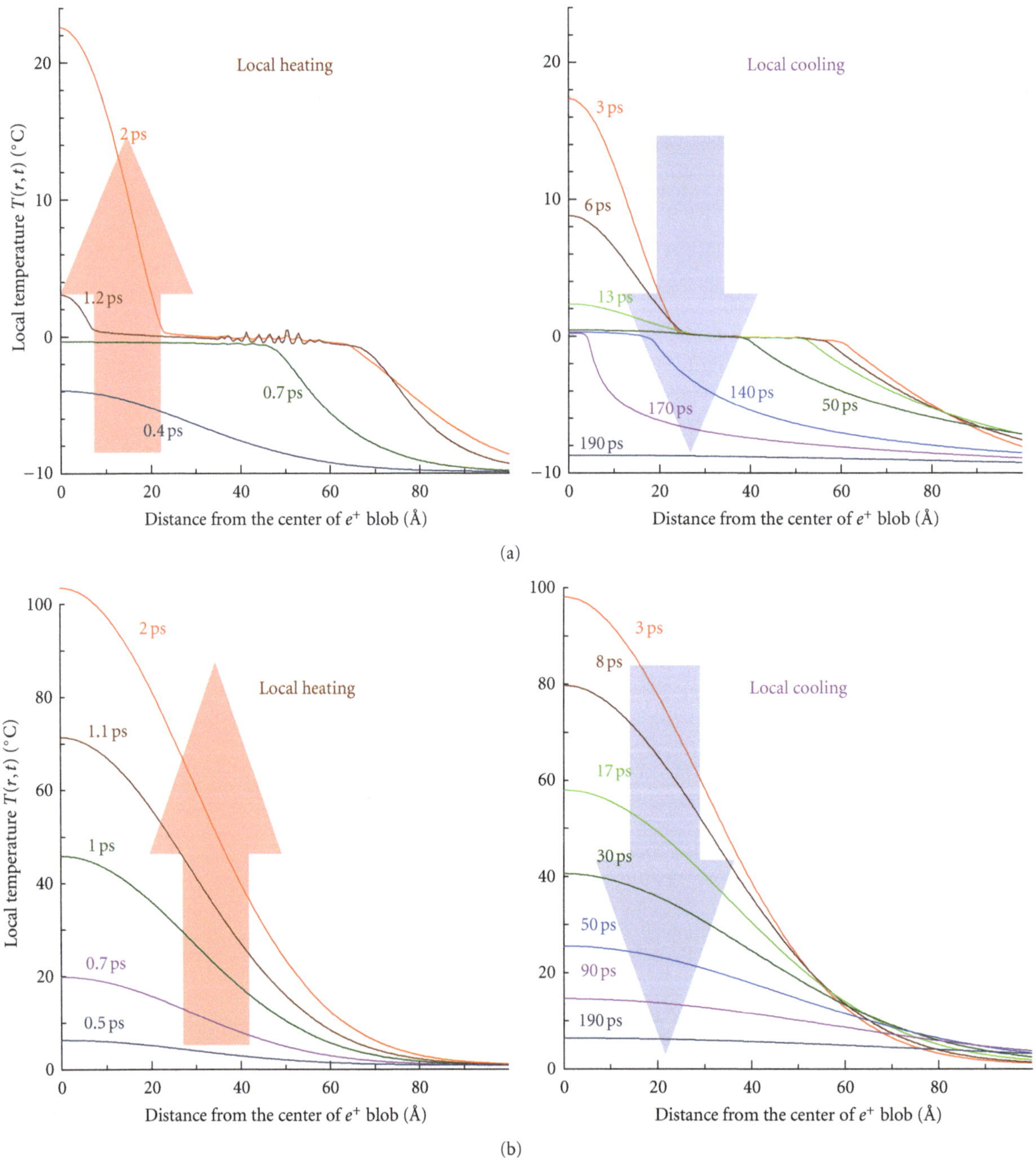

Figure 4: Calculated temperature profile in ice, $T_{\text{bulk}} = -10°\text{C}$ (on top), and in water, T_{bulk} (bottom), is slightly above 0°C. In both cases $a_{\text{bl}} = 40$ Å, $W_{\text{bl}} = 1$ keV.

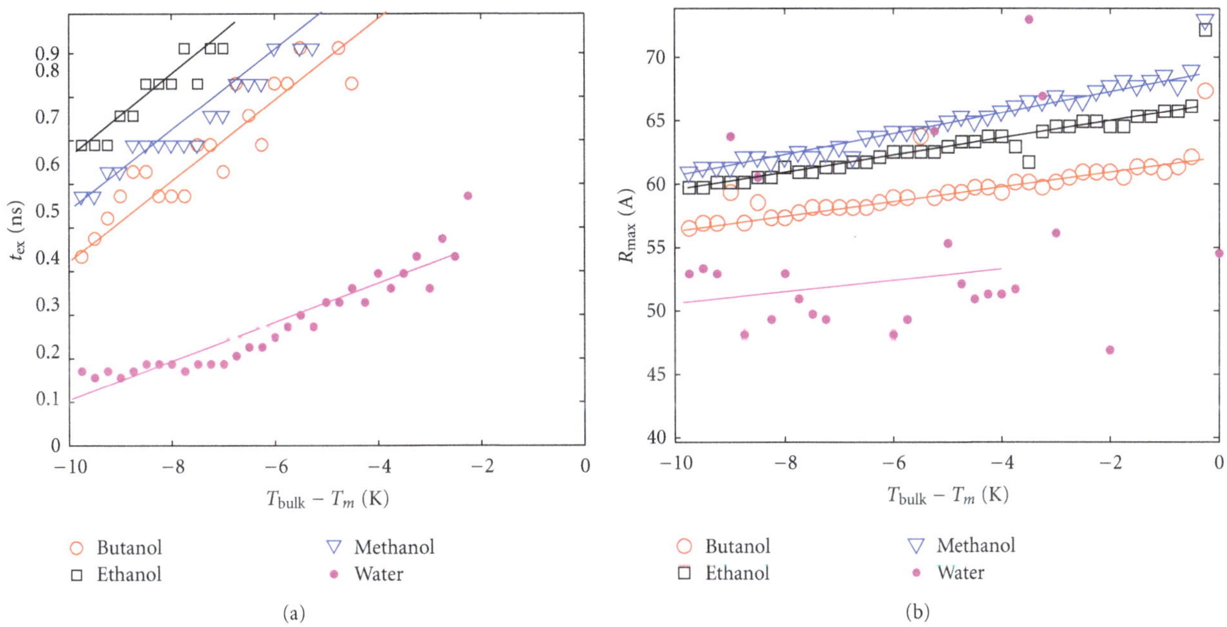

FIGURE 5: Lifetime of the molten region t_{ex} (a) and its radius R_{max} (b) for methanol, ethanol, butanol, and water are displayed as a function of the difference of the bulk temperature and the melting temperature.

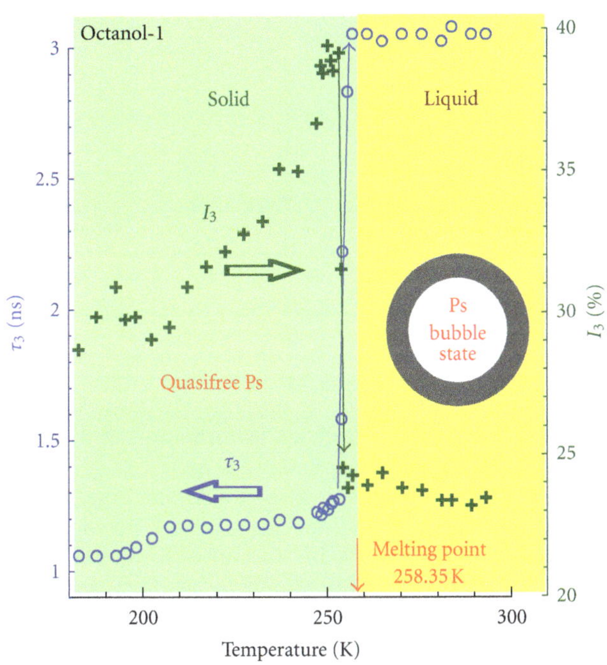

FIGURE 6: Temperature dependences of the ortho-Ps lifetime and the intensity of the corresponding lifetime component in octanol [11].

be observed experimentally as an increase of the long-lived (ortho-Ps) lifetime, τ_3, from its solid phase value to the value corresponding to the liquid state. This behavior is illustrated in Figure 6. This transient temperature regime may also affect, for example, the Ps bubble growth (through the viscosity), the mobility of intratrack species, and their reaction rates coefficients.

3. Positronium Formation in Condensed Media

3.1. The Ore Model. The Ore model was proposed for interpretation of the Ps formation in gases [25]. It implies that the "hot" positron, e^{+*}, having excess kinetic energy, pulls out an electron from a molecule M, thereby forming a Ps atom and leaving behind a positively charged radical-cation $M^{+\cdot}$:

$$e^{+*} + M \longrightarrow Ps + M^{+\cdot}. \quad (8)$$

This process is most effective when the e^{+*} energy W lies within the "Ore gap":

$$I_G - \frac{Ry}{2} < W < W_{ex} \quad (\text{or } I_G). \quad (9)$$

Here I_G is the first ionization potential of the molecule M, W_{ex} is its electronic excitation threshold, and Ry/2 = 6.8 eV is the Ps binding energy in a gas phase. If the positron energy is lower than $I_G - Ry/2$, e^+ cannot pick up an electron from a molecule. When $W > W_{ex}$, electronic excitations and ionizations dominate and Ps formation becomes less effective.

3.2. Quasifree Ps State. In the condensed phase, because of the presence of molecules and lack of free space, the final Ps state differs from that in vacuum [7]. We call this state quasifree positronium (qf-Ps). If we adopt a binding energy E_b of qf-Ps in a dielectric continuum of roughly $Ry/2\varepsilon^2$ (instead of just Ry/2 in vacuum), where $\varepsilon = n^2 \approx 2$ high-frequency dielectric permittivity is the square of the refractive index, it is seen that the Ore gap in a medium gets squeezed

$$I_L - \frac{\mathrm{Ry}}{2\varepsilon^2} < W < W_{\mathrm{ex}}^L \tag{10}$$

and may even completely disappear because of a significant decrease in the binding energy. Here, W_{ex}^L is the lower threshold of electronic excitations of a molecule in a liquid, $I_L \approx I_G - V_0^- - |U_p|$ is the liquid phase ionizing potential of a molecule, V_0^- is the energy of the ground state of an electron in a medium (sometimes it is called the e^- work function, Table 2), and U_p is the energy of polarization interaction of the positively charged ion with the medium.

3.3. Mechanism of Ps Formation. The above discussion points out the unique possibility for the Ps formation mechanism in molecular media. It postulates that Ps is formed through the combination of the thermalized particles quasifree positron and one of the intratrack electrons:

$$e_{\mathrm{qf}}^+ + e_{\mathrm{blob}}^- \longrightarrow (e^+ \cdots e^-) \longrightarrow \text{qf-Ps} \longrightarrow \text{Ps in the bubble.} \tag{11}$$

This reaction proceeds in the terminal part of the e^+ track (in the e^+ blob). If the positron is thermalized outside the blob, the only way for it to form Ps is to diffuse back and pick up one of the intrablob electrons. Otherwise it annihilates as a "free" positron. When e^+ picks up an electron, the "initial" separation between them is comparable to the average distance $(4\pi a_{\mathrm{bl}}^3/3n_0)^{1/3} \approx 20\,\text{Å}$ between intrablob species. The binding energy of such a $e^+ \cdots e^-$ pair is small, about 0.1 eV. So, the translational kinetic energies of the particles must be less than the binding energy, otherwise the pair will break up. Thus, just before Ps formation, the positron and track electron must be almost thermalized (note that the total energy of this pair at this stage is approximately the sum $V_0^+ + V_0^-$ of the e^+ and e^- work functions), Figure 7.

However, such a weakly bound pair is not at the bottom of its energy spectrum. The two particles approach each other (in average) and continue to release energy via excitation of molecular vibrations. Finally, the pair reaches an intermediate equilibrium state, which we term the quasifree Ps. Roughly, the qf-Ps binding energy is $E_b \approx \mathrm{Ry}/2\varepsilon^2$, about 1 eV. In liquids, energy gain of the $e^+ e^-$ pair is related with the rearrangement of molecules and appearance of some additional free space around it. e^+ and e^- may get closer, repelling molecules from their location and forming a Ps bubble state. A substantial decrease in their Coulombic energy is the driving force of this process.

Over the last 30 years this combination mechanism has become extremely widespread [25, 27]. It has been used to interpret numerous data on Ps chemistry and explain variations of the Ps yields (from 0 to 0.7) in very different condensed media, where parameters of the Ore gap are practically the same. It provides a natural explanation to the changes in the Ps formation probability at phase transitions. Experimentally, the observed monotonic inhibition of Ps yields (practically down to zero) in solutions of electron acceptors contradicts the Ore model but inserts well in the

Table 2: Values of V_0^- for different liquids at room temperature [26].

Liquid	V_0^-, eV
Helium; 4.2 K	1.3
n-dodecane	0.2
n-decane	0.18
n-heptane	0.12
n-hexane	0.1
Nitrogen; 77.3 K	0.05
n-pentane, c-hexane	0.01
Argon; 86.4 K	0
Benzene	−0.14
Isooctane	−0.17
Toluene	−0.22
Neopentane	−0.38
MeOH, EtOH, PrOH	−0.4
Xenon; 170 K	−0.57
Water	−1.2

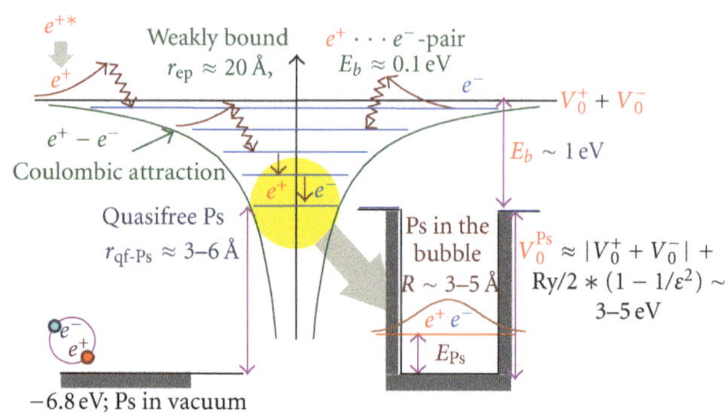

Figure 7: Mechanism of the Ps formation. Here, for simplicity, the potential in which Ps is localized is shown as a rectangular well.

recombination mechanism. It explains the anti-inhibition effect, including experiments on Ps formation in moderate electric fields in pure liquids and mixtures.

There are two models that utilize this mechanism, the spur model [28] and the blob model (or diffusion-recombination model) [29, 30]. In spite of the fact that both models answer the question about the Ps precursor in the same way, they differ as to what constitutes the terminal part of the e^+ track and how to calculate the probability of the Ps formation [7]. In the following we shall consider the blob model.

4. Intratrack Reactions in the Positron Blob

The interaction of the Ps atom with primary radiolytic products, formed in the terminal part of the positron track (in the e^+ blob) due to e^+ ionization slowing down, is a feature inherent to the positron spectroscopy of any molecular medium. Many intratrack products are strong oxidizers and/or radicals (like H_3O^+ radical cations and OH radicals in water) and their initial concentration in the

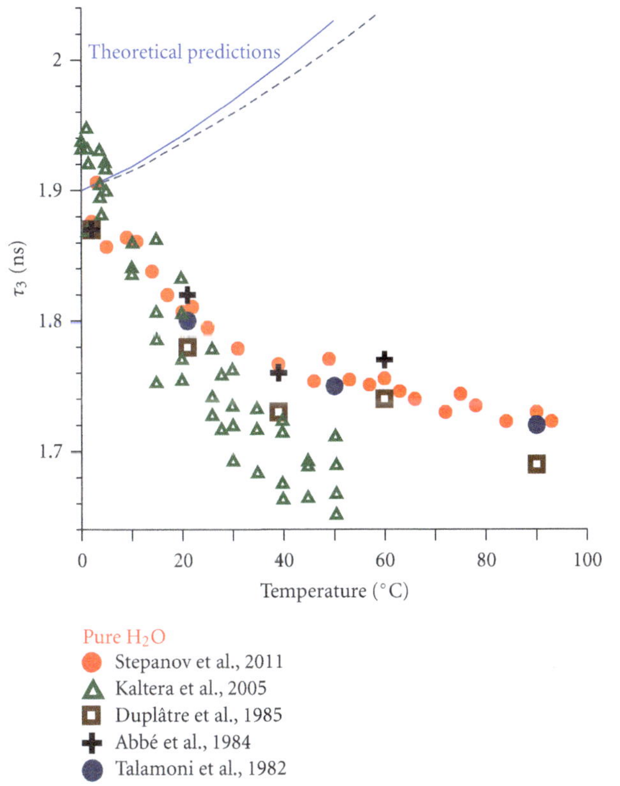

FIGURE 8: Temperature dependence of the lifetime, τ_3, of the long-lived component in pure water. Experimental data are derived from a 3-exponential deconvolution of the spectra [13–17] and theoretical predictions are based on the standard models of the Ps bubble (finite potential well: solid line; infinite potential well: dashed line) [18].

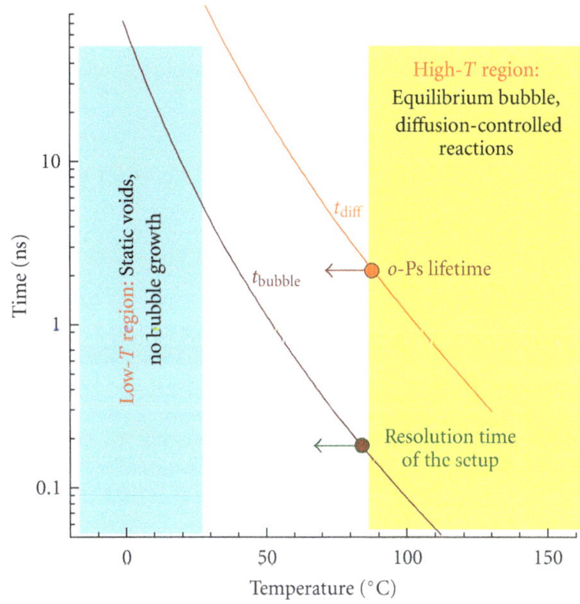

FIGURE 9: Temperature dependences of the Ps bubble formation time t_{bubble} and typical time t_{diff} of the diffusion-controlled intrablob reactions in glycerol.

e^+ blob is not small (up to 0.05 M). The average distance between intrablob species is comparable with the diffusion displacement of the Ps bubble before e^+ annihilation, so the contribution of diffusion-controlled oxidation and ortho-paraconversion reactions is quite possible.

Neglecting these reactions leads to obvious contradictions. Figure 8 shows the T-dependence of the lifetime (τ_3) of the long-lived component of PAL spectra in pure water (roughly, τ_3 is the pick-off lifetime of ortho positronium). Experimental data from different authors reasonably agree, but all are in strong contradiction with the theoretical expectations based on any Ps bubble model [18]: with increasing T, the surface tension coefficient decreases, so the size of the Ps bubble should increase, which should lead to an increase in τ_3 with T, in sharp contradiction to the PAL data. So one of the most informative ways to investigate Ps intratrack reactions is to study the temperature variation of the PAL spectra.

In liquid water, the Ps bubble growth proceeds very fast ($\lesssim$ 10 ps) and so it may be considered as an instantaneous process [31]. Further intratrack radiolytic processes are known rather well there. In glycerol at temperatures $\lesssim 50^\circ$C, formation of the Ps bubble state and further Ps interaction with intratrack radiolytic products are rather slow [32]. Of course, all these processes strongly depend on temperature, so firstly it is worth to consider two extreme cases, the cases of "high" and "low" temperatures.

4.1. High-Temperature Region. This region may be defined through the following conditions:

(1) the Ps bubble formation time, $t_{bubble}(T) \approx R\eta/\sigma$ [31] is short as compared to the time resolution of a PAL spectrometer, $\sim$ 0.2 ns. Here, $R \approx$ 3–6 Å is the radius of the Ps bubble and η and σ are the viscosity and surface tension coefficients of a liquid medium. Experimental observation of the bubble growth is impossible in this case because the equilibrium state is reached too fast (less than 0.1 ns).

(2) the diffusion length $\sqrt{6(D_{Ps}+D_i)\tau_3}$ of Ps and radiolytic products (subscript i) in the e^+ blob during the ortho-Ps lifetime, τ_3 (about few nanoseconds), must exceed the average distance between intrablob particles $\bar{r} \approx$ 10–20 Å. This can be expressed through the condition $t_{diff} < \tau_3$, where $t_{diff} = \bar{r}^2/6(D_i + D_{Ps})$; on the basis of the Stokes-Einstein relationship, $D_i \approx D_{Ps} \approx k_B T/6\pi\eta(T)R_i$, R_i is a few Å. In glycerol both conditions are satisfied at the same temperatures above 90°C (Figure 9), but in liquid water they are fulfilled at any temperature from 0 to 100°C.

In the high-T region the influence of intrablob reactions is important. The radiolytic processes initiated through ionizations induced by the slowing down of fast e^{+*} may be represented through the following basic reactions:

ionization:

$$e^{+*} + \mathrm{RH} \longrightarrow e^- + \dot{\mathrm{R}}\mathrm{H}^+ + e^+, \quad (12)$$

ion-molecule reaction:

$$\dot{\mathrm{R}}\mathrm{H}^+ + \mathrm{RH} \longrightarrow \mathrm{RH}_2^+ + \dot{\mathrm{R}}, \quad (13)$$

ion-electron recombination:

$$e^- + \dot{R}H^+ \longrightarrow \cdot\dot{R}H^*_{\text{tripl}}, RH^*_{\text{singl}}, \quad (14)$$

$$e^- + RH_2^+ \longrightarrow RH + \dot{H} \longrightarrow \dot{R} + H_2, \quad (15)$$

electron solvation (in polar media; question about e^+ solvation remains open):

$$e^- \longrightarrow e_s^-, \quad (16)$$

electronic deexcitation:

$$\cdot\dot{R}H^*_{\text{tripl}} \xrightarrow{+RH} 2\dot{R} + H_2, \qquad RH^*_{\text{singl}} \longrightarrow RH. \quad (17)$$

From reactions ((12)–(15)) one can infer that the total number of $\dot{R}$ in the e^+ blob varies weakly in time and is approximately equal to the initial number, n_0, of ion-electron pairs in the blob. However, their spatial distribution broadens in time due to outdiffusion.

As we have discussed above, Ps formation is described by reaction (11), which may be briefly written as follows:

$$e^+ + e^- \longrightarrow \text{Ps}. \quad (18)$$

The main effect of ions and radicals on Ps is oxidation and spin conversion reactions. If we denote for simplicity all these species ($\dot{R}H^+$, RH_2^+, $\dot{R}$, $\dot{H}$) by the same symbol, $\dot{R}$, and term them "radicals," these species lead to

Ps oxidation:

$$\text{Ps} + \dot{R} \longrightarrow e^+ + R^-, \quad (19)$$

ortho-paraconversion:

$$\text{para- or ortho-Ps} + \dot{R} \longrightarrow \frac{1}{4}\text{para-Ps} + \frac{3}{4}\text{ortho-Ps} + \dot{R}. \quad (20)$$

In the high-temperature region these Ps reactions are mostly diffusion controlled. With the help of the nonhomogeneous chemical kinetics approach (the "white" blob model) [6, 18] reactions (12)–(20) may be described in terms of the following equations

$$\frac{\partial c_i(r,t)}{\partial t} = D_i \Delta c_i - k_{ij} c_i c_j - \lambda_i c_i,$$
$$c_i(r,0) = c_i^0 \frac{\exp(-r^2/a_i^2)}{\pi^{3/2} a_i^3}, \quad (21)$$

where $c_i(r,t)$ is the concentration of the intrablob species of the ith type (including all the positron states; note that the initial spatial distribution of the e^+ is usually somewhat broader than that of ion-electron pairs in the blob. This is due to the appearance of the local electric field which pushes e^+ out of the blob and keeps track electrons close to the ions, see Section 2.2), k_{ij} is the rate constant of the reaction between the i and j reactants and λ_i is the decay rate of the ith particles including possible annihilation.

According to the Smoluchowski approach, the reaction rate constant $k_{\text{Ps,R}}$ (let us take as an example reactions (19)-(20)) may be written as follows:

$$k_{\text{Ps,R}}(T,t) = 4\pi D_{\text{Ps,R}}(T) R_{\text{Ps,R}} \cdot \left[1 + \frac{R_{\text{Ps,R}}}{\sqrt{\pi D_{\text{Ps,R}} t}}\right], \quad (22)$$

where $R_{\text{Ps,R}} = R_{\text{Ps}} + R$ is the reaction radius of the R + Ps chemical reaction, R_i is approximately the geometric radius of R reactant, and $R_{\text{Ps}} \approx R_U + 1/\varkappa$ is the radius of delocalization of the Ps wave function ($1/\varkappa$ is the under-barrier penetration length of the Ps center-of-mass wave function; here, we consider Ps as a point particle in a potential well) [33].

The temperature dependence of the rate constant arises through the diffusion coefficients of the reagents, $D_{\text{Ps,R}}(T) = D_{\text{Ps}} + D_{\text{R}}$. According to the Einstein relationship and the Stokes formula, the diffusion coefficients may be expressed as

$$D_{\text{R}} = \frac{T}{6\pi\eta(T)R}, \qquad D_{\text{Ps}} = \frac{T}{4\pi\eta(T)R_{\text{Ps}}}, \quad (23)$$

where η is the viscosity of the medium and R and R_{Ps} are the hydrodynamic radii of $\dot{R}$ particle and Ps bubble. Although the applicability of these expressions on the atomic scale might be questionable, our studies in neat water [18, 33] have shown that it remains valid within a reasonable accuracy.

The knowledge of the surface tension of the liquid together with the Ps bubble model allows one to estimate the equilibrium radius of the Ps bubble and calculate the pick-off annihilation rate $\lambda_{\text{po}}(t,T)$, which is of utmost importance to interpret the PAL spectra:

$$\lambda_{\text{po}}(t,T) = \lambda_p \cdot P(R_U(t,T)). \quad (24)$$

Here, λ_p is the "free" positron annihilation rate and $P(R_U)$ is the under-barrier penetration probability of e^+ into a bulk of the liquid (into space containing molecular electrons). Its calculation is discussed below.

4.2. Low-Temperature Region. In this region, the Ps bubble formation time is larger than the e^+ and Ps annihilation lifetimes. On a scale of 1 ns, the size of a preexisting void in which a quasifree Ps has been localized does not change, since $t_{\text{bubble}} \gg 1$ ns. This situation is similar to that existing in polymers. Typically at these T the diffusion time t_{diff} is also much larger than the o-Ps lifetime (~ 1 ns), so one may completely neglect the diffusion motion of radiolytic products. For example in glycerol it happens at $T < 30°$C, Figure 9.

To describe correctly the PAL spectra, one should average the pick-off annihilation kinetics over the size distribution of the preexisting voids in which Ps localization takes place. This can be done by using the theory of free volume entropy fluctuations [34], which claims that every molecule possesses some free volume v_i with the probability

$$\propto \exp\left(-\frac{v_i}{v_F}\right), \qquad v_F(T) \approx v_{\text{WS}}(T) - v_{\text{vdW}}, \quad (25)$$

where v_F is a difference between the volume of the Wigner-Seitz cell and van-der-Waals volume of a molecule. Because of the approximate character of this expression, v_F may be considered as an adjustable parameter derivable by fitting the spectra in the low-temperature region. Note that in some cases the free volume distribution in the studied substances may be more complicated (nanocavities may have some selected sizes related to a particular chemical structure and bonding) [35].

If v is the volume of a preexisting void in which the Ps atom can localize, the probability to find such a void is proportional to $\exp(-v/v_F)$. Neglecting the influence of any intratrack reactions (because of low T and not discussing their possible tunneling nature), the calculation of the Ps pick-off annihilation kinetics reduces to averaging the exponents that relate to the pick-off annihilation of individual Ps atoms located at different voids:

$$\left\langle e^{-\lambda_{po}(R_U)t} \right\rangle = \int_{v_{min}}^{\infty} e^{-\lambda_{po}(R_U)t} \cdot e^{-v/v_F} \cdot \frac{dv}{v_F}, \quad v = \frac{4\pi R_U^3}{3}. \tag{26}$$

Here, v_{min} is the volume of the minimal cavity, which may trap Ps (Section 4).

4.3. Intermediate Temperatures. The most complex task in the PAL data processing refers to the intermediate temperature region, where the bubble formation time is comparable with the e^+ lifetime in the medium (fortunately this T interval is rather narrow because of "exponential" T-dependence of viscosity). qf-Ps localizes in one of the preexisting voids, which further somewhat increases in size; the time at which the equilibrium volume is reached depends on T. Obviously, the quantum-mechanical pressure exerted by Ps on the wall and thus the growth of the void cease when e^+ annihilates.

At these temperatures, the Ps fate may also depend on the intratrack reactions. Therefore, the corresponding chemical kinetic equations must be solved by assuming that Ps initially localizes in a preexisting void of a given radius, which will increase with time until annihilation; the process must be averaged over the size distribution of the preexisting voids [32].

4.4. Interpretation of the PAL Spectra in Liquid Water at Different T. To illustrate briefly how the blob model works, we mention the solution of the problem with a puzzling temperature behavior of the long-lived component in water, Figure 8. We have rejected conventional exponential deconvolution of the spectra and explicitly took into account the intrablob processes mentioned above [13, 18]. The decrease of τ_3 versus T was ascribed to the increasing efficiency of the oxidation (19) and ortho-paraconversion (20) reactions between Ps and intratrack radicals (mainly OH radicals). All parameters included in the model have a clear physical meaning: the Ps oxidation reaction radius $R_{ox} \approx R_{Ps} + R_{OH}$ which enters the rate constant of the Ps oxidation reaction (19), relative contact density η_c in the Ps bubble state, free positron annihilation rate, the Ps formation rate constant, and some others (5 adjustable parameters in total).

It is seen from Figure 10 that the blob model explains well the experimental data in a wide range of temperatures (and magnetic fields also [13]). It does not lead to contradictions with known radiation chemistry data. The agreement between theory and experiment became even better when taking into account the time dependence of the Ps reaction rate coefficients. The temperature dependence of the Ps diffusion-controlled reaction rate constants agrees with the Stokes-Einstein law. Good fitting of the PAL data is obtained when the Ps hydrodynamic radius is equal to the radius $R_{Ps} = R_U + 1/\varkappa$ of the Ps bubble (here $1/\varkappa \approx 0.5$ Å is the under-barrier penetration depth of the Ps wave function).

5. Ps Bubble Models

Typical lifetimes of a parapositronium atom in condensed medium are about 130–180 ps. They are close to the p-Ps lifetime in vacuum (125 ps). The orthopositronium lifetime in a medium is considerably shorter (about 100 times; some ns) in comparison with that in vacuum. This is due to the so-called pick-off process-prompt 2γ-annihilation of the e^+, composing Ps atom, with one of the nearest e^- of surrounding molecules, whose spin is antiparallel to the e^+ spin. Just this property turns Ps into a nanoscale structural probe of matter. The theoretical task consists in calculating the pick-off annihilation rate λ_{po}, that is, in relating λ_{po} with such properties of the medium like surface tension, viscosity, external pressure, and size of the Ps trap.

Originally, to explain the unexpectedly long lifetime of the ortho-Ps atom in liquid helium, Ferrel [36] suggested that due to the quantum mechanical nature of the Ps atom, it forms a nanobubble around itself. This is caused by a strong exchange repulsion between the o-Ps electron and electrons of the surrounding He atoms. Ferrel approximated this repulsion by a spherically symmetric potential barrier of radius R_∞. To estimate the equilibrium radius of the Ps bubble he minimized the sum of the Ps energy in a spherically symmetric potential well, that is, $E_{Ps} = \pi^2\hbar^2/4mR_\infty^2 = (\mathrm{Ry}/2)(\pi a_B/R_\infty)^2$, Ry $= 13.6$ eV, and the surface energy, $4\pi R_\infty^2\sigma$, where σ is the macroscopic surface tension coefficient. The following relationship is hereby obtained for the equilibrium radius of the bubble:

$$\frac{\mathrm{Ry}}{2} \cdot \frac{\pi^2 a_B^2}{R_\infty^2} + 4\pi R_\infty^2 \sigma \longrightarrow \text{ min over } R_\infty \Longrightarrow R_\infty = a_B \left(\frac{\pi \mathrm{Ry}}{8\sigma a_B^2} \right)^{1/4}. \tag{27}$$

5.1. The Tao-Eldrup Model. Ferrel's idea got further development in the studies of Tao [37] and Eldrup et al. [38]. They considered the Ps atom as a point particle in a liquid, that is, in a structureless continuum, Figure 11. The repulsive Ps-liquid interaction was approximated by a rectangular infinitely deep spherically symmetric potential well of radius R_∞. In such a well, the wave function of a point particle has the following standard expression:

$$\Psi(0 \le r \le R_\infty) = \frac{\sin(\pi r/R_\infty)}{\sqrt{2\pi R_\infty} r}, \quad \Psi(r \ge R_\infty) = 0. \tag{28}$$

Here, r is the Ps center-of-mass coordinate. Because the Ps wave function equals zero at the bubble radius (and

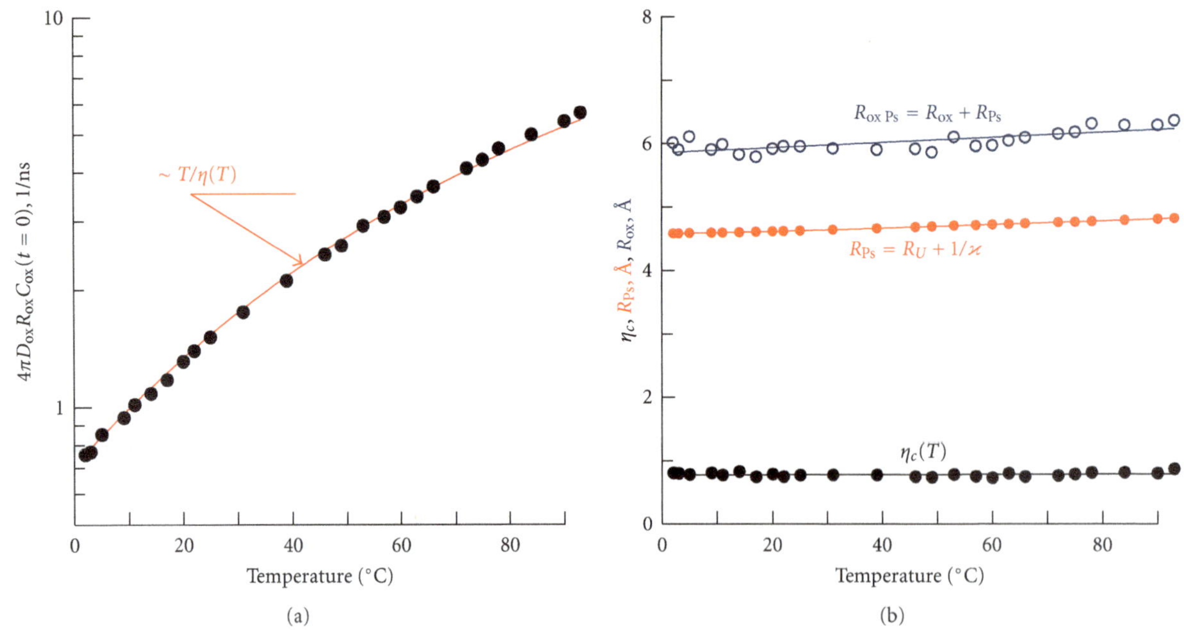

FIGURE 10: (a) Temperature dependence of the Ps oxidation reaction rate constant (at an infinite time, $t \to \infty$) multiplied by the "initial" concentration of oxidizers $c_{ox}(0)$ in the e^+ blob (OH and H_3O^+, see [13, 18] for details). D_{ox} is the sum of D_{Ps} and the diffusion coefficient of the oxidizer (OH radical, $D_{OH} \approx 1.5 \cdot 10^{-5}\,\text{cm}^2/\text{s}$). The obtained T-dependence agrees well with the Stokes-Einstein law, $\propto T/\eta(T)$. (b) T-dependences of the oxidation reaction radius, $R_{ox,Ps} = R_{ox} + R_{Ps}$, the Ps radius, $R_{Ps} = R_U + 1/\kappa$, and the relative contact density parameter $\eta_c \approx 0.8$ in Ps. R_{ox} is the radius of the oxidizer (approximately, the geometric radius of the OH radical, 1.4 Å).

outside), there is no e^+ overlapping with outer electrons of a medium. So, pick-off annihilation is absent. To overcome this difficulty it was postulated that molecular electrons, which form a "wall" of the Ps bubble, may penetrate inside the potential well. This results in the appearance of a surface layer of thickness $\delta = R_\infty - R$ having the same average electron density as in the bulk. As a result, the pick-off annihilation rate λ_{po} becomes nonzero. It is proportional to the e^+ overlapping integral with the electrons inside the bubble:

$$\lambda_{po} = \lambda_+ P_R, \quad P_R = \int_R^{R_\infty} |\Psi(r)|^2 4\pi r^2 dr = \frac{\delta}{R_\infty} - \frac{\sin(2\pi\delta/R_\infty)}{2\pi}. \tag{29}$$

This is the well-known Tao-Eldrup formula. Here, $\lambda_+ \approx 2\,\text{ns}^{-1}$ is the e^+ annihilation rate in an unperturbed medium (it is proportional to Dirac's 2γ-annihilation cross-section and the number density of valence electrons). The thickness δ of the electron layer is an empirical parameter, which may have different values in various media.

Substituting (27) for R_∞ into (29), one obtains the relationship between λ_{po} and σ with one adjustable parameter, δ. It may be easily obtained by fitting experimental pick-off annihilation rate constants with the relationship (29), Figure 12. Thus, we obtain $\delta \approx 1.66$ Å. Equation (29) with this value of δ is widely used for recalculation of the observed pick-off annihilation rate into the free volume $4\pi R^3/3$ of the cavity, where Ps atom resides and annihilates.

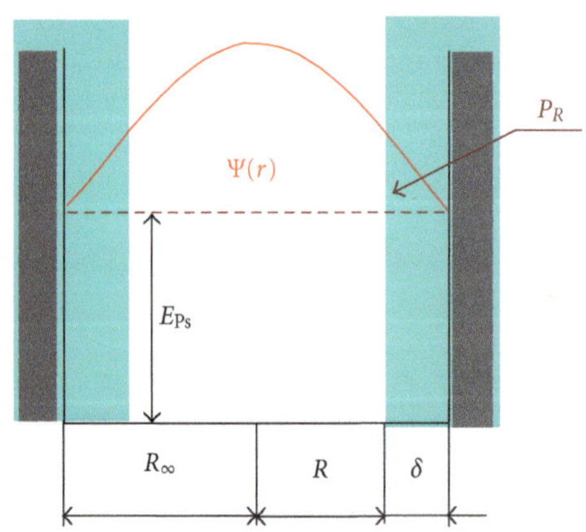

FIGURE 11: Tao-Eldrup model of Ps atom in a liquid phase. It is assumed that Ps is confined in an infinite spherically symmetric potential well of radius R_∞. $\Psi(r)$ is the center-of-mass wave function of Ps. R is the free volume radius of the Ps bubble. δ is the penetration depth of the molecular electrons into the Ps bubble.

5.2. Further Development of the Ps Bubble Models: "Nonpoint" Positronium. Along with the development of the "infinite potential well" Ps bubble model, another approach based on the finite potential well approximation was also elaborated [39–43]. However in both approaches, the Ps atom was approximated by a point particle. This leads to a significant simplification, but it is not justified from a physical viewpoint, because (1) the size of the localized state of Ps (size of the Ps bubble) does not significantly exceed the

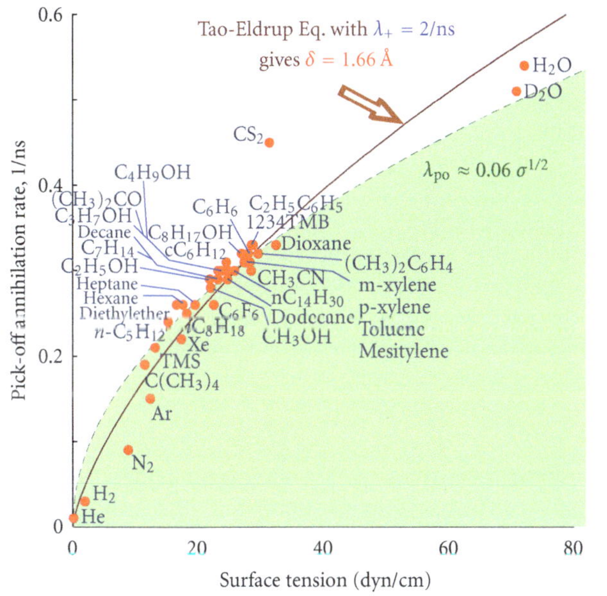

FIGURE 12: Dependence of the experimental pick-off annihilation rate constants [39] versus surface tension in different liquids. The solid curve shows the correlation given by the Tao-Eldrup at $\lambda_+ = 2\,\text{ns}^{-1}$ and optimal value $\delta = 1.66$ Å (obtained from fitting of these data by means of (29)). The dashed curve illustrates the simplest approximation $\lambda_{po} \propto \sigma^{1/2}$.

distance between e^+ and e^- in Ps (2) during the formation of the Ps bubble there is a substantial variation of the Ps internal energy (particularly of the Coulombic attraction of e^+ and e^-), which is completely ignored in the "point-like" Ps models. In a vacuum or in a large bubble, the internal energy of Ps tends to $-\text{Ry}/2 = -6.8$ eV. In a continuous liquid (no bubble) with the high-frequency dielectric permittivity $\varepsilon \approx n^2$ ($n \approx$ 2-3 is the refractive index) the energy of the Coulombic attraction between e^+ and e^- decreases in absolute value by a factor $\varepsilon^2 \approx$ 4–9. The same takes place with the total Ps binding energy, which tends to the value $-\text{Ry}/2\varepsilon^2 \approx -(0.8\text{–}1.7)$ eV (this is a simple consequence of the scaling $e^2 \to e^2/\varepsilon$ of the Schrödinger equation for Ps atom). Thus, the change in the Ps internal energy during Ps formation may reach 5 eV. Obviously, this represents an important contribution to the energetics of Ps formation. Now we shall present a more accurate estimation of this contribution, which has not been done yet.

There is only a small number of papers where the consequences of the finite size of Ps are discussed in application to positron annihilation spectroscopy. To calculate λ_{po}, the Kolkata group [44, 45] suggested to smear the Ps atom over the relative e^+-e^- coordinate exactly in the same way as it is in a vacuum. Such an approach is valid for rather large bubbles. However, they do not discuss the variation of the internal Ps energy.

In [46–48] a path integral Monte Carlo technique was used to simulate the two-particle e^+-e^- system. However, to proceed with calculations the authors need potentials describing e^--atom and e^+-atom interactions (they were taken from a variety of sources). However, the question about

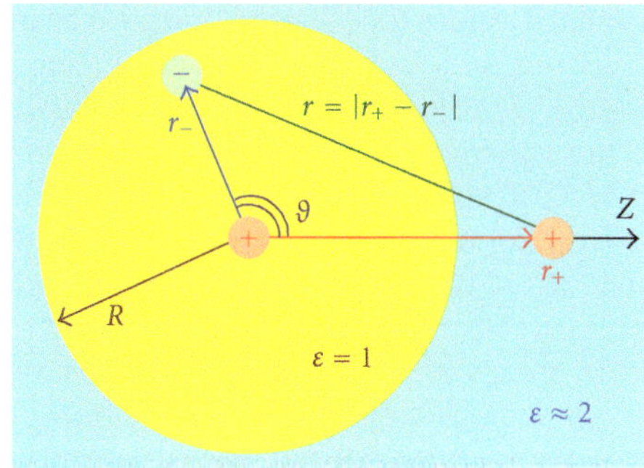

FIGURE 13: Coordinates of the e^+e^- pair.

modification of the e^+-e^- interaction because of the presence of the medium remains open (in this paper we roughly take this effect into account by means of introducing the high frequency dielectric permittivity).

In [49] the Ps atom is considered as a finite sized e^- e^+ pair, but the variation of the Coulombic interaction because of dielectric screening is not discussed. It was assumed that e^- is confined in an infinite potential well and e^+ is bound to it by means of the Coulombic attraction. The wave function of the pair was taken as a series of orthogonal polynomials, their weights being determined from a minimization procedure of the total energy of the pair.

5.3. Hamiltonian of the e^+e^- Pair in a Medium. Let the e^+e^- pair (Ps atom) have already formed in a liquid a nanobubble (spherical cavity; Ps bubble) of radius R (the onset of coordinates is taken at the center of the bubble, Figure 13). Together with the molecules surrounding the e^+e^- pair, one has to deal with a quite intricate many-body problem with a complex hamiltonian. We reduce it to the following form:

$$H \approx -\frac{\hbar^2(\Delta_+ + \Delta_-)}{2m_e} + U(r_+) + U(r_-) - U_c(r_+, r_-, R, \varepsilon). \tag{30}$$

Terms with Laplacians Δ_+ and Δ_- over r_+ and r_- (e^+ and e^- coordinates) stand for the kinetic energies of the particles. $U(r_+)$ and $U(r_-)$ describe the individual interaction of e^+ and e^- with the medium. For them we adopt the following approximation:

$$U(r_+) = \begin{cases} 0, & r_+ < R, \\ V_0^+, & r_+ > R, \end{cases} \qquad U(r_-) = \begin{cases} 0, & r_- < R, \\ V_0^-, & r_- > R. \end{cases} \tag{31}$$

Here, V_0^+ and V_0^- are the e^+ and e^- work functions, respectively, Table 2. The work function is introduced as the energy needed for an excess particle to enter the liquid without any rearrangement of its molecules and to stay there in a delocalized state, having no preferential location in a bulk. One may say that V_0^+ and V_0^- are the ground state energies of the quasifree e^+ and e^-, because their energies at rest after having been removed from the liquid to infinity are defined to be zero.

Experimental values for V_0^- are known for many liquids (Table 2). Because of a lack of experimental data on the e^+ work functions, we admit that $V_0^+ \approx V_0^-$ and $|V_0^+ + V_0^-| \approx 0$. Note that the variation of the internal energy of the pair, $\approx \text{Ry}(1 - 1/\varepsilon^2)/2 \approx 5\,\text{eV}$, related with the variation in the dielectric screening of the e^+-e^- attraction in the Ps bubble formation process, very probably significantly exceeds V_0^- and V_0^+.

In (30) U_c stands for the Coulombic interaction between e^+ and e^- in a polarizable medium. Assuming that the medium has the high frequency dielectric permittivity ε of the bulk and a spherical cavity of radius R (inside the cavity $\varepsilon = 1$), one may calculate U_c by solving the Poisson equation. Denoting the e^+ and e^- coordinates as r_+ and r_-, U_c may be written in the form of the following series via the Legendre polynomials $P_l(x = \cos\theta)$ [50]:

$$\frac{U_c(r_+ < R, r_- < R)}{\text{Ry}} = \frac{2a_B}{r} - \left(1 - \frac{1}{\varepsilon}\right)\frac{2a_B}{R} \times \left(1 + \sum_{l=1}^{\infty} \frac{(1+l)P_l(x)}{1+l+l/\varepsilon} \cdot \frac{r_+^l r_-^l}{R^{2l}}\right),$$

$$\frac{U_c(r_+ < R, r_- > R)}{\text{Ry}} = \frac{2a_B}{\varepsilon r_-}\left(1 + \sum_{l=1}^{\infty} \frac{(1+2l)P_l(x)}{1+l+l/\varepsilon} \cdot \frac{r_+^l}{r_-^l}\right),$$

$$\frac{U_c(r_+ > R, r_- > R)}{\text{Ry}} = \frac{2a_B}{\varepsilon r} + \left(1 - \frac{1}{\varepsilon}\right)\frac{2a_B}{R}\sum_{l=1}^{\infty}\frac{lP_l(x)}{l + \varepsilon + l\varepsilon} \cdot \left(\frac{R^2}{r_+ r_-}\right)^{l+1};$$

$$\frac{U_c(r_+ > R, r_- < R)}{\text{Ry}} = \frac{2a_B}{\varepsilon r_+}\left(1 + \sum_{l=1}^{\infty} \frac{(1+2l)P_l(x)}{1+l+l/\varepsilon} \cdot \frac{r_-^l}{r_+^l}\right). \tag{32}$$

Here, the argument of the Legendre polynomials is $x \equiv \cos\vartheta$, where ϑ is the angle between the z axis and the direction of r_-.

5.4. Wave Function of the e^+e^- Pair and Minimization of Its Total Energy $\langle H\rangle$. Keeping in mind further use of the variational procedure, let us choose the normalized e^+e^- wave function in the following simplest form:

$$\Psi_{+-}(r_+, r_-) = \frac{\exp(-r/2a - r_{\text{cm}}/2b)}{8\pi\sqrt{a^3 b^3}}, \quad \mathbf{r}_{\text{cm}} = \frac{\mathbf{r}_+ + \mathbf{r}_-}{2}, \quad \mathbf{r} = \mathbf{r}_+ - \mathbf{r}_-. \tag{33}$$

In both cases of a rather large bubble and a uniform dielectric continuum, Ψ_{+-} breaks into a product of two terms: the first one depends on the distance r between e^+ and e^-, and the second one depends on the center-of-mass coordinate $\mathbf{r}_{\text{cm}}$. Parameters a and b are the variational ones, over which we have minimized the energy of the e^+ e^- pair:

$$E(a, b, R) = \langle\Psi_{+-}|H|\Psi_{+-}\rangle \longrightarrow \min \Longrightarrow a(R), b(R). \tag{34}$$

The simplest verification of the calculations is to recover two limiting cases. In case of large bubbles ($R \to \infty$), one should reproduce the "vacuum" state of the Ps atom: its total energy must tend to $-\text{Ry}/2 = -6.8\,\text{eV}$, the kinetic energy to $+\text{Ry}/2$, and the Coulombic energy to $-\text{Ry}$. In case of small bubbles ($R \to 0$), the delocalized qf-Ps state must be reproduced. The Schrödinger equation for qf-Ps has the same form as for the vacuum Ps, but with the substitution $e^2 \to e^2/\varepsilon$. Then the total qf-Ps energy tends to $V_0^+ + V_0^- - \text{Ry}/2\varepsilon^2$, its kinetic part tends to $+\text{Ry}/2\varepsilon^2 = 1.7\,\text{eV}$ ($\varepsilon = 2$), and the Coulombic energy tends to $-\text{Ry}/\varepsilon^2 = -3.4\,\text{eV}$. Figure 14 displays optimal values of a and b as well as different contributions to the total energy of the e^+ e^- pair.

5.5. Relative Contact Density and Pick-Off Annihilation Rate. Using the wave function (33) it is easy to obtain the relative contact density η_c in the Ps atom:

$$\eta_c = \frac{\iint d^3r_+ d^3r_- |\Psi_{+-}(r_+, r_-)|^2 \delta(r_+ - r_-)}{\iint d^3r_+ d^3r_- |\Psi_{+-}^{\text{vac}}(r_+, r_-)|^2 \delta(r_+ - r_-)} = \frac{a_B^3}{a^3(R)}. \tag{35}$$

This quantity determines the observable Ps annihilation rate constant (including the case with applied permanent magnetic field). The resulting dependencies of η_c are shown in Figure 14 (on the left). Because, for qf-Ps, parameter a is equal to εa_B, for qf-Ps the value of η_c should be $1/\varepsilon^3 = 1/8$, which is well recovered in numerical calculations. When R increases, η_c approaches unity, because a tends to its vacuum value a_B.

Knowing the expression for the wave function (33), one may calculate the positron overlapping P_R with molecular electrons, surrounding the Ps atom, and therefore find out the pick-off annihilation rate constant:

$$\lambda_{\text{po}}(R) \approx \lambda_+ P_R, \quad P_R \approx \int_{r_+ > R} d^3r_+ \int d^3r_- |\Psi_{+-}(r_+, r_-)|^2. \tag{36}$$

Here, $\lambda_+ \approx 2\,\text{ns}^{-1}$ is the annihilation rate constant of "free" positrons. Results of calculations of $\lambda_{\text{po}}(R)$ for optimal a and b values, which correspond to the minimal Ps energy at a given R, are shown in Figure 15.

5.6. Discussion on the Nonpoint Ps Approximation. It is usually considered that Ps is a solvophobic particle, that is, it forms a bubble when entering a liquid because of exchange repulsion between e^- in Ps and the surrounding molecular electrons. If both V_0^+ and V_0^- are negative, that is, e^+ and e^- consider a cavity as a potential barrier, they are pulled to the bulk by polarization interaction with the medium. Nevertheless, even in this case the Ps bubble may be formed due to an enhancement of the Coulombic e^+e^- attraction inside the cavity (no dielectric screening inside). This feature cannot be taken into account when Ps is simulated as a point particle.

It is seen that the behavior of the total energy of the pair (red curve in Figure 14) strongly differs from the Tao-Eldrup prediction (green dashed curve; the first term in (27), where R_∞ is replaced by R), as well as from the expectation based on the finite potential well model (brown curve, Figure 14;

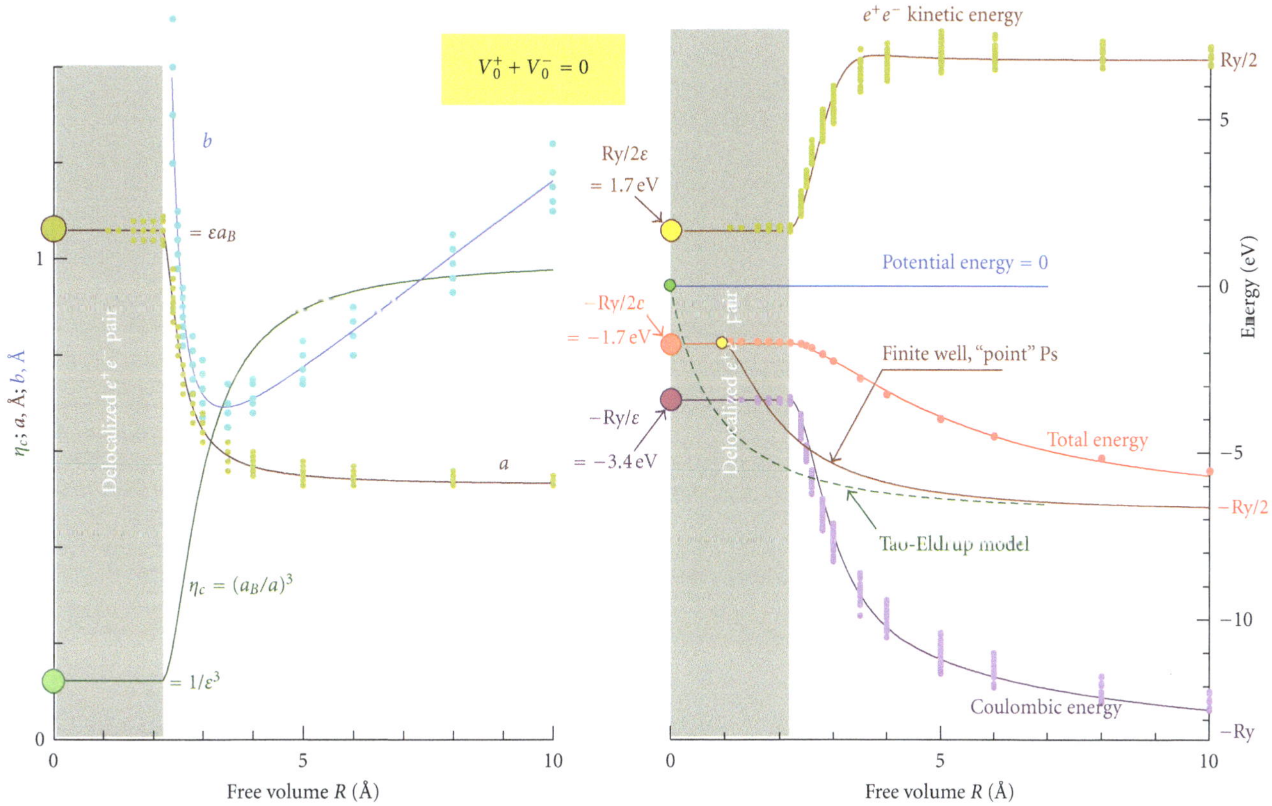

FIGURE 14: Dependences of the optimal parameters a and b versus R, the bubble radius. They enter the $e^+ e^-$ wave function and yield the minimum of the total energy $\langle H \rangle$. The relative contact density η_c and different energy contributions to $\langle H \rangle$ (at optimal a and b) are shown as well. It was assumed that $\varepsilon = 2$.

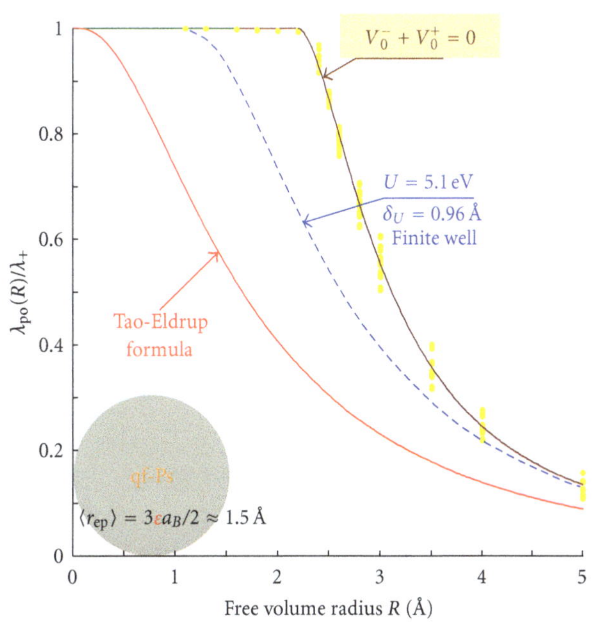

FIGURE 15: Pick-off annihilation rate constant of Ps, localized in a bubble of R when $V_0^+ + V_0^- = 0$ (brown curve). For small R ($\lesssim 2.3$ Å) the calculated values of λ_{po} are equal to λ_+. The red line shows the pick-off annihilation rate constant, calculated according to the Tao-Eldrup formula. The dashed line is the calculation according to the finite potential well model (for comparison we adopted that the depth of the well is $(1 - 1/\varepsilon^2)$ Ry/2 ≈ 5.1 eV and its radius is R; in this case the minimal radius of the well when there appears an energy level for Ps is $\delta_U = 0.96$ Å).

the Coulombic potential cannot be approximated well by a rectangular spherically symmetric potential). The same is true for the pick-off annihilation rate, Figure 15.

Calculations demonstrate one common feature: up to $R \lesssim 2.3$ Å all dependencies remain the same as in a medium without any cavity, but at larger R there are significant deviations. This is related to the known quantum mechanical phenomenon—absence of a bound state of a particle in a small finite 3d-potential well. In such cavities, Ps cannot be bound, it does not exert any repulsive pressure on their walls, and it does not stimulate their transformation towards the equilibrium Ps bubble. The possibility of finding a suitable preexisting cavity, sufficient at least for preliminary localization of qf-Ps, may be a limiting factor for the formation of the Ps bubble state.

One may find an equilibrium Ps bubble radius by minimizing the sum of the total e^+e^- energy $\langle H \rangle$ and the surface energy of the bubble. For water it turns out to be 5-5.2 Å, which is about 2 Å larger than predicted by the Tao-Eldrup model. For such a large bubble, the relative contact density is $\eta_c \approx 0.9$, Figure 14. It is somewhat higher than the experimental values (0.65–0.75 [13]). This discrepancy may indicate that e^+ and e^- really interact with a medium in a different way, for example, it could be $V_0^+ > V_0^-$. It means that the positron may be really trapped by a cavity, and e^- will be bound to the trapped e^+ by the Coulombic attraction. This scenario may be also considered in the framework of the present approach, but the expression for the trial wave function of this pair must be modified from the "symmetric" form (33) (with respect to the e^+ and e^- coordinates) to the "asymmetric" one:

$$\Psi_{+-}(r_+, r_-) \approx \frac{\exp(-|r_+ - r_-|/2a - r_+/2b)}{8\pi\sqrt{a^3b^3}}, \quad (37)$$

which explicitly takes into account the e^+ localization owing to the factor $\exp(-r_+/2b)$.

Actually, any Ps bubble model reduces the original many-body (multiparticle) problem to a simpler one, that of one or two particles in an external field, which simulates the interaction with the medium. To calculate this field one usually has to rely on some macroscopic approaches, whose validity remains uncertain. For example, in our case we have somehow related the actual arrangement of molecules around the Ps bubble with the jump of dielectric permittivity outside the bubble.

6. Conclusion

We have made an attempt to trace the fate of a positron in molecular liquids, starting from its entering the medium till annihilation with an electron. Considering the ionization slowing down of the projectile positron, one may conclude that the e^+ blob (terminal part of the e^+ track) contains several tens of ion-electron pairs, which are confined in a spherical region with a radius of several nanometers. Ion-electron pairs in the blob are bound together because of their mutual Coulombic attraction. This is why outdiffusion of the blob species obeys ambipolar diffusion law.

Subionizing e^+ can escape from the blob. It becomes thermalized either outside the blob or inside it. Energetic restrictions have been shown to leave only one possibility to form Ps atom. It is the combination of the thermalized positron with one of the intrablob electrons (the recombination mechanism or the intratrack mechanism).

Primary ion-electron pairs transform into secondary radiolytic products (typical concentration $\gtrsim 0.01$ M), which are chemically active and thus can interact with Ps: oxidize it and/or stimulate its ortho-paraconversion. Because the number of ion-electron pairs is large and the diffusion displacement of the species on a nanosecond timescale is comparable with the size of the blob (at least at a high-T diffusion region), it appears inescapable to describe these processes on the basis of inhomogeneous diffusion kinetic equations in terms of concentrations of the species, rate constants, diffusion coefficients, and so forth.

In the case of water it is clearly seen that taking into account interactions with intratrack products (Ps oxidation by OH radicals and H_3O^+) is important. Otherwise, it is not possible to explain the experimentally observed decrease of the ortho-Ps lifetime with temperature.

Calculating the pick-off annihilation rate constant is of the utmost importance. This is the main goal of the Ps bubble models. Up to now most of calculations deal with the "point-like" positronium, by using its center-of-mass wave function only. In the framework of this approximation it is not possible to take into account the variation of the internal (Coulombic) energy of the e^+-e^- pair during Ps bubble formation. However, we have shown that this energy contribution is not small and may play a decisive role.

Finally, we would like to draw attention to one of the possible applications of the positron spectroscopy: fast detection of potentially carcinogenic chemical compounds. These substances are considered as one of the major causes of human cancer. Modern technologies have led to new potential chemical carcinogens with which people may be in contact in everyday life. There is a need for a fast and cheap method of testing such compounds. Positron annihilation spectroscopy may provide such a method [51, 52].

Acknowledgment

This work is supported by the Russian Foundation of Basic Research (Grant 11-03-01066).

References

[1] J. Dryzek and P. Horodek, "The distribution of slowing-down times of positrons emitted from ^{22}Na and ^{68}Ge\$^{68}$Ga isotopes into metals," *Materials Science Forum*, vol. 666, pp. 10–14, 2011.

[2] H. G. Paretzke, "Radiation track structure theory," in *Kinetics of Nonhomogeneous Processes*, G. R. Freeman, Ed., pp. 89–170, John Wiley & Sons, New York, NY, USA, 1987.

[3] V. M. Byakov and S. V. Stepanov, "Common features in the formation of Ps, Mu, radioltic hydrogen and solvated electrons in aqueous solutions," *Journal of Radioanalytical and Nuclear Chemistry*, vol. 210, no. 2, pp. 371–405, 1996.

[4] V. M. Byakov and F. G. Nichiporov, *Intratrack Chemical Processes*, Energoatomizdat, Moscow, Russia, 1985.
[5] S. V. Stepanov, "Energy losses of subexcitation charged particles in polar media," *Radiation Physics and Chemistry*, vol. 46, no. 1, pp. 29–37, 1995.
[6] S. V. Stepanov and V. M. Byakov, "Physical and radiation chemistry of the positron and positronium," in *Principles and Applications of Positron and Positronium Chemistry*, Y. C. Jean, P. E. Mallone, and D. M. Schrader, Eds., chapter 5, World Scientific Publications, 2003.
[7] S. V. Stepanov and V. M. Byakov, "Electric field effect on positronium formation in liquids," *Journal of Chemical Physics*, vol. 116, no. 14, pp. 6178–6195, 2002.
[8] S. V. Stepanov, V. M. Byakov, C. L. Wang, Y. Kobayashi, and K. Hirata, "Electric field effect on Ps formation: black blob model," *Materials Science Forum*, vol. 363–365, pp. 392–394, 2001.
[9] S. V. Stepanov, V. M. Byakov, and Y. Kobayashi, "Ps formation in molecular media: effect of the external electric field," *Physical Review B*, vol. 72, no. 5, Article ID 054205, 2005.
[10] S. V. Stepanov and V. M. Byakov, "Ps formation in molecular media: low temperature hydrocarbons," *Physica Status Solidi (C) Current Topics in Solid State Physics*, vol. 4, no. 10, pp. 3684–3689, 2007.
[11] F. Bockstahl, I. Billard, G. Duplâtre, and A. Bonnenfant, "Positronium formation and quenching in frozen and liquid solutions in octanol," *Chemical Physics*, vol. 236, no. 1–3, pp. 393–403, 1998.
[12] M. Muhieddine, E. Canot, and R. March, "Various approaches for solving problems in heat conduction with phase," *International Research Journal of Finance*, vol. 6, no. 1, pp. 1–20, 2009.
[13] S. V. Stepanov, D. S. Zvezhinski, G. Duplâtre, V. M. Byakov, Y. Y. Batskikh, and P. S. Stepanov, "Incorporation of the magnetic quenching effect into the blob model of Ps formation. Finite sized Ps in a potential well," *Materials Science Forum*, vol. 666, pp. 109–114, 2011.
[14] K. Kotera, T. Saito, and T. Yamanaka, "Measurement of positron lifetime to probe the mixed molecular states of liquid water," *Physics Letters A*, vol. 345, no. 1–3, pp. 184–190, 2005.
[15] G. Duplâtre, A. Haessler, and J. C. Abbé, "Interactions of positronium with Co^{2+} in water: mechanisms and temperature effects," *Journal of Physical Chemistry*, vol. 89, no. 9, pp. 1756–1760, 1985.
[16] J. C. Abbé, G. Duplâtre, A. Haessler, A. Marques Netto, and D. Pilo Veloso, "Temperature and (polar) solvent effects on positronium reactions with nitroxyl free radicals," *Journal of Physical Chemistry*, vol. 88, no. 10, pp. 2071–2076, 1984.
[17] J. Talamoni, J. C. Abbé, G. Duplâtre, and A. Haessler, "Temperature effects on positronium formation and inhibition: a contribution to the elucidation of early spur processes-III. Aqueous solutions," *Radiation Physics and Chemistry*, vol. 20, no. 4, pp. 275–280, 1982.
[18] S. V. Stepanov, G. Duplâtre, V. M. Byakov, V. S. Subrahmanyam, D. S. Zvezhinskii, and A. S. Mishagina, "Influence of temperature on intratrack processes and Ps formation and behaviour in liquid water," *Materials Science Forum*, vol. 607, pp. 213–217, 2009.
[19] C. Bonacina, G. Comini, A. Fasano, and M. Primicerio, "Numerical solution of phase-change problems," *International Journal of Heat and Mass Transfer*, vol. 16, no. 10, pp. 1825–1832, 1973.
[20] R. C. Wilhoit, J. Chao, and K. R. Hall, "Thermodynamic properties of key organic oxygen compounds in the carbon range C_1 to C_4. Part 1. Properties of condensed phases," *Journal of Physical and Chemical Reference Data*, vol. 14, no. 1, pp. 1–175, 1985.
[21] C. Yaws, *Thermophysical Properties of Chemicals and Hydrocarbons*, William Andrew, New York, NY, USA, 2008.
[22] O. A. Korolyuk, "Thermal conductivity of molecular crystals of monoatomic alcohols: from methanol to butanol," *Fizika Nizkikh Temperatur*, vol. 37, no. 5, pp. 526–530, 2011.
[23] C. Yaws, *Lange's Handbook of Chemistry*, McGraw-Hill, 15th edition, 1999.
[24] D. R. Lide, *Handbook of Chemistry and Physics*, CRC Press LLC, 84th edition, 2004.
[25] O. E. Mogensen, *Positron Annihilation in Chemistry*, Springer, Berlin, Germany, 1995.
[26] T. Tabata, Y. Ito, and S. Tagawa, *Handbook of Radiation Chemistry*, CRC Press, Boca Raton, Fla, USA, 1991.
[27] Y. Ito, "Radiation chemistry: intraspur effects and positronium formation mechanism," in *Positron and Positronium Chemistry*, D. M. Schrader and Y. C. Jean, Eds., Studies in Physical and Theoretical Chemistry Series, pp. 120–158, Elsevier Scientific Publishers, 1988.
[28] O. E. Mogensen, "Spur reaction model of positronium formation," *Journal of Chemical Physics*, vol. 60, no. 3, pp. 998–1004, 1974.
[29] V. M. Byakov, V. I. Goldanskii, and V. P. Shantarovich, "About the possible role of "dry" electrons in positronium formation in a liquid," *Doklady Physical Chemistry*, vol. 219, no. 3, pp. 1090–1093, 1974, Doklady Akademii Nauk, SSSR, V. 219(3), pp. 633–636, 1974.
[30] V. M. Byakov, "The nature of the precursors of radiolytic molecular hydrogen in water, and the mechanism of positronium formation in liquids," *International Journal for Radiation Physics and Chemistry*, vol. 8, no. 3, pp. 283–288, 1976.
[31] K. V. Mikhin, S. V. Stepanov, and V. M. Byakov, "Dynamics of positronium bubble growth in liquid media and the energy dissipation problem," *High Energy Chemistry*, vol. 39, no. 1, pp. 36–43, 2005.
[32] S. V. Stepanov, D. S. Zvezhinskii, G. Duplâtre, V. M. Byakov, and V. S. Subrahmanyam, "A novel approach to the interpretation of PAL spectra in glycerol," *Materials Science Forum*, vol. 607, pp. 260–262, 2009.
[33] S. V. Stepanov, V. M. Byakov, G. Duplâtre, D. S. Zvezhinskiy, and Y. V. Lomachuk, "Positronium formation in a liquid phase: influence of intratrack reactions and temperature," *Physica Status Solidi (C) Current Topics in Solid State Physics*, vol. 6, no. 11, pp. 2476–2481, 2009.
[34] A. I. Burshtein, *Molecular Physics*, Nauka, Novosibirsk, Russia, 1986.
[35] S. K. Sharma, A. Zaydouri, G. Roudaut, and G. Duplâtre, "Effect of water on glass transition in starch/sucrose matrices investigated through positron annihilation lifetime spectroscopy: a new approach," *Physical Chemistry Chemical Physics*, vol. 13, pp. 19338–19344, 2011.
[36] R. A. Ferrell, "Long lifetime of positronium in liquid helium," *Physical Review*, vol. 108, no. 2, pp. 167–168, 1957.
[37] S. J. Tao, "Positronium annihilation in molecular substances," *Journal of Chemical Physics*, vol. 56, no. 11, pp. 5499–5510, 1972.
[38] M. Eldrup, D. Lightbody, and J. N. Sherwood, "The temperature dependence of positron lifetimes in solid pivalic acid," *Chemical Physics*, vol. 63, no. 1-2, pp. 51–58, 1981.
[39] S. V. Stepanov and V. M. Byakov, "What do positrons say about the structure and properties of fluids?" *Journal of Structural Chemistry*, vol. 43, no. 6, pp. 949–971, 2002.

[40] A. T. Stewart and C. V. Briscoe, "Positron Annihilation," in *Proceedings of the Conference International Symposium on Positron Annihilation Studies of Fluids*, A. T. Stewart and L. O. Roellig, Eds., p. 383, Wayne State UniversityAcademic Press, New York, NY, USA, 1959.

[41] A. P. Buchikhin, V. I. Gol'danskii, A. O. Tatur, and V. P. Shantarovich, "The positronium atom in organic liquids," *Journal of Experimental and Theoretical Physics*, vol. 33, no. 3, article 615, 1971, Zh. Eksp. Teor. Fiz., Vol. 60(3), p. 1136, 1971.

[42] L. O. Roellig, "Positron annihilation in liquids and condensed gases," in *Proceedings of Wayne State University Conference on Positron AnnihilationAcademic Press*, A. T. Stewart and L. O. Roellig, Eds., p. 127, New York, NY, USA, 1967.

[43] C. Dauwe, N. Balcaen, S. Van Petegem, and D. Segers, "Trapping of positronium in a size-dependent spherical square well potential," *Radiation Physics and Chemistry*, vol. 58, no. 5-6, pp. 681–685, 2000.

[44] D. Dutta, B. N. Ganguly, D. Gangopadhyay, T. Mukherjee, and B. Dutta-Roy, "Corrections to the prevalent bubble model of positronium annihilation in liquids," *Physical Review B*, vol. 65, no. 9, Article ID 094114, 2002.

[45] D. Dutta, B. N. Ganguly, D. Gangopadhyay, T. Mukherjee, and B. Dutta-Roy, "General trends of positronium pick-off annihilation in molecular substances," *Journal of Physics Condensed Matter*, vol. 14, no. 32, pp. 7539–7549, 2002.

[46] B. N. Miller and T. Reese, "Path integral simulation of positronium," *Nuclear Instruments and Methods in Physics Research B*, vol. 192, no. 1-2, pp. 176–179, 2002.

[47] L. Larrimore, R. N. McFarland, P. A. Sterne, and A. L. R. Bug, "Two-chain path integral model of positronium," *Journal of Chemical Physics*, vol. 113, no. 23, pp. 10642–10650, 2000.

[48] A. L. R. Bug, M. Muluneh, J. Waldman, and P. A. Sterne, "Positronium in solids: computer simulation of pick-off and self-annihilation," *Materials Science Forum*, vol. 445-446, pp. 375–379, 2004.

[49] T. Fülöp, Z. Farkas, A. Seeger, and J. Major, "On the inner structure of confined positronium," http://arxiv.org/abs/cond-mat/0304442v1.

[50] V. V. Batygin and I. N. Toptygin, "Problems on electrodynamics," Tech. Rep. 157, NIC RKhD, Moscow, Russia, 2002.

[51] V. M. Byakov, S. V. Stepanov, and O. P. Stepanova, "Positron method for diagnostics of carcinogens," *Materials Science Forum*, vol. 607, pp. 223–226, 2009.

[52] V. M. Byakov, S. V. Stepanov, and O. P. Stepanova, "PAL spectroscopy and testing for potential carcinogens," *Physica Status Solidi (C) Current Topics in Solid State Physics*, vol. 6, no. 11, pp. 2503–2506, 2009.

18

Quantum Instanton Evaluations of the Thermal Rate Constants for Complex Systems

Yi Zhao[1] and Wenji Wang[2]

[1] *State Key Laboratory for Physical Chemistry of Solid Surfaces and Fujian Provincial Key Lab of Theoretical and Computational Chemistry, College of Chemistry and Chemical Engineering, Xiamen University, Xiamen 361005, China*
[2] *College of Science, Northwest A&F University, Yangling 712100, Shaanxi Province, China*

Correspondence should be addressed to Yi Zhao, yizhao@xmu.edu.cn and Wenji Wang, wjwang@nwsuaf.edu.cn

Academic Editor: Laimutis Bytautas

Quantum instanton (QI) approximation is recently proposed for the evaluations of the chemical reaction rate constants with use of full dimensional potential energy surfaces. Its strategy is to use the instanton mechanism and to approximate time-dependent quantum dynamics to the imaginary time propagation of the quantities of partition function. It thus incorporates the properties of the instanton idea and the quantum effect of partition function and can be applied to chemical reactions of complex systems. In this paper, we present the QI approach and its applications to several complex systems mainly done by us. The concrete systems include, (1) the reaction of $H + CH_4 \rightarrow H_2 + CH_3$, (2) the reaction of $H + SiH_4 \rightarrow H_2 + SiH_3$, (3) H diffusion on Ni(100) surface; and (4) surface-subsurface transport and interior migration for H/Ni. Available experimental and other theoretical data are also presented for the purpose of comparison.

1. Introduction

The accurate and efficient evaluation of chemical reaction rate constant is one of prime objectives of theoretical reaction dynamics. Since rigorous quantum mechanical approaches are limited to small molecular (several atoms) reactions, a variety of approximation approaches have been proposed. Benefited from the small recrossing dynamics at not-too-high temperatures, the transition state theories (TSTs), originally proposed by Eyring [1, 2] and Wigner [3], have become a possible and popular way to estimate rate constants. Due to their practical simplicity, they have been broadly applied to numerous reactions. The TST is inherently a classical theory and suitable at sufficiently high temperatures, where the classical description of nuclear motions may be adequate. At low temperatures, especially for the reactions involving the motions of light atoms (i.e., hydrogen), however, quantum effects become quite significant. To make the TST still valid for such low temperature reactions, many approaches have been proposed to quantize it [4–14]. However, there is no absolutely unambiguous way to do it.

To develop a more accurate and less *ad hoc* quantum version of TST, with a specific focus on the tunneling regime, Miller et al. [15–17] have proposed a quantum instanton (QI) approach recently. The QI is based on an earlier semiclassical (SC) TST [18] that became known as the instanton [19, 20]. The similarity between the QI and SC instanton lies in using the steepest descent approximation to evaluate relevant integrals in the quantum rate formula, while the crucial difference is that the Boltzmann operator is evaluated by the quantum mechanics and semiclassical approximation, respectively. The QI theory thus incorporates the tunneling, corner cutting [21–24], and anharmonicity correctly and is expected to overcome the quantitative deficiency of the SC instanton model. In particular, the QI theory considers all tunneling paths and automatically gives each path its naturally weight factor from the quantum Boltzmann operator, instead of choosing a single optimal tunneling path, which is taken into account in the SC instanton and TSTs with SC tunneling corrections. Indeed, it has been numerically demonstrated that the QI predicts accurate quantum rates for one-dimensional and two-dimensional models within 20% error over a wide temperature range, from the deep tunneling to overbarrier regimes.

A lot of developments and applications [25–39] have been made since the QI theory is proposed. The original QI [15] involves the second time derivative of the flux-flux correlation at time zero. It has been further improved [29] by taking into account the higher derivatives of the flux-flux function. For the 1D and collinear reactions, the improved model is considerable accuracy, giving the rates to within 5%–10% errors. For a practical purpose, a simple and general way for choosing dividing surfaces used in the QI is suggested [31], namely, using the family of (hyper)planes normal to the minimum energy path at various distances s. A "simplest" QI model [26] has also been suggested with one dividing surface, slightly less accurate than the original QI. To reveal the relationship with conventional TSTs, the classical limit of the QI has been derived [37]. It is found that the classical TST is just a special case of the QI in high temperature limit; moreover, the quantum correction of the prefactor is more important than that of the activation energy in the TST.

Since the QI solely involves the Boltzmann operator and its relevant quantities, it can be applied to quite complex molecular systems (from gas phase [17, 37], liquid [28], to surface [38, 39]) via well-established imaginary time path integral techniques. The first implementation of QI with path integral Monte Carlo and adaptive umbrella sampling techniques is applied to the three-dimensional hydrogen exchange reaction $D + H_2 \rightarrow HD + H$ [16]. Soon, the techniques are further extended to the reaction of $H+CH_4 \rightarrow H_2+CH_3$ [17]. The thermodynamic integration with respect to the mass of the isotope and the inverse temperature is also proposed to compute the kinetic isotope effects [30] and rate constants [34] directly. To improve the convergence of the Monte Carlo simulation, the efficient "virial" estimators [32] have been derived from the logarithmic derivatives of the partition function and the delta-delta correlation functions, and it is found that their statistical errors do not increase with the number of discrete time slices in the path integral. Most recently, the QI has been compared with other conventional approaches [36] for an intramolecular proton transfer on a full-dimensional potential energy surface that incorporates high-levels *ab initio* calculations along the reaction path. The obtained kinetic isotope effects from the QI are in reasonable agreement with those from the path-integral quantum TST.

In this paper, we firstly illustrate the QI formula and its path integral representation. Then, we display several applications, which are mainly done by ourselves. The systems include two gas phase reactions $H + CH_4 \rightarrow H_2 + CH_3$ [17] and $H + SiH_4 \rightarrow H_2 + SiH_3$ [37], H diffusion on Ni(100) surface [38], and surface-subsurface transport and interior migration for H/Ni [39].

2. Method

In this section, we summarize the rate formula for the QI evaluation. The detailed derivation can be found in [15–17]. The QI model proposes the following thermal rate constant:

$$k_{\mathrm{QI}} = \frac{1}{Q_r}\frac{\sqrt{\pi}\hbar}{2}\frac{C_{ff}(0)}{\Delta H(\beta)}. \quad (1)$$

Here, Q_r is the reactant partition function per unit volume. $C_{ff}(0)$ is zero time value of the flux-flux correlation function

$$C_{ff}(t) = \mathrm{tr}\left[e^{-\beta\hat{H}/2}\hat{F}_a e^{-\beta\hat{H}/2} e^{i\hat{H}t/\hbar}\hat{F}_b e^{-i\hat{H}t/\hbar}\right], \quad (2)$$

where β is the inverse temperature $1/(k_B T)$, $\hat{H}$ is the Hamiltonian operator of the reaction system, and $\hat{F}_a$ and $\hat{F}_b$ are the flux operators given by

$$\hat{F}_\gamma = \frac{i}{\hbar}\left[\hat{H}, h\left(s_\gamma(\hat{r})\right)\right], \quad (3)$$

with $\gamma = a, b$. In (3), h is the step-side function, r represents the Cartesian coordinates of the reaction system, and $s_a(r)$ and $s_b(r)$ define two separate dividing surfaces via the equations $s_a(r) = 0$ and $s_b(r) = 0$, both $s_a(r)$ and $s_b(r)$ being positive (negative) on the product (reactant) side of the dividing surfaces.

$\Delta H(\beta)$ in (1) is a specific type of energy variance given by

$$\Delta H(\beta) = \hbar\sqrt{\frac{-\ddot{C}_{dd}(0)}{2C_{dd}(0)}}. \quad (4)$$

In order to get the correct free particle (high temperature) limit (that would be 25% too large otherwise), an ad hoc term is added to $\Delta H(\beta), \Delta H_{\mathrm{mod}}(\beta) = \Delta H(\beta) + (\sqrt{\pi} - \sqrt{2})/\beta$, which has very little effect in the low temperature regime. $C_{dd}(0)$ and $\ddot{C}_{dd}(0)$ are zero time value and its second time derivative, respectively, of the "delta-delta" correlation function

$$C_{dd}(t) = \mathrm{tr}\left[e^{-\beta\hat{H}/2}\Delta(s_a(\hat{r}))e^{-\beta\hat{H}/2}e^{i\hat{H}t/\hbar}\Delta(s_b(\hat{r}))e^{-i\hat{H}t/\hbar}\right], \quad (5)$$

where the generalized delta-function operator is

$$\Delta\left(s_\gamma(\hat{r})\right) = \delta\left(s_\gamma(\hat{r})\right)\sqrt{\sum_{i=1}^{N}\frac{1}{m_i}\left(\nabla_i s_\gamma(\hat{r})\right)^2} \quad (\gamma = a, b). \quad (6)$$

Here, N is the total number of atoms, $\nabla_i = \partial/\partial r_i$, r_i denotes the Cartesian coordinates of the ith atom, and m_i is its atomic mass.

The dividing surfaces are determined by the stationary condition

$$\frac{\partial}{\partial c_k}C_{dd}(0; \{c_k\}) = 0, \quad (7)$$

where $\{c_k\}$ is a collection of parameters that is involved in the location of the dividing surfaces. This condition originates from the SC instanton model, and the resulting dividing surfaces correspond qualitatively to the turning points of the periodic orbit that runs on an upside down PES in imaginary time (see Appendix A in [15, 18]).

Since all the relevant quantities in the QI expression (1) involved only the quantum Boltzmann operator, they can be readily evaluated using imaginary time path integral Monte Carlo (PIMC) [40] method.

We begin with the simplest quantity, $C_{dd}(0)$, which can be discretized according to the standard procedure [41–43] to give

$$C_{dd}(0) = C\int dr^{(1)}\int dr^{(2)}\cdots\int dr^{(P)}\Delta\left(s_a\left(r^{(0)}\right)\right) \times \Delta\left(s_b\left(r^{(P/2)}\right)\right)\exp\left[-\beta\Phi\left(\left\{r^{(s)}\right\}\right)\right], \tag{8}$$

where C is a multiplicative constant, P the number of imaginary time slices, and $r^{(s)} = (r_1^{(s)}, r_2^{(s)}, \ldots, r_N^{(s)})$ the Cartesian coordinates of the system associated with the sth time slice. $\Phi(\{r^{(s)}\})$ is the discretized action given by

$$\Phi\left(\left\{r^{(s)}\right\}\right) = \frac{P}{2\hbar^2\beta^2}\sum_{s=1}^{P}\sum_{i=1}^{N} m_i\left(r_i^{(s)} - r_i^{(s-1)}\right)^2 + \frac{1}{P}\sum_{s=1}^{P} V\left(r^{(s)}\right), \tag{9}$$

where $r^{(0)} = r^{(P)}$ and $\{r^{(s)}\}$ represents $\{r^{(1)}, r^{(2)}, \ldots, r^{(P)}\}$. Path integral expressions for $C_{ff}(0)$ and $\ddot{C}_{dd}(0)$ are somewhat more complicated but can be obtained in a straightforward manner. The appropriate expressions are

$$C_{ff}(0) = C\int dr^{(1)}\int dr^{(2)}\cdots\int dr^{(P)}\Delta\left(s_a\left(r^{(0)}\right)\right) \times \Delta\left(s_b\left(r^{(P/2)}\right)\right)\exp\left[-\beta\Phi\left(\left\{r^{(s)}\right\}\right)\right] f_v\left(\left\{r^{(s)}\right\}\right), \tag{10}$$

with

$$f_v\left(\left\{r^{(s)}\right\}\right) = \left(\frac{iP}{2\hbar\beta}\right)^2 \frac{\sum_{i=1}^{N}\nabla_i s_a\left(r^{(0)}\right)\cdot\left(r_i^{(1)} - r_i^{(P-1)}\right)}{\sqrt{\sum_{i=1}^{N} m_i^{-1}(\nabla_i s_a(r^{(0)}))^2}} \times \frac{\sum_{i=1}^{N}\nabla_i s_b\left(r^{(P/2)}\right)\cdot\left(r_i^{(P/2+1)} - r_i^{(P/2-1)}\right)}{\sqrt{\sum_{i=1}^{N} m_i^{-1}(\nabla_i s_b(r^{(P/2)}))^2}},$$

$$\ddot{C}_{dd}(0) = -\frac{1}{\hbar^2}C\int dr^{(1)}\int dr^{(2)}\cdots\int dr^{(P)}\Delta\left(s_a\left(r^{(0)}\right)\right) \times \Delta\left(s_b\left(r^{(P/2)}\right)\right)\exp\left[-\beta\Phi\left(\left\{r^{(s)}\right\}\right)\right] \times \left\{F\left(\left\{r^{(s)}\right\}\right)^2 + G\left(\left\{r^{(s)}\right\}\right)\right\}, \tag{11}$$

with

$$F\left(\left\{r^{(s)}\right\}\right) = -\frac{P}{\hbar^2\beta^2}\left\{\sum_{s=1}^{P/2} - \sum_{s=P/2+1}^{P}\right\}\sum_{i=1}^{N} m_i\left(r_i^{(s)} - r_i^{(s-1)}\right)^2 + \frac{2}{P}\left\{\sum_{s=1}^{P/2-1} - \sum_{s=P/2+1}^{P-1}\right\} V\left(r^{(s)}\right),$$

$$G\left(\left\{r^{(s)}\right\}\right) = \frac{2fP}{\beta^2} - \frac{4P}{\hbar^2\beta^3}\sum_{s=1}^{P}\sum_{i=1}^{N} m_i\left(r_i^{(s)} - r_i^{(s-1)}\right)^2, \tag{12}$$

with f being the total number of degrees of freedom (i.e., $f = 3N$).

In realistic calculations, we rewrite (1) as the product of several ratios

$$k_{\rm QI} = \frac{\sqrt{\pi}\hbar}{2}\frac{C_{dd}(0)}{Q_r}\frac{C_{ff}(0)}{C_{dd}(0)}\frac{1}{\Delta H(\beta)}. \tag{13}$$

The terms of $C_{ff}(0)/C_{dd}(0)$ and ΔH are directly calculated as a constrained average over the same ensemble of paths [16, 17]

$$\frac{C_{ff}(0)}{C_{dd}(0)} = \langle f_v\rangle, \qquad \Delta H^2 = \frac{1}{2}\langle F^2 + G\rangle, \tag{14}$$

with

$$\langle\cdots\rangle = \frac{\int dr^{(1)}\int dr^{(2)}\cdots\int dr^{(P)}\Delta\left(s_a\left(r^{(0)}\right)\right)\Delta\left(s_b\left(r^{(P/2)}\right)\right)\exp\left[-\beta\Phi\left(\left\{r^{(s)}\right\}\right)\right](\cdots)}{\int dr^{(1)}\int dr^{(2)}\cdots\int dr^{(P)}\Delta(s_a(r^{(0)}))\Delta(s_b(r^{(P/2)}))\exp[-\beta\Phi(\{r^{(s)}\})]}. \tag{15}$$

The evaluation of $C_{dd}(0)/Q_r$, however, meets a challenge because the $C_{dd}(0)$ is the quantity associated with the transition state, while Q_r with the asymptotic reactant domain, we evaluate it using adaptive umbrella sampling techniques [44].

The QI approximation uses the short-time information of the flux-flux correlation function. Predescu and Miller [45] demonstrate that in the classical limit, Wigner's variational principle and the quantum variational criterion based on the minimization of flux-flux correlation function produce the same optimal surface. Recently, Wang et al. [37] have shown that in the classical limit, the QI formula is the same as the classical transition state theory. These situations motivate us to write the QI formula (13) as the Arrhenius form

$$k_{\rm QI} = A_{\rm QI}\cdot e^{-\beta\Delta F}. \tag{16}$$

Here, the free energy ΔF is defined by

$$\Delta F = \frac{-1}{\beta} \cdot \log\left(\frac{C_{dd}(0)}{Q_r}\right). \tag{17}$$

This two-dimensional free energy is related to Q_r (the reactant partition function) and $C_{dd}(0;\xi_a,\xi_b)$, and it is corresponding to the quality of probability density at (ξ_a,ξ_b). $C_{dd}(0;\xi_a,\xi_b)$ has a similar property to the partition function at the transition state. The prefactor is given by

$$A_{\mathrm{QI}} = \frac{\sqrt{\pi}}{2}\frac{C_{ff}(0)}{C_{dd}(0)}\frac{\hbar}{\Delta H}. \tag{18}$$

The advantage of (16) allows us to investigate the respective quantum contributions to the rates from the quantum free energy and prefactor by comparing their quantum and classical values, since other factors such as the vibrational-rotational coupling and anharmonicity are automatically involved.

3. Results and Discussions

3.1. The Reaction of $H + CH_4 \rightarrow H_2 + CH_3$. The $H + CH_4$ reaction is a prototype of polyatomic hydrogen abstraction reaction. Quantum dynamical studies of this reaction have become possible only recently, because it involves 12 internal degrees of freedom and thus poses difficulties to quantum dynamics calculations as well as construction of the potential surface. We apply the QI methodology to this reaction using the potential energy surface constructed by Espinosa-García [46]. All calculations are performed in terms of the Cartesian coordinates of all the atoms (i.e., 18 degrees of freedom).

In the path integral simulations, the number of imaginary time slices P is chosen to be 20 and 100 at temperatures $T = 1000$ K and 200 K, respectively, while 3×10^7 Monte Carlo cycles are run to achieve <10% statistical convergence.

3.1.1. Reaction Coordinate. For this reaction, we define a generalized reaction coordinate $s(r;\xi)$, where ξ is an adjustable parameter that shifts the location of the dividing surface (defined by $s(r;\xi) = 0$). $s(r;\xi)$ is defined by a linear interpolation between two constituent reaction coordinates $s_0(r)$ and $s_1(r)$ through the parameter ξ,

$$s(r;\xi) = \xi \cdot s_1(r) + (1-\xi) \cdot s_0(r). \tag{19}$$

$s_1(r)$ is a reaction coordinate whose dividing surface is designed to pass through the top of the classical potential barrier, which is defined here as

$$s_1(r) = \max\left\{s_\alpha(r), s_\beta(r), s_\gamma(r), s_\delta(r)\right\}, \tag{20}$$

with $s_x(r)(x = \alpha,\beta,\gamma,\delta)$ being the reaction coordinate that describes the abstraction process of one of the methane hydrogens H_x by the incident one H

$$s_x(r) = r(\mathrm{C}-\mathrm{H}_x) - r(\mathrm{H}_x-\mathrm{H}) - \left[r^\dagger(\mathrm{C}-\mathrm{H}_x) - r^\dagger(\mathrm{H}_x-\mathrm{H})\right], \tag{21}$$

where $r(X-Y)$ denotes the interatomic distance between atoms X and Y and $r^\dagger(X-Y)$ is the value at the transition state geometry. $s_0(r)$, on the other hand, describes a dividing surface that is located far in the asymptotic reactant valley, which is given by

$$s_0(r) = R_\infty - |R|. \tag{22}$$

Here, R is the scattering vector that connects the incident hydrogen and the center of mass of the methane. R_∞ is an adjustable parameter which is chosen to be 9 Å in order to guarantee that the interaction potential energy between H and CH_4 is negligible.

Now, the term $C_{dd}(0)$ (5) becomes a function of two parameters, ξ_a and ξ_b, as follows:

$$C_{dd}(0;\xi_a,\xi_b) = \mathrm{tr}\left[e^{-\beta\hat{H}/2}\Delta(s(\hat{r};\xi_a)) \times e^{-\beta\hat{H}/2}\Delta(s(\hat{r};\xi_b))\right], \tag{23}$$

and thus, one seeks a stationary point of $C_{dd}(0;\xi_a,\xi_b)$ in the two-dimensional (ξ_a,ξ_b) space to obtain the corresponding "optimal" values.

3.1.2. Free Energy Surface. The quantity $C_{dd}(0;\xi_a,\xi_b)$ varies exponentially as a function of (ξ_a,ξ_b), it is convenient to define a quantum "free energy surface" as follows:

$$F(\xi_a,\xi_b) = -k_B T \log C_{dd}(0;\xi_a,\xi_b), \tag{24}$$

and locate the saddle point of $F(\xi_a,\xi_b)$ by visual inspection.

Figure 1 shows the free energy surfaces for the $H + CH_4$ reaction. It exhibits a barrier-like profile along the direction $\overline{\xi} = (\xi_a + \xi_b)/2$, while it grows approximately quadratically with the increasing of the absolute value of $\Delta\xi = \xi_a - \xi_b$. From this figure, it is seen that at a higher temperature $T = 1000$ K (Figure 1(a)), there appears only a single saddle point at $(\overline{\xi},\Delta\xi) = (1.02, 0.0)$, while at a low temperature $T = 200$ K (Figure 1(b)), the saddle point bifurcates into a distinct pair at $(\overline{\xi},\Delta\xi) = (1.1, \pm 0.35)$, which indicates the existence of nonnegligible tunneling effects in the rate constant [15, 16].

3.1.3. Rate Constants. Having obtained the "optimal" values of the (ξ_a,ξ_b) at each temperature, one can now compute the quantum instanton rate by combining various quantities as in (8). The calculated QI rates are tabulated in Table 1 as well as other theoretical and experimental ones.

Comparing the quantum instanton rates with others, we find that k_{QI} is in good agreement with the experimental data more specifically, it is closer to the rates obtained by Baulch et al. [47] than those by Sutherland et al. [48], and k_{QI} agrees with $k_{\mathrm{CVT}/\mu\mathrm{OMT}}$ within 10% for the temperature range $T = 600$–2000 K, but it becomes somewhat larger than the latter as the temperature is decreased (the deviation becomes 30% and 55% for $T = 500$ and 300 K, resp.). It should be noted that the differences between k_{QI} and $k_{\mathrm{CVT}/\mu\mathrm{OMT}}$ are much smaller than the uncertainty of the experimental data.

Figure 2 displays the Arrhenius plots of the rate constants. In Figure 2, we also plot the accurate quantum dynamics results of MCTDH (the multiconfigurational

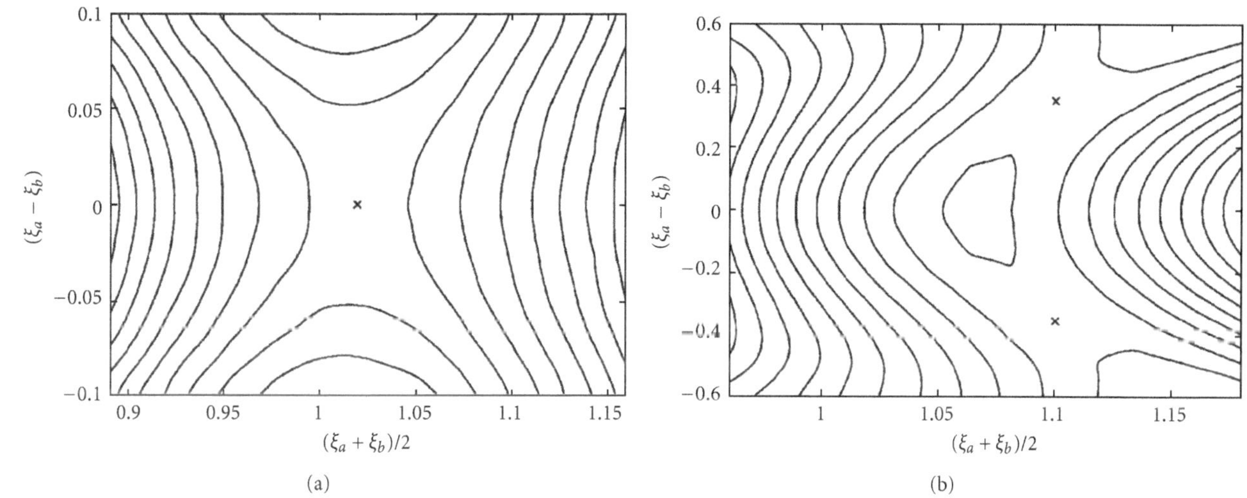

FIGURE 1: Local topography of the quantum free energy surface defined by (24) near the top of the barrier. (a) $T = 1000$ K; (b) $T = 200$ K. The cross symbols show the location of the saddle points. The values of ξ_a and ξ_b at the saddle points are used as input for computing the quantum instanton rate.

TABLE 1: Theoretical and experimental rate constants obtained for the H + CH_4 → H_2 + CH_3 reaction. (ξ_a, ξ_b) is the saddle point of the quantum free energy surface. k_{QI} is the quantum instanton rate constant. $k_{CVT/\mu OMT}$ is the rate constant of the canonical variational theory with microcanonical optimized multidimensional tunneling (CVT/μOMT) [46]. k^a_{expt} and k^b_{expt} are the experimental Arrhenius fits, $k(T) = 2.18 \times 10^{-20} T^{3.0} \exp(-4045/T)$[47] and $k(T) = 6.78 \times 10^{-21} T^{3.156} \exp(-4406/T)$ [48], respectively. Unit: $cm^3 s^{-1}$ for rates.

T (K)	(ξ_a, ξ_b)	k_{QI}	$k_{CVT/\mu OMT}$	k^a_{expt}	k^b_{expt}
200	(0.93, 1.28)	3.30 (−22)		2.87 (−22)	3.36 (−23)
300	(1.05, 1.05)	7.80 (−19)	5.03 (−19)	8.20 (−19)	1.86 (−19)
400	(1.05, 1.05)	6.00 (−17)	4.16 (−17)	5.66 (−17)	1.82 (−17)
500	(1.04, 1.04)	9.00 (−16)	6.96 (−16)	8.35 (−16)	3.33 (−16)
600	(1.03, 1.03)	5.42 (−15)	5.00 (−15)	5.56 (−15)	2.57 (−15)
700	(1.02, 1.02)	2.40 (−14)		2.31 (−14)	1.19 (−14)
800	(1.02, 1.02)	7.13 (−14)	6.72 (−14)	7.10 (−14)	4.00 (−14)
900	(1.02, 1.02)	1.85 (−13)		1.78 (−13)	1.07 (−13)
1000	(1.02, 1.02)	3.97 (−13)	3.58 (−13)	3.82 (−13)	2.43 (−13)
2000	(1.00, 1.00)	1.52 (−11)	1.70 (−11)	2.31 (−11)	1.96 (−11)

time-dependent Hartree approach) [49]. Compared to the MCTDH ones, our QI rate constants are larger by factors of about 2 to 3 over the temperature range 300–400 K. This difference may partly be due to the recrossing effect which is not considered in QI theory and partly arise from the use of the J-shifting approximation and the neglect of the vibrational angular momenta Hamiltonian in MCTDH method [49].

3.2. The Reaction of H + SiH_4 → H_2 + SiH_3. The H + SiH_4 reaction is an important step in the radical mechanism of thermal decomposition of monosilane. We calculate the rates and kinetic isotope effects (KIEs) of this reaction with the quantum instanton method in full Cartesian space, on the basis of analytical potential energy surface constructed by Espinosa-García et al. [50].

The reaction coordinate of H + SiH_4 has the same form as that of H + CH_4 (Section 3.1.1). In our QI calculations, the number of time slices, P, is set to be 20 (1000 K) and 120 (200 K) for the quantum evaluations. In our classical evaluations, the formula is the same as QI, but the number of time slices in the path integral is set to be 1. The number of Monte Carlo is about $(6 - 10) \times 10^6$ for computing a single ensemble average, and it converges the relevant quantities within 10% statistical errors.

3.2.1. Free Energy and Prefactor. We have rewritten the QI formula (8) in the Arrhenius form (16), which consists of the free energy and prefactor. In this section, we calculate the corresponding quantum and classical quantities so as to investigate the respective quantum contributions to the rates from the quantum free energy and prefactor. Our calculated results are plotted in Figures 3 and 4.

Now, we look into the quantum effects from both the free energy and prefactor. Figure 3 plots the quantum and classical free energies as a function of temperature

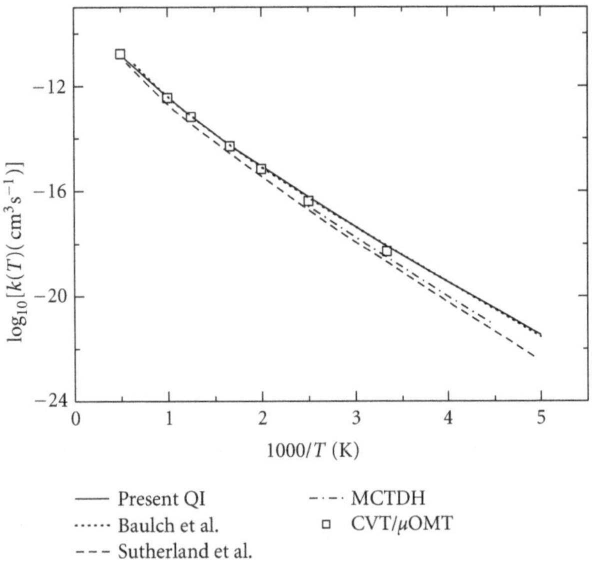

FIGURE 2: Arrhenius plots of the thermal rate constant for the $H + CH_4 \rightarrow H_2 + CH_3$ reaction. The solid line is the quantum instanton rate. Dotted and dashed lines are the Arrhenius fits of the experimental data from [47, 48], respectively. The dot-dashed line is the result of the multiconfigurational time-dependent Hartree approach (MCTDH) [49]. The open squares are the results of the canonical variational theory with microcanonical optimized multidimensional tunneling (CVT/μOMT) [46].

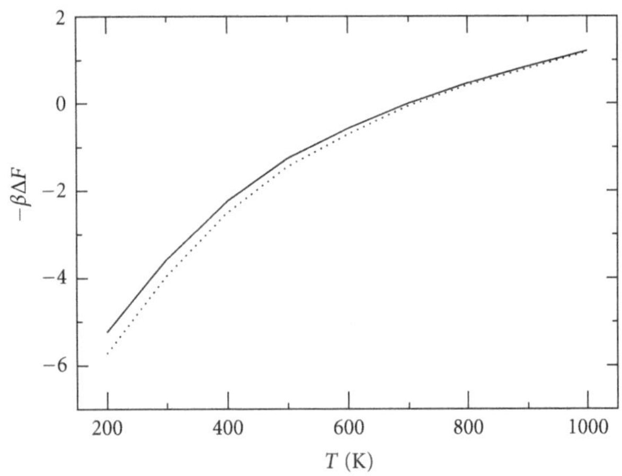

FIGURE 3: The temperature dependence of the free energy. The solid and dotted lines correspond to the quantum and classical calculations, respectively.

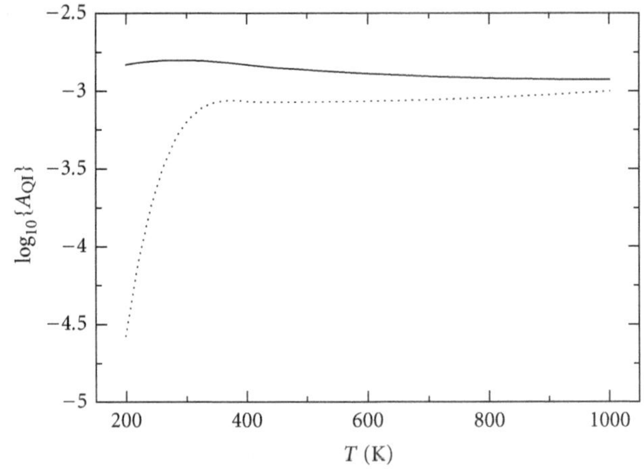

FIGURE 4: The temperature dependence of the prefactor. The solid and dotted lines correspond to the quantum and classical calculations, respectively.

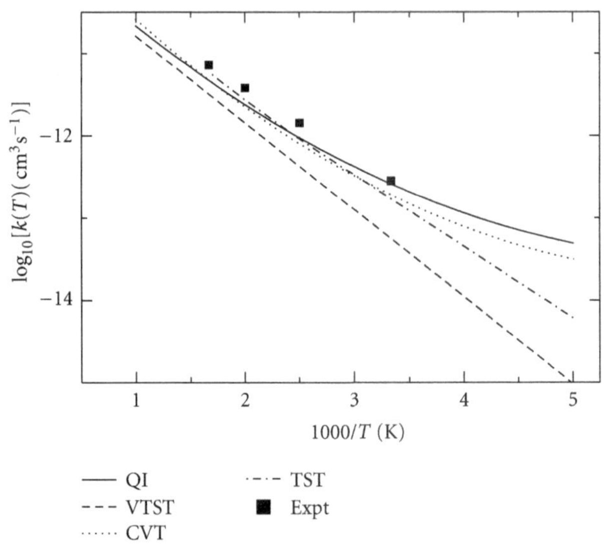

FIGURE 5: Arrhenius plot of the thermal rate constants for the $H+SiH_4$ reaction. Solid line, the QI results; dashed line, the classical VTST results; dotted line, CVT/CD-SCSAG results from [50]; dot-dashed, the values from the conventional TST with a simple Wigner tunneling factor [51]; filled squares are the experimental values from [52].

at the optimized stationary point. One immediately observes that the quantum effect becomes significant at $T < 600$ K, whereas both quantum and classical results nearly coincide at $T > 600$ K. As expected, the quantum effect always decreases the classical free energy. The difference of the contributions to the rates is about 40% at 200 K.

Figure 4 shows the temperature dependence of the quantum and classical prefactors. Compared with Figure 3, one easily finds that this quantum contribution is much larger than that for the free energy correction. Even at 1000 K, the difference is observable. At 200 K, the difference is about several orders of magnitude.

In the QI theory, it is not possible to explicitly distinguish the quantum contributions from the partition function and nuclear tunneling, but we can conclude that it is insufficient to estimate the accurate rate by only replacing the activation energy in the TST with its quantum analog, because the quantum prefactor plays more important rule in determining the quantum rate.

3.2.2. Rate Constants. In Table 2, we summarize the rate constants obtained from the QI theory and classical VTST (this kind of results are performed by setting the time slices

TABLE 2: Rate constants for the $H + SiH_4$ reaction (cm^3s^{-1}). The CVT/CD-SCSAG results from [50]. DD is the conventional TST with simple Wigner tunneling factor [51]. Expt. is the experimental Arrhenius fits, $k(T) = 1.78 \times 10^{-10} \exp(-3820/RT)$ (T = 290–660 K) [52].

T (K)	Present QI	VTST	CVT/CD-SCSAG	DD	Expt.
200	4.82 (−14)	8.88 (−16)	3.07 (−14)	6.1 (−15)	
300	2.74 (−13)	1.38 (−14)	2.09 (−13)	1.7 (−13)	2.81 (−13)
400	9.67 (−13)	4.32 (−13)	7.96 (−13)	9.1 (−13)	1.43 (−12)
500	2.37 (−12)	1.23 (−12)	2.11 (−12)	2.7 (−12)	3.78 (−12)
600	3.99 (−12)	2.55 (−12)	4.43 (−12)	5.9 (−12)	7.18 (−12)
700	7.60 (−12)	5.18 (−12)	8.00 (−12)		
800	1.18 (−11)	8.63 (−12)			
900	1.76 (−11)	1.29 (−11)			
1000	2.39 (−11)	1.99 (−11)	2.80 (−11)		

in the path integral to be 1 in the QI calculations) as well as those from the canonical variational TST with the centrifugal-dominant small curvature SC adiabatic ground-state (CVT/CD-SCSAG) approach [50], the conventional TST with simple Wigner tunneling factor [51] and the experiment [52]. The corresponding Arrhenius plots are displayed in Figure 5.

Table 2 and Figure 5 display comparable results of the CVT/CD-SCSAG to the QI values, with 36% maximal errors at 200 K and a slightly different Arrhenius slope. Both results are in good agreement with experimental data in the tested temperature range. This manifests that the PES used is reasonable accurate. However, the values from the classical VTST are always smaller than the QI results especially in the deep tunneling regions. The conventional TST results have similar tendency to those from the classical VTST. Although the classical VTST can be much improved by evaluating the partition functions quantum mechanically, we do not focus on this improvement. However, it is noted that the anharmonicity, rotational-vibrational coupling are involved in the classical simulation. It is thus expected that these large errors come from the pure quantum effects.

3.2.3. Kinetic Isotope Effects. Kinetic isotope effect (KIE) is the characteristic of chemical reactions which may reveal the quantum effect. We consider the following isotopic reactions:

$$\begin{aligned} &(\mathrm{R1}) \quad \mathrm{H} + \mathrm{SiH_4} \longrightarrow \mathrm{SiH_3} + \mathrm{H_2} \\ &(\mathrm{R2}) \quad \mathrm{D} + \mathrm{SiH_4} \longrightarrow \mathrm{SiH_3} + \mathrm{HD} \\ &(\mathrm{R3}) \quad \mathrm{H} + \mathrm{SiD_4} \longrightarrow \mathrm{SiD_3} + \mathrm{HD}. \end{aligned} \tag{25}$$

The calculated values are tabulated in Table 3.

The KIEs of $k_{QI}(\mathrm{R1})/k_{QI}(\mathrm{R2})$ and $k_{QI}(\mathrm{R1})/k_{QI}(\mathrm{R3})$ in the temperature range of 200–1000 K are summarized in Table 3. Besides the QI values, this table also displays comparable results of the CVT/CD-SCSAG [50]. It is easily found that the $k(\mathrm{R1})/k(\mathrm{R2})$ KIEs predicted by the QI are smaller than 1, in agreement with the CVT/CD-SCSAG values. The detailed comparison reveals that the CVT/CD-SCSAG predicts smaller values than those from QI theory. Again, the maximal error occurs at 200 K and is about 23%. For $k(\mathrm{R1})/k(\mathrm{R3})$ KIEs, we find that although both approaches predict the values larger than 1 at 200 to 1000 K, the QI values are smaller than those from CVT/CD-SCSAG. Espinosa-García and coworkers [50] have pointed out that the CVT/CD-SCSAG rates may have been overestimated because of the high vibrational and tunneling contributions. This manifests that the QI approach indeed correctly accounts for the quantum effects.

3.3. H Diffusion on Ni(100) Surface. Diffusion plays a fundamental role in surface process. It reveals characteristics about the underlying surface potential and is intimately involved in determining the kinetics of surface catalyzed chemical reactions. The hydrogen atom and its isotopes are ideal candidates to exhibit quantum tunneling behavior due to their small masses. We explore the evaluation of the quantum instanton approximation to the process of H diffusion on Ni(100) surface using the EAM4 potential energy surface constructed by Truong and Truhlar [53].

In the path integral calculations, the numbers of time slices, P and P_{bath}, for the degrees of freedom of H and quantized Ni atoms, respectively, are set to $(P, P_{\mathrm{bath}}) = (24\text{–}40, 6\text{–}8)$ in the temperature range of 40–300 K. The number of Monte Carlo cycle employed is $(1\text{–}10) \times 10^6$, which converges most of the relevant quantities within 10% statistical error (some of the statistical errors are within 20% below 100 K).

3.3.1. Model. Nickel crystallizes in a face-centered-cubic (fcc) lattice structure. The structural model used consists of 162 Ni atoms over four layers: 40 atoms are in each of the first and third layers, and 41 atoms are in each of the second and fourth layers.

Figure 6 shows the platform and profile chart of Ni(100) lattice. In the simulations, 89 Ni atoms (orange), in the four sides of the bulk metal and the bottom layer, are fixed. The 48 atoms (blue) closest to the fixed ones are considered to be movable and treated classically. The last 25 ones (red), surrounding the reactant and product sites and lying directly beneath the reactant and product sites, are treated quantum mechanically.

In our calculations, the lattice (see Figure 6) is rotated by 45 degrees counterclockwise, and we chose the x coordinate of H atom (along the reaction path) to be reaction coordinate.

TABLE 3: The kinetic isotope effects (KIEs) at different temperatures. CVT is the CVT/CD-SCSAG values from [50].

T (K)	QI KIE ($R1/R2$)	CVT KIE ($R1/R2$)	QI KIE ($R1/R3$)	CVT KIE ($R1/R3$)
200	0.56	0.69	30.10	33.60
300	0.73	0.71	8.57	9.81
400	0.77	0.76	4.39	5.07
500	0.85	0.81	3.11	3.46
600	0.90	0.84	2.41	2.72
700	0.92	0.86	2.15	2.32
800	0.94		1.78	
900	0.94		1.68	
1000	0.95	0.90	1.64	1.80

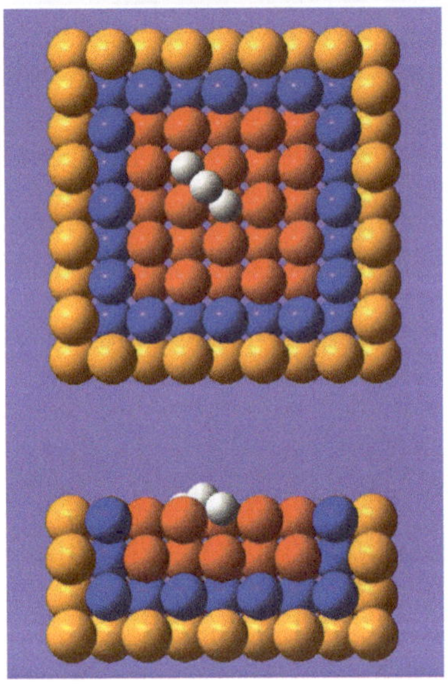

FIGURE 6: The platform and profile chart of H diffusion on Ni(100) lattice. The gray circles represent H atom on different surface sites, and the orange circles represent the fixed Ni atoms. Blue circles are Ni atoms treated classically, while the red ones are treated quantum mechanically.

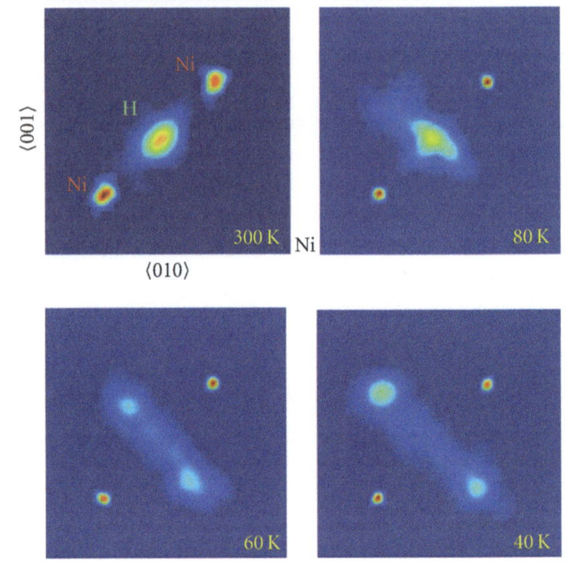

FIGURE 7: Distribution of H and Ni quantum paths on the surface with two reaction coordinate beads (x_0 and $x_{P/2}$) of H path fixed at the two dividing surfaces. Labels of <010> and <001> denote the crystal directions of nickel. The probabilities are normalized for both H and Ni.

3.3.2. Probability Distribution of Paths. The path integral has an advantage that the paths of the particles can display the character of the diffusive process. We extract the probability distributions for the paths of H and Ni atoms (two reaction coordinate beads (x_0 and $x_{P/2}$) of H path are fixed at the two dividing surfaces) for the purpose of guaranteeing the instanton property (the instanton is a periodic orbit between the two dividing surfaces). The dividing surfaces can be obtained from the free energy surface (24), their values at different temperatures are shown in Table 4.

Figure 7 displays the probability distributions for both H and Ni atoms at several temperatures. It is found that the probability distribution of H is localized at 300 K, and it becomes more and more delocalized as the temperature decreases. Below 80 K, the distribution rapidly becomes obvious in both sides of the transition state, and the path begins to continuously distribute between the two hollow sites at lower temperatures. This phenomenon indicates that H tunneling becomes remarkable at the temperatures lower than 80 K. The probability distributions of Ni atoms have small changes in the whole temperature range, and they seem to be frozen at very low temperature.

3.3.3. Diffusion Coefficients. According to the hopping model [54], the diffusion coefficient D is related to the rates through

$$D = k_{\mathrm{QI}} \cdot b^2, \tag{26}$$

where b is the hop length and is equal to 2.489 Å for the Ni(100) surface. We have calculated the diffusion coefficients at temperatures ranging from 40 to 300 K, and the results are plotted in Figure 8.

We plot two kinds of QI diffusion coefficients in Figure 8, one is the result on a rigid lattice, the other is the result on a quantized lattice. At high temperatures, these two kinds of diffusion coefficients are nearly the same; however, at low temperatures, the ones on a rigid lattice are bigger than that

TABLE 4: The positions of two dividing surfaces. (Unit in Å. The position 0.00 corresponds to the bridge site and the positions ±1.24 Å correspond to two hollow sites.)

T (K)	300	80	60	50	40
$x_a^\ddagger$	0.00	0.00	−0.75	−1.00	−1.08
$x_b^\ddagger$	0.00	0.00	0.75	1.00	1.08

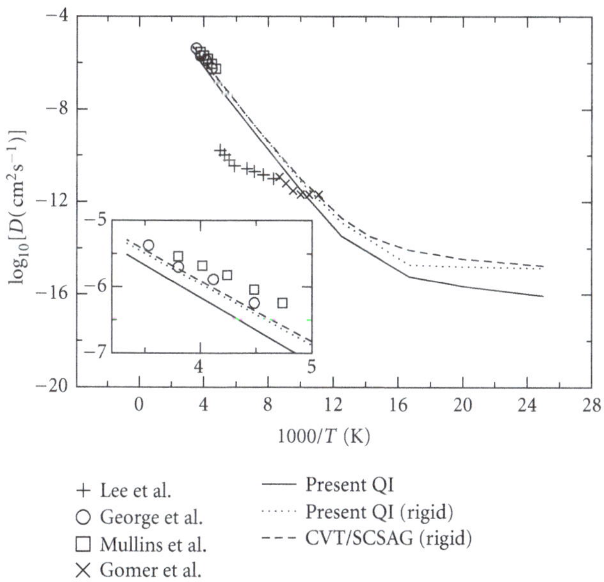

FIGURE 8: Arrhenius plots of the diffusion coefficients in the range 40–300 K. Solid line: the QI results for the quantized lattice; dotted line: the QI results for the rigid lattice; dashed line: the CVT/SCSAG results for the rigid lattice [55]. Pluses, circles, squares, and crosses are experimental results of Lee et al. [58], George et al. [56], Mullins et al. [57], and Lin and Gomer [59], respectively.

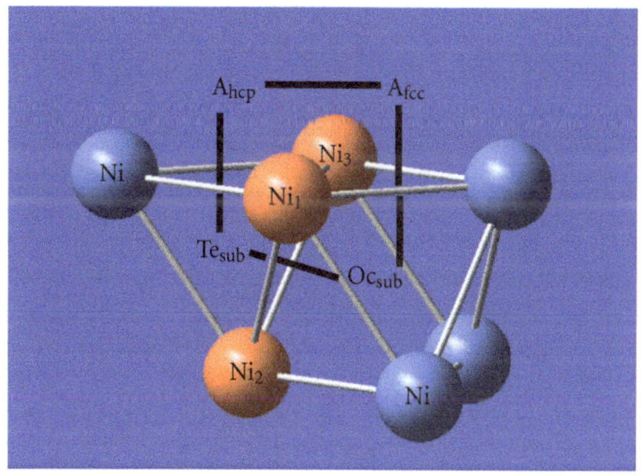

FIGURE 9: A lattice model with a few Ni atoms for the circle reaction processes of hydrogen. A_{hcp}, A_{fcc}, Oc_{sub}, and Te_{sub} are the abbreviations for a hcp hollow site, a fcc hollow site, a subsurface octahedral vacancy, and a subsurface tetrahedral vacancy, respectively. The black lines stand for the directions of the reaction paths. The atoms of Ni_1, Ni_2, and Ni_3 colored in orange are specially used to determine the reaction coordinate of the process $Oc_{sub} \rightarrow Te_{sub}$.

on a quantized lattice, and this is mainly due to the fact that the free energy on a quantized lattice is higher than that on a rigid lattice [38]. Comparing the QI diffusion coefficients with the CVT/SCSAG ones [55] on a rigid lattice, we find that our QI results on a rigid lattice are very similar to those from the CVT/SCSAG. Now, we compare the present theoretical results with experimental ones. It is found that the QI results as well as the CVT/SCSAG results are in good agreement with the experiments of George et al. [56] and Mullins et al. [57] at 200 K and 300 K. However, at low temperatures, the theoretical results are significantly larger than Lee et al. [58] and Lin and Gomer [59]. The experimental transition temperature (100 K–160 K) is also different from the present calculations. The QI predicts it to be 70 K for the quantized lattice, while it is about 66 K for the rigid lattice obtained from both the QI and the CVT/SCSAG approaches. These discrepancies between theories and experiments may be attributed to the interaction potential. It is likely that the accuracy of the potential energy surfaces needs further improvement.

3.4. Surface-Subsurface Transport and Interior Migration for H/Ni. A full-dimensional potential energy surface (EAM6), constructed by Wonchoba and Truhlar [60], using the embedded diatomics-in-molecules (EDIMs) [61] potential energy function, is used in this QI calculations.

In path integral calculations, the numbers of time slices, P and P_{bath} for the degrees of freedom of the H and quantized Ni atoms, respectively, are set to $(P, P_{\text{bath}}) = (30-40, 6-8)$ in the temperature range of 100–400 K. The number of Monte Carlo is about $(2-6) \times 10^6$ for computing a single ensemble average. It converges most of the values within 10% statistical errors (some of the statistical errors are within 20% at 100 K).

3.4.1. Model. Figures 9 and 10 show H diffusion processes in the nickel crystal with a face-centered-cubic (fcc) lattice structure. For a given rate process, we construct a lattice cell (each cell contains more than 200 Ni atoms) in which all atoms are treated to be movable to incorporate the effect of the crystal fluctuation on the rates. To be concrete, the Ni atoms in the four sides of the bulk metal and at the bottom layers are fixed, the Ni atoms surrounding the reactant and product sites and lying along the reaction path are treated quantum mechanically, and the others are treated classically. It should be noted that the lattice for the interior diffusion process has a structure of sphere and the outer layers are fixed.

In the QI calculations, we need to define the reaction coordinate operators $\hat{s}$ (6). For the systems considered in this section, the hydrogen coordinates are essentially good choices. We thus adopt the following reaction coordinates for different rate processes. In the H diffusion on Ni(111) surface

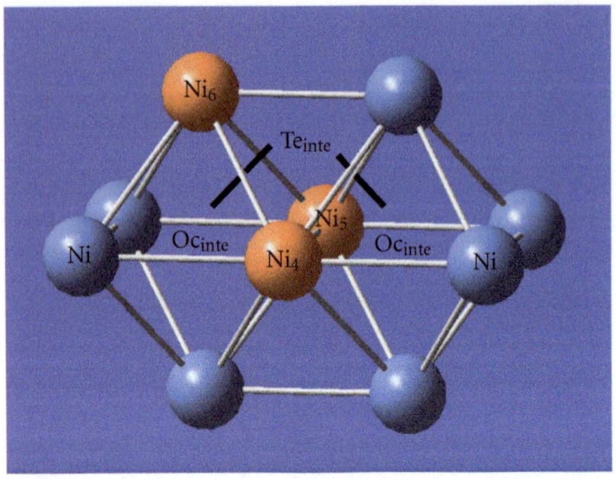

Figure 10: A lattice model with a few Ni atoms for the process of H diffusion in interior of bulk Ni. Oc_{inte} and Te_{inte} are the abbreviations for an interior octahedral vacancy and an interior tetrahedral vacancy, respectively. The black lines stand for the general directions of the reaction paths. The atoms of Ni_4, Ni_5, and Ni_6 colored in orange are specially used to determine the reaction coordinate of the process $Oc_{inte} \rightarrow Te_{inte}$.

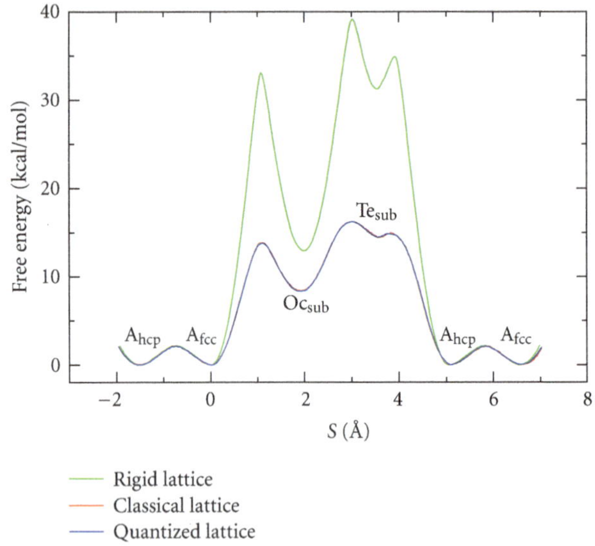

Figure 11: Free energy profiles with respect to the processes in Figure 9 at 300 K. The green, the red, and the blue lines correspond to the rigid, classical, and quantized lattices, respectively.

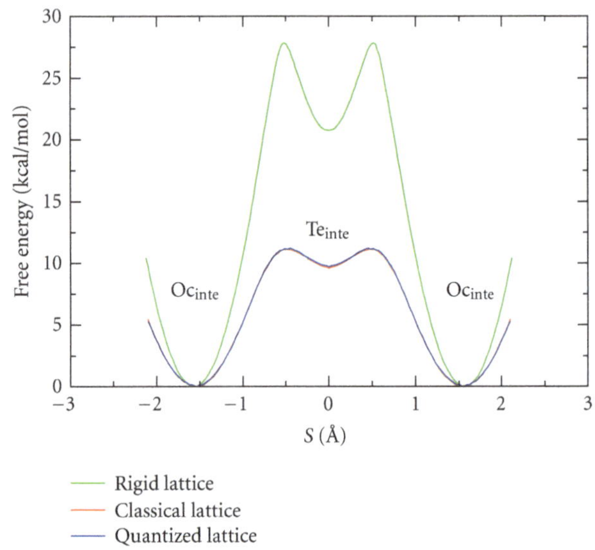

Figure 12: Free energy profiles with respect to the processes in Figure 10 at 300 K. The green, the red, and the blue lines correspond to the rigid, classical, and quantized lattices, respectively.

from a hcp site to a fcc site ($A_{hcp} \rightarrow A_{fcc}$), the x coordinate of the H atom is chosen, whose direction is showed in Figure 9 as the black line connecting the A_{hcp} site to A_{fcc} site. In the H resurfacing from a subsurface octahedral vacancy to the fcc site ($Oc_{sub} \rightarrow A_{fcc}$) and from a subsurface tetrahedral vacancy to the hcp site ($Te_{sub} \rightarrow A_{hcp}$), the z coordinate of the H atom (vertical to Ni(111) surface) is taken. In the H diffusion between the adjacent subsurface octahedral and tetrahedral vacancies ($Oc_{sub} \rightarrow Te_{sub}$) and between the adjacent interior octahedral vacancy and interior tetrahedral vacancy ($Oc_{inte} \rightarrow Te_{inte}$), the reaction coordinates are along the directions that perpendicular to the planes of Ni_1–Ni_2–Ni_3 and Ni_4–Ni_5–Ni_6, respectively.

3.4.2. Free Energy and Prefactor. In order to investigate the quantized Ni lattice effect on the rates, we have recast the QI formula in (1) into Arrhenius form, which consists of the free energy and the prefactor (17) and (18).

In the free energy calculations, firstly, we calculate the free energy profile for each step of the H hopping paths, $A_{hcp} \rightarrow A_{fcc} \rightarrow Oc_{sub} \rightarrow Te_{sub} \rightarrow A_{hcp}$ and $Oc_{inte} \rightarrow Te_{inte} \rightarrow Oc_{inte}$, with the reaction coordinate defined in Section 3.4.1. Then, we connect these free energy profiles one by one, the final free energy profiles for the whole processes are displayed in Figures 11 and 12.

Figures 11 and 12 display the calculated free energy profiles for the surface-subsurface and interior processes, respectively, with the quantized lattice, the classical lattice and the rigid one at room temperature (300 K). The corresponding free energy barriers, prefactors, and rates are tabulated in Table 5.

Figures 11 and 12 clearly show that the classical lattices always reduce the free energy barriers compared with the rigid lattices, however, the differences between the free energy barriers with the quantized lattices and the ones with the classical lattices are very small. For different processes; however, the relaxation effect on the free energies is very different. For hydrogen diffusion on Ni(111), the classical lattice only slightly lowers the free energy barrier, while it decreases the barriers by more than one-half in subsurface and interior processes. More careful analysis from Figure 11 reveals that the two preferred Ni(111) surface binding sites, that is, hcp and fcc hollows have symmetric wells, manifesting that the motions of Ni atoms in the layer beneath the surface have little influence on the surface free energies despite the fact that the Ni atoms beneath the hcp and fcc hollows have different arrangements. It is also found that the Oc_{sub} site has a deeper well than the Te_{sub} site and the well of the Te_{sub} site nearly disappears as relaxed Ni atoms are considered. It manifests that the hydrogen at the Te_{sub} site is very unstable and can easily move to the Oc_{sub} site or resurface to the hcp site. In the interior processes, although the hydrogen in the Te_{inte} site is much less stable than in the Oc_{inte} site (see Figure 12), the well at the Te_{inte} is obvious. One thus expects that a two-step reaction process can be used for the reaction from one Oc_{inte} to the other.

Next, we consider the prefactor. Table 5 shows that the prefactors of the classical lattices are much smaller than those of rigid ones except for the process of H diffusion on Ni(111), and the prefactors of quantized lattices are always smaller than that of classical ones, but their amplitudes have the same order. This may be explained by the fact that the quantum motions of the lattice atoms can induce the dissipative effect on the tunneling degrees of freedom [62, 63], because the prefactor essentially incorporates the dynamical effect. It is well known that pure dissipation in the overdamping regime always hinders the reaction rates for a given reaction barrier. The present results thus are consistent with above analysis.

The rate is determined by both the prefactor and the free energy barrier. In Table 5, the rate with the rigid lattice is smaller than the one on the classical lattice, which is because the prefactor changes a little and the free energy determines the rate. However, the rate with the quantized lattice is smaller than that of the classical lattice, which is due to the fact that quantized lattice has a smaller prefactor while the free energies are similar. Generally speaking, for the quantized lattices, the rates are lower by 20%–40% when compared to the ones on the classical lattices.

Another important feature of the free energy is its temperature dependence. Figure 13 displays the free energy profiles with both H- and Ni-treated quantum mechanically at several temperatures. Generally speaking, the free energies have a slight difference at 300 K and 400 K, whereas this difference becomes pronounced for 100 K and 200 K, and the barrier positions move to the directions of shallow well for asymmetric reactions. These properties can be explained by the hydrogen tunneling effect. At lower temperatures, the tunneling plays a more important role. Indeed, the barrier heights decrease with decreasing temperature except for $Oc_{sub} \rightarrow Te_{sub}$ among 200 K to 400 K. This special case may be due to the special structure of the lattice. The thermal average displacements of Ni_1 and Ni_3 (in Figure 9) vertical to the Ni(111) surface increase with increasing temperature. Thus, H goes through reaction bottleneck easier at a higher temperature, which makes the barrier decrease. Figure 13 also displays that the free energy barriers of $Oc_{inte} \rightarrow Te_{inte}$ change little in the temperature range of 200–400 K, while those of the $Te_{inte} \rightarrow Oc_{inte}$ become smaller and smaller with decreasing temperature. The corresponding free energy barriers are 0.47, 1.09, 1.47, and 1.66 kcal/mol at 100, 200, 300, and 400 K, respectively. It manifests that the diffusing H atom may not equilibrate in the interior tetrahedral vacancy at very low temperatures.

3.4.3. Surface-Subsurface Transport. In the resurfacing process, $Te_{sub} \rightarrow A_{hcp}$ does not show an obvious barrier (see Figure 11), as the lattice atoms are treated quantum mechanically. This step thus can be thought as a barrierless process. Table 6 tabulates the rate constants for the other resurfacing and subsurface processes in the temperature range of 100 to 600 K. We also list the available CVT/SCT results for $Oc_{sub} \rightarrow A_{fcc}$ and its reverse reaction [60]. Again, the CVT/SCT rates are close to the QI rates except at 100 K. We think our much bigger rate constant at 100 K is due to the contribution of tunneling.

3.4.4. Interior Migration. In the interior of bulk Ni, the two most stable sites to cage H are symmetric octahedral vacancies (see Figure 12). H diffusion between them has been measured experimentally [64–67]. Several theoretical calculations have also been proposed to investigate this diffusion process. Wimmer et al. [68] calculate the diffusion coefficient via two-step reactions $Oc_{inte} \rightarrow Te_{inte} \rightarrow Oc_{inte}$ by using a transition state theory together with accurate ab initio energies, while Wonchoba and Truhlar [60] consider the kinetic step as a direct process with a double maximum barrier and calculate the diffusion coefficient using the CVT/SCT. In the present QI calculations, Figure 13 has explicitly shown that the free energies have a well at the tetrahedral site from 200 K to 400 K. It is thus reasonable to assume that the diffusing H atom temporarily equilibrates in the tetrahedral site before jumping forward or backward to a neighboring octahedral site. The free energy well, however, becomes very shallow at 100 K. In this case, the direct reaction from the octahedral site to the other one may be acceptable. Here, we only calculate the diffusion coefficients via the two kinetic steps at 200–400 K.

The temperature dependence of the diffusion coefficients are commonly fitted to the Arrhenius equation

$$D(T) = D_0 \cdot \exp\left[\frac{-E_a}{RT}\right], \qquad (27)$$

where R is the gas constant and D_0 and E_a are the pre-exponential factor and the activation energy, respectively. The QI calculations predict $D_0 = 3.93 \times 10^{-3}\,\text{cm}^2\text{s}^{-1}$ and $E_a = 10.26$ kcal/mol. In the calculations, the rates of $Oc_{inte} \rightarrow Te_{inte}$ are used to obtain the diffusion coefficients for $Oc_{inte} \rightarrow Oc_{inte}$, because this process is much slower than that of $Te_{inte} \rightarrow Oc_{inte}$ and it determines the total reaction rates.

Table 7 tabulates the pre-exponential factors and activation energies coming from available experiments and theories. It is found that both D_0 and E_a from the QI calculations are close to Ebisuzaki's experimental data [65]. Further tracking down the comparisons with experiments is nontrivial, because the accuracy of the diffusion coefficients is much dependent of the potential energy surface. However, we may make a quantitative comparison for the QI and CVT/SCT results. With use of the diffusion coefficients from 295 K to 300 K obtained by the CVT/SCT, Wonchoba and Truhlar [60] predict 11.1 kcal/mol for E_a, and $1.3 \times 10^{-3}\,\text{cm}^2\text{s}^{-1}$ for D_0, respectively. These values are observably different from the QI calculations. E_a and D_0 are 0.8 kcal/mol larger and 3 times smaller than those from the QI calculations, respectively. The origin of these differences can be

TABLE 5: Free energies, prefactors, and rate constants.

Process (300 K)		ΔF (kcal/mol)	A_{QI} (s^{-1})	k_{QI} (s^{-1})
$A_{fcc} \rightarrow A_{hcp}$	a	2.19	1.58 (13)	4.07 (11)
	b	2.14	1.61 (13)	4.49 (11)
	c	2.14	1.27 (13)	3.47 (11)
$A_{fcc} \rightarrow Oc_{sub}$	a	33.06	6.87 (13)	6.22 (−11)
	b	13.79	2.85 (13)	2.64 (3)
	c	13.75	2.29 (13)	2.23 (3)
$Oc_{sub} \rightarrow A_{fcc}$	a	20.00	6.44 (13)	1.83 (−1)
	b	5.40	2.10 (13)	2.62 (9)
	c	5.40	1.72 (13)	2.15 (9)
$A_{hcp} \rightarrow Te_{sub}$	a	34.92	4.29 (13)	1.67 (−12)
	b	14.92		
	c	14.87		
$Te_{sub} \rightarrow A_{hcp}$	a	3.25	5.13 (13)	2.28 (11)
$Oc_{sub} \rightarrow Te_{sub}$	a	26.19	5.23 (13)	4.31 (−6)
	b	7.83	1.86 (13)	3.51 (7)
	c	7.86	1.45 (13)	2.74 (7)
$Te_{sub} \rightarrow Oc_{sub}$	a	7.83	8.17 (13)	1.51 (8)
	b	1.75	1.91 (13)	1.05 (12)
	c	1.73	1.49 (13)	8.22 (11)
$Oc_{inte} \rightarrow Te_{inte}$	a	27.82	5.12 (13)	2.86 (−7)
	b	11.14	2.50 (13)	1.90 (5)
	c	11.17	1.75 (13)	1.32 (5)
$Te_{inte} \rightarrow Oc_{inte}$	a	7.07	7.80 (13)	6.16 (8)
	b	1.52	2.06 (13)	1.60 (12)
	c	1.47	1.53 (13)	1.36 (12)

a: The results for a rigid lattice.
b: The results for a classical lattice.
c: The results for a quantized lattice.

TABLE 6: Rate constants (Unit: s^{-1}. Powers of 10 are in parentheses) for surface-subsurface processes. [a]The QI results with a quantized lattice. [b]The CVT/SCT results from [60].

	100 K	200 K	300 K	400 K	600 K
$A_{fcc} \rightarrow Oc_{sub}$[a]	3.81 (−14)	1.14 (−1)	2.23 (3)	5.48 (5)	2.43 (8)
$A_{fcc} \rightarrow Oc_{sub}$[b]	1.09 (−15)	1.23 (−1)	9.72 (3)	3.36 (6)	1.16 (9)
$Oc_{sub} \rightarrow A_{fcc}$[a]	1.17 (6)	4.59 (7)	2.15 (9)	1.23 (10)	1.02 (11)
$Oc_{sub} \rightarrow A_{fcc}$[b]	1.62 (5)	7.36 (7)	2.93 (9)	2.33 (10)	2.15 (11)
$Oc_{sub} \rightarrow Te_{sub}$[a]	2.78 (−4)	1.99 (4)	2.74 (7)	1.18 (9)	2.49 (10)
$Te_{sub} \rightarrow Oc_{sub}$[a]	7.24 (8)	4.68 (10)	8.22 (11)	2.00 (12)	4.13 (12)

TABLE 7: Arrhenius parameters for H diffusion in interior of bulk Ni. [a]The QI diffusion coefficients for $Oc_{inte} \rightarrow Oc_{inte}$. [b]From [60]. [c]From [68]. [d]From [64]. [e]From [65]. [f]From [66]. [g]From [67].

		D_0 (cm^2s^{-1})	E_a (kcal/mol)
Present QI[a]	200–400 K	3.93 (−3)	10.26
	295–300 K	1.3 (−3)	11.1
Wonchoba[b] (theoretical data)	300–627 K	7.8 (−4)	10.9
	627–1650 K	4.4 (−4)	10.2
Wimmer[c] (theoretical data)	273–1000 K	3.84 (−2)	10.92
Yamakawa[d] (experimental data)	220–350 K	1.9 (−3)	8.89
Ebisuzaki[e] (experimental data)	470–690 K	5.22 (−3)	9.56
Eichenauer[f] (experimental data)	660–930 K	6.73 (−3)	9.47
Katz[g] (experimental data)	670–1270 K	7.04 (−3)	9.43

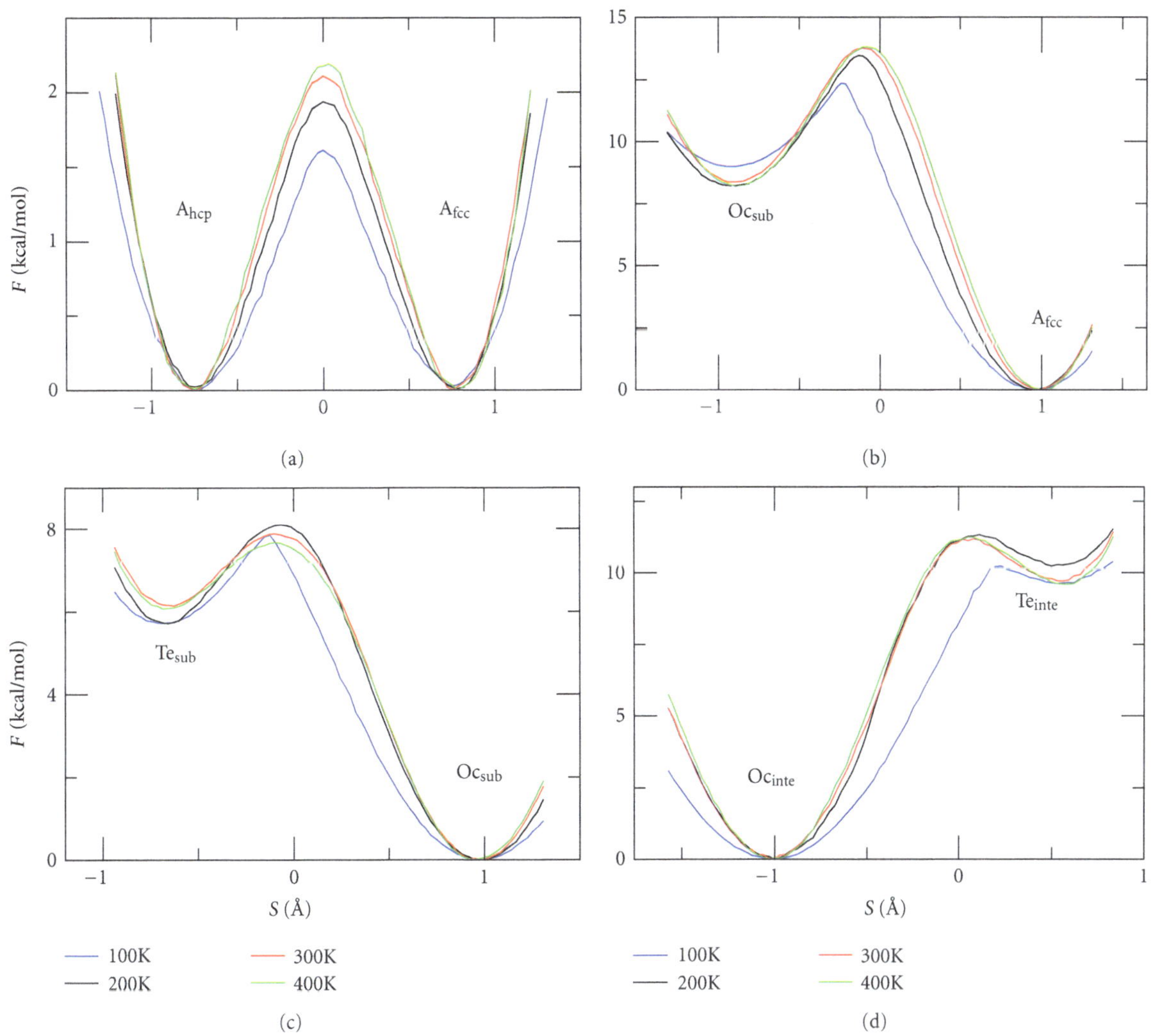

FIGURE 13: The temperature dependence of the free energies. The blue, the black, the red, and the green lines correspond to free energy profiles at 100 K, 200 K, 300 K, and 400 K, respectively.

explained by that Wonchoba and Truhlar have treated the processes $Oc_{inte} \rightarrow Te_{inte}$ and $Te_{inte} \rightarrow Oc_{inte}$ as a single kinetic step rather than as two kinetic steps, which results in a much longer tunneling path than that of the two steps. Compared with the results reported by Wimmer et al. [68], the activation energy is about 0.6 kcal/mol larger than the present one, and the pre-exponential factor is 10 times larger, which is also larger than all available experimental data. These differences may come from both the different potential energy surfaces and rate methods.

4. Conclusion

We have presented the basic principle of the quantum instanton (QI) approximation and its applications to chemical reactions from gas phase to surface systems. The applications demonstrate that the QI method makes it possible to treat hundreds of atoms because of the well-established techniques of imaginary time path integrals. For instance, more than 200 atoms have been incorporated in the present study of the H diffusion processes.

The QI approximation is a kind of "quantum transition state theory" in that there is no account of "recrossing" dynamics in the description. The recrossing effects on the quantum instanton rate constants have been quantified for several collinear reaction by Ceotto and Miller [25], and it is found especially evident for the collinear heavy-light-heavy reactions. Fortunately, the recrossing effects become generally less important in higher dimensions [69]. Therefore, the QI approach may become a suitable tool for the calculation of chemical reaction rate constants of complex systems.

Compared to conventional TST theories, the QI approximation involves two dividing surfaces, which are quantum analogs of the two turning point surfaces of the imaginary time trajectory in the semiclassical instanton theory. At high-temperature limit, these dividing surfaces coalesce into the one, the same as the dividing surface from Wigner's variational principle. In this case, the QI approximation becomes exactly the same as the classical TST. As the tunneling corrections are incorporated, the CVT rates are consistent with the QI rates except at deep tunneling regime, where QI rates are generally greater than the rates from CVT with various tunneling corrections, since the CVT method uses an optimized tunneling path, while the QI method considers all tunneling paths and automatically gives each path its natural weight by the quantum Boltzmann factor. For the reaction of $H + CH_4 \rightarrow H_2 + CH_3$, accurate quantum dynamics rate constants are obtained with the MCTDH method. Compared to the MCTDH ones, the QI rate constants are larger by factors of about 2 to 3 over the temperature range 300–400 K. This difference may partly be due to the recrossing effect which is not considered in QI theory and partly arise from the use of the J-shifting approximation and the neglect of the vibrational angular momenta Hamiltonian in MCTDH method.

Acknowledgments

This work was financially supported by the National Science Foundation of China (Grant nos. 20833004 and 21073146), National Key Basic Research Foundation Program of China (2007CB815204), and the Scientific Research Foundation of Northwest A&F University (Grant no. Z109021103).

References

[1] H. Eyring, "The activated complex in chemical reactions," *The Journal of Chemical Physics*, vol. 3, no. 2, pp. 63–71, 1935.

[2] H. Eyring, "The theory of absolute reaction rates," *Transactions of the Faraday Society*, vol. 34, pp. 41–48, 1938.

[3] E. Wigner, "The transition state method," *Transactions of the Faraday Society*, vol. 34, pp. 29–41, 1938.

[4] G. A. Voth, D. Chandler, and W. H. Miller, "Time correlation function and path integral analysis of quantum rate constants," *Journal of Physical Chemistry*, vol. 93, no. 19, pp. 7009–7015, 1989.

[5] G. A. Voth, D. Chandler, and W. H. Miller, "Rigorous formulation of quantum transition state theory and its dynamical corrections," *The Journal of Chemical Physics*, vol. 91, no. 12, pp. 7749–7760, 1989.

[6] S. Jang and G. A. Voth, "A relationship between centroid dynamics and path integral quantum transition state theory," *Journal of Chemical Physics*, vol. 112, no. 20, pp. 8747–8757, 2000.

[7] W. H. Miller, "Generalization of the linearized approximation to the semiclassical initial value representation for reactive flux correlation functions," *Journal of Physical Chemistry A*, vol. 103, no. 47, pp. 9384–9387, 1999.

[8] E. Pollak and J. L. Liao, "A new quantum transition state theory," *Journal of Chemical Physics*, vol. 108, no. 7, pp. 2733–2743, 1998.

[9] J. Shao, J. L. Liao, and E. Pollak, "Quantum transition state theory: perturbation expansion," *Journal of Chemical Physics*, vol. 108, no. 23, pp. 9711–9725, 1998.

[10] K. Yamashita and W. H. Miller, "'Direct' calculation of quantum mechanical rate constants via path integral methods: application to the reaction path Hamiltonian, with numerical test for the $H + H_2$ reaction in 3D," *The Journal of Chemical Physics*, vol. 82, no. 12, pp. 5475–5484, 1984.

[11] J. W. Tromp and W. H. Miller, "New approach to quantum mechanical transition-state theory," *Journal of Physical Chemistry*, vol. 90, no. 16, pp. 3482–3485, 1986.

[12] N. F. Hansen and H. C. Andersen, "A new formulation of quantum transition state theory for adiabatic rate constants," *The Journal of Chemical Physics*, vol. 101, no. 7, pp. 6032–6037, 1994.

[13] G. Krilov, E. Sim, and B. J. Berne, "Quantum time correlation functions from complex time Monte Carlo simulations: a maximum entropy approach," *Journal of Chemical Physics*, vol. 114, no. 3, pp. 1075–1088, 2001.

[14] E. Sim, G. Krilov, and B. J. Berne, "Quantum rate constants from short-time dynamics: an analytic continuation approach," *Journal of Physical Chemistry A*, vol. 105, no. 12, pp. 2824–2833, 2001.

[15] W. H. Miller, Y. Zhao, M. Ceotto, and S. Yang, "Quantum instanton approximation for thermal rate constants of chemical reactions," *Journal of Chemical Physics*, vol. 119, no. 3, pp. 1329–1342, 2003.

[16] T. Yamamoto and W. H. Miller, "On the efficient path integral evaluation of thermal rate constants within the quantum instanton approximation," *Journal of Chemical Physics*, vol. 120, no. 7, pp. 3086–3099, 2004.

[17] Y. Zhao, T. Yamamoto, and W. H. Miller, "Path integral calculation of thermal rate constants within the quantum instanton approximation: application to the $H + CH_4 \rightarrow H_2 + CH_3$ hydrogen abstraction reaction in full Cartesian space," *Journal of Chemical Physics*, vol. 120, no. 7, pp. 3100–3107, 2004.

[18] W. H. Miller, "Semiclassical limit of quantum mechanical transition state theory for nonseparable systems," *The Journal of Chemical Physics*, vol. 62, no. 5, pp. 1899–1906, 1975.

[19] S. Coleman, "Fate of the false vacuum: semiclassical theory," *Physical Review D*, vol. 15, no. 10, pp. 2929–2936, 1977.

[20] A. J. Leggett, S. Chakravarty, A. T. Dorsey, M. P. A. Fisher, A. Garg, and W. Zwerger, "Dynamics of the dissipative two-state system," *Reviews of Modern Physics*, vol. 59, no. 1, pp. 1–85, 1987.

[21] R. A. Marcus, "On the analytical mechanics of chemical reactions. quantum mechanics of linear collisions," *The Journal of Chemical Physics*, vol. 45, no. 12, pp. 4493–4499, 1966.

[22] E. A. Mccullough and R. E. Wyatt, "Dynamics of the collinear $H + H_2$ reaction. I. Probability density and flux," *The Journal of Chemical Physics*, vol. 54, no. 8, pp. 3578–3591, 1971.

[23] D. G. Truhlar and A. Koppermann, "Exact and approximate quantum mechanical reaction probabilities and rate constants for the collinear $H + H_2$ reaction," *The Journal of Chemical Physics*, vol. 56, no. 5, pp. 2232–2252, 1972.

[24] T. V. George and W. H. Miller, "Classical S-matrix theory of reactive tunneling: linear $H + H_2$ collisions," *The Journal of Chemical Physics*, vol. 57, no. 6, pp. 2458–2467, 1972.

[25] M. Ceotto and W. H. Miller, "Test of the quantum instanton approximation for thermal rate constants for some collinear reactions," *Journal of Chemical Physics*, vol. 120, no. 14, pp. 6356–6362, 2004.

[26] C. Venkataraman and W. H. Miller, "The Quantum Instanton (QI) model for chemical reaction rates: the "simplest" QI with one dividing surface," *Journal of Physical Chemistry A*, vol. 108, no. 15, pp. 3035–3039, 2004.

[27] Y. Li and W. H. Miller, "Different time slices for different degrees of freedom in Feynman path integration," *Molecular Physics*, vol. 103, no. 2-3, pp. 203–208, 2005.

[28] T. Yamamoto and W. H. Miller, "Path integral evaluation of the quantum instanton rate constant for proton transfer in a polar

solvent," *Journal of Chemical Physics*, vol. 122, no. 4, Article ID 044106, pp. 1–13, 2005.
[29] M. Ceotto, S. Yang, and W. H. Miller, "Quantum reaction rate from higher derivatives of the thermal flux-flux autocorrelation function at time zero," *Journal of Chemical Physics*, vol. 122, no. 4, Article ID 044109, pp. 1–7, 2005.
[30] J. Vaníček, W. H. Miller, J. F. Castillo, and F. J. Aoiz, "Quantum-instanton evaluation of the kinetic isotope effects," *Journal of Chemical Physics*, vol. 123, no. 5, Article ID 054108, pp. 1–14, 2005.
[31] Y. Li and W. H. Miller, "Using a family of dividing surfaces normal to the minimum energy path for quantum instanton rate constants," *Journal of Chemical Physics*, vol. 125, no. 6, Article ID 064104, 2006.
[32] S. Yang, T. Yamamoto, and W. H. Miller, "Path-integral virial estimator for reaction-rate calculation based on the quantum instanton approximation," *Journal of Chemical Physics*, vol. 124, no. 8, Article ID 084102, 2006.
[33] J. Vaníček and W. H. Miller, "Efficient estimators for quantum instanton evaluation of the kinetic isotope effects: application to the intramolecular hydrogen transfer in pentadiene," *Journal of Chemical Physics*, vol. 127, no. 11, Article ID 114309, 2007.
[34] M. Buchowiecki and J. Vaníček, "Direct evaluation of the temperature dependence of the rate constant based on the quantum instanton approximation," *Journal of Chemical Physics*, vol. 132, no. 19, Article ID 194106, 2010.
[35] T. Zimmermann and J. Vaníček, "Three applications of path integrals: equilibrium and kinetic isotope effects, and the temperature dependence of the rate constant of the [1,5] sigmatropic hydrogen shift in (Z)-1,3-pentadiene," *Journal of Molecular Modeling*, vol. 16, no. 11, pp. 1779–1787, 2010.
[36] K. F. Wong, J. L. Sonnenberg, F. Paesani et al., "Proton transfer studied using a combined ab initio reactive potential energy surface with quantum path integral methodology," *Journal of Chemical Theory and Computation*, vol. 6, no. 9, pp. 2566–2580, 2010.
[37] W. Wang, S. Feng, and Y. Zhao, "Quantum instanton evaluation of the thermal rate constants and kinetic isotope effects for SiH_4+H → SiH_3 + H_2 reaction in full Cartesian space," *Journal of Chemical Physics*, vol. 126, no. 11, Article ID 114307, 2007.
[38] W. Wang and Y. Zhao, "Path integral evaluation of H diffusion on Ni(100) surface based on the quantum instanton approximation," *Journal of Chemical Physics*, vol. 130, no. 11, Article ID 114708, pp. 1–10, 2009.
[39] W. Wang and Y. Zhao, "Quantum instanton evaluations of surface diffusion, interior migration, and surface-subsurface transport for H/Ni," *Journal of Chemical Physics*, vol. 132, no. 6, Article ID 064502, 2010.
[40] B. J. Berne and D. Thirumalai, "On the simulation of quantum systems: path integral methods," *Annual Review of Physical Chemistry*, vol. 37, pp. 401–424, 1986.
[41] R. P. Feynman and A. R. Hibbs, *Quantum Mechanics and Path Integrals*, McGraw-Hill, New York, NY, USA, 1965.
[42] R. P. Feynman, *Statistical Mechanics*, Addison-Wesley, New York, NY, USA, 1972.
[43] L. S. Schulman, *Techniques and Applications of Path Integrals*, John Wiley & Sons, New York, NY, USA, 1986.
[44] M. Mezei, "Adaptive umbrella sampling: self-consistent determination of the non-Boltzmann bias," *Journal of Computational Physics*, vol. 68, no. 1, pp. 237–248, 1987.
[45] C. Predescu and W. H. Miller, "Optimal choice of dividing surface for the computation of quantum reaction rates," *Journal of Physical Chemistry B*, vol. 109, no. 14, pp. 6491–6499, 2005.
[46] J. Espinosa-García, "New analytical potential energy surface for the CH_4+H hydrogen abstraction reaction: thermal rate constants and kinetic isotope effects," *Journal of Chemical Physics*, vol. 116, no. 24, pp. 10664–10673, 2002.
[47] D. L. Baulch, C. J. Cobos, R. A. Cox et al., "Evaluated kinetic data for combustion modelling," *Journal of Physical and Chemical Reference Data*, vol. 21, no. 3, pp. 411–734, 1992.
[48] J. W. Sutherland, M. C. Su, and J. V. Michael, "Rate constants for H + CH_4, CH_3 + H_2, and CH_4 dissociation at high temperature," *International Journal of Chemical Kinetics*, vol. 33, no. 11, pp. 669–684, 2001.
[49] S. Andersson, G. Nyman, A. Arnaldsson, U. Manthe, and H. Jonsson, "Comparison of quantum dynamics and quantum transition state theory estimates of the H + CH_4 reaction rate," *Journal of Physical Chemistry A*, vol. 113, no. 16, pp. 4468–4478, 2009.
[50] J. Espinosa-García, J. Sansón, and J. C. Corchado, "The SiH_4 + H → SiH_3 + H_2 reaction: potential energy surface, rate constants, and kinetic isotope effects," *Journal of Chemical Physics*, vol. 109, no. 2, pp. 466–473, 1998.
[51] K. D. Dobbs and D. A. Dixon, "Ab initio prediction of the activation energies for the abstraction and exchange reactions of H with CH_4 and SiH_4," *Journal of Physical Chemistry*, vol. 98, no. 20, pp. 5290–5297, 1994.
[52] A. Goumri, W. J. Yuan, L. Ding, Y. Shi, and P. Marshall, "Experimental and theoretical studies of the reaction of atomic hydrogen with silane," *Chemical Physics*, vol. 177, no. 1, pp. 233–241, 1993.
[53] T. N. Truong and D. G. Truhlar, "Effect of steps and surface coverage on rates and kinetic isotope effects for reactions catalyzed by metallic surfaces: chemisorption of hydrogen on Ni," *Journal of Physical Chemistry*, vol. 94, no. 21, pp. 8262–8279, 1990.
[54] J. V. Barth, "Transport of adsorbates at metal surfaces: from thermal migration to hot precursors," *Surface Science Reports*, vol. 40, no. 3, pp. 75–149, 2000.
[55] T. N. Truong and D. G. Truhlar, "The effects of steps, coupling to substrate vibrations, and surface coverage on surface diffusion rates and kinetic isotope effects: hydrogen diffusion on Ni," *The Journal of Chemical Physics*, vol. 93, no. 3, pp. 2125–2138, 1990.
[56] S. M. George, A. M. DeSantolo, and R. B. Hall, "Surface diffusion of hydrogen on Ni(100) studied using laser-induced thermal desorption," *Surface Science*, vol. 159, no. 1, pp. L425–L432, 1985.
[57] D. R. Mullins, B. Roop, S. A. Costello, and J. M. White, "Isotope effects in surface diffusion: hydrogen and deuterium on Ni(100)," *Surface Science*, vol. 186, no. 1-2, pp. 67–74, 1987.
[58] A. Lee, X. D. Zhu, L. Deng, and U. Linke, "Observation of a transition from over-barrier hopping to activated tunneling diffusion: H and D on Ni(100)," *Physical Review B*, vol. 46, no. 23, pp. 15472–15476, 1992.
[59] T. S. Lin and R. Gomer, "Diffusion of ^1H and ^2H on the Ni(111) and (100) planes," *Surface Science*, vol. 255, no. 1-2, pp. 41–60, 1991.
[60] S. E. Wonchoba and D. G. Truhlar, "General potential-energy function for H/Ni and dynamics calculations of surface diffusion, bulk diffusion, subsurface-to-surface transport, and

absorption," *Physical Review B*, vol. 53, no. 16, pp. 11222–11241, 1996.

[61] T. N. Truong, D. G. Truhlar, and B. C. Garrett, "Embedded diatomics-in-molecules: a method to include delocalized electronic interactions in the treatment of covalent chemical reactions at metal surfaces," *Journal of Physical Chemistry*, vol. 93, no. 25, pp. 8227–8239, 1989.

[62] R. Baer and R. Kosloff, "Quantum dissipative dynamics of adsorbates near metal surfaces: a surrogate Hamiltonian theory applied to hydrogen on nickel," *Journal of Chemical Physics*, vol. 106, no. 21, pp. 8862–8875, 1997.

[63] R. Baer, Y. Zeiri, and R. Kosloff, "Hydrogen transport in nickel (111)," *Physical Review B*, vol. 55, no. 16, pp. 10952–10974, 1997.

[64] K. Yamakawa, "Diffusion of deuterium and isotope effect in nickel," *Journal of the Physical Society of Japan*, vol. 47, no. 1, pp. 114–121, 1979.

[65] Y. Ebisuzaki, W. J. Kass, and M. O'Keeffe, "Diffusion and solubility of hydrogen in single crystals of nickel and nickel-vanadium alloy," *The Journal of Chemical Physics*, vol. 46, no. 4, pp. 1378–1381, 1967.

[66] W. Eichenauer, W. Loser, and H. Witte, "Solubility and rate of diffusion of hydrogen and deuterium in nickel and copper single crystals," *Z Metallk*, vol. 56, pp. 287–293, 1965.

[67] L. Katz, M. Guinan, and R. J. Borg, "Diffusion of H_2, D_2, and T_2 in single-crystal Ni and Cu," *Physical Review B*, vol. 4, no. 2, pp. 330–341, 1971.

[68] E. Wimmer, W. Wolf, J. Sticht et al., "Temperature-dependent diffusion coefficients from ab initio computations: hydrogen, deuterium, and tritium in nickel," *Physical Review B*, vol. 77, no. 13, Article ID 134305, 2008.

[69] S. Chapman, S. M. Hornstein, and W. H. Miller, "Accuracy of transition state theory for the threshold of chemical reactions with activation energy. Collinear and three-dimensional H + H_2," *Journal of the American Chemical Society*, vol. 97, no. 4, pp. 892–894, 1975.

19

Fundamental Mechanistic Understanding of Electrocatalysis of Oxygen Reduction on Pt and Non-Pt Surfaces: Acid versus Alkaline Media

Nagappan Ramaswamy and Sanjeev Mukerjee

Laboratory for Electrochemical Advanced Power, Department of Chemistry and Chemical Biology, Center of Renewable Energy Technology (NUCRET), Northeastern University, 360 Huntington Avenue, Boston, MA 02115, USA

Correspondence should be addressed to Sanjeev Mukerjee, s.mukerjee@neu.edu

Academic Editor: Milan M. Jaksic

Complex electrochemical reactions such as Oxygen Reduction Reaction (ORR) involving multi-electron transfer is an electrocatalytic inner-sphere electron transfer process that exhibit strong dependence on the nature of the electrode surface. This criterion (along with required stability in acidic electrolytes) has largely limited ORR catalysts to the platinum-based surfaces. New evidence in alkaline media, discussed here, throws light on the involvement of surface-independent outer-sphere electron transfer component in the overall electrocatalytic process. This surface non-specificity gives rise to the possibility of using a wide-range of non-noble metal surfaces as electrode materials for ORR in alkaline media. However, this outer-sphere process predominantly leads only to peroxide intermediate as the final product. The importance of promoting the electrocatalytic inner-sphere electron transfer by facilitation of direct adsorption of molecular oxygen on the active site is emphasized by using pyrolyzed metal porphyrins as electrocatalysts. A comparison of ORR reaction mechanisms between acidic and alkaline conditions is elucidated here. The primary advantage of performing ORR in alkaline media is found to be the enhanced activation of the peroxide intermediate on the active site that enables the complete four-electron transfer. ORR reaction schemes involving both outer- and inner-sphere electron transfer mechanisms are proposed.

1. Introduction

Oxygen reduction reaction (ORR) on noble and nonnoble metal surfaces remains as one of the well-investigated electrochemical processes. While under acidic conditions Pt-based systems remain the mainstay as catalyst materials for ORR, in alkaline electrolyte a wide range of nonnoble metals and their oxides are stable enough for practical applications [1, 2]. *ORR pathway rather than ORR mechanism* has typically been addressed in the literature due to the easy accessibility of the former from rotating ring-disk electrode (RRDE) studies, and the complexity in understanding the latter [3]. ORR pathway is found to be similar in both acid and alkaline media on Pt-based materials [3, 4]. Based on the initial propositions by Damjanovic et al. [4–6], rate determining step (*rds*) on Pt electrodes is widely agreed to be the first electron transfer step to the adsorbed molecular O_2 with or without rapid proton transfer [4–6]. A major alternative viewpoint to *rds* in ORR was proposed by Yeager et al. [7], wherein it was proposed that ORR on Pt surfaces is likely to involve dissociative chemisorption of molecular O_2 with the initial adsorption of O_2 (with or without an electron transfer) as the *rds*. ORR on Pt-based catalysts is understood to proceed via "parallel" routes with the $4e^-$ "direct" or "series" pathway as the predominant route and a minor route involving $2e^-$ pathway to peroxide. Various kinetic models for ORR have been developed to understand the reaction pathways. The first model was developed by Damjanovic et al. [8] following which Wroblowa et al. [9], and Anastasijević et al. [10, 11], proposed extensive models.

For electrocatalytic reactions proceeding via inner-sphere electron transfer mechanism, it is typically assumed that either molecular adsorption of reactant species (dissociatively or nondissociatively) or an electron transfer is the first step [12]. For neutral, nonpolar species like molecular O_2, direct molecular O_2 adsorption is likely to be inhibited relative to, for example, the adsorption of charged superoxide radical anion ($O_2^{\bullet -}$) unless the free energy of adsorption of O_2 molecule is very exothermic on a specific catalyst surface. This is especially true under fuel cell conditions, where the cathodic reaction typically occurs at potentials well positive of the potential of zero charge (pzc). Multistep, multielectron transfer processes like ORR that involves many adsorbed intermediates are undoubtedly classified as an inner-sphere electron transfer reaction. However, among the many elementary reaction steps involved in ORR, there could be *a surface-independent outer-sphere electron transfer component* in the overall electrocatalytic $4e^-$ inner-sphere electron transfer reaction. In that perspective, O_2 reduction by one-electron transfer to superoxide ($O_2^{\bullet -}$) is observed at $E^\circ = -0.3 \pm 0.03$ V versus SHE corresponding to $\Delta G^\circ = 30 \pm 2$ kJ mol^{-1} with both O_2 and $O_2^{\bullet -}$ remaining in the aqueous phase [13, 14]. Given the pH independence of this redox couple ($O_2/O_2^{\bullet -}$), the potential of this reaction does not change as the pH is varied from zero to fourteen [15]. Due to the occurrence of four proton transfer steps in O_2 reduction to H_2O/OH^-, its standard reduction potential changes by 0.828 V from 1.229 V to 0.401 V versus SHE as the pH value changes from zero to fourteen. This causes the overpotential for the first electron transfer step ($O_2/O_2^{\bullet -}$) to decrease from 1.53 V at pH = 0 to 0.7 V at pH = 14 indicating a sharp decrease in overpotential at alkaline pH conditions. Blizanac et al. [15] argued based on a modified Pourbaix diagram approach that the above-mentioned decrease in overpotential is the primary thermodynamic reason for the applicability of a wide range of nonnoble materials in alkaline media. Due to the high overpotential required for $O_2/O_2^{\bullet -}$ redox couple in acidic media, only certain specific catalyst surfaces such as platinum that offer high free energy of adsorption for O_2 can catalyze ORR in acidic media. While the decrease in overpotential for the first electron transfer in alkaline media is certainly significant, this argument is primarily of thermodynamic origin. The concept of involving the possibilities of outer-sphere electron transfer during ORR in alkaline media bears importance, and it was pointed out earlier by Bockris [16] and Appleby [17] that the exchange current density values in alkaline media exhibit near independence on a large number of electrode materials including silver, gold, manganese oxides, perovskites, and various carbon surfaces. So certain steps in the overall ORR process in alkaline media could proceed via a non-electrocatalytic mechanism [17].

The drive to replace expensive and scarce Pt-based catalysts for ORR has led to a class of electrocatalysts composed of first-row transition-metal ions stabilized by surface nitrogen functionalities on graphitic surfaces [2, 18–21]. The nature of the active site (obtained after heat treatment) in terms of its location on the carbon support (edge versus basal plane) [22], coordination number (Fe-N_4 versus non-Fe-N_4 environment) [23], and chemical identity of the nitrogen functional groups (pyridinic, pyrrolic, and quaternary) [24] have remained a key aspect of intense discussion. Several theories exist to explain the nature of the active site such as those proposed by van Veen et al. [25–27], McBreen et al. [28], Schulenburg et al. [29], Yeager et al. [2, 30], Scherson et al. [31–33], and Dodelet et al. [22, 34–45]. Although some authors observed that ORR is conducted by sites comprised of surface nitrogen groups devoid of any metal ion centers [46, 47], it is now widely accepted that the transition-metal ion centers coordinated to four nitrogen groups (Me-N_4) on graphitic surfaces constitute the active site [22, 23, 30, 48], whereas chelation primarily serves to prevent the metal center from passivation/corrosion under electrochemical conditions [49]. Besides the exact structure of the active site, the low active site density or the metal loading that is obtained in these catalysts eludes clear understanding. A maximum in catalytic activity is obtained at a very low metal loading (~5000 ppm by weight for inorganic precursors and ~2 wt% for macrocycle precursors) [35]. Dodelet et al. [50] suggested that only the microporosity generated during pyrolysis upon gasification of disordered carbon content acts as a host for active sites. Irrespective of whatever the exact nature of the active site is, this does not necessarily explain the fundamental origin of ORR activity in this class of heat-treated catalysts. It was recently pointed out by Bezerra et al. [51] and Garsuch et al. [52] that most studies on this class of materials have focused on the optimal synthesis conditions and structure necessary for maximum activity, whereas a more fundamental understanding will be of great help in developing innovative routes for new catalyst synthesis.

In this paper, we present a systematic study of ORR reaction mechanisms on Pt and non-Pt surfaces. Fundamental reasons for the so-called kinetic facility of ORR in alkaline media are investigated from the perspectives of the changes in the double-layer structure and the reaction mechanisms in transitioning from acidic to alkaline conditions. New evidence unraveling the influences of inner-sphere and outer-sphere electron transfer mechanisms during ORR is elucidated. Besides the efforts to understand the nature of the active site in heat-treated metal macrocycle-based catalysts, a combination of electrochemical and advanced X-ray absorption spectroscopy techniques has been utilized to understand (1) the fundamental origin of electrocatalytic ORR activity upon heat treatment and (2) the reasons for the low density of active metal sites. To this end, a comparative study of ORR on Pt and non-Pt surfaces in both acid and alkaline media has been performed to (1) analyze the influence of both inner-sphere and outer-sphere electron transfer mechanisms, (2) elucidate the ORR reaction mechanisms, and (3) understand the importance of stabilization of the reaction intermediates on the active site.

2. Experimental

2.1. Electrochemical Characterization. All electrochemical measurements were made at room temperature using a rotating ring-disk electrode (RRDE) setup from Pine Instruments connected to an Autolab (Ecochemie Inc., model-PGSTAT 30) bipotentiostat. Alkaline (0.1 M NaOH) and acidic (0.1 M $HClO_4$) electrolytes were prepared using sodium hydroxide pellets (semiconductor grade, 99.99%, Sigma-Aldrich) and double-distilled 70% perchloric acid (GFS Chemicals), respectively. 30% Pt/C catalyst from BASF-ETEK (Somerset, NJ) was used as received. Ru/C catalyst was synthesized in-house via aqueous route at a metal loading of 20% by weight [53]. Iron(III) meso-tetraphenylporphyrin chloride (FeTPPCl) was procured from Alfa Aesar. FeTPPCl was mixed with Black Pearl carbon in the mass ratio 1 : 4 and ball milled for 2 hours at 400 rpm followed by pyrolysis at 800°C for 2 hours under argon atmosphere. Catalyst inks were prepared by dispersing 25 mg of the catalyst in 10 mL of 1 : 1 millipore H_2O : isopropyl alcohol mixture along with 100 μL of 5 wt% Nafion$^{(R)}$ solution as a binder. 10 μL aliquot of the catalyst ink was dispensed on Glassy Carbon (GC) disk of 5.61 mm dia. Gold ring electrode was held at 1.1 V versus RHE in alkaline electrolyte and at 1.3 V versus RHE in acidic electrolyte to detect stable peroxide intermediate. Collection efficiency of the disk-ring electrode was 37.5%. All potentials are referred to reversible hydrogen electrode (RHE) scale prepared from the same solution as the bulk electrolyte unless otherwise stated.

2.2. X-Ray Absorption Spectroscopic (XAS) Measurements. The *in situ* XAS studies at Fe K-edge (7112 eV) were performed at X19A beamline of National Synchrotron Light Source (NSLS, Brookhaven National Laboratory, NY). Detailed information on the spectroelectrochemical cell design are given elsewhere [54]. Spectra at Fe K-edge were collected in fluorescence mode using a PIPS detector. Argon- or oxygen-saturated 0.1 M NaOH was used as the electrolyte. Complete details on data analysis of X-ray absorption near edge structure (XANES) and extended X-ray absorption fine structure (EXAFS) are available elsewhere [54]. Typical k-range window during EXAFS fit was 2.500–12.500 Å^{-1} (Kaiser-Bessel). Data analysis for Delta-mu ($\Delta\mu$) studies at Fe K-edge involved specific normalization procedures detailed elsewhere [54, 55]. Briefly, this involves careful calibration of edge energy (Fe K-edge 7112 eV), alignment to standard reference scan to account for any drift in the beam energy. A postedge normalization procedure was then applied to the aligned scans via a cubic spline function which normalizes the oscillations over a specific energy range (typically 25 to 200 eV with respect to E_0) on a per-atom basis. Difference spectra were obtained using the equation $\Delta\mu = \mu(\text{V}) - \mu(0.1\,\text{V})$, where $\mu(\text{V})$ is the XANES spectra of the catalyst at various potentials and $\mu(0.1\,\text{V})$ is the reference XANES signal at 0.1 V at which potential no evidence for electrochemical adsorbates (H_{upd}, O_{ads}, OH_{ads}) were found on iron-based catalysts. Theoretical delta-mu curves ($\Delta\mu_t$) were constructed using the FEFF 8.0 code [56]. This was accomplished using the relationship $\Delta\mu_t = \mu(O_{ads}\text{-Fe-N}_x\text{-C}) - \mu(\text{Fe-N}_x\text{-C})$, where the oxide species (O_{ads} or OH_{ads}) is in a specific binding site on Fe.

3. Results and Discussions

3.1. ORR on Carbon-Supported Platinum Nanoparticles: Acid versus Alkaline Media. Figure 1 provides a snapshot of the cyclic voltammetry (CV) and ORR on 30% Pt/C in 0.1 M NaOH and 0.1 M $HClO_4$ electrolytes. As shown in Figure 1(a), CV of 30% Pt/C catalyst features the typical hydrogen underpotential deposition/stripping region below 0.5 V versus RHE in both of the electrolytes followed by oxide formation on Pt at potentials above 0.7 V versus RHE. Oxide formation on Pt in acidic media is due to oxidation of the solvent water molecules (water activation) [57] and in alkaline media is due to specific adsorption of hydroxide anions from the supporting electrolyte [58, 59]. The onset potential of Pt-OH formation is similar in both electrolytes, although in alkaline media oxide formation current exhibits a characteristic peak shape whereas in acidic media oxide formation current is relatively more flat. Half-wave potential ($E_{1/2}$) of Pt-OH in 0.1 M NaOH is 0.775 V whereas in 0.1 M $HClO_4$ $E_{1/2}$ of Pt-OH formation is shifted slightly more positive to 0.810 V. Figure 1(b) shows the ring current measured during ORR on Pt/C at 900 rpm in 0.1 M $HClO_4$ and 0.1 M NaOH. The corresponding ORR polarization curves are shown in the inset of Figure 1(b). The ORR onset potential in both of the electrolytes is ~1 V versus RHE which is followed by mixed kinetic-diffusion region between the potentials 0.7 V and 1 V. Well-defined diffusion-limited current density region is observed below 0.6 V versus RHE. The marginally higher Pt-OH coverage in alkaline media (between the potential region of 0.9 V to 0.8 V versus RHE) causes a penalty of ~25 mV higher overpotential at a kinetic current density of $i_k = 1\,\text{mA cm}^{-2}{}_{\text{geo}}$ for ORR in 0.1 M NaOH relative to 0.1 M $HClO_4$. The most interesting observation in this experiment using Pt/C is in the ring currents measured during ORR. As seen in Figure 1(b), in 0.1 M $HClO_4$ the ring current due to peroxide oxidation is lower in the entire potential region and does not show any significant increase until the disk potential enters the H UPD region. On the contrary, the ring current due to peroxide oxidation in 0.1 M NaOH electrolyte shows a sharp increase at 0.9 V in the cathodic scan. The sharp increase in ring current at 0.9 V during ORR in alkaline media is closely related to Pt-OH formation from specific adsorption of hydroxide anions as seen from the CV in 0.1 M NaOH. As shown in Figure 1(a), in 0.1 M NaOH electrolyte, Pt-OH formation commences at ~0.7 V versus RHE and reaches a peak current at 0.81 V and a plateau at ~0.9 V. On the cathodic direction, the peak potential for Pt-OH reduction is ~0.75 V versus RHE. As seen in Figure 1(b), on the cathodic ORR scan, the increase in ring current commences at 0.9 V and reaches a peak potential of 0.75 V versus RHE. This clearly establishes the interaction between molecular oxygen and the hydroxyl species on the surface. This peroxide formation in the mixed kinetic-diffusion region in alkaline media is more evident at a higher electrolyte concentration of 1.0 M NaOH as shown in Figure 1(c). At a higher concentration of alkaline electrolyte, that is, at 1 M NaOH, peroxide is detected at the ring electrode in both the anodic and the cathodic scan [60].

In order to explain the link between the specifically adsorbed hydroxide anions and the peroxide detected in the mixed kinetic-diffusion region in alkaline media, it is important to understand the double-layer structure of the electrode-electrolyte interface during ORR at high-pH

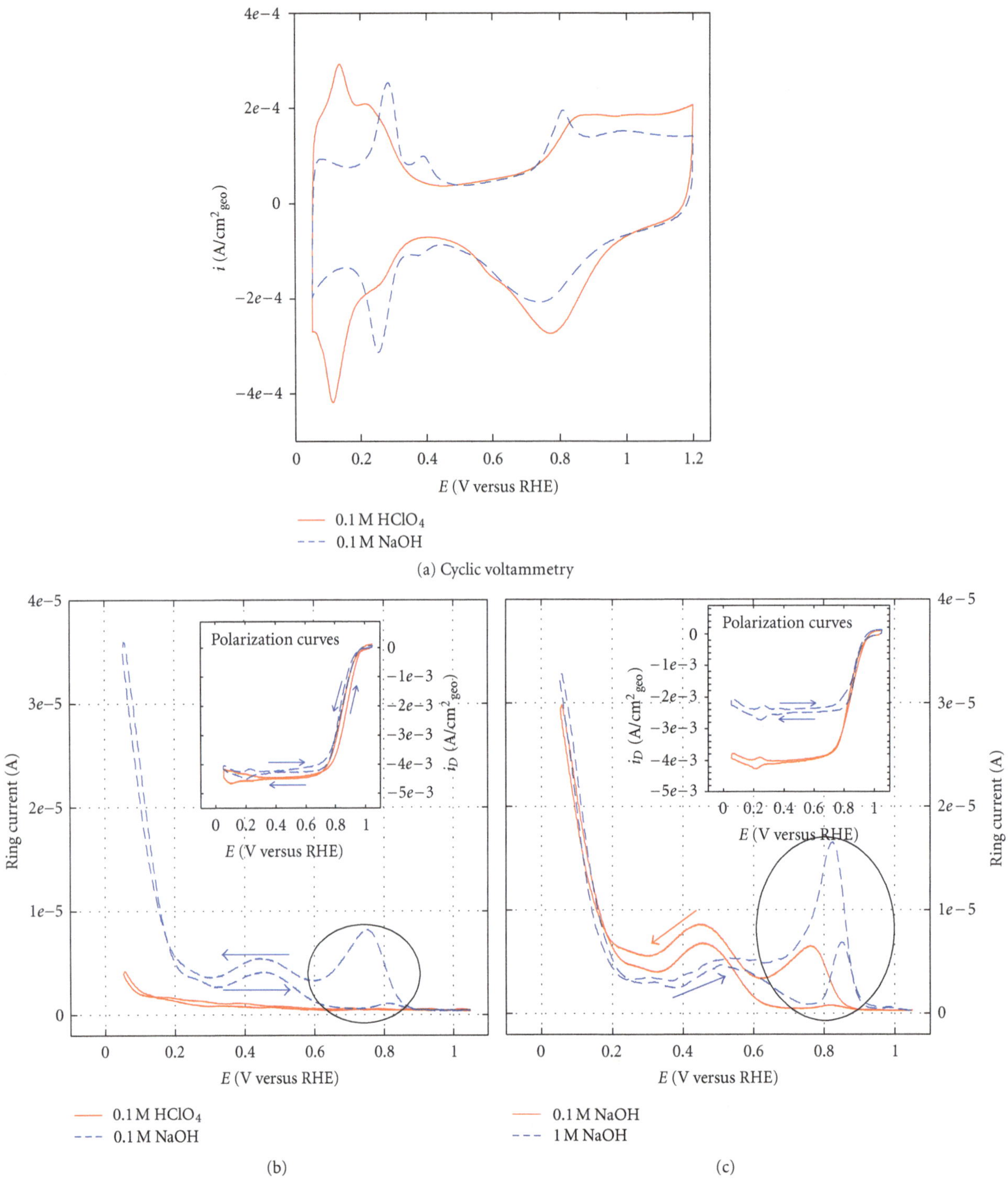

FIGURE 1: Comparison of electrochemical characteristics of BASF-ETEK 30% Pt/C in 0.1 M $HClO_4$, 0.1 M NaOH, and 1.0 M NaOH electrolytes. (a) CV in argon-saturated electrolytes, (b, c) Ring currents measured during ORR study at 900 rpm in O_2-saturated electrolytes. Insets in (b, c) show the corresponding ORR polarization curves at 900 rpm. Scan rate: 20 mV/s. E_{Ring} = 1.1 V versus RHE in 0.1 M NaOH and 1.3 V versus RHE in 0.1 M $HClO_4$. Catalyst loading: 15 $\mu g_{Pt}/cm^2$ on 5.61 mm Glassy Carbon disk electrode.

environments. A schematic illustration of the double-layer structure is shown in Figure 2. In an operating fuel cell, cathodic ORR process typically occurs at potentials well positive of the potential of zero charge (pzc). In alkaline media, water molecules not only act as solvent but also serve as the source of protons required in ORR. At these conditions in alkaline media, inner Helmholtz plane (IHP) is populated by specifically adsorbed hydroxyl species, solvent water dipoles (oriented with the oxygen atom towards the electrode surface), and chemisorbed O_2. Alkali metal ions

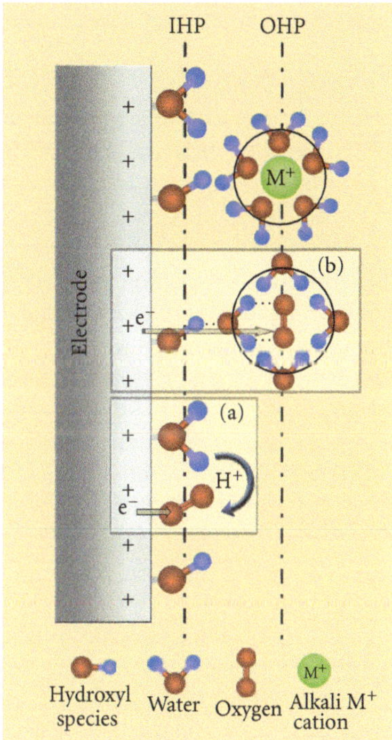

FIGURE 2: Schematic illustration of the double-layer structure during ORR in alkaline media. Insets (a, b) illustrate the inner- and outer-sphere electron transfer processes.

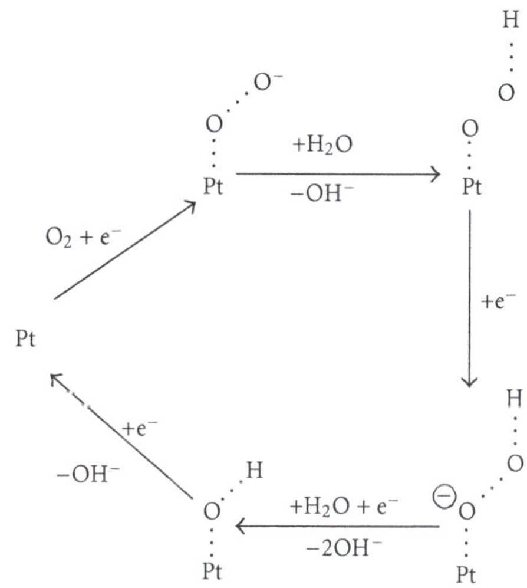

FIGURE 3: Electrocatalytic inner-sphere electron transfer mechanism for ORR on Pt in alkaline media assuming end-on adsorption of O_2 on a single Pt active site.

are typically well solvated and are classically expected to populate the outer Helmholtz plane (OHP). In ORR, the case of a typical electrocatalytic inner-sphere electron transfer mechanism involves strong chemisorption of molecular O_2 (with or without bond breaking) on oxide-free Pt active sites followed by $4e^-/4H^+$ transfer. This inner-sphere mechanism depicted in Figure 2 (inset (a)) is common to both acidic and alkaline media, wherein all the reaction intermediates remain adsorbed on the active site until the final product is desorbed into the bulk electrolyte. Also shown in Figure 2 (inset (b)) is the outer-sphere electron transfer mechanism that appears to be unique to the alkaline conditions. The solvated molecular O_2, represented here as a cluster $O_2 \cdot (H_2O)_n$, could interact with the surface hydroxyl species (OH_{ads}) via a hydrogen bond between H atom in OH_{ads} and O atom in the solvent water molecule. Such hydrogen bond energies (<35 kJ mol^{-1}) are typically much lower than the energy associated with covalent bond strength such as in the case of direct chemisorption of O_2 on Pt (>300 kJ mol^{-1}) [61]. This hydrogen bond formation stabilizes the solvated molecular oxygen $O_2 \cdot (H_2O)$ cluster in the OHP and promotes an outer-sphere electron transfer to form the superoxide species.

Following this new evidence on the involvement of outer-sphere electron transfer in alkaline media, two ORR mechanisms at high-pH environment are proposed here. The first is the well-known electrocatalytic inner-sphere electron transfer mechanism, where molecular O_2 undergoes direct chemisorption on *oxide-free* Pt site leading to a direct/series $4e^-$ pathway without the desorption of reaction intermediates (such as peroxide) from the surface according to the following well-known reaction scheme (also depicted in Figure 3) [3]:

$$O_2 \longrightarrow O_{2,ads}, \tag{1a}$$

$$O_{2,ads} + H_2O + 2e^- \longrightarrow (HO_2^-)_{ads} + OH^-, \tag{1b}$$

$$(HO_2^-)_{ads} + H_2O + 2e^- \longrightarrow 3OH^-. \tag{2}$$

The second is the outer-sphere electron transfer mechanism, where solvated molecular O_2 cluster $O_2 \cdot (H_2O)_n$ weakly interacts with adsorbed hydroxyl species to promote a $2e^-$ reaction pathway to HO_2^- anion as a reaction product which desorbs from the surface and is eventually detected at the ring electrode. This reaction is formulated as follows:

$$\begin{aligned} &\text{M-OH} + [O_2 \cdot (H_2O)_n]_{aq} + e^- \\ &\quad \longrightarrow \text{M-OH} + (HO_2^{\bullet})_{ads} + OH^- + (H_2O)_{n-1}, \end{aligned} \tag{3}$$

$$(HO_2^{\bullet})_{ads} + e^- \longrightarrow (HO_2^-)_{ads} \tag{4}$$

$$(HO_2^-)_{ads} \longrightarrow (HO_2^-)_{aq}. \tag{5}$$

The first step in the previous reaction shown in (3) involves electron transfer (or tunneling) from the electrode surface across a thin oxide film and at least one layer of solvation shell to solvated O_2. Equation (3) involves several elementary steps as written in the following:

$$\begin{aligned} &\text{M-OH} + [O_2 \cdot (H_2O)_n]_{aq} + e^- \\ &\quad \longrightarrow \text{M-OH} + [O_2^{\bullet -} \cdot (H_2O)_n]_{aq}, \end{aligned} \tag{6a}$$

$$[O_2^{\bullet -} \cdot (H_2O)_n]_{aq} \longrightarrow (O_2^{\bullet -})_{ads} + nH_2O, \tag{6b}$$

$$(O_2^{\bullet -})_{ads} + H_2O \longrightarrow (HO_2^{\bullet})_{ads} + OH^-. \tag{6c}$$

First electron transfer to $O_{2,aq}$ forms $(O_2^{\bullet -})_{aq}$ which then undergoes desolvation and subsequent *adsorption on the oxide substructure* of the Pt surface to form $(O_2^{\bullet -})_{ads}$, followed by proton transfer to form adsorbed hydroperoxyl radical, $(HO_2^{\bullet})_{ads}$. Second electron transfer to $(HO_2^{\bullet})_{ads}$ yields $(HO_2^{\bullet -})_{ads}$. The binding energy of $(HO_2^{\bullet -})_{ads}$ on the oxide substructure of Pt is likely to be lower than that on oxide-free Pt site. This leads to the facile desorption of HO_2^- anion into the electrolyte which is eventually detected at the ring electrode. Also hydrogen peroxide anion formed via the homogeneous chemical disproportionation of superoxide anion would also be detected at the ring electrode. *The interaction between the $O_2 \cdot (H_2O)_n$ cluster and the surface hydroxyl species causes certain nonspecificity to the identity of the underlying electrode metal. This nonspecificity opens the gate to use a wide-range of nonnoble metals and their oxides as electrode materials for ORR in alkaline media.* On noble metals such as Pt, this nonspecificity is observed only in the oxide formation region, and depending on the extent of $-OH$ coverage both inner- and outer-sphere electron transfer mechanisms coexist in alkaline media. On nonnoble metals such as the first-row transition elements that are completely passivated by a layer oxide film upon immersion into the aqueous electrolyte, this outer-sphere electron transfer process is likely to be the dominant mechanism. Furthermore as identified in Figures 1(b) and 1(c), we consider the peak-shaped ring current in the potential range of 0.6 V to 0.9 V in alkaline media to be a characteristic signature of the outer-sphere electron transfer reaction mechanism. In acidic media, the adsorbed OH_{ads} species from water activation primarily serve only to block/inhibit the adsorption of molecular O_2 and other reaction intermediates via the well-known site-blocking effect [62, 63]. However, as shown here in alkaline media, the OH_{ads} species not only block the direct adsorption of O_2 but also serve to promote the $2e^-$ outer-sphere electron transfer reaction to peroxide. There are several precedents for electrochemical reactions which are mediated and/or promoted by specifically adsorbed anions and surface functional groups [64–67]. The anions or the surface groups act as an *outer-sphere bridge* between the reactant and the electrode surface. Furthermore as seen in Figures 1(b) and 1(c), peroxide intermediate is detected at the ring electrode at 0.5 V and also below 0.3 V versus RHE, and the reason for this was explained in detail in an earlier publication [53, 60]. Briefly, in alkaline media, the peak in ring current at 0.50 V is due to the carbon support and the increase in ring current below 0.3 V is due to reorientation of the water molecules at the electrode surface [53, 60].

In light of this new evidence of involvement of outer-sphere electron transfer mechanism during ORR in alkaline media, it is noted that, while this mechanism allows the possibility of nonnoble metals to be used as electrode materials, it predominantly seems to yield the $2e^-$ peroxide intermediate as the final product. So, for efficient $4e^-$ conversion of O_2, it is important to facilitate direct adsorption of desolvated molecular O_2 on the OH-free active site and avoid the precipitous reaction of solvated molecular O_2 with the OH-covered active site. In agreement with Bard [12], this implies the necessity to promote the electrocatalytic inner-sphere electron transfer mechanism for efficient $4e^-$ reduction of oxygen.

3.2. ORR on Heat-Treated Carbon-Supported Ru Nanoparticles. Figure 4 shows the ORR on Ru/C catalyst in O_2-saturated 0.1 M NaOH and 0.1 M $HClO_4$ electrolytes recorded at 900 rpm. The ORR onset potential for Ru/C in 0.1 M $HClO_4$ is 0.78 V versus RHE, whereas in 0.1 M NaOH it is 0.9 V versus RHE. In the alkaline media, mixed kinetic-diffusion region between 0.9 V and 0.5 V is ensued by a well-defined limiting current region below 0.5 V versus RHE. In the acidic electrolyte, ORR is kinetically controlled as indicated by the absence of any clear diffusion-limited current region even at very high overpotentials. As shown in Figure 4(b) at a kinetic current density of 0.1 mA $cm^{-2}{}_{geo}$, Ru/C exhibits ~125 mV lower overpotential in 0.1 M NaOH than in 0.1 M $HClO_4$. Primary reasons to this lower overpotential in alkaline media are delineated here. As seen in Figure 4(c) between the potential regions of 0.7 V to 0.8 V, the shoulder in ring current in 0.1 M NaOH electrolyte is characteristic of the outer-sphere electron transfer mechanism on oxide-covered Ru/C catalyst in alkaline media. This outer-sphere reaction mechanism was depicted in (3)–(5) giving rise to the peroxide intermediate (HO_2^-) as the final product at oxide-covered Ru sites. This $(HO_2^-)_{aq}$ formed is detected at the ring electrode and appears as a shoulder in ring current between 0.7 V and 0.8 V as shown in Figure 4(c). In acidic media, peroxide detected is very minimal and about an order of magnitude lower than that in 0.1 M NaOH. It should be noted that, in 0.1 M NaOH, this outer-sphere electron transfer occurs only at oxide-covered Ru sites, whereas at oxide-free Ru sites direct molecular O_2 adsorption should take place leading to efficient $4e^-$ reduction of O_2 to OH^- via an inner-sphere electrocatalytic pathway. So, in alkaline media, a combination of both inner-sphere and outer-sphere electron transfer mechanisms is operative. The consequence of this outer-sphere electron transfer in alkaline media is that this mechanism leads to higher concentration of HO_2^- to be generated near the electrode surface, that is, the double layer; however, no evidence for such an outer-sphere reaction is observed in acidic media. Higher activity of HO_2^- effectively shifts the potential of the electrode from that of the O_2/HO_2^- couple to that of the HO_2^-/OH^- redox couple by carrying out HO_2^- reduction to OH^- on oxide-free Ru sites. This is shown in Figure 4(d), where ORR and hydrogen peroxide reduction reaction (HPRR) are shown on Ru/C catalyst in both 0.1 M NaOH and 0.1 M $HClO_4$. $2e^-$ reduction of hydrogen peroxide in acidic (H_2O_2/H_2O) and alkaline media (HO_2^-/OH^-) are written as follows [13]:

$$\text{Acidic Media: } H_2O_2 + 2H^+ + 2e^- \longrightarrow 2H_2O \qquad E^\circ = 1.763\,\text{V versus SHE}, \tag{7}$$

$$\text{Alkaline Media: } HO_2^- + H_2O + 2e^- \longrightarrow 3OH^- \qquad E^\circ = 0.867\,\text{V versus SHE}. \tag{8}$$

As is well known, the standard reduction potentials of the above reactions are well positive of the $4e^-$ reduction of

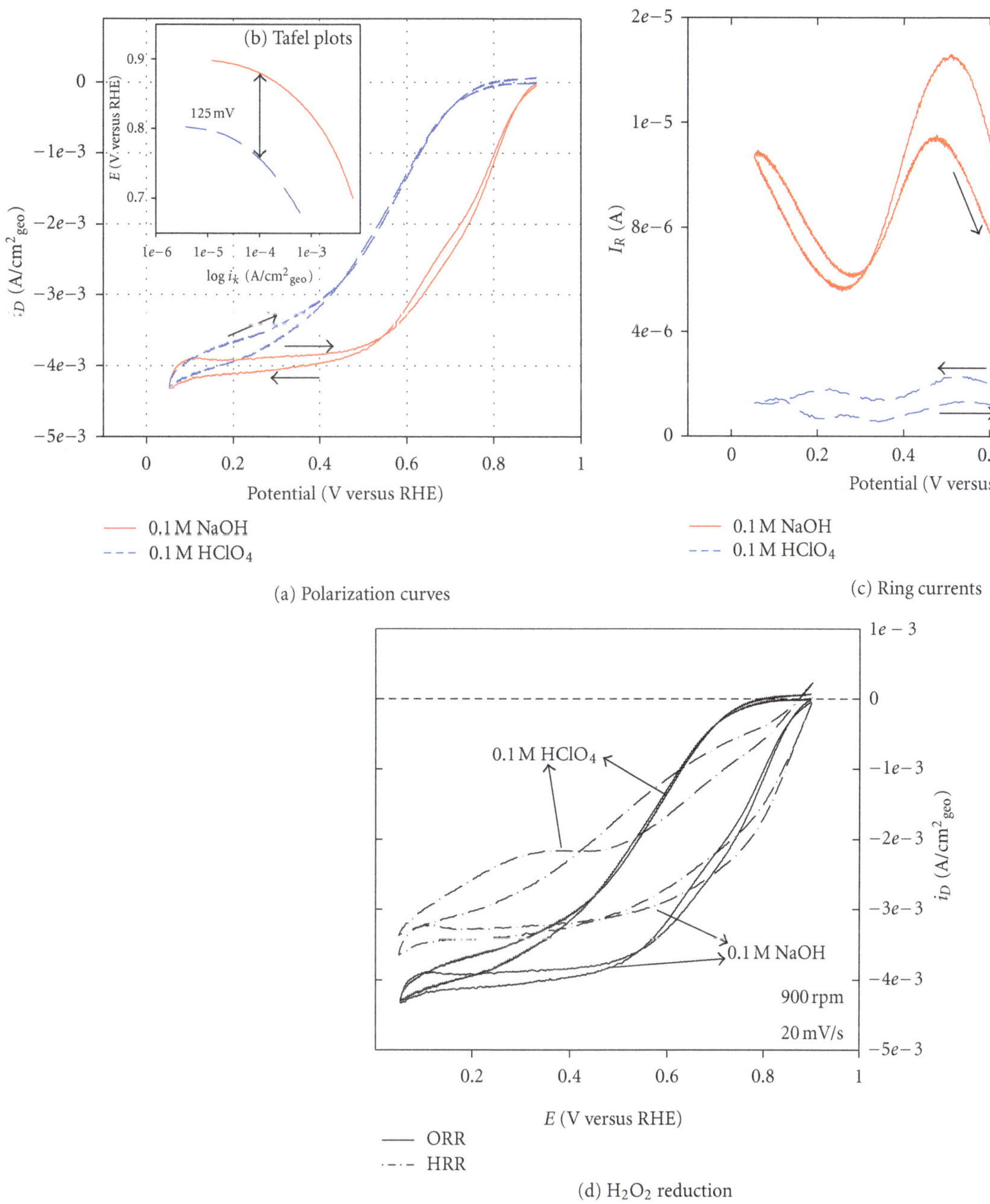

FIGURE 4: Oxygen reduction reaction on Ru/C heat-treated (HT) catalyst in 0.1 M NaOH and 0.1 M $HClO_4$ electrolytes. (a) ORR polarization curves at 900 rpm and 20 mV/s, (b) mass-transport-corrected Tafel plots, (c) ring current measured during ORR at 900 rpm. E_{Ring} = 1.1 V versus RHE in 0.1 M NaOH and E_{Ring} = 1.3 V versus RHE in 0.1 M $HClO_4$, and (d) hydrogen peroxide reduction reaction (HPRR) in comparison to ORR in 0.1 M $HClO_4$ and 0.1 M NaOH. HPRR is shown in oxygen-free electrolytes containing externally added H_2O_2 at a concentration of 3.5 mM.

molecular O_2 in both acidic and alkaline electrolytes. So, from a thermodynamic perspective, any peroxide intermediate formed should be immediately reduced further. This is also kinetically true on ruthenium as shown in Figure 4(d). Half-wave potential ($E_{1/2}$) of HPRR on Ru/C in 0.1 M NaOH is 40 mV positive compared to that of the $E_{1/2}$ of ORR [68]. So the kinetics of the reaction in alkaline media favors the immediate reduction of any peroxide intermediate generated during ORR. This is also true in acidic media although the shape of the HPRR profile in acidic media on Ru/C requires more explanation. The onset potential of HPRR on Ru/C in 0.1 M $HClO_4$ is as high as 0.88 V which is only 20 mV lower than ORR and HPRR onset potentials in 0.1 M NaOH electrolyte. As is known previously [69], H_2O_2 undergoes decomposition to O_2 and H_2O in acidic electrolyte at Ru/C surface. This decomposition reaction that generates O_2 near the electrode surface skews the HPRR profile in acidic media to higher overpotentials characteristic of ORR in acidic media. However, the aspect of relevance to the discussion here is that the kinetics of the system favors further reduction of hydrogen peroxide intermediate in both acidic and alkaline media. Once the H_2O_2/HO_2^- stable intermediate is generated, this species undergoes adsorption at oxide-free Ru sites and further reduces to H_2O/OH^- according to (7)/(9). So, any *in situ* generation of hydrogen peroxide intermediate should shift the potential to more positive values, and indeed this is what is observed in alkaline media. The *in situ* parallel generation of HO_2^- anion intermediate via the outer-sphere electron transfer reaction scheme shown in (3)–(5) serves to shift the ORR potential to more positive values in alkaline media. On the contrary, this excess parallel generation of H_2O_2 via the outer-sphere reaction mechanism is not observed in acidic media and, hence, the higher overpotential for ORR in acidic media. In general, this can also be further extended to other catalyst systems that exhibit lower ORR overpotential in alkaline media compared to that in acidic electrolytes. The prospects of modifying the carbon-supported ruthenium nanoparticle catalyst by chalcogen groups (S/Se) in order to suppress oxide formation and promote the inner-sphere electrocatalytic process by facilitating direct O_2 adsorption was discussed recently as a separate treatise [70].

3.3. Pyrolyzed Metal Macrocycle for Promoting Inner-Sphere Electron Transfer. Figure 5(a) shows a comparison in dilute acidic and alkaline electrolytes of the ORR activity of FeTPP/C catalyst pyrolyzed at 800°C. As observed in Figure 5(a), the onset potential for ORR in 0.1 M NaOH is 0.95 V versus RHE, whereas in 0.1 M $HClO_4$ electrolyte it is 0.80 V versus RHE. This 150 mV lower overpotential in alkaline media is clearly reflected over the entire mixed kinetic-diffusion region. In 0.1 M NaOH electrolyte, the mixed kinetic-diffusion region is ensued by a well-defined diffusion-limited region. In 0.1 M $HClO_4$, no clear diffusion-limited region could be discerned, which is indicative of kinetic control in acidic media even at high overpotentials. At a potential of 0.80 V versus RHE, the ORR kinetic current density of FeTPP/C (pyrolyzed at 800°C) is clearly four orders of magnitude higher in 0.1 M NaOH electrolyte than that in 0.1 M $HClO_4$. For a given catalyst, the four orders of magnitude difference in kinetic activity between acidic and alkaline media are intriguing. Figure 5(b) shows the HPRR activity of FeTPP/C catalyst (pyrolyzed at 800°C) in both acidic and alkaline media in comparison to the corresponding ORR polarization curves. The onset potential for peroxide reduction in 0.1 M $HClO_4$ is 0.84 V versus RHE, whereas in 0.1 M NaOH it is 1.01 V versus RHE. Besides this onset potential difference, in 0.1 M NaOH electrolyte, the mixed kinetic-diffusion region for peroxide reduction is more anodic compared to that of ORR in the same electrolyte which is then followed by a reasonably discernable diffusion-limited current density region. This clearly indicates that peroxide reduction in alkaline media is kinetically favored such that any peroxide intermediate formed during ORR in 0.1 M NaOH will be immediately reduced to the $4e^-$ product. On the contrary, the reduction of hydrogen peroxide in acidic media is kinetically unfavorable due to weak binding of H_2O_2 on the active site leading to its desorption into the bulk electrolyte and/or catalytic decomposition to molecular O_2. *This clearly demonstrates that stabilizing the peroxide intermediate on the active site is important in effectively carrying out $4e^-$ ORR process, and the higher stability of peroxide intermediate on the active site leads to the so-called kinetic facility in alkaline media but not in acidic media.* Given the pK_a values for the first and second ionization of H_2O_2 at 25°C (pK_1 = 11.69 and pK_2 = ~20), the predominant peroxide species for pH < 12 is H_2O_2 whereas at pH > 12 it is HO_2^- [58]. This leads to the formation of a stable Lewis acid-base adduct in alkaline media (Fe^{2+}-HO_2^-) that enables complete $4e^-$ transfer. On the contrary, the lack of stabilization of (Fe^{2+}-H_2O_2) adducts in acidic media leads to desorption of H_2O_2 into the bulk electrolyte.

Figure 5(c) shows the ring current due to peroxide oxidation measured during ORR in both acidic and alkaline electrolytes. Clearly, the peroxide yield during ORR on FeTPP/C (pyrolyzed at 800°C) is higher in acidic media by an order of magnitude compared to that in alkaline media. This observation is opposite to what was observed on the Pt/C and Ru/C catalysts discussed previously. The onset potential for ORR in 0.1 M NaOH is 0.95 V versus RHE, whereas the corresponding peroxide oxidation current does not begin until 0.8 V. In 0.1 M $HClO_4$, the onset potential for both ORR and peroxide oxidation is 0.8 V. This is further proof for the instability of peroxide intermediate on the active site in acidic media because the weak binding of the H_2O_2 intermediate on the Fe^{2+} active site facilitates its desorption and/or decomposition into the bulk electrolyte. This desorption of H_2O_2 into the bulk electrolyte due to its weak binding on the active site is the primary source of peroxide detected at the ring in acidic media. At very high overpotentials (<0.3 V), the ORR process in acidic electrolyte becomes very efficient and consequently the peroxide detected decreases [71, 72]. The ring current profile in 0.1 M NaOH is more complex. Briefly, in alkaline media, the peak in ring current at 0.50 V is due to the carbon support and the increase in ring current below 0.3 V is due to reorientation of the water molecules [60, 73]. More importantly, a weak shoulder in the ring current in 0.1 M NaOH is observed between 0.6 V and 0.7 V. Also shown in Figure 5(c) is the ring current profile of Pt/C for comparison. As discussed previously, the ring current peak due to Pt/C around 0.80 to 0.90 V is the signature for the peroxide formation via outer-sphere electron transfer mechanism at –OH-covered Pt sites. On Fe-N_x-based catalyst, only a weak shoulder in ring current at more negative potentials of 0.60 to 0.70 V is observed indicating that the outer-sphere electron transfer mechanism is significantly muted. This indicates the facilitation of direct adsorption of O_2 on the Fe^{2+} active site. This is because the specifically adsorbed OH^- species on the Fe^{2+} site acts as a labile ligand that is easily displaced by the molecular O_2. This is ensued by the successful $4e^-$ reduction of O_2 with minimal occurrence of outer-sphere electron transfer mechanism.

The nature of the active site obtained upon pyrolysis and the fundamental origin of the activity is investigated here in detail. Figure 6(a) shows the square wave voltammetry (SWV) of non-heat-treated Fe(III)TPPCl/C in

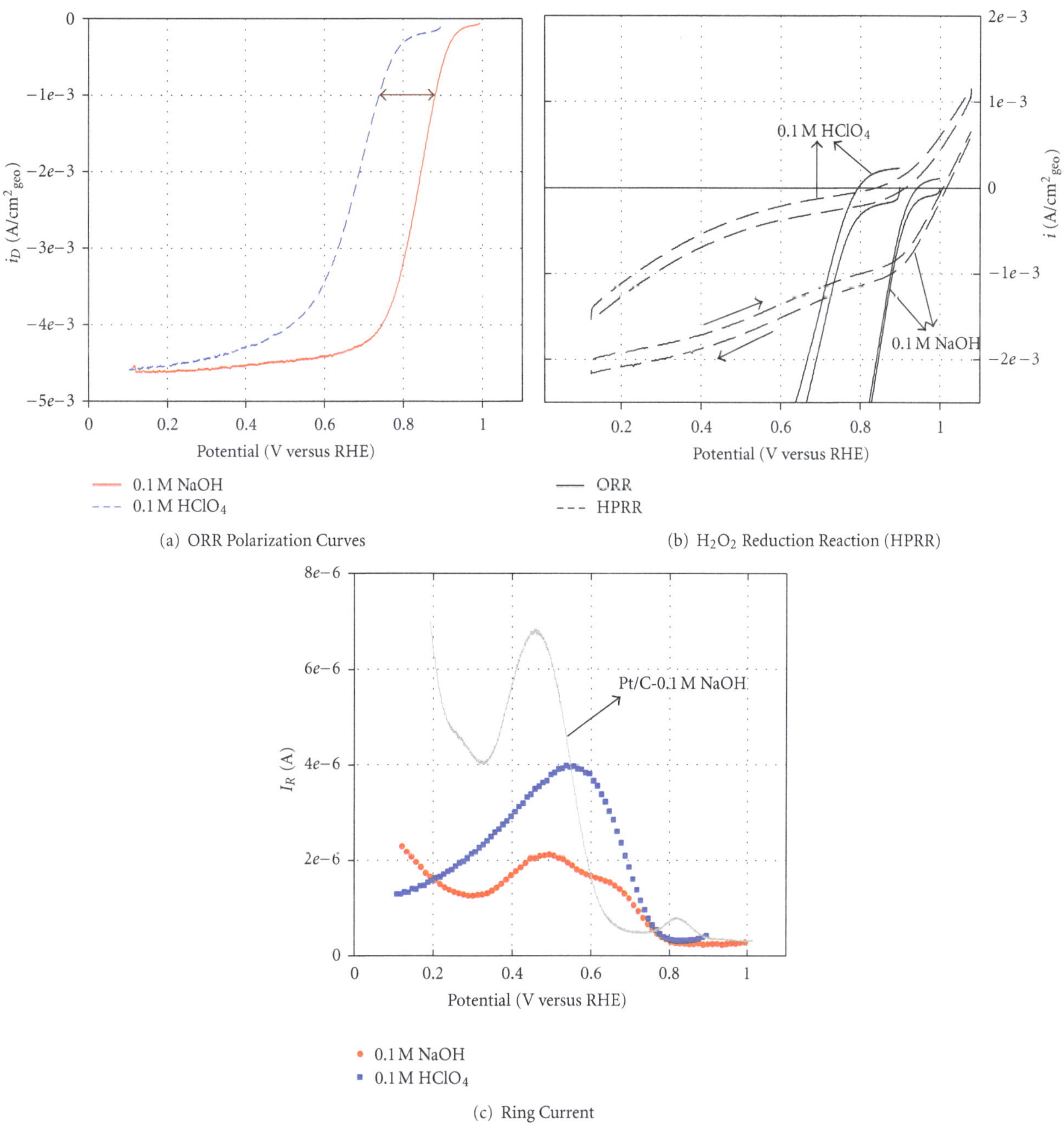

FIGURE 5: ORR activity of FeTPP/C catalyst (pyrolyzed at 800°C) in O_2 saturated acidic and alkaline electrolytes. (a) ORR polarization curves, (b) hydrogen peroxide reduction reaction (HPRR) in comparison to ORR, and (c) ring current. All measurements were performed at 900 rpm rotation rate and 20 mV/s scan rate. E_{Ring} = 1.1 V versus RHE in 0.1 M NaOH and E_{Ring} = 1.3 V versus RHE in 0.1 M $HClO_4$. HPRR is shown in oxygen-free electrolytes containing externally added H_2O_2 at a concentration of 3.5 mM. Also shown in (c) is the ring current profile of 30% Pt/C in 0.1 M NaOH for comparison.

argon-saturated 0.1 M NaOH and 0.1 M $HClO_4$ electrolytes, and Figure 6(b) shows the corresponding CV. As seen in both the SWV and the CV profiles, the redox transition involving the metal center $Fe^{2+/3+}$ is observed at 0.314 V versus RHE in 0.1 M NaOH and at 0.155 V versus RHE in 0.1 M $HClO_4$ [30, 31, 74]. As seen in the SWV profile, the peaks at 1.260 V in 0.1 M $HClO_4$ and 1.508 V in 0.1 M NaOH correspond to the one-electron redox transition involving the delocalized π-electron system in the macrocycle ligand [74]. As observed in the literature [30], CV of the heat-treated catalysts does not yield useful information, since in most cases after heat treatment the double-layer charging current overwhelms the Faradaic currents. In order to overcome this limitation, SWV was performed as shown in Figure 7 in order to understand the evolution of the active site in these catalysts with increasing heat treatment temperatures. Figures 7(a)

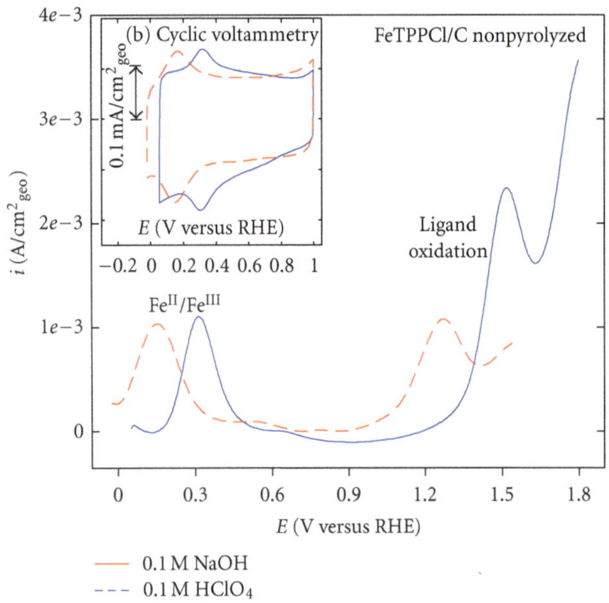

(a) Square wave voltammetry

FIGURE 6: (a) Square wave voltammetry (SWV) and (b) cyclic voltammetry (CV) of as-received iron(III) meso-tetraphenylporphyrin chloride (FeTPPCl/C) supported on Black Pearl carbon. Experiments were performed in argon-saturated 0.1 M NaOH and 0.1 M $HClO_4$ electrolytes. SWV parameters: 5 mV step potential, 20 mV amplitude, and 10 Hz frequency. CV: 20 mV/s.

and 7(b) are divided into two panels each due to different potential ranges that are required for baseline correction of SWV profiles. As seen in Figure 7(a) in alkaline media, after 300°C pyrolysis, the $Fe^{2+/3+}$ redox couple at 0.31 V and the ligand oxidation peak at 1.5 V are similar to the case of non-heat-treated catalyst. After heat treatment at 500°C and 600°C, the magnitude of the peak currents decrease significantly due to the sublimation of some fraction of the metal macrocycle complex. More importantly, the $Fe^{2+/3+}$ peak potentials have shifted to more anodic potentials. As seen in Panel 1 of Figure 7(a), after pyrolysis at 500°C and 600°C, the peak potential (E_p) of the $Fe^{2+/3+}$ couple has shifted to 0.405 V and 0.427 V, respectively. Correspondingly, the ligand oxidation peak at ~1.5 V, shown in Panel 2 of Figure 7(a), decreases in magnitude significantly due to the sublimation and/or destruction of the macrocycle. The ligand oxidation peak is not observed after heat treatment at temperatures greater than 600°C indicating that the delocalized π-electron system of the macrocycle does not survive such high temperatures on the carbon support. After 800°C pyrolysis, $Fe^{2+/3+}$ couple has shifted more anodically to $E_p = 0.48$ V. As shown in Figure 7(a) (Panel 2), the most interesting observation is that after pyrolysis at 600°C a new shoulder begins to emerge at ~1.2 V. This shoulder resolves into a clear peak after 800°C pyrolysis with a peak potential of 1.25 V versus RHE in alkaline media. This new peak is attributed to an anodic shift in the $Fe^{2+/3+}$ redox couple based on the *in situ* XANES experimental results shown later (*vide infra*). This indicates the presence of two $Fe^{2+/3+}$ redox couples, one at a low potential (~0.48 V) and another at a high potential (1.25 V). The anodic shift in $Fe^{2+/3+}$ redox peak potentials shown in Figure 7(a) (inset) clearly indicates that after 600°C pyrolysis the metal center exists in two redox environments. Qualitatively, similar behavior was observed in acidic media as shown in Figure 7(b). (Note that the ligand oxidation peak in acidic media that occurs above 1 V versus RHE is not shown in Panel 2 of Figure 7(b).) It is interesting to note that after heat treatment at 600°C the $Fe^{2+/3+}$ redox peak potential in 0.1 M $HClO_4$ was shifted to an anodic potential of only 0.80 V versus RHE compared to the 1.25 V in alkaline media. Thus, there are two reasons for the lower ORR overpotential for FeTPP/C catalyst in alkaline media: (1) the higher redox potential of the $Fe^{2+/3+}$ metal center in alkaline media and (2) the improved stability of the peroxide intermediate on the active site. This translates into efficient $4e^-$ ORR reaction with lower overpotential in alkaline media compared to acidic media.

3.3.1. Delta-Mu ($\Delta\mu$) Studies. Both the extended X-ray absorption fine structure (EXAFS) and X-ray absorption near edge structure (XANES) being bulk averaged techniques overlook the critical electrochemical reactions occurring on the catalyst surface. Delta-mu ($\Delta\mu$) is a surface sensitive, spectral subtraction technique, where the bulk structure of the catalyst is effectively removed leading to information on the nature and site specificity of the surface adsorbates [54, 55, 75, 76]. Figure 8(a) shows the XANES region of FeTPP/C catalyst pyrolyzed at 300°C taken at two different potentials of 0.1 V and 0.9 V versus RHE. At 0.1 V, the metal center exists in the reduced Fe^{2+} state with no adsorbates (neither hydrides nor oxides) at the axial position, and immediate coordination environment is reminiscent of the iron(II) phthalocyanine complex, where the preedge peak is muted. At 0.9 V, the metal center is oxidized to Fe^{3+} state with an oxygen atom at the axial position, and the metal coordination environment is similar to that of the original porphyrin complex, where the preedge Fe(1s → 3d) forbidden transition at 7112.5 eV is observed. The delta-mu spectra is obtained by subtracting the XANES regions according to the equation $\Delta\mu = \mu(0.90\,\text{V}) - \mu(0.10\,\text{V})$. In the delta-mu spectra of Figure 8(a), the positive peak feature (boxed portion) indicates the difference in absorption probability at the preedge energy (7112.5 eV). This positive peak feature could be safely assigned as a signature for the existence of the metal center in a centrosymmetric environment undergoing a transition from Fe^{2+}-N_4 coordination geometry at 0.10 V to O-Fe^{3+}-N_4 coordination geometry at 0.90 V. Ensuing this positive peak is a steep negative dip featuring a split peak below 20 eV. This negative dip characterizes charge transfer from the metal center to the adsorbed oxygen species [77]. As shown in Figure 8(b), the XANES spectra of the 800°C pyrolyzed catalyst is predominantly characteristic of metallic iron at both 0.10 V and 1.10 V that precludes proper analysis of the active site. However, careful analysis of the corresponding delta-mu spectra clearly indicates the positive peak feature at the preedge energy indicative of the fact that the active site is Fe^{2+}-N_4 where the metal center is in a centrosymmetric environment which is mildly disrupted by the presence of

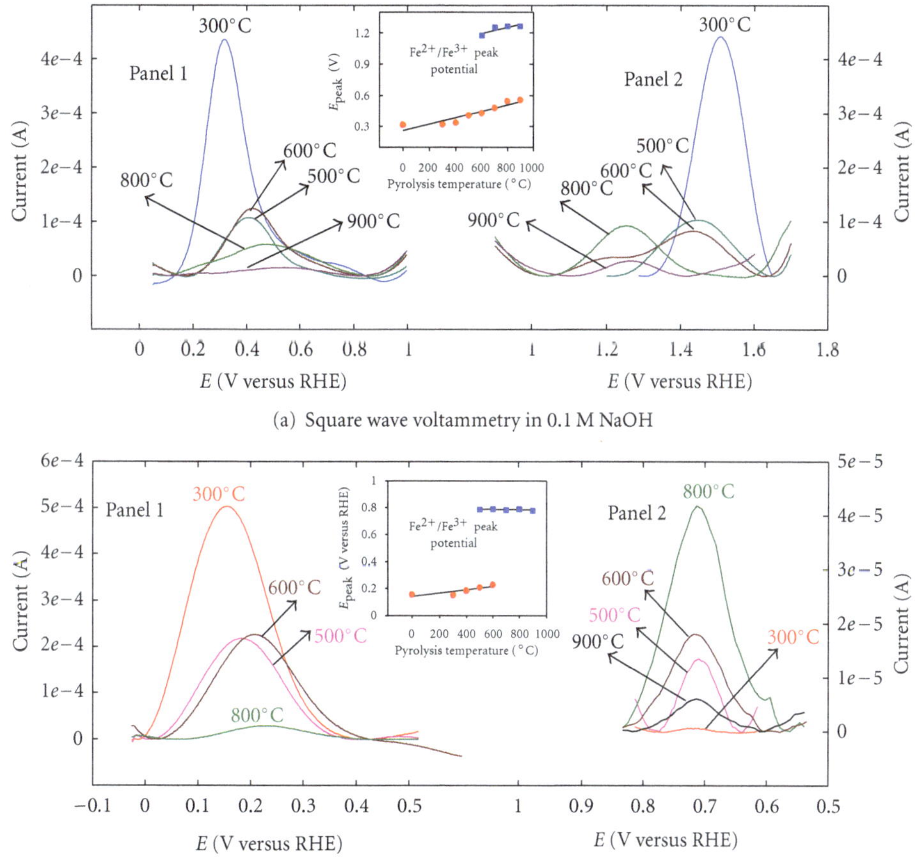

(a) Square wave voltammetry in 0.1 M NaOH

(b) Square wave voltammetry in 0.1 M HClO$_4$

FIGURE 7: SWV profiles of FeTPP/C catalyst as a function of heat treatment temperature in (a) 0.1 M NaOH and (b) 0.1 M HClO$_4$ electrolytes. Insets show the peak potential of the $Fe^{2+/3+}$ redox couple as a function of the heat treatment temperature. SWV experiment parameters: step potential −5 mV, amplitude −20 mV, and scan frequency −10 Hz.

an axial oxygen atom. This clearly indicates that the Fe^{2+} metal center surrounded by four nitrogen atoms is the active site that binds oxygen in the axial position and the redox transition from Fe^{3+} to Fe^{2+} triggers oxygen adsorption according to the redox mechanism [78]. Figure 9 shows the theoretical delta-mu ($\Delta\mu_t$) spectra obtained using FEFF8.0 code [56]. The structural models used are shown in the insets of Figure 9. Only the atoms encircled were used in the theoretical FEFF8.0 modeling. These theoretical spectra were calculated from Fe-N$_4$-C$_x$ models derived from prior crystallographic data adjusted to the EXAFS fitting results according to the relation $\Delta\mu_t = \mu(\text{O}_{\text{ads}}\text{-Fe-N}_4\text{-C}) - \mu(\text{Fe-N}_4\text{-C})$, where the oxide species (O$_{\text{ads}}$ or OH$_{\text{ads}}$) is in a specific binding site. In all cases, the positive peak feature at the preedge energy was observed only when the adsorbed oxygen atom was placed in the axial position of the metal center. No successful theoretical delta-mu fits could be obtained for Fe-N coordination numbers less than four or for oxygen adsorption modes other than at the axial position. As shown in the inset of Figure 9(a), after 300°C pyrolysis the immediate coordination environment of the original precursor porphyrin macrocycle is clearly retained. This corresponds to an FeN$_4$C$_{12}$ cluster, where the metal center is coordinated to four nitrogen atoms and each nitrogen atom in turn bonded to two carbon atoms. Finally, inclusion of the four methine carbon bridges gives a tally of C$_{12}$. Figure 9(b) shows the theoretical delta-mu spectrum that likely mimics the line shape of the experimental delta-mu shown in Figure 8(b) for the 800°C heat-treated catalyst. As shown in the inset of Figure 9(b), the molecular cluster used to simulate the theoretical delta-mu spectrum consisted of FeN$_4$C$_{10}$. While compositionally this cluster is not very different from the 300°C pyrolyzed sample, the immediate coordination environment of the metal active site after 800°C heat treatment is found to be reminiscent of the crystallographic atomic defects such as the divacancy on the graphitic surfaces. This is schematically depicted in Figure 9(c) [79, 80]. Atomic defects such as monovacancy and divacancy on microporous carbon and carbon nanotubes are known to thermodynamically exist or can be induced via various chemical or physical processes [81–85]. The presence of a monovacancy in carbon atom creates three dangling bonds whereas divacancies create four dangling bonds. These dangling bonds give rise to unsaturated valences that then become favorable for nitrogen doping [80]. Consequently, these nitrogen-doped sites constitute defective pockets for metal coordination. Such atomic vacancies are either already present on the graphite surfaces or can be created during the

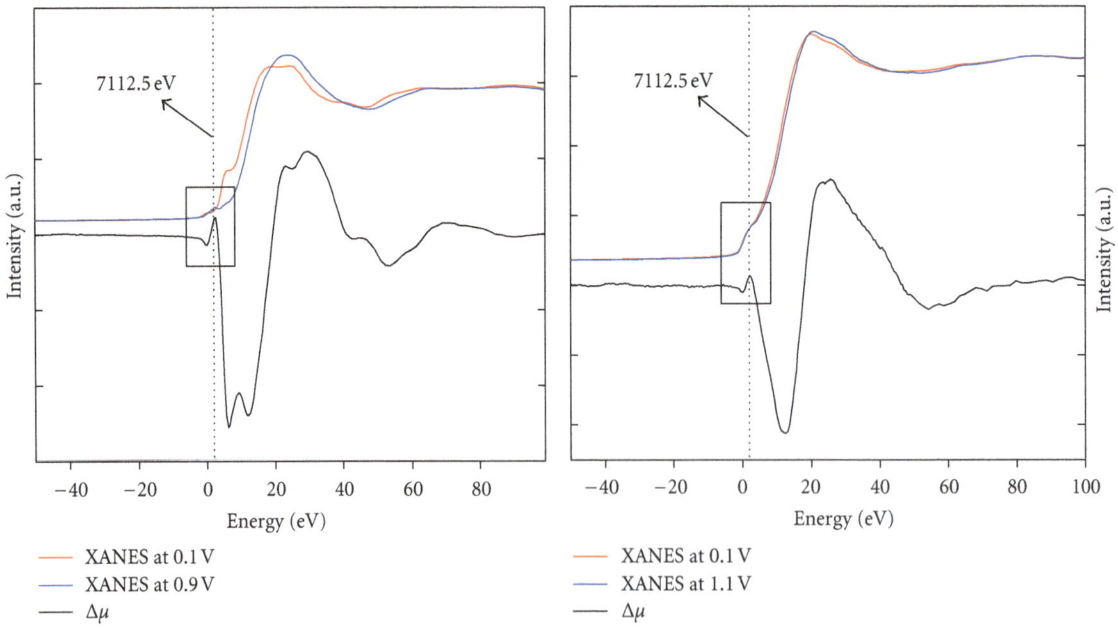

(a) 300°C Pyrolysis

(b) 800°C Pyrolysis

FIGURE 8: Experimental XANES and delta-mu signatures of FeTPP/C catalyst heat treated at (a) 300°C and (b) 800°C. Delta-mu signatures were obtained by subtracting the XANES signatures according to $\Delta\mu = \mu(0.90 \text{ (or) } 1.10\,V) - \mu(0.10\,V)$. Experiments were conducted at Fe K-edge under *in situ* conditions in argon-saturated 0.1 M NaOH electrolyte. Vertical dotted line indicates the preedge position at 7112.5 eV and the boxed region focuses the preedge region. Delta-mu spectra have been multiplied by a factor of 5 for visual comparison of the line shapes.

heat treatment step [81, 83]. During heat treatment under inert atmosphere, carbothermic reaction causes desorption of oxygen functional groups along with creation of vacancy defects [83]. It has also been observed earlier that heat-treated Fe-N_x catalysts showed higher activity when supported on carbon that was previously treated in concentrated inorganic acids [52]. Since such acid treatment steps lead to oxygen functional groups on carbon support, it is likely that this yields higher number of defective sites during subsequent heat treatment. These defective pockets are likely the favorable zones for anchoring FeN_4 active sites. Presumably, the low concentration of the defective sites limits the active site density. Therefore, increasing the defect density likely holds key to increasing the metal loading in this class of catalysts.

The following observations are made in an attempt to correlate the SWV and the delta-mu results. The low-potential $Fe^{2+/3+}$ redox couple is characteristic of the original FeN_4 porphyrin moiety, whereas the high-potential $Fe^{2+/3+}$ redox couple is characteristic of the FeN_4 cluster found in the microporous carbon defect sites. At a pyrolysis temperature of 600°C, both of these sites seem to coexist and the higher ORR activity obtained for heat treatment temperatures ≥600°C is due to the high-potential $Fe^{2+/3+}$ redox couple seated in the defective pockets. Based on the above experimental results, the following reaction scheme is proposed for ORR in dilute alkaline media on heat-treated FeTPP/C catalyst. Similar set of reactions can be developed, *mutatis mutandis*, for dilute acidic media [22, 71, 72, 86]. Equation (9) shows the redox reaction involving the metal center that is a prerequisite for adsorption of molecular oxygen on the active site $[N_4\text{-}Fe^{II}\text{-}OH^-]$:

$$\left[N_4\text{-}Fe^{III}\text{-}OH\right] + e^- \longrightarrow \left[N_4\text{-}Fe^{II}\right] + OH^- \quad (9)$$

While the Fe^{II} valence state favors a square-planar tetra-coordinate environment, the high potential of this redox reaction causes the OH^- species to poison the active site at the axial position. This poisonous OH^- species prevents direct molecular adsorption of O_2 on the active site. Further, the adsorbed OH^- species mediates the $2e^-$ outer-sphere electron transfer reduction of solvated O_2 molecule as shown in the reaction schemes (3)–(5) although this outer-sphere mechanism appears to be extremely weak in this class of pyrolyzed catalysts as explained previously. It should be noted that the case of outer-sphere mechanism does not arise in acidic media for reasons explained previously.

Finally, the $4e^-$ electrocatalytic inner-sphere electron transfer mechanism is shown in Figure 10, where the molecular O_2 displaces the OH^- species and chemisorbs directly on the Fe^{II} active site. Based on our experimental results, the electrocatalytic process in Figure 10 is shown to take place via a redox mechanism involving $Fe^{II/III}$ couple [78]. Once molecular O_2 adsorbs on the Fe^{II} active site, the reaction proceeds to the ferrous-hydroperoxyl adduct via the superoxo and the ferric-hydroperoxyl states. The ferrous-hydroperoxyl adduct is very critical since its stability determines the product distribution. Peroxide anion (HO_2^-) is a stable intermediate. So, weak binding of HO_2^- on the

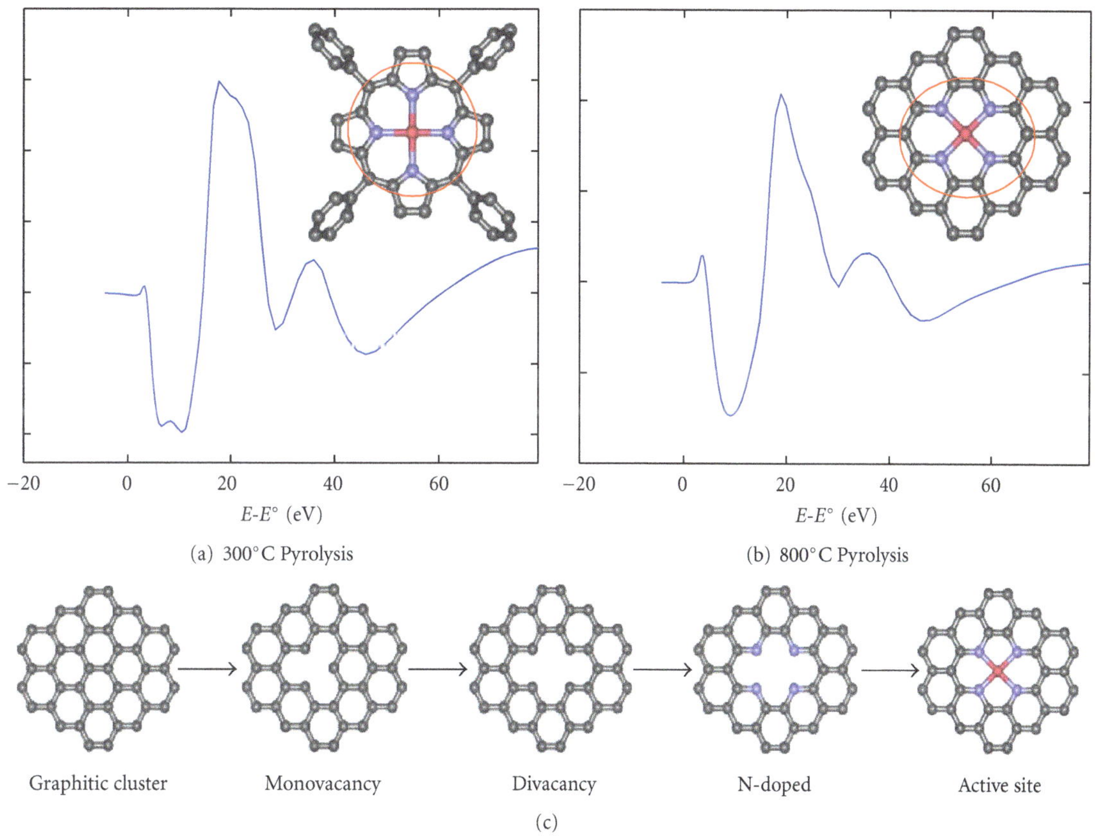

FIGURE 9: Theoretical FEFF8 $\Delta\mu = \mu(\text{Fe-N}_4\text{-C}_x\text{-O}_{\text{ads}})-\mu(\text{Fe-N}_4\text{-C}_x)$ signatures obtained for (a) 300°C and (b) 800°C pyrolysis conditions. The insets in (a, b) show the corresponding structural models utilized. Only the atoms encircled in these structural models were used for FEFF8 simulation. (c) Schematic illustration of the mono- and divacant defective pockets in amorphous carbon acting as anchors for active site formation during pyrolysis. Color codes: pink: Fe, blue: N, and black: C.

active site will lead to its desorption into the bulk electrolyte. As mentioned previously, the stable acid-base adduct formed in alkaline media stabilizes the ferrous-hydroperoxyl adduct. This ensures that the catalytic cycle shown in Figure 10 regenerates the active site via the formation of ferric-hydroxyl species. However, in acidic media the analogous ferrous-hydrogen peroxide adduct Fe^{II}-(OHOH) is unstable and causes desorption and/or disproportionation of the peroxide intermediate (H_2O_2) into the bulk electrolyte. This leads to higher peroxide yield in acidic media as shown previously.

4. Conclusions

The case of ORR in alkaline media shows interesting mechanistic differences compared to acidic media. In acidic media, the presence of OH_{ad} due to water activation blocks the direct adsorption of molecular O_2 on the active site. In alkaline media, the presence of OH_{ad} due to specific adsorption of hydroxide species not only blocks the O_2 adsorption but also promotes outer-sphere electron transfer mechanism to yield the $2e^-$ peroxide intermediate as the product. The interaction between the $O_2 \cdot (H_2O)_n$ cluster and the surface hydroxyl species causes certain nonspecificity to the identity of the underlying electrode metal. This nonspecificity is a result of kinetic differences in alkaline media that opens the gate to use a wide-range of nonnoble metals and their oxides as electrode materials, while in acidic media the absence of this outer-sphere mechanism necessitates the use of highly catalytic surfaces based on Pt. On noble metals such as Pt, this nonspecificity in alkaline media is observed only in the oxide formation region, and depending on the extent of OH_{ad} coverage both inner- and outer-sphere electron transfer mechanisms coexist in alkaline media. On nonnoble metals such as the first-row transition elements that are completely passivated by a layer oxide film upon immersion into the aqueous electrolyte, this outer-sphere electron transfer process is likely to be the dominant mechanism. While the outer-sphere mechanism primarily seems to yield peroxide intermediate as the final product, this could be beneficial in cases like Ru/C catalyst, where the peroxide intermediate generated in the outer-sphere process could be readsorbed on the oxide-free Ru site to reduce further to form hydroxide as the final product. However, in cases like nonnoble metal surfaces, where the electrode surface is incapable of reducing peroxide intermediate any further, only $2e^-$ reduction products are likely to predominate in alkaline media. So, it is important to promote the inner-sphere electron transfer mechanism in alkaline media by facilitating the direct molecular

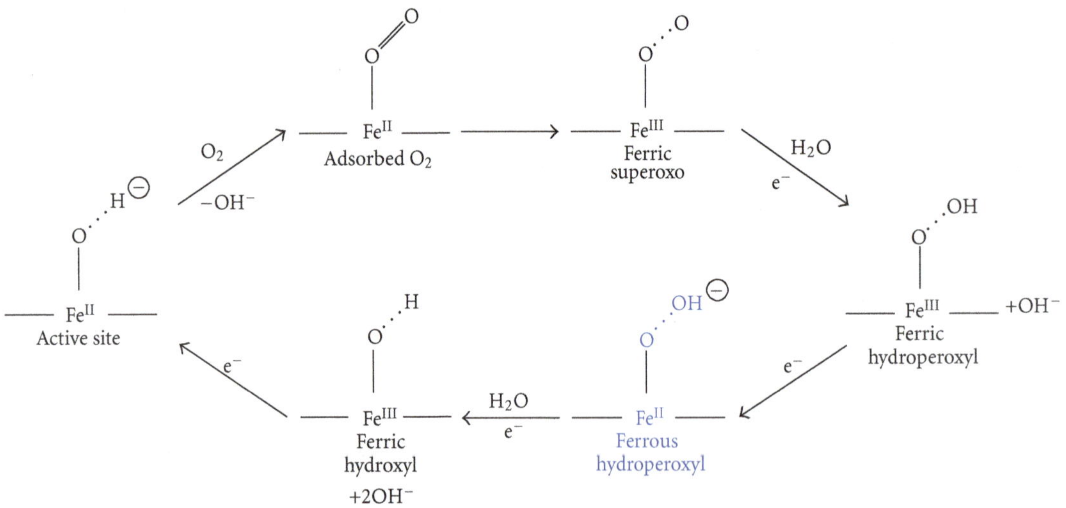

Figure 10: Catalytic cycle showing the redox mechanism involved in ORR on heat-treated iron porphyrin macrocycles in dilute alkaline medium. Nitrogen atoms in the square planar positions have been omitted for clarity.

O_2 adsorption and promoting complete $4e^-$ reduction process.

On catalyst surfaces such as the Fe-N_4 systems, the operation of redox mechanism allows the direct molecular O_2 adsorption on the Fe^{2+}-N_4 active site and prevents the precipitous outer-sphere reaction of solvated O_2 molecule with the OH_{ad}-covered metal site. This is due to the OH_{ad} species acting as a labile ligand in the Fe-N_4-based catalysts that helps to promote the $4e^-$ inner-sphere electron transfer mechanism by efficiently reducing O_2 to OH^- in alkaline media. Further, in the class of Fe-N_4-based catalysts, it is important to perform ORR studies with a thin film of catalyst surface ($\leq 25\,\mu m$) rather than thick catalyst films that are typically used in the literature, where the peroxide formed is completely quenched within the catalyst layer and hence not being detected at the ring electrode. This is especially true while Koutecky-Levich plots for calculating the number of electrons transferred since such calculations were derived for smooth electrode surfaces. Also, it is noted that if a catalyst is considered to perform $4e^-$ oxygen reduction, then it is imperative to show the reduction of externally added hydrogen peroxide in the absence of O_2. Finally, in conclusion, the presence of the outer-sphere electron transfer mechanism in alkaline media is a kinetic effect. The only thermodynamic advantage of performing ORR at high pH is the improved stability of the hydroperoxide anion intermediate on the active site that enables complete $4e^-$ transfer. So, in order to stabilize the H_2O_2 intermediate in acidic media, the use of dimeric active sites that simultaneously bind both oxygen atoms are required.

Acknowledgments

The authors deeply appreciate financial assistance from the Army Research Office under the Single Investigator Grant. The authors also gratefully acknowledge the supply of platinum electrocatalysts from BASF fuel cells (Somerset, NJ, USA). Use of the National Synchrotron Light Source (NSLS) at Brookhaven National Laboratory (BNL) was supported by the US Department of Energy, Office of Basic Energy Sciences. Support from NSLS beamline personnel Drs. Kaumudi Pandya (X11A), Syed Khalid (X19A), and Nebojsa Marinkovic (X18B) is gratefully acknowledged.

References

[1] J. S. Spendelow and A. Wieckowski, "Electrocatalysis of oxygen reduction and small alcohol oxidation in alkaline media," *Physical Chemistry Chemical Physics*, vol. 9, no. 21, pp. 2654–2675, 2007.

[2] E. Yeager, "Electrocatalysts for O_2 reduction," *Electrochimica Acta*, vol. 29, no. 11, pp. 1527–1537, 1984.

[3] R. Adzic, "Frontiers in electrochemistry," in *Electrocatalysis*, J. Lipkowski and P. N. Ross, Eds., vol. 197, Wiley-VCH, New York, NY, USA, 1998.

[4] D. B. Sepa, M. V. Vojnovic, and A. Damjanovic, "Kinetics and mechanism of O_2 reduction at Pt in alkaline solutions," *Electrochimica Acta*, vol. 25, no. 11, pp. 1491–1496, 1980.

[5] D. B. Sepa, M. V. Vojnovic, L. M. Vracar, and A. Damjanovic, "Different views regarding the kinetics and mechanisms of oxygen reduction at Pt and Pd electrodes," *Electrochimica Acta*, vol. 32, no. 1, pp. 129–134, 1987.

[6] A. Damjanovic, A. Dey, and J. O. M. Bockris, "Kinetics of oxygen evolution and dissolution on platinum electrodes," *Electrochimica Acta*, vol. 11, no. 7, pp. 791–814, 1966.

[7] E. Yeager, M. Razaq, D. Gervasio, A. Razaq, and D. Tryk, "The electrolyte factor in O_2 reduction electrocatalysis," *Proceedings of the Electrochemical Society*, vol. 92-11, pp. 440–473, 1992.

[8] A. Damjanovic, M. A. Genshaw, and J. O'M Bockris, "Distinction between intermediates produced in main and side electrodic reactions," *The Journal of Chemical Physics*, vol. 45, no. 11, pp. 4057–4059, 1966.

[9] H. S. Wroblowa, Y.-C. Pan, and G. Razumney, "Electroreduction of oxygen a new mechanistic criterion," *Journal of Electroanalytical Chemistry*, vol. 69, no. 2, pp. 195–201, 1976.

[10] N. A. Anastasijević, V. Vesović, and R. R. Adžić, "Determination of the kinetic parameters of the oxygen reduction reaction using the rotating ring-disk electrode. Part I. Theory," *Journal of Electroanalytical Chemistry*, vol. 229, no. 1-2, pp. 305–316, 1987.

[11] N. A. Anastasijević, V. Vesović, and R. R. Adžić, "Determination of the kinetic parameters of the oxygen reduction reaction using the rotating ring-disk electrode. Part II. Applications," *Journal of Electroanalytical Chemistry*, vol. 229, no. 1-2, pp. 317–325, 1987.

[12] A. J. Bard, "Inner-sphere heterogeneous electrode reactions. Electrocatalysis and photocatalysis: the challenge," *Journal of the American Chemical Society*, vol. 132, no. 22, pp. 7559–7567, 2010.

[13] A. J. Bard, R. Parsons, and J. Jordan, Eds., *Standard Potentials in Aqueous Solution*, 1985.

[14] H. H. Yang and R. L. McCreery, "Elucidation of the mechanism of dioxygen reduction on metal-free carbon electrodes," *Journal of the Electrochemical Society*, vol. 147, no. 9, pp. 3420–3428, 2000.

[15] B. B. Blizanac, P. N. Ross, and N. M. Markovic, "Oxygen electroreduction on Ag(111): the pH effect," *Electrochimica Acta*, vol. 52, no. 6, pp. 2264–2271, 2007.

[16] J. O. Bockris and J. Appleby, "Alkaline Fuel Cells," in *Assessment of Research Needs for Advanced Fuel Cells*, S. S. Penner, Ed., vol. 11, p. 95, 1986.

[17] A. J. Appleby, "Electrocatalysis," in *Comprehensive Treatise of Electrochemistry*, B. E. Conway, Ed., vol. 7, pp. 173–239, Plenum, New York, NY, USA, 1983.

[18] R. Jasinski, "A new fuel cell cathode catalyst," *Nature*, vol. 201, no. 4925, pp. 1212–1213, 1964.

[19] J. H. Zagal, "Metallophthalocyanines as catalysts in electrochemical reactions," *Coordination Chemistry Reviews*, vol. 119, pp. 89–136, 1992.

[20] "Macrocycles," in *Handbook of Fuel Cells-Fundamentals*, J. H. Zagal, Ed., pp. 544–554, John Wiley & Sons, Chichester, UK, 2003.

[21] B. Wang, "Recent development of non-platinum catalysts for oxygen reduction reaction," *Journal of Power Sources*, vol. 152, no. 1-2, pp. 1–15, 2005.

[22] J.-P. Dodelet, "Oxygen reduction in pem fuel cell conditions: heat- treated non-precious metal-N4 macrocycles and beyond," in *N4-Macrocyclic Metal Complexes*, J. H. Zagal, F. Bedioui, and J.-P. Dodelet, Eds., p. 83, Springer, Berlin, Germany, 2006.

[23] J. M. Ziegelbauer, T. S. Olson, S. Pylypenko et al., "Direct spectroscopic observation of the structural origin of peroxide generation from Co-based pyrolyzed porphyrins for ORR applications," *Journal of Physical Chemistry C*, vol. 112, no. 24, pp. 8839–8849, 2008.

[24] K. Artyushkova, S. Levendosky, P. Atanassov, and J. Fulghum, "XPS Structural studies of nano-composite non-platinum electrocatalysts for polymer electrolyte fuel cells," *Topics in Catalysis*, vol. 46, no. 3-4, pp. 263–275, 2007.

[25] A. L. Bouwkamp-Wijnoltz, W. Visscher, and J. A. R. van Veen, "The selectivity of oxygen reduction by pyrolysed iron porphyrin supported on carbon," *Electrochimica Acta*, vol. 43, no. 21-22, pp. 3141–3152, 1998.

[26] A. L. Bouwkamp-Wijnoltz, W. Visscher, J. A. R. van Veen, E. Boellaard, A. M. van der Kraan, and S. C. Tang, "On active-site heterogeneity in pyrolyzed carbon-supported iron porphyrin catalysts for the electrochemical reduction of oxygen: an in situ mössbauer study," *Journal of Physical Chemistry B*, vol. 106, no. 50, pp. 12993–13001, 2002.

[27] A. L. Bouwkamp-Wijnoltz, W. Visscher, J. A. R. van Veen, and S. C. Tang, "Electrochemical reduction of oxygen: an alternative method to prepare active CoN_4 catalysts," *Electrochimica Acta*, vol. 45, no. 3, pp. 379–386, 1999.

[28] J. McBreen, W. E. O'Grady, D. E. Sayers, C. Y. Yang, and K. I. Pandya, "An EXAFS study of pyrolyzed metal macrocyclic electrocatalysts," S. Srinivasan, S. Wagner, and H. Wroblowa, Eds., vol. 87-12, p. 182, The Electrochemical Society, Pennington, NJ, USA.

[29] H. Schulenburg, S. Stankov, V. Schünemann et al., "Catalysts for the oxygen reduction from heat-treated iron(III) tetramethoxyphenylporphyrin chloride: structure and stability and active sites," *Journal of Physical Chemistry B*, vol. 107, no. 34, pp. 9034–9041, 2003.

[30] A. A. Tanaka, S. L. Gupta, D. Tryk, C. Fierro, and E. B. Yeager, "Electrochemical and spectroscopic aspects of heat-treated transition metal macrocycles as electrocatalysts for oxygen reduction," in *Structural Effects in Electrocatalysis and Oxygen Electrochemistry*, D. Scherson, D. Tryk, M. Daroux, and X. Xing, Eds., vol. 92-11, p. 555, Proceedings of the Electrochemical Society, 1992.

[31] I. T. Bae, D. A. Tryk, and D. A. Scherson, "Effect of heat treatment on the redox properties of iron porphyrins adsorbed on high area carbon in acid electrolytes: an in situ Fe K-edge X-ray absorption near-edge structure study," *Journal of Physical Chemistry B*, vol. 102, no. 21, pp. 4114–4117, 1998.

[32] S. Kim, I. T. Bae, M. Sandifer et al., "In situ XANES of an iron porphyrin irreversibly adsorbed on an electrode surface," *Journal of the American Chemical Society*, vol. 113, no. 24, pp. 9063–9066, 1991.

[33] S. Kim, D. A. Tryk, I. T. Bae et al., "In situ extended X-ray absorption fine structure of an iron porphyrin irreversibly adsorbed on an electrode surface," *Journal of Physical Chemistry*, vol. 99, no. 25, pp. 10359–10364, 1995.

[34] M. Lefèvre, J. P. Dodelet, and P. Bertrand, "Molecular oxygen reduction in PEM fuel cells: evidence for the simultaneous presence of two active sites in Fe-based catalysts," *Journal of Physical Chemistry B*, vol. 106, no. 34, pp. 8705–8713, 2002.

[35] M. Lefevre, J. P. Dodelet, and P. Bertrand, "Oi reduction in PEM fuel cells: activity and active site structural information for catalysts obtained by the pyrolysis at high temperature of Fe precursors," *Journal of Physical Chemistry B*, vol. 104, no. 47, pp. 11238–11247, 2000.

[36] M. Lefèvre and J. P. Dodelet, "Fe-based catalysts for the reduction of oxygen in polymer electrolyte membrane fuel cell conditions: determination of the amount of peroxide released during electroreduction and its influence on the stability of the catalysts," *Electrochimica Acta*, vol. 48, no. 19, pp. 2749–2760, 2003.

[37] M. Lefèvre, E. Proietti, F. Jaouen, and J. P. Dodelet, "Iron-Based catalysts with improved oxygen reduction activity in polymer electrolyte fuel cells," *Science*, vol. 324, no. 5923, pp. 71–74, 2009.

[38] G. Lalande, R. Côté, D. Guay, J. P. Dodelet, L. T. Weng, and P. Bertrand, "Is nitrogen important in the formulation of Fe-based catalysts for oxygen reduction in solid polymer fuel cells?" *Electrochimica Acta*, vol. 42, no. 9, pp. 1379–1388, 1997.

[39] G. Lalande, G. Faubert, R. Côté et al., "Catalytic activity and stability of heat-treated iron phthalocyanines for the electroreduction of oxygen in polymer electrolyte fuel cells," *Journal of Power Sources*, vol. 61, no. 1-2, pp. 227–237, 1996.

[40] M. C. M. Alves, J. P. Dodelet, D. Guay, M. Ladouceur, and G. Tourillon, "Origin of the electrocatalytic properties for O_2 reduction of some heat-treated polyacrylonitrile and

phthalocyanine cobalt compounds adsorbed on carbon black as probed by electrochemistry and X-ray absorption spectroscopy," *Journal of Physical Chemistry*, vol. 96, no. 26, pp. 10898–10905, 1992.

[41] G. Faubert, R. Côté, J. P. Dodelet, M. Lefèvre, and P. Bertrand, "Oxygen reduction catalysts for polymer electrolyte fuel cells from the pyrolysis of FeII acetate adsorbed on 3,4,9,10-perylenetetracarboxylic dianhydride," *Electrochimica Acta*, vol. 44, no. 15, pp. 2589–2603, 1999.

[42] G. Faubert, G. Lalande, R. Côté et al., "Heat-treated iron and cobalt tetraphenylporphyrins adsorbed on carbon black: physical characterization and catalytic properties of these materials for the reduction of oxygen in polymer electrolyte fuel cells," *Electrochimica Acta*, vol. 41, no. 10, pp. 1689–1701, 1996.

[43] P. He, M. Lefèvre, G. Faubert, and J. P. Dodelet, "Oxygen reduction catalysts for polymer electrolyte fuel cells from the pyrolysis of various transition metal acetates adsorbed on 3,4,9,10-perylenetetracarboxylic dianhydride," *Journal of New Materials for Electrochemical Systems*, vol. 2, no. 4, pp. 243–251, 1999.

[44] C. Médard, M. Lefèvre, J. P. Dodelet, F. Jaouen, and G. Lindbergh, "Oxygen reduction by Fe-based catalysts in PEM fuel cell conditions: activity and selectivity of the catalysts obtained with two Fe precursors and various carbon supports," *Electrochimica Acta*, vol. 51, no. 16, pp. 3202–3213, 2006.

[45] F. Jaouen, E. Proietti, M. Lefèvre et al., "Recent advances in non-precious metal catalysis for oxygen-reduction reaction in polymer electrolyte fuel cells," *Energy and Environmental Science*, vol. 4, no. 1, pp. 114–130, 2011.

[46] D. P. Amalnerkar, S. Radhakrishnan, H. Minoura, T. Sugiura, and Y. Ueno, "Origin of the photocathodic effect at CdS + CuCl electrodes," *Journal of Electroanalytical Chemistry and Interfacial Electrochemistry*, vol. 260, no. 2, pp. 433–440, 1989.

[47] G. Gruenig, K. Wiesener, S. Gamburzev, I. Iliev, and A. Kaisheva, "Investigations of catalysts from the pyrolyzates of cobalt-containing and metal-free dibenzotetraazaannulenes on active carbon for oxygen electrodes in an acid medium," *Journal of Electroanalytical Chemistry*, vol. 159, no. 1, pp. 155–162, 1983.

[48] J. A. R. van Veen, J. F. van Baar, and K. J. Kroese, "Effect of heat treatment on the performance of carbon-supported transition-metal chelates in the electrochemical reduction of oxygen," *Journal of the Chemical Society, Faraday Transactions 1*, vol. 77, no. 11, pp. 2827–2843, 1981.

[49] P. Vasudevan, Santosh, N. Mann, and S. Tyagi, "Transition metal complexes of porphyrins and phthalocyanines as electrocatalysts for dioxygen reduction," *Transition Metal Chemistry*, vol. 15, no. 2, pp. 81–90, 1990.

[50] F. Jaouen, M. Lefèvre, J. P. Dodelet, and M. Cai, "Heat-treated Fe/N/C catalysts for O_2 electroreduction: are active sites hosted in micropores?" *Journal of Physical Chemistry B*, vol. 110, no. 11, pp. 5553–5558, 2006.

[51] C. W. B. Bezerra, L. Zhang, K. Lee et al., "A review of Fe–N/C and Co–N/C catalysts for the oxygen reduction reaction," *Electrochimica Acta*, vol. 53, no. 15, pp. 4937–4951, 2008.

[52] A. Garsuch, A. Bonakdarpour, G. Liu, R. Yang, and J. R. Dahn, "Time to move beyond transition metal–N–C catalysts for oxygen reduction," in *Handbook of Fuel Cells*, vol. 5, Springer, New York, NY, USA, 2009.

[53] N. Ramaswamy, R. J. Allen, and S. Mukerjee, "Electrochemical kinetics and X-ray absorption spectroscopic investigations of oxygen reduction on chalcogen-modified ruthenium catalysts in alkaline media," *Journal of Physical Chemistry C*, vol. 115, no. 25, pp. 12650–12664, 2011.

[54] T. M. Arruda, B. Shyam, J. S. Lawton et al., "Fundamental aspects of spontaneous cathodic deposition of Ru onto Pt/C electrocatalysts and membranes under direct methanol fuel cell operating conditions: an in situ X-ray absorption spectroscopy and electron spin resonance study," *Journal of Physical Chemistry C*, vol. 114, no. 2, pp. 1028–1040, 2010.

[55] T. M. Arruda, B. Shyam, J. M. Ziegelbauer, S. Mukerjee, and D. E. Ramaker, "Investigation into the competitive and site-specific nature of anion adsorption on Pt using in situ X-ray absorption spectroscopy," *Journal of Physical Chemistry C*, vol. 112, no. 46, pp. 18087–18097, 2008.

[56] A. L. Ankudinov, B. Ravel, J. J. Rehr, and S. D. Conradson, "Real-space multiple-scattering calculation and interpretation of x-ray-absorption near-edge structure," *Physical Review B*, vol. 58, no. 12, pp. 7565–7576, 1998.

[57] V. S. Murthi, R. C. Urian, and S. Mukerjee, "Oxygen reduction kinetics in low and medium temperature acid environment: correlation of water activation and surface properties in supported Pt and Pt alloy electrocatalysts," *Journal of Physical Chemistry B*, vol. 108, no. 30, pp. 11011–11023, 2004.

[58] M. R. Tarasevich, A. Sadkowski, and E. Yeager, "Kinetics and mechanisms of electrode processes," in *Comprehensive Treatise of Electrochemistry*, B. E. Conway, J. O. M. Bockris, E. Yeager, S. U. M. Khanand, and R. E. White, Eds., vol. 7, pp. 301–398, Plenum Press, New York, NY, USA, 1983.

[59] N. M. Marković, H. A. Gasteiger, and P. N. Ross, "Oxygen reduction on platinum low-index single-crystal surfaces in alkaline solution: rotating ring diskPt(hkl) studies," *Journal of Physical Chemistry*, vol. 100, no. 16, pp. 6715–6721, 1996.

[60] N. Ramaswamy and S. Mukerjee, "Influence of inner- and outer-sphere electron transfer mechanisms during electrocatalysis of oxygen reduction in alkaline media," *Journal of Physical Chemistry C*, vol. 115, no. 36, pp. 18015–18026, 2011.

[61] E. Janin, H. Von Schenck, M. Göthelid, U. O. Karlsson, and M. Svensson, "Bridge-bonded atomic oxygen on Pt(110)," *Physical Review B*, vol. 61, no. 19, pp. 13144–13149, 2000.

[62] G. Jerkiewicz, G. Vatankhah, J. Lessard, M. P. Soriaga, and Y. S. Park, "Surface-oxide growth at platinum electrodes in aqueous H_2SO_4 Reexamination of its mechanism through combined cyclic-voltammetry, electrochemical quartz-crystal nanobalance, and Auger electron spectroscopy measurements," *Electrochimica Acta*, vol. 49, no. 9-10, pp. 1451–1459, 2004.

[63] N. M. Markoví and P. N. Ross, "Surface science studies of model fuel cell electrocatalysts," *Surface Science Reports*, vol. 45, no. 4–6, pp. 117–229, 2002.

[64] P. Chen and R. L. McCreery, "Control of electron transfer kinetics at glassy carbon electrodes by specific surface modification," *Analytical Chemistry*, vol. 68, no. 22, pp. 3958–3965, 1996.

[65] P. Chen, M. A. Fryling, and R. L. McCreery, "Electron transfer kinetics at modified carbon electrode surfaces: the role of specific surface sites," *Analytical Chemistry*, vol. 67, no. 18, pp. 3115–3122, 1995.

[66] N. Wakabayashi, F. Kitamura, T. Ohsaka, and K. Tokuda, "Effect of adsorbed anions on the outer-sphere electron-transfer reactions of cobalt complexes at platinum single-crystal electrodes," *Journal of Electroanalytical Chemistry*, vol. 499, no. 1, pp. 161–168, 2001.

[67] T. Wandlowski and R. de Levie, "Double-layer dynamics in the adsorption of tetrabutylammonium ions at the mercury—water interface Part 4. The reduction of hexammine-cobalt

(III) through tetrabutylammonium films," *Journal of Electroanalytical Chemistry*, vol. 380, no. 1-2, pp. 201–207, 1995.

[68] N. A. Anastasijevic, Z. M. Dimitrijevic, and R. R. Adzic, "Oxygen reduction on a ruthenium electrode in alkaline electrolytes," *Journal of Electroanalytical Chemistry and Interfacial Electrochemistry*, vol. 199, no. 2, pp. 351–364, 1986.

[69] N. A. Anastasijevic, Z. M. Dimitrijevic, and R. R. Adzic, "Oxygen reduction on a ruthenium electrode in acid electrolytes," *Electrochimica Acta*, vol. 31, no. 9, pp. 1125–1130, 1986.

[70] N. Ramaswamy, R. J. Allen, and S. Mukerjee, "Electrochemical kinetics and X-ray absorption spectroscopic investigations of oxygen reduction on chalcogen-modified ruthenium catalysts in alkaline media," *The Journal of Physical Chemistry C*, vol. 115, no. 25, pp. 12650–12664, 2011.

[71] R. Boulatov, "Billion-Year-Old Oxygen Cathode that Actually Works: Respiratory Oxygen Reduction and Its Biomimetic Analogs," in *N4-Macrocyclic Metal Complexes*, J. H. Zagal, F. Bedioui, and J.-P Dodelet, Eds., p. 1, Springer, Berlin, Germany, 2006.

[72] R. Boulatov, "Metalloporphyrin Catalysts of Oxygen Reduction," in *Fuel Cell Catalysis*, M. T. M. Koper, Ed., p. 637, John Wiley & Sons, New York, NY, USA, 2008.

[73] N. Ramaswamy and S. Mukerjee, "Electrocatalysis of oxygen reduction on nonprecious metallic centers at high pH environments," *ECS Transactions*, vol. 33, no. 1, pp. 1777–1785, 2010.

[74] K. M. Kadish, E. van Caemelbecke, and R. Royal, "Electrochemistry of metalloporphyrins in nonaqueous media," in *The Porphyrin Handbook*, K. M. Kadish, K. M. Smith, and R. Guilard, Eds., vol. 8, chapter 55, pp. 1–97, Academic Press, San Diego, Calif, USA, 2000.

[75] M. Teliska, V. S. Murthi, S. Mukerjee, and D. E. Ramaker, "Correlation of water activation, surface properties, and oxygen reduction reactivity of supported Pt-M/C bimetallic electrocatalysts using XAS," *Journal of the Electrochemical Society*, vol. 152, no. 11, pp. A2159–A2169, 2005.

[76] M. Teliska, W. E. O'Grady, and D. E. Ramaker, "Determination of O and OH adsorption sites and coverage in situ on Pt electrodes from Pt L23 X-ray absorption spectroscopy," *Journal of Physical Chemistry B*, vol. 109, no. 16, pp. 8076–8084, 2005.

[77] E. A. Lewis, C. U. Segre, and E. S. Smotkin, "Embedded cluster Δ-XANES modeling of adsorption processes on Pt," *Electrochimica Acta*, vol. 54, no. 28, pp. 7181–7185, 2009.

[78] F. Beck, "The redox mechanism of the chelate-catalysed oxygen cathode," *Journal of Applied Electrochemistry*, vol. 7, no. 3, pp. 239–245, 1977.

[79] S. R. P. Silva, "Properties of Amorphous Carbon," in *Properties of Amorphous Carbon*, EMIS Datareviews Series 29, 2003.

[80] Y. J. Cho, H. S. Kim, S. Y. Baik et al., "Selective nitrogen-doping structure of nanosize graphitic layers," *Journal of Physical Chemistry C*, vol. 115, no. 9, pp. 3737–3744, 2011.

[81] J. R. Hahn and H. Kang, "Vacancy and interstitial defects at graphite surfaces: scanning tunneling microscopic study of the structure, electronic property, and yield for ion-induced defect creation," *Physical Review B*, vol. 60, no. 8, pp. 6007–6017, 1999.

[82] S. L. Mielke, D. Troya, S. Zhang et al., "The role of vacancy defects and holes in the fracture of carbon nanotubes," *Chemical Physics Letters*, vol. 390, no. 4–6, pp. 413–420, 2004.

[83] Y. Luo, Y. Heng, X. Dai, W. Chen, and J. Li, "Preparation and photocatalytic ability of highly defective carbon nanotubes," *Journal of Solid State Chemistry*, vol. 182, no. 9, pp. 2521–2525, 2009.

[84] T. Kondo, T. Suzuki, and J. Nakamura, "Nitrogen doping of graphite for enhancement of durability of supported platinum clusters," *Journal of Physical Chemistry Letters*, vol. 2, no. 6, pp. 577–580, 2011.

[85] P. Nemes-Incze, Z. Koínya, I. Kiricsi et al., "Mapping of functionalized regions on carbon nanotubes by scanning tunneling microscopy," *Journal of Physical Chemistry C*, vol. 115, no. 8, pp. 3229–3235, 2011.

[86] A. B. Anderson and R. A. Sidik, "Oxygen Electroreduction on FeII and FeIII Coordinated to N4 Chelates. Reversible Potentials for the Intermediate Steps from Quantum Theory," *Journal of Physical Chemistry B*, vol. 108, no. 16, pp. 5031–5035, 2004.

Permissions

The contributors of this book come from diverse backgrounds, making this book a truly international effort. This book will bring forth new frontiers with its revolutionizing research information and detailed analysis of the nascent developments around the world.

We would like to thank all the contributing authors for lending their expertise to make the book truly unique. They have played a crucial role in the development of this book. Without their invaluable contributions this book wouldn't have been possible. They have made vital efforts to compile up to date information on the varied aspects of this subject to make this book a valuable addition to the collection of many professionals and students.

This book was conceptualized with the vision of imparting up-to-date information and advanced data in this field. To ensure the same, a matchless editorial board was set up. Every individual on the board went through rigorous rounds of assessment to prove their worth. After which they invested a large part of their time researching and compiling the most relevant data for our readers. Conferences and sessions were held from time to time between the editorial board and the contributing authors to present the data in the most comprehensible form. The editorial team has worked tirelessly to provide valuable and valid information to help people across the globe.

Every chapter published in this book has been scrutinized by our experts. Their significance has been extensively debated. The topics covered herein carry significant findings which will fuel the growth of the discipline. They may even be implemented as practical applications or may be referred to as a beginning point for another development. Chapters in this book were first published by Hindawi Publishing Corporation; hereby published with permission under the Creative Commons Attribution License or equivalent.

The editorial board has been involved in producing this book since its inception. They have spent rigorous hours researching and exploring the diverse topics which have resulted in the successful publishing of this book. They have passed on their knowledge of decades through this book. To expedite this challenging task, the publisher supported the team at every step. A small team of assistant editors was also appointed to further simplify the editing procedure and attain best results for the readers.

Our editorial team has been hand-picked from every corner of the world. Their multi-ethnicity adds dynamic inputs to the discussions which result in innovative outcomes. These outcomes are then further discussed with the researchers and contributors who give their valuable feedback and opinion regarding the same. The feedback is then collaborated with the researches and they are edited in a comprehensive manner to aid the understanding of the subject.

Apart from the editorial board, the designing team has also invested a significant amount of their time in understanding the subject and creating the most relevant covers. They scrutinized every image to scout for the most suitable representation of the subject and create an appropriate cover for the book.

The publishing team has been involved in this book since its early stages. They were actively engaged in every process, be it collecting the data, connecting with the contributors or procuring relevant information. The team has been an ardent support to the editorial, designing and production team. Their endless efforts to recruit the best for this project, has resulted in the accomplishment of this book. They are a veteran in the field of academics and their pool of knowledge is as vast as their experience in printing. Their expertise and guidance has proved useful at every step. Their uncompromising quality standards have made this book an exceptional effort. Their encouragement from time to time has been an inspiration for everyone.

The publisher and the editorial board hope that this book will prove to be a valuable piece of knowledge for researchers, students, practitioners and scholars across the globe.

List of Contributors

Daniel Kozma
Institute of Enzymology, Research Centre for Natural Sciences, Hungarian Academy of Sciences, Karolina ´ut 29, Budapest 1113, Hungary

Gergely Toth
Institute of Chemistry, E¨otv¨os Lor´and University, P.O. Box 32, Budapest 1518, Hungary

Sadhana Senapati
PG Department of Chemistry, Utkal University, Odisha, Bhubaneswar 754004, India

S. P. Das and A. K. Patnaik
PG Department of Chemistry, Ravenshaw University, Odisha, Cuttack 753003, India

K. Ramesh, S. Shylaja, K. C. Rajanna, P. Giridhar Reddy and P. K. Saiprakash
Department of Chemistry, Osmania University, Hyderabad, Andhra Pradesh 500 007, India

Mona A. Shouman, Nady A. Fathy, Soheir A. Khedr and Amina A. Attia
Surface Chemistry and Catalysis Laboratory, National Research Centre, El-Dokki, Cairo 12622, Egypt

J. Espinosa-Garcia, M.Monge-Palacios and J. C. Corchado
Departamento de Qu´ımica F´ısica, Universidad de Extremadura, 06071 Badajoz, Spain

John A. Bumpus
Department of Chemistry and Biochemistry, University of Northern Iowa, Cedar Falls, IA 50614, USA

Ruggero Barni, Stefano Zanini, and Claudia Riccardi
Dipartimento di Fisica Occhialini, Universit`a degli Studi di Milano-Bicocca, Piazza della Scienza 3, 20126, Milano, Italy

K. Indira and L. Rajendran
Department of Mathematics, The Madura College, Madurai 625011, India

Haitao Ma, Chunfang Zhang, Zhijun Zhang and Wensheng Bian
Beijing National Laboratory for Molecular Sciences, Institute of Chemistry, Chinese Academy of Sciences, Beijing 100190, China

Xiaojun Liu
Key Laboratory of Luminescence and Optical Information, Ministry of Education, Institute of Optoelectronic Technology, Beijing Jiaotong University, Beijing 100044, China

Satoshi Maeda
The Hakubi Center, Kyoto University, Kyoto 606-8302, Japan

Satoshi Maeda and Keiji Morokuma
Fukui Institute for Fundamental Chemistry, Kyoto University, Kyoto 606-8103, Japan

Koichi Ohno
Toyota Physical and Chemical Research Institute, Nagakute, Aichi 480-1192, Japan

Keiji Morokuma
Department of Chemistry and The Cherry L. Emerson Center for Scientific Computation, Emory University, Atlanta, GA 30322, USA

P. Giridhar Reddy, K. Ramesh, S. Shylaja, K. C. Rajanna and S. Kandlikar
Department of Chemistry, Osmania University, Hyderabad 500001, India

Ricardo Infante-Castillo
Department of Physics and Chemistry, University of Puerto Rico, Arecibo Campus, P.O. Box 8152, Arecibo, PR 08152-00613, USA

Samuel P. Hernández-Rivera
ALERT-DHS Center of Excellence/Center for Chemical Sensors Development, University of Puerto Rico, Mayag¨uez Campus, P.O. Box 9000, Mayaguez, PR 00681-9000, USA

Sk. Fakruddin, B. R. Venkateswara Rao and K. Narendra
Department of Physics, Velagapudi Ramakrishna Siddhartha Engineering College, Andhra Pradesh, Vijayawada 520007, India

Ch. Srinivasu
Department of Physics, Andhra Loyola College, Vijayawada 520008, India

Vahid Moeini and Mahin Farzad
Department of Chemistry, Payame Noor University, P.O. Box 19395-3697, Tehran, Iran

Martin Alberto Masuelli
Instituto de F´ısica Aplicada-CONICET, Facultad de Qu´ımica, Bioqu´ımica y Farmacia, Universidad Nacional de San Luis, Area de Qu´ımica F´ısica, Chacabuco 917, 5700 San Luis, Argentina

Ali Ebrahimi, Mostafa Habibi-Khorassani, Farideh Badichi Akher and Abdol karim Farrokh zadeh
Department of Chemistry, University of Sistan and Baluchestan, P.O. Box 98135-674, Zahedan, Iran

Sergey V. Stepanov, Vsevolod M. Byakov, Dmitrii S. Zvezhinskiy, Roman R. Nurmukhametov and Petr S. Stepanov
Institute of Theoretical and Experimental Physics, Moscow 117218, Russia

Vsevolod M. Byakov
D. Mendeleev University of Chemical Technology of Russia, Miusskaya Square 9, Moscow 125047, Russia

Gilles Duplatre
Institut Pluridisciplinaire Hubert Curien, CNRS/IN2P3, BP 28, 67037 Strasbourg Cedex 2, France

Yi Zhao
State Key Laboratory for Physical Chemistry of Solid Surfaces and Fujian Provincial Key Lab of Theoretical and Computational Chemistry, College of Chemistry and Chemical Engineering, Xiamen University, Xiamen 361005, China

Wenji Wang
College of Science, Northwest A&F University, Yangling 712100, Shaanxi Province, China

Nagappan Ramaswamy and Sanjeev Mukerjee
Laboratory for Electrochemical Advanced Power, Department of Chemistry and Chemical Biology, Center of Renewable Energy Technology (NUCRET), Northeastern University, 360 Huntington Avenue, Boston, MA 02115, USA